The Humaniverse Guide to Better Reasoning and Decision-Making

Keith A. Seland

NEWMAN SPRINGS PUBLISHING
320 Broad Street
Red Bank, NJ 07701

First originally published by Newman Springs Publishing 2019

ISBN 978-1-64531-092-1 (Paperback)
ISBN 978-1-64531-093-8 (Digital)

Printed in the United States of America

To you, the reader, for your willingness and curiosity to inquire and learn.

Contents

Acknowledgments 7

Introduction...9

Chapter 1: The Hypothesis 21

Chapter 2: Critical Thinking and the Scientific Method.............30

Chapter 3: What Is Science?...49

Chapter 4: The Inductive and Deductive Methods of
Scientific Reasoning 79

Chapter 5: The Yin and Yang of the Science and
Ufology Argument 101

Chapter 6: Who Has Seen Them? 119

Chapter 7: Man's Technology Explosion....................................167

Chapter 8: Why Here? Why Not Somewhere Else? 180

Chapter 9: How to Move My UFO 200

Chapter 10: Ancient Texts of High Strangeness 233

Chapter 11: Ancient Structures of High Strangeness Part 1 263

Chapter 12: Ancient Structures of High Strangeness Part 2 295

Chapter 13: Ancient Paraphernalia of High Strangeness...........340

Chapter 14: Physical Evidence of UFOs and Aliens Abound.....366

Chapter 15: What Scientists Are Saying 392

Chapter 16: What People (Many at One Time) Are
Observing The Public...415

Chapter 17: What People (a Lot of Them at One Time)
Are Seeing The Especially Trained Public 451

Chapter 18: What People (a Lot of Them at One Time)
Are Seeing The Military 490

Chapter 19: What a Lot of People Are Seeing Plus
a Lot More 524

Chapter 20: Disclosure: An Overview of What We
Don't Know..576

Chapter 21: Why Would You Become Interested in the
UFO/UAP/Alien Hypothesis 622

Chapter 22: Prove ET Ways to Provide Evidence for
Possibility of Existence 648

Chapter 23: The Final Chapter 679

Appendix A: Shapes of UFOs..725

Appendix B: Types of Reasoning and Fallacies Dictionary.........726

Acknowledgments

I would like to thank the efforts and guidance of the following people who made creation of *The Humaniverse Guide to Better Reasoning & Decision-Making* overarchingly easier to accomplish:

To my family, whose constant encouragement and enthusiasm were beacons of inspiration.

To my line editor, Karen Szudzik, who stood beside me throughout polishing and buffing the ideas and providing insight for my writing.

To the Roswell UFO Museum and Research Library for their continuing support and access to meaningful knowledge that helped envision and shape the story.

To the staff at Newman Springs Publishing, including Sadie McLaughlin, for her expertise, guidance, and wisdom in providing the framework for better consequential decisions; in conjunction with Sean and the entire editorial staff, all whom championed the cause of *The Humaniverse Guide to Better Reasoning & Decision-Making* project and aligned its compass through a successful launching that enabled it to take flight.

Introduction

Extraordinary claims require extraordinary evidence.

—Simon-Pierre LaPlace
Mathematician, requited by Carl Sagan, science communicator

LaPlace and Sagan forgot one thing.

These claims also demand extraordinary study in order to properly answer them and prove their existence. They are notably worthy of it. Would you demand the same of the many questions and problems in your life?

The Humaniverse Guide to Better Reasoning & Decision-Making will provide for you a glimpse of how nature and the environment operate and how people talk about, interact with, and make decisions from being immersed in them.

You will engage in and think enthusiastically about how the discipline of ufology is viewed and used in society. This is a subject where virtually every societal community has some input.

If only your life wasn't so busy and each day filled with so many activities. If it weren't for this reality, you could better develop and shape your theories about unidentified flying objects and extraterrestrial life. This could be accomplished by your range of ability to assimilate a lot of facts and data and be able to sort out the hoaxes and false information that exist in the literature.

The Humaniverse Guide to Better Reasoning & Decision-Making is all about *facts*. Questions about these themes can only be answered and reality uncovered by a rigid adherence to the collection, analy-

9

sis, and determination of how facts and data will best answer those inquiries and how truth and reality are discovered.

A word I've used in informal conversations where questions are raised and answers are sought by use of the best methods available is *pragmascience*. It is a paraphrase for *science inquiry* and *pragmatism*. Today, there is too much proliferation of impulsive opinions and unreasoned explanations offered to answer many of the questions raised in these debates. *The Humaniverse Guide to Better Reasoning & Decision-Making* cares about this impulsiveness in recognition of the thematic summary hypotheses known as "The UFO Hypothesis" and "The Extraterrestrial Hypothesis." You will see acronyms for these, such as UFOH or ETH throughout the text. Pragmascience is defined as a reasoned and thoughtful inquiry into a science topic that includes elements of scientific knowledge and exploration. Even if it is regarded by some as "pseudoscience," it is still given due analysis and consideration. This consideration is only concerned with facts and data obtained from episodes of real phenomena that occur in nature and the environment.

The Humaniverse Guide to Better Reasoning & Decision-Making will present a lot of facts and data about the history, evolution, and current state of affairs while engaging you in the UFOH and the ETH. You will immerse yourself in the roles of scientist, forensic investigator, and jury while studying all the evidence. When you think about it, if a scientist, a forensic investigator, and a juror practice their methodologies in the correct manner, each performs the same functions and critical thinking in the same ways. I will have a bit more to say about this later.

At this point, suffice to say, some people accurately practice the methodologies as principled in the sciences, the social sciences, and the law respectively. Some people do not. The research literature is populated, however, with rebuttals from some journalists, scientists, debunkers, and members of the general public. These critics give unreasoned opinions, bordering on the irrational, about questions of inquiry without much forethought or factual explanation. *The Humaniverse Guide to Better Reasoning & Decision-Making* is asking you to be the chief investigator as you read and offer your own critical thinking and insight into the study of the UFOH and the ETH.

After a thorough history and introduction to the ways the philosophy and science disciplines study phenomena in nature, you will fly into the theater of the UFOH and the ETH. The history epistemology will take you into studies of things such as Gobekli Tepe, a twelve-thousand-year-old settlement that is rewriting many archaeological and scientific theories about ancient humankind and its place in it.

You will learn about the many glyphs and apparatus left by all the humankind ancient civilizations since before the origin of written language. This history will be thought-provoking for you in that you might start hypothesizing about where and how language was itself created and developed.

The earth is a big place with which to harbor our species. In the tens of thousands of years since accurate dating of discovered artifacts have been applied to, there have existed thousands of civilizations. Literally, all these societies have communicated some form of anomalous experiences of nature and the environment down to their descendants. They have passed them onto us today to explore. It seems amazing that these ancient paraphernalia have existed for this long and that we are discovering ever more and more of them even now.

You will read and think about the Egyptian Pyramids and the Great Sphinx and wonder how and why they came into existence. Ancient China and India have similar structures that have been enigmatic to modern-day researchers of all trades. The ancient Mayan and Meso-American cultures and the Incas and their ancestors of ancient South America also left us their great monolithic and megalithic artifacts to study. From Stonehenge to Angkor Wat and Puma Punku to Machu Picchu humankind has left us so many puzzles that stretch our bewilderment and critical thinking to heights not present in most any other subject with which to study.

Moving along this timeline, humankind's thinking was transformed by the writings of the ancient Greek philosophers. These historians wrote the first texts for the discovery of new knowledge and ways to learn and reflect upon by both philosophical and scientific principles. From Thales of Miletus to Socrates, Aristotle, Plato, and others, you will come to know how human thought was so beautifully and critically organized and advanced. Because by these times writing as a linguistic tool was fully developed, other ancient journalists and prophets wrote and archived such literature as the *Egyptian Pyramid Texts*, *The Hebrew and Christian Bible*, *Indian Mahabharata*, and the *Mayan Codexes*.

You will ascertain that the intervening two thousand years are as full of texts, structures, glyphs, megaliths, and events of high strangeness as that of the many prior millennia. In the times since the beginning of the Neolithic Era, descendant peoples from all walks of life have voiced their encounters of events and phenomena that were and continue to be difficult to explain. This explanation obstacle to truth occurs even when studied by the most sophisticated technologies and when they are given a thorough and dedicated investigative treatment. Unfortunately, this does not happen often enough.

Homo sapiens have occupied the earth for perhaps one hundred thousand years or more. In the last fifteen percent or so of it, their world started to congregate to form villages, cities, and nations that represent communal ways of living. The history of the noted hundreds of societies that have existed is replete with scripture of their adventures of travel, exploration, and conquest. The metrics of all this represents humankind's propensity to survey other lands, many of them far from their home.

The Humaniverse Guide to Better Reasoning & Decision-Making will inspect these notions and traits as they may have occurred through the unique lenses of different forms of intelligent life. You will become aware of a context and perspective that explores the metrics not only from your own anthropological eye but also that of another motivated potential representative of intelligent life. As we are at the pinnacle of intelligent life species on Earth, this must mean that these potential representatives would emerge from the cosmos. As such, a particular means of travel must be utilized for any potential visitors to traverse that cosmos to get here from there and interact. Both of these criteria must be accompanied by a motivation for other intelligent life to want to come here. As humankind wrestles with questions such as our uniqueness as an intelligent life species and Earth's uniqueness as a requisite planetary body, *The Humaniverse Guide to Better Reasoning & Decision-Making* probes these problems in correlation with the body of evidence that time has left for us to pursue and peruse.

Humankind's liturgical record is full of terminology with which to define such things as extraterrestrial beings and UFOs. Here is a short list among the many, many definitions all authors have used to depict extraterrestrial life-forms. *ALF* means "alien life-form." *Bedroom Visitors* used to describe those alleged abduction originators; *Blondes* used to describe a type of Nordic extraterrestrial being; *Blues*, a benevolent being that made contact and nurtured natives of the Hopi Indian tribe of the Southwest United States; *Celestial Beings*; life-forms, assumed to be intelligent species, not native to Earth; *Cosmic Beings*, which is synonymous with Celestial Beings; *EBE's*, extraterrestrial biological entities; *ET*, extraterrestrial.

Here are some phrases used to describe UFOs: *AAP*: anomalous atmospheric phenomena, originated use from USSR covert military intelligence; *AFO*: alien flying object (acronym); *Daylight Disc*: term for a distant UFO seen in the daytime, originating from J. Allen Hynek; *EFO*: extraordinary flying object (acronym); *Flying Saucer*: a disc-shaped UFO, assumed by some to be an AFO; *IFO*: a UFO that has been definitively identified; *INFO*: identified nonflying object; *Mothership*: a particular type of UFO that is the interstellar core vessel that other "scout" ships have safe harbor; *Orb*: a spherical airborne object that emits some form of light in the visual range of the electromagnetic spectrum; *OVNI*: UFO (Spanish, "Objecto Volador No Identificado," or similar Italian, French, or Portuguese acronym); *UAP*: unidentified aerial phenomenon, originated use from Great Britain covert military intelligence.

A variety of archives exists that depict shapes of UFOs that have been reported in the past. In "Appendix A," there is an imagery list of a variety of these terms. Among the shapes include: disks, flattened like an ice hockey puck or a thick coin, cigars, lenticular, spherical, point-light, saucer-shaped (either concave or convex), triangular-shaped, diamond-shaped, mushroom-shaped, elliptical (or egg-shaped) orbs, cubical, conical, crescent-shaped, boomerang, pyramidal (a form of triangular), hexagonal, and donut-shaped.

You will also become familiar with a couple other terms used to summarize a group of related UFO sightings. These two terms denote a temporal relationship to the group. A *flap* is a series of UFO sightings within a loosely-defined block of time. A *wave* is a series of experiences, similar in scope but not in time frame to a flap.

This is not just a book about UFOs and aliens—far from it. If one were to absolutely attribute one overarching thematic to *The Humaniverse Guide to Better Reasoning & Decision-Making*, I would refer you back to the opening sentence of this "Introduction." *The Humaniverse Guide to Better Reasoning & Decision-Making* will provide for you a glimpse of how nature and the environment operate and how people talk about, interact with, and make decisions from being immersed in them. The endogenous and internal meanings from this sentence point you in the direction of an impending inspection about understanding your life and world and how you fit into and interact with it.

Oh, by the way, *The Humaniverse Guide to Better Reasoning & Decision-Making* will also not give you the apprehensive feeling you remember when you read from science textbooks. There will be discussions that will show you that the way much of the science you were taught created some of those unpleasant experiences. These discussions may help you understand how those impressions may have originated within you. The readings are meant to appeal to you and to offer reflection about those times in a different context. I hope you can, just as you should, give it another chance.

Some of the chapters will provide a nuanced history of science and philosophy as the basis of how they exist today. In this way, *The Humaniverse Guide to Better Reasoning & Decision-Making* may be partially characterized as a "quasi-textbook." The history will offer you the advantages of learning about science and philosophy topics that most people, including most scientists, are not taught in their educational training. It will also help you to come to a new understanding about how science is not evil and impossible to get the hang of. After all, we are immersed in nature every day of our lives, and science attempts to discover an understanding of why, how, when, where, and if something exists.

There will be an opportunity for you to acquire a lot of scientific knowledge in what are referred to as the natural disciplines that science topics cover. These include chemistry, biology, geophysics, physics, archaeology, anthropology, and biotechnology. Please do not feel the least bit intimidated or frightened by the breadth of topics that will be discussed. This information has significant links to the study of ufology. Also in thinking about this immersion into science right now, please erase or at least store those thoughts and visualizations into short-term memory. You can reflect upon this later.

You will be thoroughly introduced to all the concepts and ideas presented in such a way that should help you readjust your attitude about the sciences. These concepts are all basic to any of the sciences and do not require any prerequisite knowledge. They will be presented in a fashion, which will give you vivid visualizations no matter what perspective lens you are viewing them from. There will be no attempts made to place any of this within a myopic methodology. But there will be cases where myopia appears to dominate the landscape anyway. They are all, however, related to a main theme of *The Humaniverse Guide to Better Reasoning & Decision-Making*. This theme will discuss ufology, another discipline that has not achieved the status of being called mainstream science.

There are issues within the subject that some people have reasoned do not qualify for this classification. As you will see in some of the readings, when a body of knowledge is unacceptable for some of the science communities as proven according to their specifications, then an attribution of pseudoscience is given to it. You will also read of many, many situations in the history of the sciences where this classification is given to just those attributes at their time of involvement. The recipients of these dishonors include Galileo, Charles Darwin, Einstein, Alfred Wegener (discoverer of the earth's geophysical process known as plate tectonics), and Gregor Mendel (discoverer of the biological gene and founder of most disciplines related to genetics) just to mention a very few.

This is a time when you will read that there is some new terminology and decorum to learn both here and in the main chapters. The science research communities communicate with each other officially about their new hypotheses, theories, and knowledge discoveries in a respectful and courteous, if somewhat rigid, decency. To many of us, as science students, we may have been exposed to a brief example of this. If you have ever read a science research paper, it is a very formal and precise document. The paper is sectioned to include reasons the research study is important, what new knowledge areas it hopes to discover, what the results were, and what direction future research teams should take to follow-up on their research.

When you read through these pages, you may get a general visualization similar to what I have just explained. While *The Humaniverse Guide to Better Reasoning & Decision-Making* is not pretending or desiring to fall within the precise rigid and tortuous confines of what you would experience when reading such a paper, it is a read to give you a glimpse of what thoughtful analysis looks like on and in such a paper. Therefore, there will be a lot of new terminology interspersed throughout each chapter. You will be tipped off at the beginning of each with "significant terminology" pointers. Although each of the terms will be explained when you first read them in the chapter, feel free to look them up before you read.

Part of the formal nature of a science research paper lies in the terminology used. There is a dictionary subset of terms frequently used in science and educational research literature. To continue with this preliminary discussion about terminology, a *List of 128 Words Often Used by Science and Scientists* has been provided for you:

absolute	*dynamics*	*observation*	*temperature*
adhesion	*elasticity*	*organism*	*theory*
algorithm	*element*	*paradigm*	*time*
alternative	*energy*	*particle*	*transmitter*
amplitude	*environment*	*perceive*	*unit*
analogous	*equation*	*phase*	*validity*

analysis	*equilibrium*	*placebo*	*variable*
angular size	*evidence*	*polarize*	*vector*
angular velocity	*evolution*	*population*	*volatile*
anomalous	*exigent*	*precipitation*	*volume*
apparatus	*expansion*	*pressure*	*weight*
apparent	*experiment*	*principle*	
approach	*fact*	*process*	
area	*fluidity*	*procedure*	
argument	*force*	*protocol*	
atom	*frequency*	*radiate*	
axis	*friction*	*reaction*	
cohesion	*grid*	*refute*	
compound	*hybrid*	*regimen*	
concept	*hypothesis*	*reliability*	
conclusion	*impact*	*research*	
constant	*inference*	*resonance*	
control	*inquiry*	*respiration*	
convergent	*intensity*	*response*	
core	*interpretation*	*sample*	
credibility	*interval*	*saturation*	
critical	*intervention*	*scientific method*	
cross section	*laboratory*	*tension*	
data	*laws*	*test*	
deduction	*magnitude*	*sensitized*	
density	*mass*	*significant*	
derive	*matter*	*space*	
deviation	*measure*	*spectrum*	
diagram	*mechanism*	*stasis*	
diffusion	*medium*	*state function*	
dimension	*motion*	*static*	
discipline	*myopia*	*stimulus*	
distribution	*normal*	*study*	
divergent	*objective*	*surface*	

Don't worry; you will *not* see some of these terms throughout *The Humaniverse Guide to Better Reasoning and Decision-Making*!

Your immersion into the universe and its environmental system does take on one all-encompassing theme when it comes to being capable of understanding. All people use a thought process that helps enable them to come to an answer to questions about events that occur in their daily lives. This thought process is not unique but in fact is a characteristic human trait. Our entire civilization is structured around this practice.

This process takes on different levels of nuance for each of them. Some inquiries involve only one person. The individual makes inquiries about events and tries to answer them using any of the different levels of this process. All these levels involve collecting and analyzing facts and data. After an analysis, the inquirer answers the question (hypothesis). In everyday life, the inquirer takes this answer and acts on it. That would then end the process. To summarize, the inquiry is made, facts and data are studied, and conclusions or answers are made.

Some inquiries involve more than one person. One popular way to describe a process of inquiry or the statement of the problem and form a conclusion as an answer in this fashion is via an argument or discourse. The inquiry-answer process in this example is the same as is in the thought process for an individual, except one additional step in this different overall process is taken. This step is the use of a form of critique or rebuttal to the answer. Other synonymous terms frequently used to describe this rebuttal are refutation, appraisal, commentary, disapproval, debunk, and discrediting to name only a few of them. To summarize, the inquiry is made, facts and data are studied, conclusions and answers are made, then *additional* arguments or refutations are made.

When the discourse involves a debate in more of a professional realm, the definition and use of the tools and methods that make up the inquiry-answer-rebuttal process become more formal and rigid. This is not a bad thing. There is a real efficient and organized benefit to adoption and use of this process. Known in some terms as *decorum, etiquette,* or *courtesy,* I assert that the process should definitely be formally taught to everyone in school and everyone should practice it. This process is known as the "scientific method."

Everyone, including the trained professionals, is guilty of not using this methodology all the time. As human beings, our thought process gets interrupted by some extraneous factors. Sometimes, we fall short of answering ours or others' inquiries in the most correct, rational, and reasoned manner. This is part of being human. There are rules in play when it comes to a reasoned and civil discourse about an inquiry. The scientific method asks the question(s), collects facts, inspects, and investigates them on the bases of validity, reliability, and credibility. If the inquirer needs more facts, then he collects them. The rules that govern this part of the process include healthy skepticism and an open mind. If the inquirer does not possess and use all these rules, his conclusion will be incorrect.

Validity means "the quality of being logically or factually sound; soundness or cogency." Reliability means "the extent to which an experiment, test, or measuring procedure yields the same results on repeated trials." Credibility means "the quality of being trusted and believed in." Along with the rules that govern skepticism and open-mindedness, none of these rules are bestowed with any special entitlements nor can anyone claim entitlements from them. Everyone should use these rules in a reasonable and rational manner and not in an excessive one or for any other ad hoc purpose. This becomes increasingly more important, the more important the inquiry and its consequences become.

When it comes to people making a hypothesis (the inquirer) and those who are responding to those claims, the claimant's inquiry or responder's conclusion/answer should not be agreed upon unconditionally or accepted solely on his reputation or some sense of entitlement nor should anybody's answer. The conclusion/answer should be based on facts and data in conjunction with the validity, reliability, and credibility of those facts in equally significant proportions. The thought process one should take in an investigation of a hypothesis moves first from initial skeptic and having an open mind about the situation being studied to an inspection of the body of facts and data that result from testing of the hypothesis. There should be no preconceived conclusions or notions present. Let the process play itself out by dedicated study, testing, and analysis. Validity and reliability are tested and approved when the facts can be established as true (validity) and can be replicated, observed, and experienced by others. Then conclusions can be formulated based on the established body of valid and reliable facts and data. The situation and its occurrence in nature are the major prerequisites for conclusion, not the credibility of the claimant or responder. They must all earn their credibility with every situation in which they provide a discourse. A claim must be proved by the same rigor of evidence as a response or refutation.

What you are then left with is the approach of twenty-three chapters full of facts and data that represent symptoms of something that has been happening in nature and the environment for an extremely long time. With your new tool set, you can study this case in the same way that all the scientific experts drive their own investigative inquiries. The process moves first from hypothesis to fact and data analysis, then to rational reflection, and lastly to a conclusion in the same way as theirs. You will be the forensics investigator, the research scientist, and the expert juror. It is now time for you to apply all this to the next two problems of inquiry.

CHAPTER 1

The Hypothesis

*Thinking about "aliens" is a good way to understand,
and appreciate, what it means to be human.*

—Guy Consolmagno SJ[1]

Key words: alternative hypothesis, theory, scientific method, scientific law, prima facie, interaction (in medical research), ufology, exobiology

Hypothesis 1: Unidentified flying objects do not exist.

Hypothesis 2: ET does not exist.

The above hypotheses are two separate claims. In addition to a significant number of people, critical thinking and philosophical and scientific inquiry would reach the above conclusions at the end of an investigation. In other constructs, these are exactly the statements that come at the beginning of an investigation. When we are involved in situations where an inspection about an assertionis made, the claimants can structure our hypothesis in either of these two trains of thought. One statement says something "is," and for others, something "is not."

This is how a statement of hypothesisis summarized in science literature. After some observations and reflection on a phenomenon, scenario, or problem, the researcher or claimant builds his case about forming the sort of hypothesis he is learning to address. Once the scientist feels it is warranted, the hypothesis is formalized and extended to the representative research communities. Hypotheses are the leading questions that allow the researcher or investigator to proceed with the study.

Here, the term *communities* is often undefined. A community usually consists of a scientist's network of peers exclusively or may represent the society at large.

Nevertheless, once the hypothesis is formulated, the task at hand traverses a series of paths with the intent of disproving a claim based on its theory. Next, more observations are made, and experiments are conducted if possible. For the researcher, if laboratory experiments cannot be performed or tested, this negates any possibility that it can be disproved. So the hypothesis is dismissed.

Consider this quote from famous cosmologist, theoretical physicist, and futurist Michio Kaku: "No one knows who wrote the laws of physics or where they come from. Science is based on testable, reproducible evidence, and so far, we cannot test the universe before the big bang." This is an example of the criteria by which the science communities may dismiss any hypotheses or theories. Here, they treat and think of the universe at a specific time around the occurrence of the big bang as an experiment. They feel that because they cannot replicate this experiment and its results, the hypothesis cannot be tested and, therefore, should not be stated or acknowledged.

With hypotheses that can be tested and results reproduced, experiments are conducted, and tests are performed to collect data that will be measured and analyzed. This path then leads to drawing conclusions about whether the hypothesis has, in fact, been disproven.

The first step is often referred to as the null hypothesis. In these cases, the data is analyzed and measured to produce statistical information. A conclusion is reached based on the data. If there are no differences among what the null hypothesis is claiming and what the data infers, then the null hypothesis is sustained.

The second case, when a null hypothesis cannot be sustained is often referred to as an *alternative hypothesis*. If the theory cannot be disproven and the data indicates significant divergence from the null hypothesis (Ho), it is rejected. In the case of our two hypotheses, this rejection of the Ho null hypothesis suggests that it has not been proven that UFOs or ET do not exist. Subsequently, more data is then obtained to continue studying the null and alternative hypotheses as they are stated.

Eventually, if repeated testing is not rejected, the hypothesis can no longer be refuted. Once a hypothesis receives sufficient support, it becomes a proven *theory*.

A theory that has been upheld and thus proven repeatedly attains the highest status in the *scientific method* process; it is now described as a *scientific law*. Some theoretical philosophers and scientists assert that scientific law cannot exist in nature. In one way, they are correct. This is because all things in the universe will eventually change from their current state. Nothing remains the same forever. Another context with which to accept this notion is to say everything will change sooner or later. If, on the other hand, a hypothesis has been rejected, it must be restated or reformulated if it needs to be pursued further.[2]

This is a brief synopsis of the process known as the scientific method. Cosmologist and astrobiologist Carl Sagan described science this way: "Science is a way of thinking much more that it is a body of knowledge." Given this brief introduction to the scientific method and knowing that there are a multitude of details about the process, some of which will be discussed later, at first glance, it appears that there is a lot of work that goes on within testing a hypothesis. The big question is, will the payoff only be a minute increase in knowledge? This captures the essence of Sagan's quote and is, in fact, reality within the landscape of scientific inquiry.

There is a lot of confusion as to whom and where this methodology was created. The "modern" system's history is connected to both Muslim scholars of the tenth to fourteenth centuries and, in Europe, to Roger Bacon in the 1200s (inductive reasoning). Earlier foundational architecture goes back even farther. The most basic philosophical construct goes back to Aristotle of Greece in the third century BCE and Thales of Miletus in the early 500s BCE. During that era, the notions of scientific thought and inquiry were just being created. The study of philosophy started the study of scientific thought and science inquiry. This is where it all began, as a segment of Western thought.

There is no question determining whether any of these responders—refuters, skeptics, and debunkers—are guilty of any fallacious, faulty, or other dismissive thinking. Insight into both the problem of studying UFOs and extraterrestrials—aka extrabiological entities (EBEs), aliens, or exobiological organisms and applications of the methodology with which to study them—comes from accomplished researcher Richard Dolan. In his book, *UFOs and the National Security State: Chronology of a Coverup, 1941–1973*, he talks about "what constitutes proof but who is authorized to deem it so."[3]

Then there is the matter of commentary and editorial opinion. It is a constant struggle between humankind and nature in trying to capture the best explanations for what is happening on Earth and in the universe. The best argument to support this hypothesis comes from theoretical physicist Richard P. Feynman who feels, "Reality must take precedence over public relations, for nature cannot be fooled." Apparently, Dr. Feynman also meant this to apply to the government and the military.

Or consider this quote by noted neurologist and psychiatrist Sigmund Freud: "It is a mistake to believe that science consists in nothing but conclusively proved propositions, and it is unjust to demand that it should. It is a demand only made by those who feel a craving for authority in some form and a need to replace the religious catechism by something else, even if it be a scientific one."

Taken on its initial impression or on a *prima facie* basis of being correct, it would be hard to decide who is authorized to determine what could be considered proof of UFOs and/or extraterrestrials. Authority, as a term, implies a context of power and quintessential rebuttal against the process of rebuttal itself: "Who is going to refute authority?" When you read the later chapters about "Disclosure," you may develop an impression that the government and military act like an absolute authority about UFOs and extrabiological entities and that no one can offer a refutation. Nevertheless, whoever has the authority to decide cannot detract from or negate what is or is not real and what does or doesn't exist in nature.

In his article, *UFOs: The Physical Evidence-Overwhelming-but as Elusive as Ever*, author Michael Jordan asserts that, "The best prospect for achieving a meaningful evaluation of relevant hypotheses is likely to come from the examination of physical evidence."[4]

All of us have used some logical version of the scientific method in our everyday lives. In this step-by-step procedure, we make a statement of claim as the hypothesis after some initial observations are made about a phenomenon or situation. Then we use examples to collect facts and data. Thus, information is measured and tested and will either prove or disprove our claim. Most often in our lives, we will try to prove the claim. This is a different thinking process from what many researchers and philosophers would pursue in their arguments.

The user of the scientific method often tries to disprove the null hypothesis claim (Ho). Some others, a majority of which are social scientists and not professionals in the natural science, will set up his or her study to try to prove the null hypothesis for multiple reasons.

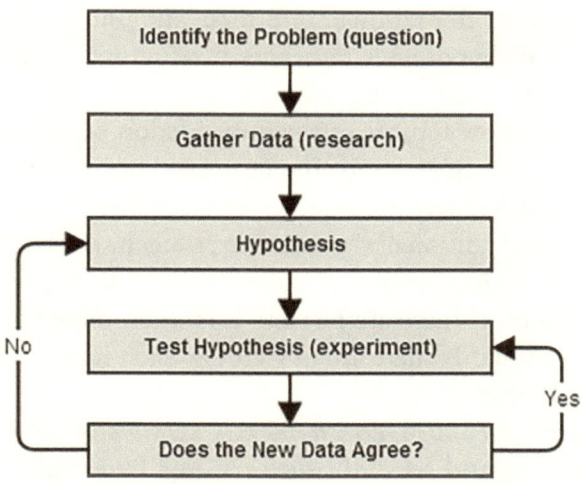

THE SCIENTIFIC METHOD

Most of us use the stepwise diagram of the scientific method the way a social scientist or educator does. The questions and problems in our lives that we try to prove or disprove involve things that do not relate to physics, chemistry, or geology such as "proving to your spouse that Emma does not take the bus home from school every day." This is an example of a hypothesis that would be investigated the way a social scientist would.

Consider this quote by Enrico Fermi, one of the nuclear physicist involved in developing the atomic bomb that ended World War II: "There are two possible outcomes: if the result confirms the hypothesis, then you've made a measurement. If the result is contrary to the hypothesis, then you've made a discovery." This enables us to see how this either agrees with or contradicts one or the other formulations for a hypothesis statement and study to either prove or disprove it.

The scientific method is taught to high school and college students in the hope that they then take those principles and apply them to their careers to offer rational, informed, and unbiased conclusions about studies of hypotheses rather than refutation with unabashed opinions. At least that is the ongoing hope.

Our focus now returns to the introduction of the two hypotheses stated at the beginning of the chapter. They are two profound hypotheses just because of what is represented in the underlying subject matter. Additionally, characterizations have been created by every facet of society. In some logical sense, one hypothesis could lead to the other. If there are not any ETs, then there are no UFOs. How could this be? If there are not UFOs, then how could there be ETs? Or could there?

The terms *ET, alien, extraterrestrial, extrabiological entity, exobiological organism*, and *celestial visitor* are used frequently throughout this book. They will be treated as interchangeable, exactly like ufology literature does.

While there may be a correlation between the two hypothetical claims, on some other levels, there are not, especially if research has found that one or the other or both of the two claims cannot be upheld by the burden of proof that applies to all investigations.

As a consequence, these two hypotheses will initially be addressed separately. Theories that have not been proven or disproven should still be treated independently until one or the other can be conclusively determined one way or the other. When conclusions can be made considering each of them individually, then the synergistic effects of the two taken together can occur.

There is a term used in medical research called an *interaction*. An interaction means there is a positive relationship between two variables in an experimental intervention or test. Drug interactions, for example, are two drugs that when used together cause a different effect on a patient's body in addition to the effects each drug has on its own. Within the examination of two hypotheses, this interaction can have some severe consequences. It is, therefore, prudent to fully investigate them individually before making any attempt at a synergistic analysis.

This book will address these two hypothetical statements at length from many perspectives. The two subtitles defining the epistemology of these two statements are *ufology* and *exobiology*.

Ufology is defined as "the study of reports, visual records, physical evidence, and other phenomena related to unidentified flying objects (UFO). UFOs have been subject to various investigations over the years by governments, independent groups, and scientists, but ufology as a field of study has yet to be embraced by academia."[5]

Exobiology is "a branch of biology concerned with the search for life outside the earth and with the effects of extraterrestrial environments on living organisms."[6]

Another reason for the separate treatment of UFOs and extraterrestrial life is described within these definitions. It should be clear that an unidentified flying object is literally a visual sighting of something with which the witness cannot establish definitive identity. There is no explicit or implicit differentiation of the origin of a UFO. Indeed, at least a simple majority of UFO objects have earthly origins. In addition, the UFO does not explicitly or implicitly prove that extraterrestrial life exists.

Further, extraterrestrial life does originate from outside of planet Earth (or does it; how is that for a "monkey wrench!"). But it would be a fallacy to assume that ET always needs a UFO to get here.

Hence, there will be some separate treatment of UFOs and extraterrestrials and of ufology and exobiology in this text. Also there will be abundant examples of a synergistic treatment of both concepts. Both will be investigated separately and together. Evidence, experiments, and observations will be tested and studied with the objective of offering an application to try to disprove or fail to disprove the statements that UFOs and extraterrestrials do not exist.

A lot of facts, events, situations, science knowledge, and data of a contemporary spatial and temporal nature will also be available to you, the reader, for analysis as you peruse these pages. Also mentioned will be some perspectives of a trivial nature; it is for you to decide to influence the outcome of the two hypotheses presented here.

The most important input on the analysis of the two hypotheses will be you as the researcher. The facts, events, situations, scientific knowledge, and contemporary and historical data will be here for your examination, inference, and deduction.

You will be the scientist, forensic investigator, and juror; and you will be asked to perform the same "input through put output" methods in your everyday lives as they do. You are skilled at it, whether knowingly or not. With this treatise, you will also have fun, learn, and make hypothetical decisions and self-conclusions about two very profound topics that are a lage part of society today.

UFOs do not exist.

ET does not exist.

Were you expecting something different? Please have fun and proceed.

Reference

[1] Consolmagno, SJ, 2005. *Intelligent Life in the Universe*. Catholic belief and the search for extraterrestrial intelligent life. Published and copyrighted by The Incorporated Catholic Truth Society.

[2] Gimbel, Steven, 2011. *Exploring the Scientific Method: Cases and Questions*. Edited by Steven Gimbel. Copyright 2011 by University of Chicago. Published 2011 by University of Chicago Press.

[3] Dolan, Richard, 2002. *UFOs and the National Security State: Chronology of a Coverup, 1941–1973*. Copyright 2002 by Richard Dolan. Published January 2002 by Hampton Roads Publishing Company, Inc. Originally published in 2000 by Keyhole Publishing Company.

[4] Jordan, Michael, 2002. *UFOs: The Physical Evidence-Overwhelming-but as Elusive as Ever*. Journal of Alternative Realities, January 2002; Volume 10, Issue number 1, p. 2.

[5] Definition of ufology. Wikipedia. Copyright 2016 by Wikimedia Foundation Inc.

[6] Definition of exobiology. Copyright 2016 by Merriam-Webster Incorporated.

CHAPTER 2

Critical Thinking and the Scientific Method

The most incomprehensible thing about the world is that it is comprehensible.

—Albert Einstein

Key words: scientific method, human factors, affirming the consequent fallacy, peer review, validity, credibility, process, social constructionist, pedagogy, reasoning, hypothetical-deductive, null hypothetical, alterative hypothetical

As I start the survey about our world and how people interact with it, there are a few major principles I need to define in order to move forward for three reasons. First, to provide useful knowledge to use as new or additional tools to assist in looking at, analyzing, and solving problems that occur in our future everyday lives. Second, to help us gain a better understanding of how science and its member science researchers are encouraged to think. Third, to integrate these topics into the nature of the science versus ufology debate.

I will talk about the first two of these major principles next. These are the *scientific method* and *critical thinking*. The third includes types of *thought reasoning*. This is the topic of chapter 4.

The first of these principles is one discussed previously referred to as the scientific method. It has evolved into a *process* of stepwise analysis and conclusion that science and the science working communities utilize to establish a consistent orderly investigation of a problem or statement of hypothesis. There is no one generally accepted precise definition of the scientific method in use today. If we were to consult ten different sources for their precise working methodology, we would obtain ten different explanations of the procedures embodied in this type of investigation.

That ambiguity is only part of the story. In real life and throughout science history, there are many examples where scientists, laboratories, and other science fields do not precisely follow the steps of the scientific method as defined in their discipline. There are numerous contemporary examples in the literature that exemplify the impreciseness of the practice of this principle. The subject examples range widely through and come from all the science disciplines. In other words, all disciplines are culpable parties.

Two examples of research literature from the recent past for reference include two timeless research books. The first is Bruno Latour and Steve Woolgar's book, *Laboratory Life: The Construction of Scientific Facts.*[1] The second, by Sharon Traweek, is titled, *Beamtimes and Lifetimes: The World of High Energy Physics.*[2]

Bruno Latour is a science sociologist, philosopher, and anthropologist. Steve Woolgar is a science sociologist. *Laboratory Life: The Construction of Scientific Facts* was the culmination of their three-year research study of an in-house investigation at the Jonas Salk Institute. The institute researches biological and biogenetic applications for human benefit.

Anthropologist Sharon Traweek undertook a similar research study of her own. This investigation lasted three years. She studied the life and environment at two high-energy physics particle accelerator campuses: The Stanford Linear Accelerator (SLAC) and the KEK, a similar campus in Japan. Both works explored the inner communities of two entirely different science disciplines and discovered and proved modern science does not work in the realm of what scientific method protocol defines as "what science should be." Both books discovered utilization of the same protocol patterns and paradigms characteristic of a proliferation of human factors. These factors are present in the collection of data, facts, and the disposition of any discoveries made. Conclusions were also the direct result of these human factors.

This means that experiments produce a body of facts used by any science research team to discover and uncover more and new facts. Then these teams use various human factors (office politics if one may) to determine which will be used in the analysis of the study questions and hypotheses. Analysis and conclusions are made by the research team and findings published. The last step toward the objective of widespread approval and application of the discovery is the *peer review*. It is a verification step or a type of checks and balances that assists in determining whether these findings are in fact a true representation of nature and reality.

A sort of social politicking reduces the scientific methodology in importance or even replaces it entirely. A reason for this is to expedite a movement toward a conclusion. Final results are then published in the science research papers. There could also be fiscal factors that influence some conclusions to problems and/or the ultimate disposition of those findings for later practical applications in science. All or some of these factors could be present to allow the science community to produce their results. They are not always the correct or even the best findings nor does the general public gain knowledge of these results all the time.

So what is the scientific method? *Merriam-Webster* defines it as "a method of procedure that has characterized natural science since the seventeenth century, consisting in systematic observation, measurement, and experiment, and the formulation, testing, and modification of hypotheses."[3]

A contemporary formulation of the principles of the scientific method is discussed in Peter Godfrey-Smith's book, *Theory and Reality: An Introduction to the Philosophy of Science*. To summarize, the *hypothetical-deductive* method is a type of scientific method. The scientific inquiry continues by formulating a hypothesis in a way that could allow it to be proven false. The hypothesis testing that could and does run contrary to predictions of the hypothesis is taken as a falsification of the hypothesis. A test that could but does not run contrary to the hypothesis strengthens the argument in favor of the theory. It is then proposed to compare the explanations of the competing hypotheses by further testing to determine the merits of the strength of either or any of the hypotheses under consideration.[4]

This is not a new principle but a reemergence of its interpretation. In his book titled, *The System of the World*, Sir Isaac Newton proposed the process of science inquiry and methodology this way:

1. Use your experience. Consider the problem and try to make sense of it. Gather data and look for previous explanations. If this is a new problem to you, then move to step 2.
2. Form a conjecture (hypothesis). When nothing else is yet known, try to state an explanation to someone else or to your notebook.
3. Deduce predictions from the hypothesis. If you assume 2 is true, what consequences follow?
4. Test (or experiment). Look for evidence (observations) that conflict with these predictions in order to disprove 2. It is a logical error to seek 3 directly as proof of 2. This formal fallacy is called affirming the consequent.[5]

Notice the mention of the concepts relating to disproving and falsification. These principles were reinterpreted during a mid-twentieth century movement by some science philosophers. Many of the proponents of this grass-roots movement were known as being part of the social constructivists or post-positivism movements. One of the founders of the revolution was Karl Popper. In his thesis, *The Logic of Scientific Discovery*, Popper brings the notions of disproving and falsification of hypotheses to the forefront. Popper is "formalizing the attempt to disprove hypotheses rather than prove them."[6]

A good working definition is this: a procedure is done to answer a question about something pertaining to science and nature. Suppose we observe something or an event that piques our curiosity or interest. We usually wish to learn more about what we observed or experienced. This happens to us often in our daily lives. This could be considered a freeze-frame moment.

Now at this point, we may not wish to wait for or seek another example of this phenomenon before forming an educated conclusion about what we observed or experienced. Human nature assures the majority of us will not wait to observe further at this point. To summarize, in the first part of this sequence, we made observations about something. We then formed a hypothesis about what we could conclude about the observation(s) or event(s). These are the first general steps as explained in the template of the scientific method. I emphasize that all of us utilize this procedure in our everyday lives, though we do not consciously think of the terminology or procedure. We just "do it."

Next, more observations and data are collected as evidence to be compared with our hypothesis in a pro-or-con fashion. When establishing a hypothesis, we are looking for evidence similar to what was obtained from our first observations. The more observations collected of this nature, the stronger our hypothesis becomes when making later conclusions. We often find many positive "pro" evidence facts and sometimes "con" evidence facts.

Having collected enough of what we feel to be evidence facts, we can then move toward analyzing the data evidence and begin to form conclusions based on the collected factual evidence. In summary, the steps in using the scientific method include these: make an initial observation of an event. Next, make more observations. If they appear similar in example and scope, a hypothesis can be formed. Next, more observations are made and then analyzed. Finally, make some sort of conclusion based on the collected evidence. These are the general steps in the initial process of scientific method investigation.[7]

There are more nuanced details involved in practicing the scientific method than the shell template just noted. To explain these, let us start again at the first step of making an initial observation. This observation will lead us into one way of thinking or another on the pro-versus-con dimension. In this example, Stephen has a belief that there were no such things as UFOs or aliens. He has had a sighting or experience that contains enough evidence with which to question the matter. Up to then, Stephen's background and life experience (remembering our history discussion in the last chapter) formed his reality. However, some skeptical persons would call this a belief, and they would argue that beliefs are disqualified from scientific thought and argument and that UFOs or aliens do not exist. To them, a belief and reality may be different things. With this new experience, however, Stephen's original belief is brought into question.

An event occurred that caused Stephen to question or maybe refute his original conclusion. This observation has caused Stephen to make a new hypothesis, which includes what he just saw or experienced. What Stephen is actually making in this instance is not one but two hypotheses about the same problem. In reality, there are at least two positions or alternative answers to a situation. He can conclude that something may happen in this way given these experiences or it may happen one way sometimes and another way at other times.

Stephen is making a hypothesis based on his observation or experience. His original conclusion before the experience was that he thinks UFOs or aliens do not exist. This is his default hypothesis, also often called the *null hypothesis* in the sciences and mathematics/statistical sciences. Therefore, Stephen has just made an anomalous observation to his original belief set. Therefore, he then makes an *alternative hypothesis* to the null hypothesis that states, "Because of this observation, I now cannot prove the null hypothesis. Some other conclusion must be made about this problem. I must obtain more data and evidence to make a better analysis. Then I can think about a conclusion given this new evidence."

He then gathers more evidence, which could include, if he is fortunate, yet another experience and/or sighting, or seeks others who can provide similar examples of factual evidence. Stephen gathers and then factors these into his analysis. After this analysis, he returns to his two hypotheses to determine which of these has been better proven. He then forms a conclusion from this procedure. Stephen could state as a new conclusion that he still does not believe in UFOs or aliens or that he does indeed now believe or, as a third outcome, he may need still more data with which to analyze. In the third case, he returns to the experiment venue to collect evidence that is more factual. At that time, the analysis and conclusion cycle is repeated.

This is the practice of the scientific method as it is taught today and how it is presented in school textbooks. The entire subject is discussed in only the first chapter. There are investigative research situations and practiced paradigms that contain more nuanced situations about the procedures involved which are not explained in textbooks.

For instance, in the analysis and conclusion steps of the procedure, it is never explained that in real life there is a lot of debate in the laboratory over interpretations of experimental analysis and conclusions. Karl Popper described this sequence of events in his book, *Conjectures and Refutations: The Growth of Scientific Knowledge* (Popper, 1963).[8] According to Popper and other authors in the literature, members of a research group will argue and often debate over many fine details/facts/evidence about the experiments being performed, the evidence data, and the analysis of it. They are all human beings, and as noted, in any community, there will be human factors, which some have come to describe as office politics. Somewhere within the evidence data collection and analysis efforts, there will (usually) be a debate. This debate will reconcile the differences among the community members. This happens not only in a science environment but in everyday life as well.

To briefly mention how extreme this particular notion of scientific debate can and has occurred in science history, there is another fine 1988 reference book titled, *Great Feuds in Science: Ten of The Liveliest Disputes Ever*, by Hal Hellman.[9] This factual account of scientific infighting involves twenty such great scientists in history including Galileo, Newton, Wegener, Einstein, etc. The ten debates Hellman describes delve into the human factors that have been present in science discourse for millennia. In each of the debates all sense of civility, the science communities have practiced throughout history fell into chaos. Each of these debates offered new knowledge that represented a challenge to the structure of the paradigm that the existing science community was practicing. When the revolution ended, a new paradigm was created embodying the new knowledge and practices accepted from this debate and the revolution.

The scientific method is not just a method of steps that is followed in specific order to obtain a conclusion about the problem. It is in reality a process in which first an anomalous event of some kind is experienced. Next, initial statements about the true nature of if, why, how, when, and where it is occurring are made. More data evidence is collected, analyzed, and debated. Then after the human factors are worked into the process, conclusions of some sort are derived and presented. These conclusions could allow for the rejection of the null hypothesis, whatever it is. One could alternatively accept the null hypothesis. As a third option, the evidence facts could allow for a conclusion where the alternative hypothesis cannot be rejected. Lastly, you could accept the alternative hypothesis as the closest conclusion to the reality of the situation.

The conclusions are presented in a science research paper. Written by team members, it is an invitation for the science community at large to further experiment, collect more data and facts, and prove (or fail to prove) the original research conclusions as having *validity* and some sort of *credibility*. The validity of the conclusions relies on the experimental aspects of the investigation. This refers partly to the physical evidence collection process and partly to the human factors.

An experiment is thought to be valid if others can replicate the parameters of the experimental design in the same manner to produce data evidence. Here, the data facts may not be an exact match for the data facts obtained by the other researchers. In reality, depending on the nature and design of the experiment, often, the data facts obtained will not be exactly as the other experiments produced. However, the data must be in accordance and correspondingly attributable to the experimental design being a replica of the original design. This is so there is assurance that the same things are being measured.

The credibility of the conclusions also relies on the human factors involved in the process. These include consideration of the reputations of the lead researchers and writers, the social politicking, and the fiscal factors mentioned earlier along with other factors. Validity and credibility are two important dimensions used in argumentation and critical thinking studies to add weight to a debater's claims and conclusions as being right and/or better than that of his opponents. These concepts of validity and credibility will be discussed later.

The peer review is the critique element of the entire scientific method process. Peer reviews serve somewhat as the "bias equalizers" whose partial objective is to eliminate the human elements of conclusion formulation from the process. The reviews will either provide more positive evidence, which strengthens the originator's claims (hypothesis or conclusion), or counter it with anomalous data evidence.

When and if enough peers review and corroborate the originator's conclusions, those conclusions can become more factually accepted. Remember, these conclusions will not be irrefutable! Science by the most realistic definition can never conclude something to be absolute all the time. This is because the experimental designs contain one or more faults and most peer reviews do not actually repeat the experiments for fiscal reasons. That is, they cost money to perform.

Eventually, after the results have been repeated or more peers corroborate that they do, many times, the hypothesis would move toward being accepted as a proven theory. A theory is also not irrefutable! Taken separately the peer review has a function of being a "bias equalizer." Together with the scientific research paper, these two procedures are the scientific method's fail-safe mechanism, which should produce correct and meaningful science, in a perfect world.

For a hypothesis in science to move toward proven theory, much evidence is needed and supported by most of a scientific community. Scientists say the most important part of their investigations is the amount of one-sided evidence they can gather. There is no industry standard to indicate "how much is enough." This is a very subjective notion which in and of itself does not contain its own conclusion. A required amount of evidence to one scientist differs significantly to that of another.

This is also a cause for scientists and science communities to move very slowly when working through an investigation. Often, a research study and peer review can take years or decades to move toward some definitive course of a useable conclusion. Two recent examples of this discourse involve the study of the global warming/climate change debate and the evolution of a pharmaceutical product study from conception-to-market.

You know now the scientific method is a process that takes an initial observation or series of observations and causes the researcher to make an initial prediction about a problem. He states his hypothesis and assumes responsibility for and introduction of its alternative hypothesis simultaneously into the study. The research team then collects data and gathers facts and data. Analysis of that experimentation or data collection is undertaken. Then the research team engages in a discourse in which the human factors characteristic of social collaboration, critical analysis and argumentation, social politicking, credibility establishment, and fiscal factors are present. An effective outcome should produce, in an ideal situation, critical and correct conclusions published in a science research paper. Upon peer review and critique and utilizing all the preceding factors to restudy the problem, only if and when the findings of all reviewers (or most and the most influential) are in corroboration, the hypothesis can either be upheld as stated, refuted as stated, or restated, then further investigation would transpire. At this point, the process would continue with another round of evidence acquisition, analysis, and so on. When many successive experimental procedures are undertaken so that results are continually replicable and verifiable, the experimental design has then attained enough validity and the hypothesis enough credibility to move closer to becoming known as a theory.[10]

The most important idea to take from this process of scientific investigation is it has a procedural stepwise arrangement and a very significant and influential sociocultural arrangement within the protocol. The nature of science in reality is not only a perception of the rigid choreographed "this step follows that step" laboratory of practice that you may have imagined in class. There is a just-as-significant sociocultural process within a process that in reality and to a very large degree defines what ultimately comes out as science knowledge, be it correct or not. Refer to Hal Hellman's *Great Feuds in Science: Ten of the Liveliest Disputes Ever* and Bruno and Latour's *Laboratory*

Life: The Construction of Scientific Facts for enrichment on the importance of this part in a more open-minded scientific method.

A concept important to remember when discussing the scientific method is how the hypothesis is structured. The null hypothesis is the default statement brought into an investigation. In our previous example, for Stephen, the null hypothesis states he did not believe in the existence of UFOs or aliens. Knowing that any scientific investigation may entail a debate involving two contrary hypotheses or statements, an *alternative hypothesis* statement needs to be introduced for the investigation to move forward. Stephen's alternative hypothesis statement is "because I have had a sighting of something anomalous and/or a more profound experience, I bring my disbelief about the existence of UFOs and/or aliens into question." The investigation then proceeds from there.

Some scientists would question the use of those two statements in the fashion and order with which they were presented. A major belief of some science researchers regarding hypothesis formation is such that a default null hypothesis statement must be made to *not* include the words "does not," "has not," "did not," or any other structure with a negative connotation. This is because the investigation and use of experimentation and the structure of a protocol in the science world usually seeks to disprove, refute, or negate the default null statement. In reality, there is no law or rule of thumb, which would cause a different connotation of hypotheses statements to be any less warranted or faulty, or one, which produces less accurate conclusions than the other does. The viewpoint centers on the possibility for refutation of the hypothesis and better introduces the notion of science being incapable of concluding something absolutely that favors the former construct.

Here the dilemma to the scientist would be that the positioning of the two default statements is not to their satisfaction. In the example, Stephen is using the "not" connotation in the default statement. So in his alternative statement, Stephen is not trying to disprove something; rather, he is trying to prove something. Some scientists would argue and actually prefer that science protocol demand that the default null statement make a "not negative" connotation so that it is allowable to disprove it.

This is just an argument on the use of the inductive method of critical reasoning versus the deductive method of critical reasoning. These notions of human thought and discovery of a rational reasoning trace back to Aristotle in ancient Greece. The topics of inductive and deductive reasoning deserve more of an explanation but are deferred until the next chapter. Then it will become clearer that these arguments are mostly ambiguity issues within the science studies.

These arguments are part of a larger disagreement present in the landscape of contemporary science and thought. Some of these issues have been presented. The two adversaries in the disagreement over science (and critical thinking), thought, and design are the *positivists* and the *social constructionists*.

Merriam-Webster defines *positivism* as "a philosophical system that holds that every rationally justifiable assertion can be scientifically verified or is capable of logical or mathematical proof, and that therefore rejects metaphysics and theism."[11] The key words are *philosophical* and *logical* or *mathematical proof*.

Science takes its epistemology from the study of philosophy. The evolution of both philosophy and science has been briefly hinted at. In this example, philosophy is the higher discipline from which positivism took its ideology in a hierarchal sense. When science studies a problem or situation, they assert that each situation is to be studied objectively. Positivism maintains that all "authentic" knowledge requires verification and that scientific knowledge obtained scientifically is the only such knowledge available. Some major proponents of positivism include August Comte and Pierre-Simon Laplace.[12]

A situation—which itself cannot be studied to specifications using the above rigid and exclusionary format—is rejected as pseudoscience, metaphysics, or theism among other disqualifying attributions. Positivism is the objective, empirical side of science thought yet still allows some philosophical ideology to be considered.

The positivist movement of the 1800s was popular until the 1920s when a branch of positivism, the Logical Positivism Movement, began to flourish. The stance of the logical positivists was even more hard-lined than that of the positivists. Logical positivism rejected anything that came from the philosophy hierarchy. Advocates practiced their paradigm in even more of a closed system of epistemology than did the original positivists. You can see the progression of science and thought moving increasingly toward exclusivity from society in which the objective seemed to be to isolate their system from the rest of the world of thought and maybe reality. For the communities that made up these fields, the only correct train of thought and hence the only correct conclusions came from this ideology of thought. Metaphysics, theism, and any other types of thought were deemed meaningless.

Social constructionism was and is the opposing ideology that has gained popularity in recent decades. Social constructionists allow for what I have been referring to as the human factors present in any investigation or collaborative study including those involving philosophy and the sciences. I have introduced some of the works of Karl Popper, Thomas Kuhn, Bruno Latour, and Steve Woolgar among others. These philosophers and scientists recognized that every investigation and study of something in science and nature involves a social group. They acknowledged the existence of varying visualizations, viewpoints, and perspectives that each member of that group brought to the investigation. This ideology alone, when practiced, makes science investigations, indeed any investigation, more open-minded and open to ideas being refuted and disagreed on but at least introduced to reduce the possibilities of "missing out" on a more correct description of reality and nature than what existed previously.

I have presented a synopsis of some issues that have shaped science's modern ideology and dogma and critically ties into the discussion. I encourage you to look for additional literature, which further discusses the two opposing systems of thought to satisfy any additional curiosity.

Speaking of "critically," the discussion of scientific thought and how science is conducted and learned is not possible without knowing (having knowledge of) how the training templates of learning science are utilized. This ties into how all human beings think.

Whatever roles we play in life, work or otherwise, our brains are wired the same way as every other human being. What makes anyone's way of thinking apparently more efficient and productive is through "brain training" of some kind. In addition to just memorizing facts and such, our brains can be trained to assimilate and understand so much more about the world we live in. Critical thinking is one such brain-training technique. It is not inherited or part of the hard wiring of our brains. It is taught and learned and has been extensively studied and practiced for millennia.

Critical thinking is the intellectually disciplined process of actively and skillfully conceptualizing, applying, analyzing, synthesizing, and/or evaluating information gathered from or generated by observation, experience, reflection, reasoning, or communication as a guide to belief and action. In its exemplary form, it is based on universal intellectual values that transcend subject matter divisions: clarity, accuracy, precision, consistency, relevance, sound evidence, good reasons, depth, breadth, and fairness.[13]

Critical thinking is a technique, which can be used productively by anyone. The science disciplines like to describe critical thinking as "just another" way to describe what they call "scientific thinking." This is a way the disciplines like to provide degrees of separation from that of other studies. Consider these two quotes from science professor Steven D. Schafersman.

The first one is from his scholarly paper, *An Introduction to Science: Scientific Thinking and the Scientific Method*: "Since critical thinking and scientific thinking are, as I claim, the same thing, only applied for different purposes. If one learns scientific thinking in a science class, one learns critical thinking. This, to my mind, is perhaps the foremost reason for college students to study science no matter what one's eventual major, interest, or profession."[14]

The second is from a scholarly paper titled, *An Introduction to Critical Thinking*. Critical thinking can be described as the scientific method applied by ordinary people to the ordinary world. This is true because critical thinking mimics the well-known method of scientific investigation: a question is identified, a hypothesis formulated, and relevant data sought and gathered. Subsequently, the hypothesis is logically tested and evaluated, and reliable conclusions are drawn from the result.[15]

These are two more examples of the wont of the science disciplines to obtain and retain a measure of separation of thought, epistemology, and *pedagogy*, which are the method and practice of teaching, especially as an academic subject or theoretical concept, in their field from others in society. It is another reason, as previously noted, people have difficulty fully engaging with the science disciplines. In fact, the term *scientific thinking* is essentially synonymous with *critical thinking* with no substantive separation of application.

The steps of critical thinking and the scientific method are equivalent in design and application if not always in scope. An observation(s) is (are) made to create the curiosity in both designs. The thought process is the same and begins in the same way. Data, facts, and information are collected under both schemes. When enough of these are obtained, a statement of hypothesis within the scientific thinking method or an expression of the main idea under a critical thinking method is made.

Next, experiments are conducted within a scientific thinking/method process or an assessment of information/fact sources within the critical thinking process is made. Both methods reference the sources of fact or data gathering. They are, therefore, equivalent though they may not be viewed immediately in exactly the same way. Here, synonymous with facts and data is the concept of evidence.

After all the facts and data or evidence are gathered, the analysis and assessment of these can be made. Again, they are equivalent under both scientific thinking/method and critical thinking schemes. When the analysis is completed within the scientific method, conclusions concerning the acceptance or rejection of both hypotheses (the null hypothesis and the obligatory alternative hypothesis) are made. In critical thinking, the main idea is revisited, as well as an alternative idea, with the same vigor as the scientific method.

The results are generally equivalent on almost all dimensions under both schemes. There is really no difference in whether the situation involves experimental procedures in a science laboratory, observations and data/fact/evidence gathering in the field, jurors being asked to deliberate the guilt or innocence of the defendant, or personal decisions on how to get to and from work the fastest way during rush hour with traffic reports on the radio.

The philosophical disciplines place a high degree of concentration on the debate mechanism as it applies to arguments. The science discipline pedagogies have historically been subordinate to the philosophy disciplines. Therefore, it should be noted that logically thinking, critical thinking ideology naturally and reasonably focuses on a debate mechanism. This is why when with friends or family, we engage in arguments about whatever topic contains dialogue with which we disagree. In summary, critical thinking contains the same processes as the scientific method. In reality, there is no substantive degree of separation between the two, save for the individual perspective and life experience the individual brings to the process.

There is a very informative practice instruction set created by the Learning Development Service of Queen's University of Belfast, United Kingdom. The title of the instruction set is, *Tips for Critical Thinking, Reading, and Writing.* It is a ten-point list of stepwise instruction to improve critical thinking skills. The ten titles are listed below. While the set focuses somewhat on writing and essay skills, the process is the same whether in a situation that does or does not involve writing. The list is portable and encourages regular practice exercises. Details are available from the link in the reference section of this chapter.

The ten steps are:

1. *Understand the question, problem, or situation.* The material in this book is meant for this purpose.
2. *Compose a list of reading reference pertaining to the question if the situation permits.* This obviously applies primarily to critical reading or writing situations.
3. *Obtain the reading materials.*
4. *Try to analyze and use the critical thinking skills needed for the situation.* This includes identifying and evaluating evidence, such as what is presented in this book.
5. *Make applicable notes.* These notes are main facts, data, and evidence for points of analysis.

6. Organize the facts, data, and evidence.

7. *Become aware of your own unique perspectives that others may not have if they were in this situation.* Your visualizations should be objective to minimize types of bias that could cloud the reality and conclusions drawn at the end of the process.

8. *Analyze and merge the facts, data, and evidence with your reflections (your lens) on the experience.* It is vitally important that these analyses are rational, reasoned, relevant, logical, sound, valid, and reliable. These last four attributes are formal concepts that are extensively studied and discussed in any textbook on logic and philosophy.

9. *Present the conclusions you have formed to the situation whether in debate form with another person or through your writing.*

10. *Finally, apply this step only to written analyses and conclusions.* It could also be a step that allows your open mind to respond to debating discourse and/or new evidence presented in step #9 or elsewhere in the environment of the situation. Often, this is the most difficult step for anybody to master and conscientiously use. An open-minded attribute allows you to adjust for changes to elements of the situation in an ever-changing and fluid world. Critical thinking adopts and requires an open-minded attitude to make proper and correct decisions and modifications when needed.[16]

Within the constitution of being a critical thinker, there are a few principles involved pertaining to how a train of thought is contrived. Remember that thinking, whether critically, scientifically, or emotionally is contrived by our brains, as opposed to existing and occurring in or as a state of nature or of the universe. A train of thought ideally takes facts and coordinates them into a form of *reasoning*, which is the subject of the next chapter.

References

1   Latour, Bruno and Woolgar, Steve. *Laboratory Life: The Construction of Scientific Facts.* 1979. Sage Publications – Beverly Hills, CA.

2   Traweek, Sharon. *Beamtimes and Lifetimes: The World of High Energy Physics.* 1992. Harvard University Press.

3   Definition Ref 3: *Merriam-Webster Dictionary.* Copyright 2016. Published by Merriam-Webster Incorporated.

4   Godfrey-Smith Ref 4: Godfrey-Smith, Peter. 2003. *Theory and Reality: An Introduction to the Philosophy of Science.* Copyright 2003 by the University of Chicago Press. ISBN 0-226-30063-3.

5   Newton Ref 5: Newton, Issac. 1687, 1713, 1726. *Philosophiae Naturalis Principia Mathematica, Third Edition.* Book 3; *The System of the World: Rules for the Study of Natural Philosophy,* p. 794–6. Published 1999 by Princeton Press. ISBN 0-691-04334-5. From Bernard Cohen and Anne Whitman's 1999 translation.

6   Popper Ref 6: Popper, Karl. *The Logic of Scientific Discovery.* Copyright 1959 by Karl Popper. Originally published in a paper retrieved April 12, 2016, from: http://www.cosmopolitanuniversity.ac/library/LogicofScientificDiscoveryPopper1959.pdf) p. 17–20, 249–252, 437–8.

7   Scientific Method Ref 7: Gimbel, Steven; as editor. 2011. *Exploring the Scientific Method: Cases and Questions.* Copyright and published 2011 by the University of Chicago.

8   Popper, Karl. *Conjectures and Refutations: The Growth of Scientific Knowledge.* 1962. Published by Basic Books, New York and London.

9   Hellman, Hal. *Great Feuds in Science: Ten of the Liveliest Disputes Ever.* 1998. John Wiley & Sons, Inc. Publishers.

10 Wolf, S. January 22, 2013. *Introduction to the Scientific Method.* Copyright 2013 by S. Wolf. Retrieved April 12, 2016 from: http://teacher.nsrl.rochester.edu/phy_labs/appendixe.html

11 Definition Ref 11: *Merriam-Webster Dictionary.* Copyright 2016. Published by Merriam-Webster Incorporated.

12 Positivism Ref 12: Larrain, Jorge (1979). *The Concept of Ideology.* London: Hutchinson. p. 197. One of the features of positivism is precisely its postulate that scientific knowledge is the paradigm of valid knowledge, a postulate that indeed is never proved nor intended to be proved.

13 Definition Ref 13: Defining Critical Thinking. The Critical Thinking Community. Foundation for Critical Thinking. Retrieved August 28, 2016, from: http://www.criticalthinking.org/pages/defining-critical-thinking/766

14 Definition Ref 14: Schafersman, Steven D. 1997. *An Introduction to Science: Scientific Thinking and the Scientific Method,* p. 2. Copyright 1997 by Steven D.

Schafersman. Retrieved August 28, 2016, from: http://www.geo.sunysb.edu/esp/files/scientific-method.html

[15] Definition Ref 15: Schafersman, Steven D. 1991. *An Introduction to Critical Thinking*, p. 3. Copyright 1991 by Steven D. Schafersman. Retrieved August 28, 2016, from: http://facultycenter.ischool.syr.edu/wp-content/uploads/2012/02/Critical-Thinking.pdf

[16] Definition Ref 16: Learning Development Science. Tips for Critical Thinking, Reading, and Writing. Queen's University Belfast. Retrieved August 28, 2016, from: https://www.qub.ac.uk/directorates/sgc/learning/FileStore/Filetoupload,628271,en.pdf or http://www.qub.ac.uk/lds

CHAPTER 3

What Is Science?

Yet, in holding scientific research and discovery in respect, as we should, we must also be alert to the equal and opposite danger that public policy could itself become the captive of a scientific/technological elite.

—President Dwight D. Eisenhower
Presidential Farewell Address (January 17, 1961)

Key words: pseudoscience, Occam's razor, peer review, rhetoric, dialectic, ethics, physics, skepticism, inductive reasoning, deductive reasoning, scientific method

What Is Science?

A transition from the general path of this investigation of the existence of UFOs and extraterrestrial life to a new topic is warranted. The reason for this is some arguments, battles, debates, debunking, and misconceptions have continued between the two sides of the UFO/Alien argument—the "nays" and the "yeas." Indeed, a battle such as this has existed for many years. As you may know, many science professionals doubt such existence and reality. Ufologists sit on the other side of the disagreement. There are some from both sides that are more skeptical, and some are more pragmatic. This book intends to disprove a hypothesis that either of these does not exist.

This chapter and the next few will provide an understanding of and unravel some of the causes and the evolution of multiple levels of disagreement between the two sides. Many explanations have been presented about different aspects of the subject called *science*. The explanations will address some misconceptions that most people have regarding science. Chemistry, physics, biology, and so on are the individual disciplines that comprise a study of the sciences and are the subjects that will frequently be referred to in this text.

To most of us who are not scientists and to some who are, the generic term invokes a kind of "mystique." In our youth, we are exposed to a required set of school subjects introducing the various disciplines that make up the more generic term *science*. Many perceived science to be this subject matter that extended to the other subjects we were/are taught in school, and the same holds true for mathematics. Mathematics and the sciences are taught as two distinct subject areas and are necessarily perceived by us to be on two separate platforms. There is at least one major reason for this. Mathematics is a language in and of itself. Science has maintained a separate language of terminology, syntax, and notions for use within the construct of science instruction and discourse about science. Each of the science disciplines has its own language, nuances, and unique notions that help define that science. The science disciplines consist of anthropology, archaeology, astronomy, biology, chemistry, geology, and physics. There are subset subjects within all these disciplines. In addition, the language of each discipline defines how it operates in the field and how it is taught in the classroom. This language protocol is a reason many in society are discouraged from making an effort to study or perhaps pursue a career in the science or mathematics fields.

The mystique of science is this perception that because of how science is constructed and conducted that the subject behaves differently from the other subjects taught in school. Here, *differently* means "foreign." Science seems foreign to most of us because these subjects are taught from a lot of notions and concepts that are abstract and inferences of other things. Examples of these concepts involve things in nature not visible, inaudible, or intangible at least directly such as individual atoms or a quasar in outer space. Most other school subjects are taught less abstractly. Our life experiences and backgrounds help shape and comprehend any subjects' content in meaningful ways.

Science does not do this as easily or effectively. We are taught a new language of terms and facts in an attempt to help explain things we are learning in that class or being taught "to a test." While there are some facts and content in science that are pretty good representations of how a phenomenon behaves in nature, most notions in science and nature need a perspective from our own life experiences and background to make the most sense and meaning for us.

For example, in physics, all the introductory courses start with a very brief introduction to the science path to investigation of anything in nature. This is called the *scientific method* and will be discussed extensively in later chapters. Then the student is lead directly into an extensively abstract mathematical interpretation of an explanation of gravity. Most of the time, there are little to no examples of real-life experiences to aid students in understanding the more abstract elements of science concepts.

We proceed through our school years with this type of abstract learning and instruction, which can be very frustrating. Eventually, because of these and other issues related to learning, many of us become discouraged from choosing science as a career. Current education policy and textbook administration are often cited as reasons for most of these problems.

Those who work within the science communities are versed, at least on some intuitive level, of how a new science subject or discipline evolves. An example of the creation of a new science subject is noted here. It may be helpful to think about an example of this evolution in action. Here, the example will be the development of the scientific discipline of genetics.

Gregor Mendel discovered the epistemological foundations of the core concepts of genetics. One of the concepts he discovered is that of the gene. From his collective experimentation and study with pea pods and other plants, he was able to hypothesize and prove (deduce and induce) that living matter on Earth possessed the substances that he eventually labeled genes. That term was universally accepted.

In his time, genetics was not a subject according to the science professions. In fact, Mendel himself, and his work, was ridiculed and shamed because he was not a scientist according to the scientific community. These scientific communities explained that Mendel's work was incredible, as opposed to credible.

The term *credible* is a vitally significant one which researchers use as a basic convention in determining how scientific writing and literature is to be considered and accepted by the communities. The main attribute of *credibility* resides in the person who is making the statement claim and the scientific writing. Credibility, therefore, is a complement bestowed upon the author by the reader and the community at large. It is supposed to serve as a type of vetting step, which helps direct other readers to the acceptability of the writing.

The term *incredible* is one which you can reflect on, using your logic and reasoning skills, to assess given the explanation of what credible is in this context. Incredible is not really a polar opposite of credible, though. It is possibly to be thought of as a continuation of the scope of what credible is. Incredible is an outlier when considered on the same measurement scale.

When using this context, incredible can be discussed with what is termed *pseudoscience*. This is a favored term use within science communities to describe phenomena and concepts that are too anomalous to fit into their practice and experiment protocols, too incredible for them to be offered the complement of credibility. These protocols dictate that experiments will only happen when a researcher challenges insight into one specific, often minor, aspect of a topic being investigated. The tests are conducted, data collected and analyzed, and conclusions are made. If the new knowledge is confirmed by the data, it is peer-reviewed and, if agreed upon, finally accepted as fact. Pseudoscience, to this culture, is too much of a leap of faith, which they feel requires too many topics to be tested. Sort of like algebra protocols, if there are more than one unknown in a problem, then the answer cannot become real or fact.

The attribution was what the science communities assigned to Mendel's body of work and his studies. He was a Franciscan monk by trade and studied genetics as a hobby. These two factors conspired significantly against acceptance of the man and his work. His meticulous note-taking, though, was an instrumental factor in the ultimate foundational groundwork toward the successful evolution of the first texts about genetics. But all of it was ignored and ridiculed by the scientists and the like for over thirty-five years as pseudoscience. An extreme attribute used by some critics during that time was to explain away his work and concepts of genes as magic.

There are two additional situations presented here but left to you to explore on your own. One is the work of Alfred Wegener, a meteorologist and polar researcher. The scientific communities ridiculed his discovery of the geophysical theory of the continental drift for decades. Because his work was so unusual and did not follow the paradigm of incremental knowledge acquisition but instead leapfrogged existing knowledge, he and his work were ridiculed and ignored. It was thus branded as pseudoscience. The other situation that occurred precisely as the Alfred Wegener situation did was Albert Einstein and his theories of space-time and relativity. And so this is how an evolution of the creation of a science discipline is allowed by the industry to function.

Coming up next, you will learn of more of these situations that occurred in science history. To summarize, a science discipline does not become recognized as a discipline in the field of science until it evolves through a stage of pseudoscience, sometimes deemed as magic. However long that stage takes depends on many other factors aside from our main focus. The main point is that through history, this is how all science disciplines began and evolved.

It is important to gain a better understanding of what science is in order to do so; we need to explore the main science versus ufology arguments. The science disciplines could embody the learning aids of life experience and educational background, but to understand what science is, we should go back in history and learn about the progression of the subject. From this, we will know that in our society, the current education policy's inadequacies and shortcomings are not just newly created phenomena but are in fact those that have lasted for a long time. Current and past generations were taught science in the same way. The mystique of science is, thus, being perpetuated through society in this way. The notion of the mystique of science also contributes to the disagreements the field of science has with ufology and the nature of our two subjects of discourse: the existence of UFOs and of aliens as extraterrestrials.

The purpose of presenting this is to help you to decide the nature of the two hypotheses in this book from your own unique life perspective and cultural experiences, science, and thought. As you will see, science is not a lengthy catalog of irrefutable fact, terminology, and content that is force-fed to you for your memorization in a classroom, though you may have embodied of this impression in your school or other learning years. It is a main reason most of us would not pursue science as a career. The procedure you would use to determine for yourself what your conclusions would be for the two hypotheses presented is grounded in thought that relates to what science is about and how it has been taught for thousands of years.

Before tackling the question of what science is, we should discover what science is *not*. This discussion is presented as following an example of a course of investigation protocol first theorized and used by English science philosopher and Franciscan friar, William of Ockham (1287–1347). His principle of investigation and thought is known as *Occam's Razor*. The definition is such that if two or more courses of action or conclusions can be drawn from a hypothesis, the simplest one with which to pursue or to answer is most often the correct one. A contemporary translation of this principle would be, "Keep it simple." Use of this principle will greatly simplify the discourse that follows.

What Science Is *Not*

1. Science is not able to conclude an unequivocal or undeniable *fact* about any particular thing being studied. No absolute certainty can be concluded about anything in science or nature. A thing or event is only more probable to occur or not to occur than an alternative to it.

2. Science is not an undeniable truth or unequivocal fact. Science has only one objective: to gain a better understanding of how things in the natural world work. Your experi-

ences help to achieve the answer to succeeding questions that arise about science and nature.

3. Science is not able to prove anything. As you will see shortly, a methodology of science is of how to *disprove* an idea that is proposed. Additionally, if some "thing" cannot be disproven, it does not mean that it is proven. It only means that the "thing" has a stronger base of an explanation than existed before the attempt to disprove it.

4. Science is not able to be totally objective. Though science methodologies, such as the *peer review* process, are used in attempting to achieve some level of objectivity to conclude that science can be absolutely objective about anything is a fallacy. Science is comprised of human beings practicing their professional discipline as scientists. Without humans, there would be no science. So humans are necessary in order to conduct science. Humans also have biases. These biases are precisely results of and ramifications from the very same unique personal life experiences and knowledge backgrounds that shape scientists and our perceptions and viewpoints about the nature of the world around us.

5. Science is not as fashioned from reason #4 above; science cannot be entirely relied upon to be consistently objective. There are mechanisms structured into the practiced format of science investigation that offers help in deflecting some of the biases, subjectivities, and bad science practices. These are not all used effectively all the time.

6. Science is not a catalogue of the best answers to a problem or issue. There may be steps taken to help determine an answer that holds correctness for a time. Because science deals only with the natural world and that natural world is always changing or fluid, it too will have a different answer to an issue either at the same time as the other answer is concluded or at a later time. This also indicates science is changing or fluid. As follows from reason #1, a fact in science is not irrefutable.

7. Science is not a study of anything in nature in which the conclusion is based on than belief or faith. This includes religion.

8. Science is not currently able to address issues of nature or inquiry beyond those that occur in three dimensions without much difficulty. As you have seen elsewhere, answers to questions concerning matters involving the fourth and additional dimensions, the ones that Einstein and other theoretical physicists and cosmologists have been hard at work researching for centuries, are very elusive and create even more questions for each one that is answered. A large part of this example and more generally any such matter of science inquiry demands the requirement for inventing new measuring instruments and other technologies to be able to address the questions being asked. In the theoretical physics field, the CERN Hadron Collider in Switzerland needed to be invented and built before any such research could be undertaken.

9. Science is not technology in and of itself. The term *technology*, as defined, is an apparatus that science uses to study something in nature. Indeed, technology, when describing new machines or tools, has been unequivocally useful in discoveries of virtually all new knowledge in all the science disciplines through the centuries. All sciences, from astronomy to zoology, benefitted from new technology.[1,2]

There are other more available nuanced notions of what science is not. The ones listed here are those we recognize when we think about what science means and how it affects us.

What Science Is

Science is a process of methodology and a study of the investigation of things and issues in nature.

This is the best and shortest definition, although there is more to it. Understanding that nature includes the environment in which nature is found. Here, environment does not imply only the outside environment. It more fully means the environment around the thing or issue in nature that is being studied. There is an external environment when studying something as small as a human cell, as large as a solar system, DNA, igneous rock deep inside the earth, archaeological remains excavated in South Africa, an atom of iron, or the centrifugal force of an object in space within the physical sciences.[3]

Many scientists unfortunately place their way of thinking about nature within a closed system. Many scientists only think of nature as being of and around planet Earth. Nature does not exist on any other planet or satellite except for that of the earth.

This is a fallacy of egocentric thinking. They simply ignore any occurrences of natural events or and exclude them from the discourse on the subject.

Science can be intimidating to many of us. Its methodology of study includes many steps we do not use elsewhere in our daily lives. The subject uses too many abstract concepts, languages, and epistemologies to describe phenomena we cannot often *see* happening. It uses awkward and ambiguous terminology which is not used elsewhere in our society and is taught like a "mystical" subject. Each of the disciplines of science, the various subjects that comprise the curriculum defined as "science," have their own language of terminology, syntax, their own way of talking, their own paradigms, their own substantive structure, and their own culture of conducting science in that course of study that is different from the others. From this, we can add two additional characteristics to this schematic.

Science could do a better job in explaining their knowledge to us. While most scientists disagree with this notion, it is their responsibility, professionally and ethically, to do this. Here is another reason why they should not disagree with this notion.

The general public, being end users and learners of science content, have been asking for many decades at least that all researchers and other science community participants be more aware of and practice more, better, more effective, transparent, and readily available communication. This is empirically logical in that funding for most research study papers, articles, and other published materials are dependent upon and have their origins from the public in the form of government grants. Without those funding sources, their work is not possible.

The study of the sciences has recently been itself studied voraciously via educational research. There is plenty of research literature. Incidentally, science research papers are one of the "steps" used in any study of a scientific phenomenon. From before the 1900s and continuing right through to today, many scholars, scientists, and philosophers have weighed in on the educational aspect of science. Specifically, a lot of conclusions and proposals have been put forth calling for future radical change to how children are taught this subject.

The current pedagogy or theory and practice of science education has unexplicably been in use since before the 1900s. It is the practice we experienced in our classroom instruction. Your science courses consisted of memorizing facts and theories, learning new terminology, and trying to construct concepts from abstract visualizations. Most often, these visualizations were from only your own mind and not with the help of any digital technologies available today as learning aids. Learning this way from unreadable textbooks, uninspiring lectures, and participating in the occasional hands-on experimental laboratory activity was often unengaging and unmotivated. Our life experiences, background knowledge, and cultural history were excluded from science course content. Because of the significant influence from education policy and the textbook publishing industry, teachers found it easiest, given these inputs, to "teach to the test." This pedagogy allowed them to better maintain classroom discipline throughout each semester. The rigidness discouraged most from being interested in studying science further than the required amount to graduate.

Even with the wealth of educational research literature from such gifted twentieth and twenty-first century contributors such as John Dewey, Thomas Kuhn, Karl Popper, John Schwab, Jay Lemke, and Karen Gallas, science education still plods along in its long-existing paradigm with some hope for only incremental change and development of new and more efficient education policy.

As discussed elsewhere, science can be thought of as a process that evolved from a separate advancement of the subject of philosophy and philosophical thought. In the advancement of philosophy, the concepts of nature and knowledge embody the process of how humans observe and interpret the world around us. Science application and protocol flow from this philosophical process.

As previously noted, throughout history man's brain is hardwired to communicate and explain by way of analogy. All forms of language—including written text, symbolism, artifact, or structure—use analogies to describe things in nature.

Science, in turn, is but one application of this philosophy. Science explains through analogy also. In his book *Intelligent Life in the Universe*, author Guy Consolmagno, astronomer for the Vatican Observatory, describes the role and outcome of science that was borne, as you will read, in ancient Greece: "All language and all explanation or scientific is by analogy. Certainly, science explains and describes by analogy." In the context of this book, in surveying possibilities of intelligent extraterrestrial life, Consolmagno continues, "It is difficult to talk about 'life as we don't know it.' We don't have the words to describe it." The ancient witnesses to this type of encounter had no language with which to express themselves, so they were unable to describe what they were observing. They did their best to describe what they saw, just as we do today. Consequently, Consolmagno further defines the boundaries of science in this manner.

> Even a scientific equation is metaphor for behavior in nature. No equation is ever a perfect description, but an approximation of reality that is always simplified and incomplete. Science never gives us absolute complete knowledge of the truth.[4]

We take from this that maybe the process of science inquiry is incapable of explaining everything all the time. From the application of the process of the scientific method, in reality, this process does take a very long time to evolve hypotheses into theories and laws. They do not explain this to us.

Do we place too much faith in science and scientists to be the only application or tool of the philosophy of nature and knowledge that allows us to describe the correct recognition of reality? A plethora of examples of science's shortcomings can be seen in these areas through history throughout Consolmagno's book.

With regard to the hypotheses in this book, maybe science is not the only discipline that should be involved in any comprehensive study of the existence of UFOs and extraterrestrial life, whether intelligent or not. These hypotheses of discovery are and will be among the most important for humankind. If you were to look up the definition of life in a dictionary, a biology text, a chemical text, and in a text of any of the other natural sciences, you would find significantly different explanations and details. Each science subject has a different perspective on the term *life*, which is from their trained and biased point of view. There has to be a better attempt at consensus which could include other perspectives.

Consequently, maybe science, at least not any one particular science discipline, should not be the only discipline that can honestly explain and make comprehensive conclusions that are an accurate representation of reality and nature for the study of ufology and its objectives.

Now you may have some understanding of why you might feel the way you do about learning about science, so let's move toward being able to apply some of your learning to the battle between science and ufology and the purpose of this book. First, though, it is important to describe how science is structured. The best way to accomplish this is by starting with a look at the historical heritage of science. Incidentally, it has been widely agreed upon in science education research that education policy has caused some of the current problems with science learning. Students from before the twentieth century until now are not made aware or introduced to any meaningful history of science nor of any history of the individual science disciplines. Once you are aware of this history, you will start to gain a clearer picture of how science, scientists, and philosophers act and react to an issue of science in nature and to our two hypotheses under study. Here next is a look at some of science history and structure of the disciplines of the sciences.

The field of science was first studied following the study of philosophy. The study and structure of the subject of the sciences today rely heavily on the epistemology of philosophical research and its history. The study of philosophy dates back to ancient Greece of over 2,500 years ago.

Thales of Miletus (625–546 BCE) of Ionia, eastern Greece, is agreed to be the founder of Greek philosophy. Also a mathematician, Thales was the first recognized to apply the notion of *deductive reasoning* to a problem under study. This was a product of his mathematical background, especially in geometry. Deductive reasoning is a very basic tenet of scientific reasoning today. Thales applied his background and experiences to the study of the nature of matter. He proclaimed water to be the basis of all things. This is important in that it became the first referenced instance of the notion of something being explained as matter the way we have come to understand it today. The notion of water being the basis of all things was expanded to later include fire, air, wind, and earth.

Thales's advocates expanded on the study of the nature of matter and things in nature. It would not be until the 400 BCE that Leucippus and Democritus of Abdera attributed the use of the term *atoms* to that of matter. These atoms were the indivisible bodies, infinite in number that could be combined in infinite ways. This became the basis of another tenet of science that of form. Atoms were deduced to determine the form of something according to how these atomswere arranged. These notions sound and look familiar to those we hold today, including the study of chemistry.

Thus, the notions of matter and form, the "practical" life, are put to use in further learning about the nature of things. The origins of the subject of philosophy emerged in Athens and Samos simultaneously with the emergence of the study of nature. Pythagoras of Samos (582–504 BCE) and Anaxagoras of Clazomenae (500–428 BCE) were credited with the creation of the structural aspects of philosophy as a course of study using new ideas and of divine reason to establish an ordering principle of matter, form, and their generation and disappearance as what happens in nature. These pioneers of science concepts allowed ancient society to become aware of their lives, their environment, and their interactions.

From the breakthroughs in the development of the notions of reasoning and of human knowledge and understanding came an additional principle of human thought and not just scientific. Advocates who descended from Pythagoras and Anaxagoras became what was known as "Sophists." They were educators who specialized in using an epistemology of *rhetoric*. Rhetoric is a basic component of the study of critical thinking today; language that is intended to influence people and that may not be honest or reasonable. It is the art or skill of speaking or writing formally and effectively, especially as a way to persuade or influence others.[5] This is also a "tool" used in scientific debates.

A new notion was introduced around this time in Greek history. The Sophists also charged money for their lessons of learning. Consequently, only those who could afford their lessons could participate in them. These educators were the elite of the Greek city-states. This is a very significant principle that had a part in laying down the origin of the notion of science evolving as an elitist function and was one of President Eisenhower's focuses in his presidential farewell address as quoted at the beginning of this chapter.

Unfortunately, access was not available to those who could not afford the lessons, which was most Greek city-states. This series of events may aid in rationalizing the notions you've concluded as to why and how the study of the sciences seems to be out on the peripheral thought platform as the mystique referred to earlier. If studying science seems to be some distance away from other core subjects of study and difficult to comprehend, the evolution of this elitist practice is a major piece of evidence indicating how it all began.

Charging fees or tuition for teaching content also angered the next succeeding great Greek philosopher, Socrates (469–399 BCE). He was generally disgusted with the idea of inaccessibility of these teachings by his mainstream society. Socrates's accomplishments followed along the lines of the philosophy of man. He added other elements as important learning tools of some practices in the legal profession. Socrates's main claim to fame was that he compiled an encyclopedia of the views of the "virtues of people." Those virtues included such legal notions as justice, courage, morality, and reasoning. He spent his years acting like a cross-examination attorney attacking people's claims and perspectives on any subject.

Socrates was ultimately executed for his continued preaching along these paths and of "corruption of the minds of the youth of Athens." In any society, there is much validity and sentiment attributed to the legacy of a person who is persecuted, especially for these reasons. Martyrdom precipitates being remembered and revered. As a result of these reasons among others, it appears evident that the legacy of Socrates continues uncompromised and unaltered.

Plato (428–348 BCE) was fortunate to be an established student of Socrates before his teachers' untimely assassination. He was credited with organizing all the scattered efforts of earlier philosophers into a comprehensive system of study. There are three branches within this central system. They are the *dialectic, ethics*, and *physics*. The dialectic—which consists of the dictionary, language, and discourse of philosophical argument—is generally recognized as the first of all the science disciplines. In addition, the notions of critical thinking were also borne here.

The dialectics, together with ethics and physics, formed Plato's school curriculum called *The Academy*. You can see that the platform of human thought, reasoning, rationality, and argumentation was very basic to Plato's academy. All the early Greek philosophers took thoughts and ideas and started to build a curriculum of civics and civil affairs as a precursor to the discipline of the social sciences. More importantly, they created a curriculum of the study of nature, matter, and form. What emerged was the first comprehensive study of the sciences for humankind. These essentially were the origins of the first philosophies of scientific thought.

A generation down from Plato, one of his students, lived Aristotle (384–322 BCE). It was Aristotle and his followers, known as the Peripatetics, which separated themselves from any residual connections to the notions of *metaphysics* as they may apply to anything mystical or mythological in society. Metaphysics is defined as "the branch of philosophy that deals with the first principles of things, including abstract concepts such as being, knowing, substance, cause, identity, time, and space. A study of abstract theory or talk with no basis in reality."[6]

People in the early Greek societies still and would continue for a while after lived their lives according to the mythos of the universe. So philosophical thought was a large break away from the societal views of those times. This is a central motivation for why Socrates was executed for his "crimes."

Aristotle's main claim to legend was not of any great contribution to science but of developing a system of logical discourse and argumentation in a debate format. His creation of distinct argument constructs such as critical thinking included the classic syllogistic logic processes that are still used in today's society. You can find Aristotle ideology in most all subjects from science to the socio-sciences, law, business, politics, military, and so on.

Aristotle and the later Peripatetics gave great effort to popularize their doctrines with the masses of their society. As referenced earlier, philosophers in Plato's time and also before and after suffered the recurring problem of acting like elitists. The profession of philosophy in its entirety also suffered from characteristically sheltering themselves from mainstream society in their elitist practices and attitudes. This paradigm of the "mystique" of philosophy and later the science disciplines is yet another characteristic that traces practices and doctrines of the sciences back to these origins. Scientists up through today's time have been taught by this same paradigm structure for the last 2,500 years. They will continue to be taught in this way into the future for some amount of time unless changes are made.

One last piece of historical Greek lineage that links ancient science thought and learning, as they were practiced, with today is the notion of *skepticism*. In Aristotle's later years, the origins of the Skeptical School were being conceptualized by Pyrrho of Elis (365–275 BCE). Know that the philosophy profession was trying to reconcile the many arguments and refutations in the attempt to construct a "theory of everything." This everything included the existence of man, his environment, and the nature of it all. Matter, form, and truth were being shaped into an epistemology with an end point of a successful future of humankind.

So the basis of the Skeptical School and Pyrrho's doctrines was predicated on these attributes: what is matter and form; how are we, as humans, related to it; and what should be our attitude and perspective toward these things. A definition of skepticism in philosophy and science is "the theory that certain knowledge is impossible. A skeptical attitude; doubt as to the truth of something."[7,8,9]

If one holds that nothing in science can be concluded as a fact or ultimate truth without the ability to be refuted, disclaimed, or unproven, then skepticism can exist also and is a basis for explaining the nature of things.

This is a basic tenet of contemporary science thought and study. It is also a state of mind which is blended into contemporary scientific dialogue. This is another point of the mystique of science. Other courses of study and learning are not made as requisite for skepticism as the sciences are to children in the classroom. As former students, this has also shaped your attitude toward the sciences as being mystical or mysterious. You may not have been conscious of this but recognized something like it from intuition.

Before moving away from the historical background of the pedigree of the science disciplines, the preceding discussion will be summarized. When one picks apart the underpinnings of the paradigm of the subjects or the disciplines of the philosophy and the sciences, the origins of most of this subject of thought are traced back to ancient Greece of the period from 2,700–2,200 years ago. When you interact with a scientist, a science teacher, or a science researcher, you will now know the origins of a good part of his or her train of thought. Beside the scientists' life experiences and unique background, their education in the sciences—all in combination—will determine how they view, study, analyze, respond to, and conclude a problem or issue in science presented to them. This is not to say anything as to the degree of validity of how, why, or if a scientist correctly studies a problem or issue in science. If everything in nature is changing, then how can there be a final correct answer or truth to a problem in science? This is one way in which a person of the science professions is motivated to think.

Let me next bring together some concepts we have just covered in the last few pages. These concepts represent methods of thought processes and are not unique to the study of the sciences. They were originally conceived and first used by the philosophers of ancient Greece. Soon thereafter, there were slow, incremental additions to these bodies of doctrine and study by the ancient Roman philosophers in the centuries immediately after the Golden Age of ancient Greece. The period lasted from 800–250 BCE. Other cultures such as ancient India, Western Asia, Mesopotamia, and even Egypt made important contributions to the philosophy and science schools of thought.

With this basis in reasoning and thought, critical thinking, argumentation, and skepticism the Greek philosophers created and refined next, you will learn of a couple of principles about the human mind, reasoning, and the construction of thoughts. These principles of thought are titled *inductive reasoning* and *deductive reasoning*. Let me also introduce a concept that merges these two principles into a framework of scientific thought and process.

This concept is titled the *scientific method*. To review, the scientific method is a process of thought and investigation to an issue in science which tries to explain something in nature. Remember that the notions of discovery, of thought, and how they relate to the natural world was very basic to the Greek philosophers' argument set. This is the "umbrella" that helps to govern how a scientist reasons through a problem he is researching. In reality, it is only a part of the entire process of scientific reasoning or at least how it should ideally be investigated and used in the study of a hypothesis or problem. There are more elements to the entire process such as analysis and refutation, which are also parts of the total process of scientific investigation known to scientific reasoning.

I mentioned before that the notion of critical thinking is immersed through the entire general process of scientific reasoning. The Greek philosophers (remember they were primarily marketed then as philosophers or thinkers and as developers of the early science disciplines: the physics, astronomy, and biology) created, studied, developed, and taught these notions and started trying to tie them together as their "theory of everything" of nature.

The design and use of the scientific method, however, did not originate with ancient Greece; it was developed much later. There is some amount of reference to Aristotle and even as far back as Thales of Miletus for utilizing certain specific traits of a methodology that resemble those of elements of such a method. But there was no "science" discipline of study that existed separately from ancient mythos and deity study in ancient Greece. There were some reference texts in existence that discussed metaphysics, some astronomy, and some biology. All the divine processes that governed the natural world were still immersed into the "soup" that was the metaphysical and mythos of the ancient societies. Later, some uncovered works found in ancient Egypt and Babylonia paralleled the efforts of the Greeks. There were no real science disciplines to acquire and develop the meaning and scope of the methodology they proposed and founded. There were no formal structures of either science disciplines or of the method which could be synergized into a construct.

The birth of a much more organized and recognizable science system is attributed to Ibn Al-Haytham of Iraq. In 1021, he published *The Book of Optics*. It was the first recognized popular use of *experimentation, observation*, and rational thought process used within a scientific discipline. These are some of the "steps" used today in any scientific method study. This was followed by the contributions of a Roger Bacon in about 1265. He was inspired by an immediate predecessor Robert Grosseteste. Grosseteste provided commentary on the subject of *posterior analytics* and gave Bacon a catalytic inspiration to publish his three treatises: the *Opus Majus*, *Opus Minus*, and *Opus Tertium* to Pope Clement IV. These treatises explained the steps in a basic scientific method investigation which we recognize today. This new method bore the most resemblance to today's system than any previously uncovered.

About 350 years later, Francis Bacon (no family relation to Roger Bacon) published *Novum Organum* in 1620. Francis Bacon's work refined the earlier works of the Renaissance period scientists/

philosophers. *Novum Organum* contained the most recognizable written comparison to modern versions of the scientific method system to that date. Other later great scientists/philosophers such as Descartes, Galileo, and Newton among others contributed perspectives that added only incrementally to the construct of what is used today as the modern scientific method.

Hopefully, it is now more evident to you that there were a lot of great inventors and thinkers, spanning the millennia, which contributed to the creation and design of philosophy and science. A lot was developed from the efforts of those geniuses in those historical times. However, it still leaves plenty more epistemology to be introduced, going forward, to make these subjects more effective and efficient in their stated and objective goals of making sense of reality, truth, and understanding for our civilization.

With all this design architecture and development having been explained, it should not confuse anyone as to why, if UFOs and extraterrestrials did exist then, there could not have been better descriptions of them in terms and visualizations we use today and which we would have wanted our ancients to have used back then. It is human nature for anyone to seek out and use contemporary language, concepts, and visualizations most familiar to us and which we are aware of to describe and explain events and things of nature. This is using today's language to provide perspective on those events.

Remember, from any read of ancient texts, that this exact ideology was also used by our ancestors. The language of that particular time that was the only one in which scribes and writers knew of and could use. It is only future translations of those texts passed down through the millennia that have corrupted, more than helped, the true meaning of what the ancients saw, heard, felt, and experienced.

In terms of the terminology, language, and concepts just discussed for our ancestors, the dictionary they had was more basic and a work in progress. The printed dictionary was far thinner than what we have today. So they had fewer tools in their toolbox to use with which to analyze and conclude precise and true definitions of things they observed, encountered, or experienced than we do today.

While the development of the philosophy and science disciplines and the scientific method would continue after Francis Bacon's time, sufficient conceptual framework had been laid down by then to allow us to proceed with more description of the basics of philosophical and scientific thought and process which we use today. Next, we will continue with more discussion of critical thinking, the scientific method, and types of thought and reasoning. You will also learn some new ones which will soon become popular.

This was an abridged version of the history of the evolution of philosophy and the natural sciences. You now have an image of how the sciences came to be the way they appear in our modern society. There are a lot of details absent from here which would be better served if explained in a future book. There are, however, a few major concepts that still need to be further defined to then better integrate this topic into the nature of the science versus ufology debate, which will be discussed in chapter 5.

Let me continue with a specific discussion on the concepts of the scientific method. It has evolved into a methodology, which science and practicing scientists utilize to establish an orderly investigation of a problem analysis. There is no one generally accepted precise definition of "the scientific method." If you were to look this up in ten different places, you would obtain ten different explanations of the procedures embodied in this type of investigation.

That ambiguity is only part of this story. In real life and throughout science history, there are too many examples where scientists, laboratories, and other science study venues do not follow the "steps" of the scientific method as defined in their investigations to begin to mention here. Refer to Bruno and Latour's *Laboratory Life* and Traweek's *Beamtimes and Lifetimes* for examples I introduced in chapter 2.[10,11]

Laboratory Life was the culmination of a three-year in-house research study by Latour and Woolgar at the Jonas Salk Institute. The institute researches biological and biogenetic applications for human benefit. Sharon Traweek undertook a similar three-year research study of her own at two high energy physics particle accelerator campuses: The Stanford Linear Accelerator (SLAC) and a similar campus in Japan, the KEK Institute. Both works explored the inner human communities of researchers working in two entirely different science disciplines. They discovered and proved that modern science does not at all work within the realm of what the ideological protocols define "what science should be." Both books discovered the same protocol patterns (paradigms) in which many human factors were present in the research team communities and which influenced research decisions. These human factors help define the office politics and dictated what conclusions were made public.

Another way to explain this is that too often the methodology preached about in the scientific method is reduced in importance or replaced by a sort of social politicking within that scientific community. This cause and effect take the path to the publication of conclusions in which those human factors have dominated many critical aspects of the research study.

There could also be fiscal factors which influence science studies. Many studies in medicine, for example, are funded by corporate grants and not from governmental or academic sources. All or some of these influences could factor in to the construction of the final published reports. From the Bruno and Latour and the Traweek books just cited and many other examples in these areas, human factors are often a dominating force in published research results. That bias detracts from and eliminates a published explanation of some of the science facts, truths, and discoveries which were really made.

So what is the scientific method? A short definition may be that it is a process which is undertaken to help answer a question about something in science and nature. Suppose you observe something or an event that piques your curiosity or interest. You wish to learn more about what you observed or experienced. This happens to everybody in their everyday lives. This would be a freeze-frame moment.

Now at this point, you often may not wish to wait for or seek out another example of this phenomenon before forming an educated guess about concluding what you observed or encountered. Human nature assures that the majority of us will not wait to observe further at this point in order to make a conclusion.

To reiterate, in the first part of this sequence, you make and record an observation about an event or phenomenon. Next, you form a hypothesis about what you could reason from the observation(s) or event(s). These are the first general steps in practicing the scientific method. All of us utilize this procedure often in our everyday lives, though we do not consciously think of the terminology or the steps within the process. We just "do it."

Next, more observations and data are collected as evidence points. When you make the hypothesis, you are looking for more evidence examples of similar finding as you obtained from your first observations. The more observations of similar features you collect of this nature, the stronger your hypothesis becomes when you then make conclusions about the hypothesis. Often, you find many positive "pro" evidence examples and sometimes you also find some "con" evidence examples.

When you have collected up enough of what you feel are evidence points, you can then move toward *analysis* of the data evidence and then begin to form conclusions based on the evidence you obtained. To summarize the steps in using the scientific method, you make an initial observation of an event. Next, you either make more observations, and if they appear similar, then you form a hypothesis. You continue to make even more observations and then analyze all them. The last step in your process is to focus your analysis to form some sort of conclusion based on the evidence you have obtained.

There are more nuanced details involved in practicing the scientific method than the overview I just presented. To show you some of these, now let's start again at the beginning step of making an initial observation. That observation will lead you into one track of thinking or another or the "pro versus con" path. For example, let's say that Stephen holds a belief that there are no such things as UFOs or aliens. Stephen now has just had a sighting of or encounter with such which is evidence enough for him to ask questions. Up to that day in his life, Stephen's previous background and life experience helped form his conclusion (some skeptical persons would call this a "belief," and they would argue that beliefs are disqualified from scientific thought and argument) that there is no such thing as UFOs or aliens. With this new experience, however, Stephen's conclusion set is brought into question.

At this point, an anomalous experience happened to cause Stephen to question or maybe refute his original conclusion. This observation or experience now causes Stephen to reevaluate his previous position and to form a hypothesis about what he saw or experienced. What Stephen is actually making here is not one but two hypotheses about the same problem. In reality, there are at least two positions or alternative answers to a problem such as this. One can state that he concludes something may happen in this way or in another way.

Stephen is making a hypothesis based on his observation or experience. His original conclusion before the experience is that he thinks UFOs or aliens do not exist. This is his default hypothesis and is also called the *null hypothesis* in the science literature. But Stephen has just made an observation which is different from his original belief. The encounter contradicts his original belief set. So in addition, he also makes an alternative hypothesis, which states, "I cannot prove the null hypothesis from this new observation. Some other conclusion must be made about this problem. I must obtain more data and evidence to make a better analysis and to help form a better conclusion."

Stephen then gathers more evidence that could include another sighting and/or experience or to seek out others who can provide more examples of similar evidence. He obtains and then weighs this evidence. After analysis, Stephen returns to his two hypotheses to determine which of these has been better proven. He finally makes his conclusion from using this procedure. He could state that he still does not believe in UFOs or aliens, or that he does indeed now believe, or that he may still need more data to analyze. In the third case, Stephen then works to collect more evidence. The entire process would then be repeated.

So it goes with the practice of the scientific method as it is used today and as how it is presented in school textbooks. The discussion of the scientific method is never more than a portion of one of the very first chapters of texts. There are also more nuanced details about the procedures involved that present situations which do not get explained there either. For example, in the analysis and conclusions steps of the procedure, it is never explained that in real life there is a lot of debate that occurs in the laboratory over interpretations of these analyses and conclusions in experiments. This is a major thematic of Bruno's *Laboratory Life*. Karl Popper described this notion as the title of his book *Conjectures and Refutations: The Growth of Scientific Knowledge* (Popper, 1963).[12] Popper describes that the members of a group will argue and debate often over many fine details about the experiments they are performing, the evidence data, and the analysis itself.

As I noted before, they all are human beings and that in any community there will be human factors, which some have come to describe as office politics. Somewhere within the data collection and analysis efforts, there will (usually) be a period of debate to reconcile differences in perspective and interpretation among the community members. These social dynamics happen not only in a science environment but also in everyday life as well.

To briefly mention how extreme this notion of scientific debate can and has occurred in science history, there is another fine 1998 reference book titled *Great Feuds in Science: Ten of the Liveliest Disputes Ever* by Hal Hellman.[13] This factual account of scientific "infighting" involves such great scientists as Galileo, Newton, Wegener, Einstein, etc. The ten feuds Hellman describes delve into the human factors present in those science situations. They describe how all sense of decorum, which the adversaries and all members of science communities are especially trained to maintain and practice, falls into disarray. Personal attacks dominate argument of the facts and data that is the real evidence needed to make correct and reliable conclusions.

The scientific method is thus not just a method of steps that is specifically followed in order to obtain a conclusion about the problem. It is in reality a process of experiencing an anomalous event of some kind and forming initial guesses about the true nature of "if, why, how, when, and where" it is occurring. More data evidence is collected, analyzed, and argued about. Then after the human factors are worked into the process, conclusions of some sort are derived and presented.

The presentation of these conclusions comes in the publication of the science research paper. Written up by the science team members, they are an invitation for the science community at large to further experiment and prove (or fail to prove) the conclusions as having validity and some sort of credibility and reality. Here, the validity of the conclusions references back to the experimental aspects of the investigation, the physical evidence data collection parts of the process. The credibility of the conclusions references back to the human factors involved in the process. These factors often include the reputations of the lead researchers and writers, the social politicking, and the fiscal factors mentioned earlier. Validity and credibility are two important dimensions used in argumentation and critical thinking studies to add weight to a debater's claims and conclusions as being "right" and/or better than that of his opponents.

The peer review is the critique element of the entire process. Peer reviews are supposed to act as the "bias equalizers" that try to eliminate the human elements of science discourse from the process. The reviews will either provide more positive evidence to strengthen the originator's hypothesis or counter it with anomalous data evidence. When and if enough peers verify and corroborate the originator's conclusions, then those conclusions become more and more generally accepted as factual. (Remember, they are not irrefutable facts!) Eventually, after numerous peer reviews are published to verify original findings, the hypothesis would move on the dimension toward being accepted as a proven theory. (Remember, a theory is also not irrefutable!) Taken separately, the peer review has a function of being a bias equalizer. Together with the scientific research paper, these two functions are the scientific method's "fail safe" mechanism that they hope would produce meaningful science and an accurate explanation of the reality and truth of that phenomenon in nature.

For a hypothesis in science to move toward proven theory, a lot of evidence is needed for most of a scientific community to become supporters. As I mentioned in the last chapter, to put it this way, most of the "nays" would be convinced to align with the "yeas" of the science versus ufology debate. Scientists say the most important part of their investigations is the amount of evidence they can gather. There is no industry standard as to mean "how much is enough." This is very subjective in nature. A requisite amount of evidence to one scientist differs significantly from that of another. It is also why scientists and a scientific community can appear to move very slowly when working through an investigation to discover new knowledge. Often, an investigation can take years or decades or more to move toward some definitive course of conclusion. Two recent examples of this discourse would be the study of the global warming/climate change and the evolution process of a pharmaceutical product study.

Now you know that the scientific method is a process that takes a prediction about a problem and collects a lot of evidence data. Next, the methodology entails analysis of that experimentation or evidence collection. Then the community engages in a discourse in which the human factors characteristic of social collaboration, critical analysis and argumentation, social politicking, credibility establishment, and fiscal factors play themselves out in the process. An effective outcome should produce, in an ideal situation, critical and correct conclusions that are then published in a science research paper. Upon peer review and critique—utilizing all the preceding factors to restudy the problem—if the findings of all reviewers (or most and the most influential of them) are in agreement, then the hypothesis can either be upheld as stated, refuted as stated, or restated in which then further investigation would take place. The process would then continue with another round of evidence acquisition, analysis, etc.

The most important idea to take from this process of scientific investigation is that it has a procedural stepwise arrangement and also a very significant and influential sociocultural arrangement also. The nature of science in reality is not only or just your perception of the rigid, choreographed "this step follows that step" laboratory of practice that you tried hard to understand from that mystical "thought platform" ideology each science subject portrayed to you in school. There is a just-as-significant sociocultural process within a process that, in reality, mostly defines what ultimately comes out as science knowledge, be it correct or not. Please refer to Hellman's *Great Feuds in Science* and Bruno and Latour's *Laboratory Life* for an extensive enrichment of the importance of this part of the more broad-based scientific method.

A concept important to remember when discussing the scientific method is how the hypothesis is structured. The null hypothesis is the default statement brought into the investigation. For Stephen, the null hypothesis was that he did not believe in the existence of UFOs or aliens. Knowing that any scientific investigation entails a "debate" involving two contrary hypotheses or statements, an alternative hypothesis statement is required for the investigation to move forward. Stephen's alternative hypothesis statement is that "because I have had a sighting of something anomalous and/or a more profound experience, I bring my disbelief about the existence of UFOs and/or aliens into question." The investigation proceeds from that point.

Some scientist types here would, in their practice and application, question the use of the two statements in the fashion and order in which they were presented. A practice of some scientists is such that a default null hypothesis statement "must" be made which does not include the words "does not," "has not," "did not," or any other structure with a negatively connotative meaning. In other words, use of "it is," "it does," or "it has" is the proper way to describe the null hypothesis. This is because the investigation and use of experimentation, in the science world, often is structured to try and disprove, refute, or negate the default (null) hypothesis statement. The alternative hypothesis, therefore, would, in their minds, properly include the "nots." I refer you to many of Karl Popper's works for a better discussion of the structure of scientific hypotheses and refutations.

Here, the rub would be that the positioning of the two default statements is not "right" to satisfy some scientific types. Here, Stephen is using the "not" connotation in the null hypothesis statement. So in his alternative hypothesis, Stephen is not trying to disprove something; rather, he is trying to *prove* something. Some scientists would argue that science protocol demands that the default null statement make a "not negative" or "it is" connotation so that it is allowable to then try and disprove it.

This argument over the structure of hypotheses is just an argument of the use of the inductive method of critical reasoning and thought versus the deductive method of critical reasoning and thought. These notions of human thought and discovery of a rational train of that thought originate from the teachings of Aristotle and the Peripatetics in ancient Greece, as noted. This topic deserves more of an explanation. Then it will become clearer that this issue over the proper structure of hypotheses is ambiguous within the studies of the sciences. It is the subject of the next chapter.

Reference

[1] What Science Is Not Ref 1: University of Indiana, 2011. What Science Is Not: Teaching the Nature of Science. Retrieved June 1, 2016, from: http://www.indiana.edu/~ensiweb/lessons/unt.not.html

[2] What Science Is Not Ref 2: University of Georgia, 2016. What Science Isn't. Retrieved June 1, 2016, from: http://www.gly.uga.edu/railsback/1122science3.html

[3] What Science Is Ref 3: University of Georgia, 2016. What Is Science? Retrieved June 1, 2016, from: http://www.gly.uga.edu/railsback/1122science2.html

[4] Consolmagno, Guy, 2005. *Intelligent Life in the Universe?* Catholic belief and the search for extraterrestrial intelligent life, p. 24. Copyrighted and published by The Incorporated Catholic Truth Society.

[5] Definition Ref 5: *Merriam-Webster Dictionary.* Copyright 2016. Published by Merriam-Webster Incorporated.

[6] Definition Ref 6: *Merriam-Webster Dictionary.* Copyright 2016. Published by Merriam-Webster Incorporated.

[7] Definition Ref 7: *Merriam-Webster Dictionary.* Copyright 2016. Published by Merriam-Webster Incorporated.

[8] Philosophy Ref 8: Graham, Jacob N. *Ancient Greek Philosophy.* Copyright 2016 Internet Encyclopedia of Philosophy and its Authors. ISSN 2161-0002. Retrieved August 27, 2016, from: http://www.iep.utm.edu/greekphi/

[9] Philosophy Ref 9: Cohen, S. Marc; Curd, Patricia; and Reeve, C.D.C., 2011. *From Thales to Aristotle, Fourth Edition.* Copyright 2011 by Hackett Publishing Company, Inc.

[10] Latour, Bruno and Woolgar, Steve. *Laboratory Life: The Construction of Scientific Facts*, 1979. Sage Publications, Beverly Hills, CA.

[11] Traweek, Sharon. *Beamtimes and Lifetimes: The World of High Energy Physics*, 1992. Harvard University Press.

[12] Popper, Karl. *Conjectures and Refutations: The Growth of Scientific Knowledge*, 1996. Published by Basic Books, New York and London.

[13] Hellman, Hal. *Great Feuds in Science: Ten of the Liveliest Debates Ever*, 1998. John Wiley & Sons, Inc. Publishers.

The Inductive and Deductive Methods of Scientific Reasoning (And All Types of Reasoning)

Thus it is uncertain which of these impressions are true or false; for one kind is no more true than the other, but equally so. And hence Democritus says 1 that either there is no truth or we cannot discover it.

—Aristotle from "Metaphysics"[1]

Key words: deductive and inductive reasoning, premise, conclusion, logic, syllogism, validity, soundness, probable, verifiable, credible, epistemology, transcendental arguments, world disclosure reasoning, human factors, pseudoscience

UFOs do not exist.

Aliens do not exist.

The forms of *reasoning* known as logic, deduction, and induction have been used by humans in everyday life since before the ancient Greek philosophers scripted them over 2,500 years ago. Whether you believe that the Greeks adapted earlier versions to use for their verbal or written dialogue, as there is evidence from cuneiform texts attributed to ancient Babylonia,[2] it was their principles, which were passed downand, are still utilized today. The discipline of reasoning that was learned from them was very organized and effective. The disciplines that define reasoning and logic were used to organize one's thoughts. This was done to help, in the ideal world, achieve accurate decisions and conclusions about the situation being studied. Based on what we have learned about the ancient Greek society, we know that the birth of other learning disciplines—first mathematics then the sciences of physics, astronomy, and biology—engaged in the use of the thought reasoning principles that are also still practiced today.

Observations are made of phenomena in nature and perceived and studied by the observer. But there needed to be a mechanism that would aid in understanding with the objective to discover its correct interpretation. It is argued that any conclusion must have a foundation of facts and/or perceptions recorded and reported by the observer before being able to make one. In other words, a conclusion has to have been derived from a set of facts or observations that are uncovered from these observations and perceptions. These can be defined as *premises*.

In a way, the conclusions could be described as reason, and the sensory perceptions or observations made when nature is allowed, without interference to act on its environment, are the facts or truths. It is human nature to seek the truth. Can reason exist without truth? Can truth exist without reason?[3]

Everyone utilizes some form of the two methods of logic and thought reasoning in their everyday lives. As previously noted, these are the methods of inductive reasoning and deductive reasoning. They both exist because there is no overwhelming efficiency to using one over the other in critical thinking to the point of eliminating the use of the other. Sometimes, there are situations where one form of reasoning is required and that using the other would fail in making correct decisions about the question at hand in either or both a relative or absolute sense. In other words, there are uses for both forms of reasoning, which it is why both are still practiced. These are not two opposing methods. Most situations, in fact, involve solving a problem and entail use of both forms in a part of the process from the question/problem to the answer/solution. This happens no matter if the situation is philosophical, mathematical, scientific, or societal in nature.

It is important to be aware that while the deductive and inductive reasoning methods are commonly used in all professions and personal situations, they are not used consistently in all examples. Additionally, there are other forms of reasoning processes that, while sometimes employed in argument discourse over a broad range of subjects, predominately the sciences, are known as subsets of the two main forms of deductive and inductive reasoning.

Deductive reasoning is more commonly known as top-down reasoning. A statement or statements, known as *premises*, is structured and used to later form a *conclusion* based on the principles of logic and critical thought. Many of the principles of logic are the same principles the ancient Greek philosophers contemplated and then developed. This form of logic is applied to the question for an interpretative analysis of the problem. If all the statements are true, then the conclusion must be true according to the rules of *logic*.

Rules of logic are woven into an argument. These rules are the catalog of argument and debate protocol philosophers created, most notably, by Aristotle over 2,500 years ago and have been adapted for use in the science disciplines.

To look at another definition of deductive reasoning is to say the process takes a set of factual statements and deduces other facts in a logical manner. These situations happen to everyone every day. Taking the example from Aristotle and his creation of the principles of *syllogisms*, suppose "A" defines some factual statement. Additionally, "B" defines another factual statement. Likewise, "C" is a third factual statement. The deduction proceeds as: "if A equals B, and if B equals C, then by logic and deduction A equals C."[4]

What is essential in distinguishing the deductive reasoning method from any other is that the logic behind the process does not create or provide any new information. The logic process here only takes known facts and deduces more facts based solely on those facts already known. This could be construed as a "closed system" of factual development.

When forming a problem that uses deductive reasoning, the premise statements (the known facts as input to the problem) can consist of more than the three statements used in the last example. So a good way to visualize a study of logic and deductive reasoning is to picture a bucket that could contain many more than three factual premise statements. In any particular problem, one can use three, four, or eleven factual statements. Instead of using "A and B" then "C," one could use, "A+B+C...+J" then "K." The logical analysis is attempting to make new facts from the known bucket of facts. These new facts are defined from the arrangement of the premises factual statements. The main point here is an argument problem can have more than two logical facts to then define a last new fact. The bucket is the closed system of facts that are drawn upon to construct new realities. It is this process that has been used in the sciences since ancient Greek times.

The purpose of reasoning by deduction is to take these premise statements and to rationalize a measure of *validity* and *soundness* to the conclusion statement. In a 2016 paper, philosopher Wei-Ming Wu defined validity in this way:

> A deductive argument is valid if its conclusion indeed necessarily follows from its premises. If the premises are true/acceptable, then the conclusion must also be true/acceptable. The premises must be one hundred percent supporting the conclusion.[5]

This is necessary for validity to be established. Soundness can be defined this way: "For a deductive argument to be sound, it has to meet two conditions. First, it has to be valid. Second, every one of its premises has to be either true or acceptable."[6] Note that within the principle of deduction, both concepts of validity and soundness require one hundred percent truthfulness of all its logical statements. This is not necessarily the case for other forms of reasoning, as you will see.

Deductive reasoning argumentation constructs these premises to be parts of a *closed realm* of thought discourse. The process is like taking individual bits (premises) of something out of a container of a limited number of these bits and trying to validate whatever conclusion you are claiming. Then you construct the soundness of your argument or conclusion from this intermediate step. If all the presented premises are true, then the outcome must be true and would, therefore, be valid and sound. If the conclusion, however, turns out *not to be* true, the argument thus will not be valid or sound. For the argument to be valid, all statements of logic must be true. If this occurs, it logically follows then that the conclusion will be true according to deductive reasoning. If all the premises are true and the argument is valid, then the argument is, thus, also sound.

An essential point in studying deductive reasoning and thought is to remember that the premise statements are part of a closed system or *closed mind* of statement appeals. This means that a premise is a very specific statement that states only one thing. From here, it should be noted that the more premise statements that are introduced into the dialogue, the harder for the argument to be proven to be true, valid, and sound. If "a" equals "b" and "b" equals "c," then "a" must equal "c" is the classic if/then syllogism Aristotle created for us. A small statement cluster is easy to see and then used to validate a successful logical conclusion.

"A" and "B" were taken from such a closed system container. If there are, let's say, seventeen statements and of them are needed to prove the conclusion, it becomes much harder to validate the truthfulness of seventeen different logical statements. Remember, for deductive reasoning to work, all seventeen premises must be true in order to substantiate and prove the conclusion, its validity, soundness, and its argument. When applied to an actual scientific research study, you can appreciate the vastly increased complexities of working exclusively under such a thought pattern.

A substantial (there is no way to use and prove use of the term *majority* here) number of scientists argue for the exclusive use of deductive reasoning in their work. They are claiming an absoluteness of truth to make scientific discoveries, gain new knowledge, and obtain new facts about science and nature. The logic of deductive reasoning provides an opportunity to gain some certainty about the correctness of this new knowledge and those new facts. It is how this group of scientists thinks when they work on problems in their fields.

You will learn next of the other, though not diametrically opposite, way of reasoning and way of thinking. Inductive reasoning takes a bottom-up approach to the reasoning and critical thought process. Logical statements are also used in the inductive reasoning process. They are used, however, to support the credibility of the conclusion or claim. In this thought discourse, if the premises statements are true, then it becomes more *probable*, *verifiable*, and *credible* that the conclusion is also true.

The definitions of these terms according to *Merriam-Webster* are as follows. Probable is, "Supported by evidence strong enough to establish presumption but not proof; *a probable hypothesis."* *Verifiable is defined as,* "Something is scientifically verifiable if it can be tested and proven to be true. Verifiable comes from the verb verify, authenticate, or prove, from the Old French verifier, "find out the truth about."[8] Finally, the definition of credible is, "The quality of being believable or worthy of trust: usage; 'After all those lies, his credibility was at a low ebb.'"[9]

A key attribute to the term *credible* refers to the "worthy of trust" reference. This is attributable to a significant part of this definition to the person who is offering the conclusion and is a critical concept and practice within the sciences. Very often the believability and acceptance of a conclusion is supported and even determined from "who is offering it"; not necessarily from the facts, observations, evidence, and analysis. This is a practice that, when played out, often leads to incorrect and sometimes untrue conclusions. Tied into this thought is the notion of reputation, whereby the conclusion statement is upheld only due to the reputation of the person making the conclusion and ignoring the other parameters.

These concepts are usually classified as an inductive reasoning principle of thought. Use of a logical inductive reasoning method means to take a set or a bucket of factual statements and discover a pattern among the factual statements based on some amount of common characteristics or traits that defines the pattern in question. The conclusion statement thus creates a generalization of the relationship between the factual statements of logic.

An additional characteristic mechanism that the inductive reasoning method uses is observation. When considering the definitions of *observation*, they all are used in most scientific experiment studies, especially those that are participatory in nature. That is, subjects of some kind are used to obtain data evidence for analysis by nature of the inductive reasoning method. Observation is a vital step in the process of the scientific method whether using either deductive or inductive reasoning or both protocols.

It is noted that, within a study question or problem, the broader use of deductive reasoning (top down) may require construction of an individual fact using inductive reasoning (bottom up). Let us say the points being used to make a deductive reasoning are "A" facts. When a single "A" fact is being deduced, it may require the use of a set of, say "B" facts. These "B" facts, the way that they are observed and constructed may necessitate use of the inductive reasoning method. When that single "A" fact is constructed, it then becomes part of the body of facts used to deduce the answer/solution to the question/problem.

Use of inductive reasoning in this way introduces the possibility of obtaining new information from the set of facts under consideration. The deductive method takes facts and creates new ones only; it does not allow for such a possibility. Those deductive facts can and are used to make a more generalized statement which is the precise inductive reasoning train of thought.

The inductive reasoning method can accommodate not only a larger bucket of factual statements than the deductive reasoning method to work, but also it can also work well in an "open system" or "open-minded" environment of subject study. Inductive reasoning, thus, uses observation, experiences, analogies, and other comparative statements in its analysis of all the evidence to provide conclusions in ways that allow for an open system in which the study is taking place. The deductive reasoning method cannot provide this throughout.

Another way to summarize the points thus far is to say that when using the deductive reasoning method to study a problem, whether in science or elsewhere, there should be a one hundred percent chance of being accurate, given the closed minded or closed system of factual premises statements. When using inductive reasoning, there may be a less than one hundred percent chance of being factual (or truthful). But the more open-minded nature of and open system of logical and accurate statements allows for more insight. In essence, these forms of reasoning are a significant part of the human learning process. This concept of learning is called *epistemology*.

An important point is that, in utilizing the inductive reasoning method, a conclusion is not categorically *certain* to be true. Here, the notions of thought are those of probability, verifiability, and credibility. So in going back to the statement "bin" from earlier, if two statements A and B are taken to be true, then there is a good possibility of C being true; but this is not absolutely sure as in the deductive reasoning example. If, however, the same seventeen statements are used and all are true, then the conclusion has a lot more truthful sampling data that all agree. The conclusion, thus, has a lot more probability, verifiability, and credibility.

All seventeen may not, however, agree with each other. Inductive reasoning may still be able to verify the claim and conclusion, whereas, deductive reasoning the argument would be categorically discounted. If the statement bin were a more open system and contained 260 statements or greater, for an argument to use as logic, then the inductive reasoning method would make better use of them than deductive reasoning because of the lessening of the rigid, restrictive elements present in the deductive method.

Another restriction on the deductive method is it is assumed all the premises statements are entirely true all the time for the argument to be valid and sound. Remember, we are using science discourse as it applies to things in nature and events in the science disciplines. Systemic events and behaviors of the sciences, nature, and the universe are fluid, dynamic (not static), and part of an open system forever. The study of the cosmos and how nature acts and operates, going to an extreme, is an example of the most open system there is in science and nature.

Nature is always changing. Nature and the tangible matter that exists in life is continually evolving. Therefore, knowledge in all the sciences constantly changes. This means that eventually the specific statements of logic used to construct theories that are then used in studying problems in scientific arguments will eventually become untrue. Another way to say this is that science cannot prove forever that something is the way it is or is not. A scientific theory is always subject to change or rebuttal. This is how new knowledge and scientific discoveries and facts are born. This happens not only in science but in all aspects of our society and life as well.

Another example is a concept in chemistry known as *entropy*. Entropy is,

> The measure of such randomness and disorder in the universe is called entropy. In chemistry, entropy (represented by the capital letter S) is a thermodynamic function that describes the randomness and disorder of molecules based on the number of different arrangements available to them in a given system or reaction.[10]

Inherent in this definition is the thought that recognizes there is a lack of order or organization of matter within the universe. Thus, a theory can never be permanent because the inherent disorder in the universe and nature causes the organizational construct of a theory; its observations, data evidence, thought, analysis, and conclusion set is to break down eventually.

In this regard, it follows that a substantial number of the scientific community recognizes the usefulness and advantages of inductive reasoning. It is used predominately in the social sciences, legal, business professions, and by us in the public realm more often than we realize.

This is not to be amenable to explaining the results of any competition between the two methods to determine a winner. There would be no winner in such a debate. The inductive reasoning method uses a more open-minded system ideology that may work better in numerous situations when science and nature are the subjects of investigation. Alternately, the deductive reasoning method works better in many situations too. It would be more damning to a member of the scientific community if he took a polarized viewpoint in this argument and used only one of them. His career as a scientist would be a short one. They are meant to be used together.

In some research studies, the inductive method is being used even though the science team may not have intended it. After a hypothesis is stated, the experiments are conducted to obtain data as evidence. Sometimes, the data produces many more undesired or negative results (premises) when compared to the hypothesis statement (conclusion). Therefore, the conclusion cannot be taken to be true, valid, or sound. The conclusion must be modified or abandoned. So in reality, experiments are using both the deductive and inductive methods.

This analysis is leading to its conclusion. The two methods of logical, critical, and reasoned thought processes—adopted by the mathematics and science disciplines—are useful in their own right but not perfect, especially when taken exclusively one over the other. Remember that these principles of thought were, in fact, created by those imperfect philosophers back in the day. Does this beg for an evolution of the industry that is the study of logical, critical, and reasoned thought analysis? Or is the application process at fault here? After all, the system is over 2,500 years old, and we still uphold its dogmas, especially in the scientific communities. Nature is a real system and an open one. Nature is a dynamic, fluid, constantly changing open system subject to the effects of entropy. The events in nature are the ones that science studies and attempts to gain more perfect knowledge from. Those occurrences are real. There may be a sense of "disattachment" with the sciences when they confront problems of nature that are real events. Do they try to gain knowledge from studying the reality of those events, or is the argument discourse the operative standard that the sciences have held onto while out on their mystical thought platforms? Are reality and truth most important or is the argument? This is another example of *human factors* interacting with the scientific method and science thinking processes.

This is an appropriate place to mention other younger methods of logical reasoning. One additional major form of this ideology is called the *abductive reasoning* method. Abduction is mostly a logical branch from the primary study of inductive reasoning. Where the epistemology of inductive reasoning charts a path toward a more open-minded dimension, the abductive reasoning method goes even further in this respect. Abductive reasoning can utilize the inferential methodology in separate and more creative ways that can lead to a "best explanation" of a situation. Additionally, this form takes relevant evidence and explains the situation using the best hypothesis conjecture to proceed then to refute or not to be able to refute that hypothesis.

Abductive reasoning is most significantly a sort of hybrid method that merges aspects of both deductive and inductive reasoning. It allows for an even more open-minded and open-system use of its environment than inductive reasoning but demonstrates some of the strengths of the deductive reasoning method in that it eliminates unlikely or weak arguments as possibilities of correct or likely hypotheses.

A scholarly definition of abduction is taken from the Stanford Encyclopedia of Philosophy.

> Abduction or, as it is also often called, Inference to the Best Explanation is a type of inference that assigns special status to explanatory considerations. Most philosophers agree that this type of inference is frequently employed, in some form or other, both in everyday and in scientific reasoning. However, the exact form as well as the normative status of abduction are still matters of controversy.[11]

An example of abductive reasoning looks something like this. The type and size of the bucket of facts for this type of reasoning are similar to the one used for the inductive reasoning. The premises statements of fact usually consist of one "major premise" statement and sometimes more than one of these can be used. The abductive argument also uses one or more "minor premise" statement to move, as the other methods, toward making a "concluding" statement.

Major premise statement: A coffee urn is full minus exactly one cup.

Minor premise statement: Carl has a cup of coffee in his hand. The cup is full.

Conclusion: The coffee in Carl's cup came from the coffee urn.

The logical analysis is consistent and flows to a valid, sound, and credible conclusion in the way it is presented, though the abductive (and inductive) reasoning methods dictate that it is not one hundred percent certain that the conclusion is correct. The coffee in Carl's cup may have come from elsewhere. The deductive reasoning method would dictate that, if this determination is indeed correct, the coffee *must have come from* the coffee urn.

Abductive reasoning was borne from a more contemporary philosophy in the literature. Karl Popper's *Conjectures and Refutations* is a seminal contribution to this form.[12] This work discusses very thought-provoking evidence on the nature, activities, and definitions of important terminology within the philosophy, science, and education disciplines. Not only do *conjectures and refutations* portray a landscape of modern scientific practices, but it highlights emerging utilization of a portion of the abductive reasoning method along with others. There is a place for all three methods of logic, argumentation, and discourse in life and within our civilization.

Some other principles of reasoning have been recycled through recent times as occurring in waves of interest and then fall into disinterest. Two of these are mentioned here. The first is the principle of *transcendental arguments*. This form of thought reasoning is a favorite tool of skeptics. Here, the nature of the premises or the nature of the evidence required to prove a conclusion is held to an even higher stringency than those used under a principle of deductive reasoning. It is the form with which transcendental argumentation shares most of its characteristics.

This notion is explained best in a quote one of its contemporary proponents, philosopher, and professor, Charles Taylor, made in his book *Philosophical Arguments*. Taylor once said, "Transcendental arguments have to formulate boundary conditions we can all recognize. Once they are formulated properly, we can see at once that they are valid. The thing is self-evident. But it may be very hard to get to this point, and there may still be dispute. For although a correct formulation will be self-evidently valid, the question may arise whether we have formulated things correctly."[13]

This means the following. Take some variable "X" and let us define it as "UFOs and/or extraterrestrials exist." To make a statement such as "Y" the "X" is a required condition that makes the occurrence in a spatial or temporal manner possible. For "Y" to be a possible reality, "X" must not only be possible but must have existed.

Say that "Y" is, "there has been a UFO or extraterrestrial sighting" or some evidence artifacts like radar returns from a military installation, physical evidence from a fallen ship, or a person who had an encounter and now has serious radiation burns or other physical evidence. If "Y" is the case and is real and that has occurred, then logically, "X" must also be the case.[14]

The transcendental argument begins at the point where the skeptic or debater refutes the contender's claim. This way, the burden of proof is never on the skeptic in the argument. The contender must always reclaim or commence arguing in a progressive or regressive process, a sort of a cause-and-effect process.

While the principle of the transcendental argument may seem flexible on one level, it is an increasingly rigid form of deductive reasoning with the burden of proof being harder to confirm. If the contender can satisfy the argument using transcendental arguments, it is an accomplishment. The refuter is offered a lot of refutation strategies that have made this principle less popular. A main approach is that, in an extreme example, the skeptic can simply refuse to accept anything his opponent claims. Therefore, no closure can come of the argument. This approach is often seen in the UFO and extraterrestrial hypothesis discourse.

The notion of argument is a necessary aspect of any principle of thought reasoning. One other type of reasoning that shares a more open-minded attribution with all types of such thinking is *world disclosing reasoning*. World disclosing reasoning can address problems involving logic and debate that most significantly deductive reasoning cannot. This is because deductive reasoning—the one most often frequented by the science disciplines—is on the closed-system end of the system dimension; whereas inductive, abductive, and transcendental reasoning contribute a more open-system consideration that allows for resolution of an argument or debate and adds to increased knowledge.

A typical example of where world disclosing or the other non-deductive types argumentation would occur in situations where the investigation has discovered unusual observations and other evidence that could not be resolved in the currently-practiced paradigm in which the investigation is taking place. Thought reasoning has elements of a philosophical theory Thomas Kuhn wrote about in his book, *The Structure of Scientific Revolutions.*

World disclosing reasoning and all the other more open-minded approaches to "reasoning to resolution," to use Kuhn's famous phrase, a paradigm shift enables possibilities for new insight and the discovery of new facts that contribute to a resolution of the investigation. One of the prominent followers of these principles was philosopher and theorist Nikolas Kompridis. In his book titled *Critique and Disclosure*, Kompridis said, "Since we are not dealing with deductive or inductive styles of reasoning (which are truth preserving, not possibility disclosing), we cannot know in advance what form (they) will take."[15] This allows for arguments that cannot be reconciled under the classical method(s) to offer consideration with possibilities for new insight, knowledge, and/or resolution to the debate.

In summary, one other way of describing epistemological ways of learning that are characteristic of and interlock forms of reasoning are to differentiate an argument from an explanation. Arguments are undertaken to prove or convince someone that something "is," "was," or "will be." An explanation is centered on why or how something was, is, or will be. An investigation will proceed through its process to a conclusion. The resolution will state that "this situation was, is, or will be." The explanation will be the evidence of the how and why of the argument resolution part of the conclusion. The most necessary and required prerequisite of any argument, investigation, study, problem, or the question is that a thorough and all-inclusive fact-gathering procedure be included for analysis. This is one of the most fallible aspects of any inquiry a human being can undertake. As noted elsewhere in this book, human beings are fallible. Critical thinking, reasoning, and logic are not hardwired into our brains. While training—such as what is undertaken by philosophers, mathematicians, scientists, and others—can provide an understanding of such fallibilities, constant practice is required to achieve the most acute understanding of the truth they are possible of attaining.

It looks more and more certain that the dogma of science thought and reason collides with the subject of ufology in an unproductive manner for both. Some science communities work within their specifically designed paradigm and with a certain mind-set of critical thought. Others do not. This pattern is usually structured within a closed-minded and closed system of practice.

Depending on the circumstances of the problem under study, they may be required to practice in this manner, especially if the problem being studied pertains to a matter of life or death. Some of the biological sciences predict certain procedural behaviors on the part of laboratory research scientists designing a new pharmaceutical for control of or cure for a disease. The necessity of working in such a controlled environment, complete with all the procedural and sociocultural biases, dictates the course of the procedure. Sometimes, it is necessary for this to occur. Each study design and paradigm is unique. This uniqueness must be accounted for.

This example also recalls the notion of human factors. Human factors always influence many parts of a thought or experimental study. They are present in every community whether in science or others within society. They often influence the reasoning process and conclusions in very significant ways.

Most of the time, the scientists and those of us who have taken science courses in the past were taught precisely the way that would force their working practices and procedures into their very uniquely slotted paradigm. This is dictated from the historical epistemology of the science discipline they work in. This is not meant to absolve them from blame or consequence when the research study results in little or no progress. In a publish or perish world of the scientist as laboratory worker, research paper writer, or academic professional, the human factors and political pressures built into the science discipline often propel their work in these ways.

From the examples given here about the tools scientists use to construct facts and analyze data, there often seems to be a misuse of these reasoning tools by scientists in their investigations. Also perceptions about a set of facts a scientist trained in a particular field may make often differ from those perceived by another scientist in a different field. This is attributable to many of the study issues that arise in the field of ufology.

Some of the dichotomies can be visualized by a crevasse descending through an otherwise flat platform or mesa region around it. This crevasse is formed over time and represents any particular scientist's working protocol within his science discipline. Depending on the scientist's particular discipline and that he or she works within a closed system of logical study. Such a closed-minded protocol does not always function perfectly within matters related to nature. The scientists' ways of thinking are well defined but quite rigid. This was formed to a significant degree by the specific training they received in their field.

For example, say there are seven people sharing a seashore cottage on a beautiful summer long weekend. Three of them are sighted, and four are blind. For every day of their vacation, the sky remains blue and cloudless. The sun rises and sets every day and is continuously visible throughout. Every night, the moon and stars rise and set and are observable throughout. The three sighted people *see* this situation play out. The four who are blind do not. On the fourth day of their vacation, a researcher interviews all seven of the attendees individually. The main question asked of each participant states, "Do the sun, moon, and stars exist?" This would be the statement of hypothesis in a research study experiment.

Scientists from the different disciplines would approach and conclude their answers differently. The physicist would analyze the variabilities, motions, and physical interactions of all the objects under consideration and form a conclusion ignoring the fact that four of the participants are blind. The astronomer would use some of the physicists' concepts but also consider light as a vital premise because without the existence of light the astronomer could not create and study a research question in his field. In light of this, would the astronomer offer a different conclusion about the existence of the sun, moon, and stars because four of the participants are blind and cannot see these?

The chemist would concentrate less on the macro environment described as the entire tangible landscape or substrate of the universe's "theater of operation" and more on a micro environment where individual atoms and molecules exist and in which the chemist's training and perspective were meant to apply. The study of chemistry is vital to understanding whether something exists or not; whether something is real or not. Individual atoms cannot be seen; molecules can if they become large enough. When these are the molecules, they are usually called substances or compounds as defined by the chemistry community. From an invisible atom arranged in concert with many other atoms to form molecules, compounds, and substances, creation is given to the realm of visible matter. The corresponding physical properties are created and then attributed to the construction of invisible matter so that they become large enough to form visible matter. This then allows for a merger of the two science disciplines' ideologies of reality in nature. Without the help of the physicist and astronomer, though, could the chemist even provide a study analysis that would allow him to rationally conclude these objects all exist? It becomes a matter of perspective when a discussion such as this is undertaken.

A biologist would be another scientist to consider exploring the blind vacationers in the problem analysis. The biologist would have the most significant understanding of what it means to be blind. She could best relate the reality of the seven participants' situations in the experimental interviews to the problem stated. She knows a blind person can sense elements and characteristics in a physical environment at an enhanced sensitivity level that accompanies such an affliction. This is accomplished because the human body has a magnificent ability to compensate with enhanced system functioning (in this example a heightened sense of hearing, touch, smell, and/or taste) when they are blind. The brain has been evolutionarily wired to accomplish this. This is evidence that a scientist from another field studying the question does not have. In turn, this different tool set could, in turn, lead to the biologist forming a different conclusion from the other scientists.

An important point here is to have the awareness to recognize that different scientists may view things differently. This is partially due to their training as we have noted. More generally, humans have different views on the same things throughout our daily lives. Hence, not everyone will agree that "some thing" is exactly that something. Enough people, though, will agree that the "something is exactly that thing" so that it is generally accepted to be true. Going back to the example of the experiment involving the blind participants, so, in turn, do the sun, moon, and stars exist? This is the stuff philosophers work on throughout their entire careers and have done so for thousands of years.

When it comes to science and ufology, it is not a clash of disciplines that exists. The debate is more individualized. From some indications, ufology has been placed in a sort of preliminary science discipline. A *pseudoscience* would be the descriptive response from another scientist. A pseudoscience would become a mainstream science discipline when (and if) a theory demonstrating a positive existence of UFOs and aliens is concluded to be proven by some preponderance of enough of those in agreement.

Science history is filled with such examples of preliminary status of a science discipline. Often, this preponderance is meant to be enough for "the right people." This is an example of the human factors present in all investigations as they are practiced. It is thus known as the pseudoscience just noted. A great example of this embodies the history and career of Gregor Mendel. He was a nineteenth-century Augustinian friar whose work on crossbreeding pea plants in his field continued for fifty years. It spanned the entire second half of the nineteenth century. His career (a self-described hobby to him) of experimenting with the pea plants and others propelled him to discover the concept of the gene in biology. He single-handedly created the entire subject of genetic biology and jump-started several other science disciplines that evolved only in recent generations. Mendel was declared to be no scientist" by the science fraternities. Subsequently, his work was completely ignored and ridiculed by the science community for over thirty-five years, sadly, until after his death in 1884.[16] His hobby was declared pseudoscience by all them. Did science make an enormous mistake here? Hmmm.

To provide an outline of this discussion, the three main formats of logical reasoning are:

1. Deductive: Formal Deductive Reasoning
   Informal Deductive Reasoning
2. Inductive: Formal Inductive Reasoning
   Informal Inductive Reasoning
3. Abductive: Formal Abductive Reasoning
   Informal Abductive Reasoning

Formal reasoning is based on the *validity* of premises or truths and the validity of conclusions or reasons. (See the opening passages of this chapter for an analogy.) As noted, these provide no new information, just a rearrangement of facts. Validity is most often attributed to the deductive reasoning principle but could have attributed to all three types. Informal reasoning can include all the elements of formal rationale and additionally include truths and probabilities about the premises and concluding arguments themselves. Often, this is crucial information that can assist in obtaining the correct answer to the question, investigation, or representation of something in nature. By definition, these are things deductive reasoning cannot address. It is often classified within the realm of either inductive or abductive reasoning.[17]

An expanded list of the *types of reasoning* for you to inspect is located in the appendix. The list includes links to the main types being discussed here and more.[18]

In summary, it has long been a practice of human thought, debate, and discourse to explain, connect, and manipulate logic and reason. Many students of the philosophy become enamored with the functions of the mechanisms used to obtain logic and reason; that is, the argument. Often, what gets left out is the truth.

In his 2011 *Discover Magazine* article, *Is Reasoning Built for Winning Arguments, Rather Than Finding Truth?* author Chris Mooney states,

> Reasoning is generally seen as a means to improve knowledge and make better decisions. Much evidence shows that reasoning often leads to epistemic distortions and poor decisions. This suggests that the function of reasoning should be rethought. Skilled arguers, however, are not after the truth but after arguments supporting their views. This explains the notorious confirmation bias. Reasoning so motivated can distort evaluations and attitudes and allow erroneous beliefs to persist.[19]

Perhaps, Agent Mulder from the *X-Files* television show had it right when he argued continuously for finding the truth; he knew that "the truth is out there."

References

[1] Quote Ref 1: Aristotle. Metaphysics. Pg. 1009b. Retrieved July 29, 2016, from: http://www.perseus.tufts.edu/hopper/text?doc=Perseus%3Atext%3A1999.01.0052%3Abook%3D4%3Asection%3D1009b

[2] Introduction Ref 2: Before Nature: Rochburg, Francesca. 2016. Cuneiform Knowledge and the History of Science. Copyright 2016 by University of Chicago Press.

[3] Quote Ref 3: Aristotle. Metaphysics. Pg. 1009b. Retrieved July 29, 2016, from: http://www.perseus.tufts.edu/hopper/text?doc=Perseus%3A-text%3A1999.01.0052%3Abook%3D4%3Asection%3D1009b

[4] Aristotle Ref 4: Crane, Gregory R. Tufts University, Perseus Collection of Aristotle, Metaphysics. Perseus Digital Library; Perseus 4.0. Perseus Hopper. Retrieved July 29, 2016, from: http://www.perseus.tufts.edu/hopper/collection?collection=Perseus%3Acorpus%3Aperseus%2Cwork%2CAristotle%2C%20Metaphysics

[5] Deduction Ref 5: Wei-Ming, Wu. 2016. Butte College. This work is licensed under a Creative Commons Attribution-Noncommercial-No Derivative Works 3.0 United States License. Retrieved July 29, 2016, from: http://www.butte.edu/resources/interim/wmwu//iLogic/1.3/iLogic_1_3.html

[6] Deduction Ref 6: Wei-Ming, Wu. 2016. Butte College. This work is licensed under a Creative Commons Attribution-Noncommercial-No Derivative Works 3.0 United States License. Retrieved July 29, 2016, from: http://www.butte.edu/resources/interim/wmwu//iLogic/1.3/iLogic_1_3.html

[7] Induction Ref 7: *Merriam-Webster 2016 Collegiate Dictionary*. Copyright 2016 Merriam-Webster Inc. Retrieved July 29, 2016, from: http://www.merri-am-webster.com/dictionary/probable

[8] Induction Ref 8: Vocabulary.com Dictionary. 2016. Copyright 2016 by Dictionary.com. Retrieved July 29, 2016, from: https://www.vocabulary.com/dictionary/verifiable

[9] Induction Ref 9: Vocabulary.com Dictionary. 2016. Copyright 2016 by Dictionary.com. Retrieved July 29, 2016, from: http://www.dictionary.com/browse/credibility

[10] Induction Ref 10: Study.com. 2016. Entropy in Chemistry: Definition & Law. Copyright 2003-2016 by Study.com. Retrieved July 29, 2016, from: http://study.com/academy/lesson/entropy-in-chemistry-definition-lesson.html

[11] Douven, Igor. 2011. Abduction, Pg. 1. March 9, 2011. Copyright 2011 by Igor Douven. Published 2011 later copyrighted 2014 by The Metaphysics Research Lab, Center for the Study of Language and Information (CSLI), Stanford University. Retrieved July 29, 2016, from: http://plato.stanford.edu/entries/abduction/

[12] Popper, Karl R. 1962. *Conjectures and Refutations*. Copyright 1962 by Karl Popper. Published by Routledge Publishing.

[13] Taylor, Charles M. 1997. Philosophical Arguments: The Validity of Transcendental Arguments, Pg. 32. Copyright 1995 by Charles Taylor. Published March 1997 by Harvard University Press.

[14] Stern, Robert. 2015. Transcendental Arguments, Pg. 1. Published first February 25, 2011, and revised April 17, 2015. Copyright 2015 by Robert Stern. The Metaphysics Research Lab, Center for the Study of Language and Information (CSLI), Stanford University. Retrieved July 29, 2016, from: http://plato.stanford.edu/entries/transcendental-arguments/

[15] Kompridis, Nikolas. 2006. *Critique and Disclosure: Critical Theory between Past and Future*, Pg. 174. Copyright 2006 by Massachusetts Institute of Technology. Published 2006 by Cambridge: MIT Press.

[16] Wikipedia. 2016. Gregor Mendel. Retrieved July 29, 2016, from: https://en.wikipedia.org/wiki/Gregor_Mendel

[17] Unknown author(s). Fibonacci. Logical Reasoning. © 2016 Fibonicci by Fiboni V.O.F., Kastanjelaan 3, 6176DB Spaubeek, The Netherlands. Retrieved July 29, 2016, from: https://www.fibonicci.com/logical-reasoning/

[18] Changing Minds. Types of Reasoning. Copyright 2002–2016 by Changing Works. Retrieved July 29, 2016, from: http://changingminds.org/disciplines/argument/types_reasoning/types_reasoning.htm

[19] Mooney, Chris. 2011. "Is Reasoning Built for Winning Arguments, Rather Than Finding Truth?" *Discover Magazine*. April 25, 2011. Retrieved July 29, 2016, from: http://discovermagazine.com/intersection/2011/04/25/is-reasoning-built-for-winning-arguments-rather-than-finding-truth/#.V73tDI-cG9o

CHAPTER 5

The Yin and Yang of the Science and Ufology Argument

Everyone has a different idea of what proof really is. Some people think we should accept a new model of an airplane after only five or ten hours of flight testing. Others wouldn't be happy unless it was flight tested for five or ten years. These people have set an unreasonably high value in the word proof. *The answer is somewhere in between.*

—Excerpt from Edward J. Ruppelt book, *The Report on Unidentified Flying Objects*[1]

Key words: peer review, false memory, psychotropic

Frequently, when a scientist enters into a discussion with a ufologist about a topic related to UFOs and the like, a level of disagreement quickly surfaces. The ufologist is faced with opposition or a debate and mainly just the need to get a sincere and realistic acknowledgment and explanation from the science professional in response to a question.

The ufology field has been struggling to receive assistance regarding this topic. The paradigm has grown since the beginning of the accepted modern age of ufology, which was around 1947. This is due to the lack of acknowledgment and cooperation from the sciences on anything to do regarding any aspect of the field or related inquires.

Many science professionals do not believe any evidence is real, whether physical or circumstantial, to prove UFOs and extraterrestrials existence. The argument thus becomes intense. The disagreements that ensue usually produce absurd conclusions and refutations from scientists and other debunkers such as the "swamp gas" explanation, the Kenneth Arnold UFO sighting being a flock of pelicans,[2] the "mind playing tricks" definition of a *false memory* explanation, the *psychotropic* explanation, and the fallacious "don't bother me with the evidence because my mind is made up" explanation. Scientists say it is a good to be able to offer constructive and accurate refuted information and data to satisfy the objective of finding answers. The book *Ten Great Feuds in Science* by Hal Hellman reveals how pointless and extreme the matter of debunking and skepticism can be.[3] This proves the reality and long history of inefficiency and controversy away from the truth that has existed in the science disciplines for millennia.

Ten Great Feuds in Science also documents, highlights, and proves the existence of the notion that too often science is not about advancing knowledge and truth in nature. Instead, it is all about personal affronts and meaningless disagreements carried to an extreme. According to the author, "Skilled arguers, however, are not after the truth but after arguments supporting their views. This explains the notorious confirmation bias; reasoning so motivated can distort evaluations and attitudes and allow erroneous beliefs to persist."[4] In *Ten Great Feuds in Science*, approximately, twenty of the most acclaimed scientists in history took center stage for ten one-on-one debates that ended in what could be called social lunacy.

Some of the great debates in *Ten Great Feuds in Science* involved Pope Urban VIII versus Galileo (the earth or the sun as center of the universe); Sir Issac Newton versus Leibniz (who really invented calculus?); Charles Darwin versus "Soapy Sam"; T.H. Huxley (evolution wars); and Johanson versus the Leakeys (the missing link between apes and man). Each debate became viciously contentious and lasted years.

Each of these episodes caused what Thomas Kuhn, science educator and philosopher, categorized as a revolution within the debaters' particular science discipline. This is another example of the presence of human factors negatively influencing the scientific method.

Those great debates detracted from the body of positive attributions of scientific investigation. But what is disturbing is that they keep happening. Let's take a look at how a scientist and ufologist (or any student of the UFO and extraterrestrial subject) look at the phenomenon from different lenses that any of them, as humans, could.

We have begun to examine ways in which differences and disagreements arise within the body of the scientific community. In fact, if they do not feverishly attempt to debunk, discourage, and offer conclusions without inspecting much data and evidence, then why alternatively, aggressively retreat from being involved in any discourse whatsoever?

The first significant reasons have just been discussed. In the academic or scientific world of publish or perish, according to their terminology, it would be a matter of professional suicide if they were discovered to have been researching or encouraging investigation of the UFO phenomenon. It is a form of employment suicide. Noted author Upton Sinclair once said, "It is difficult to get a man to understand something when his salary depends upon his not understanding it."[5] There should be a measure of empathy directed toward the science educators. A lot of their resumes relate to examples of how, if they do not participate in professional and office political discourse, risk not only ridicule but discouragement from advancement in their profession and fraternity. This is and has been a very real condition applicable to those disciplines.

The second set of reasons enters into more of a rhetorical demonstration of excuses rather than allowing for making progress in real science discourse within ufology. Some rebuttals include the following:

- There is not enough evidence for me to say yes to the questions of evidence and existence of UFOs and extraterrestrials.
- The existence hypotheses are not testable.
- Scientists have to be able to physically handle their evidence. If one cannot handle parts of a spacecraft and be able to test them, this cannot be classified as evidence.
- They do not follow the typical science discourse.

- It is scientifically impossible for either phenomenon to exist. These responses offer some such impossibilities, including the physics behind travel to Earth, not showing themselves to us, and a perception indicating no motivation on their part.
- Evolution of life is unique to Earth and the uniqueness of this could only allow for a path to where and how we exist. (Regarding this point, are we trapped into the thinking the same way our ancestors did before the time of Copernicus when Earth was regarded as being flat? In accordance with the rules of those societies, believing otherwise puts their freedom and their lives risk. It would mean that in Europe, the Vatican was the governing authority. An abundance of evidence already existed during that time period and before that would have proven the earth was not flat. Would it then be considered or reduced to a matter of timing when the person who offers the proof is finally believed? Or maybe because the belief of a non-flat earth was so real in the minds of most people back in those times that they would just be more comfortable believing the status quo of the time. Relate this mind-set to the supposed reality of UFOs and aliens. If the machines and life-forms behind these alleged realities are more than a few hundred years ahead of man's technological realities, this notion may be too "out of the box" to be believed and accepted by today's society.
- Often, the scientist's job would literally be in jeopardy if they publicly expressed belief.
- The ufology subject is pseudoscience. Pseudoscience, depending on what definition you come across, consists of methods, theories, or systems that are considered as having no scientific basis. It is also defined as a set of practices or beliefs that are considered mistakenly based on the scientific method. Astrology and topics about the extra-dimensional are examples. A main argument for this conclusion is that these phenomena cannot be handled or physically tested.

Science is not equipped to practice in these areas. Most often, the phenomena of these issues cannot be proven or tested in and of themselves. This contradicts those who practice the scientific method, according to the classical, utilitarian definition you read about in science texts, and human factors cannot be included. Often, a subject that was initially deemed to be pseudoscience graduates into such widespread scientific recognition and acclaim that it is then officially considered as science. Some examples include the invention of genetics from Gregor Mendel's revolutionary work, evolution according to Darwin, the space-time continuum from Einstein, and geological plate tectonics according to Alfred Wegener.

Consider this quote from NASA in the FAQ section of the article, "Warp Drive, When?":

That whole subject (UFOs) is really irrelevant to our own human quest to travel to space. If we, humans, are going to figure out how to build space vehicles, then we have to build our own space vehicles. It doesn't matter if it has or has not been done by someone else. For example, if someone in the previous century saw a film of a 747 flying past, it would not tell them how to build a jet engine, what fuel to use, or what materials to make it out of. Yes, the wings are a clue, but just that, a clue.[6]

This sounds like a statement reflective of "office speak" within a science community that is stuck in an example of the typical science referenced by Thomas Kuhn, the great science educator and philosopher in his book *The Structure of Scientific Revolutions*.[7] This is an example of the definition of pseudoscience. The closed system of the standard at NASA, at least in a public persona, offers this scenario of an observer from the nineteenth century who is necessarily stuck in the embodiment of this example of human aerial flight. Referring back to the witness of the 747, he knows that it works because he just saw it! It is real.

A paradigm is a pattern of work protocol and procedure within a science community. According to Kuhn, a paradigm practices normal science that is designed to "maintain the status quo of the science community." A paradigm in a science community is practiced within a closed system. When someone steps out of the closed system via finding or doing something that is inconsistent or is disagreed with by everybody else, a crisis is formed. This is followed by an unsettled science community. Eventually, a revolution, according to Kuhn, forms. This revolution produces a new paradigm or way of practice in the community. The end result is that new scientific knowledge is discovered.

Here, NASA has demonstrated its arrogance and ignorance and its reliance on their closed system of the standardscience paradigm, at least publicly. From their anthropocentric perspective, it is more apt to dismiss the notion of warp drive science primarily if it has come from sources other than those of humankind's invention.

Here are some of the characteristics present and excuses given within any particular scientific paradigm:

- There is no funding for academic research in this field, which is similar to the notion of 'publish or perish' that has been present in academia since ancient Grecian times.
- A pack mentality that is akin to some of the basic social attitudes human's hold. If a maverick breaks from the herd, he is now held solely responsible for upholding his conclusions.
- A confirmation bias exists as another fundamental social attitude. The person tends to seek out evidence that only confirms his belief or his expectations. Any contrary evidence, no matter how extensive, elicits a response of, "Don't bother me with this information. My mind is made up."

- A progression from the previous two reasons demonstrates unwillingness on the scientist's behalf to conflict with the anti-ufology community that has been in existence since the late 1940s. An anti-ufology propaganda war has exploded in the literature since this time. Key constituency includes various federal government agencies (CIA, FBI, U.S. military departments, and so on) and academia, among others.
- Many scientists have accepted the preliminary explanation that UFO and alien event reports are the result of the observers misinterpreting identified objects and phenomena or that there are too many reports of hoaxes about such events.
- A good synopsis of the skeptical nature of scientists' conclusions regarding the existence of UFOs and alien existence was featured in a presentation given by J. Allen Hynek, PhD at the annual MUFON Symposium in 1983.[8] Hynek discussed seven reasons why he—as astronomer, astrophysicist, professor, ufologist, and chairman of Northwestern University's Astronomy department—would not or could not reach a positive conclusion about what is known as the "extraterrestrial hypothesis." This is basically the conclusion that some unidentified flying objects are piloted by extraterrestrials visiting Earth from other planetary systems.
- The seven reasons Hynek gave were:

1) A failure of manmade surveillance systems to detect incoming or outgoing UFO
2) Concerns about the exotic, unknown physics behind the notions of antigravity
3) What he described as "statistical considerations": elusive, evasive, and absurd behavior of UFOs and any occupants
4) Astronomical distances from the planetary origins of extraterrestrials who visited Earth
5) The aerodynamic unworthiness of UFO design factors

6) What Hynek described as "the Cheshire cat effect." It is a phenomenon, which analogizes that the quantum physics notion of the same particle of matter can be present in two different places at the same time is like the Cheshire cat that can physically separate itself from its smile so that they reside in two separate locations at the same time.

Dr. Hynek further explained each of these points respectively as follows:

1) Despite the most advanced radar and satellite tracking systems in use prior to 1983, UFOs are reported to move very fast in and out of the earth's atmosphere.

2) Extraterrestrials, though having been cataloged as mostly having humanoid phenotypes as physical traits and features, can exist on Earth and have adapted to the atmosphere with little difficulty. Reports of alleged abductions indicate many were not wearing space suits at the time of the event. This has an initial conflict with what knowledge man has gained about exoplanets, aerial systems, gravity, and adaptation of life to such unearthly systems as presented in 1983.

3) What Hynek meant by statistical considerations was that due to the sheer number of UFO and alien events reported in history (and the majority of these events went and continue to go unreported today) that a statistical analysis of this variable compared with the number of alien civilizations that could have visited Earth is an anomaly in itself.

4) The behavior of extraterrestrials during some of these events is not characteristic of what man has come to know about the expected actions given the circumstances of the incident.

5) The impractical nature of interstellar travel based on the then current knowledge is due to not understanding the mechanism of the various astrophysical characteristics

required for operating spacecraft with such design characteristics and the physical distances needed for such journeys.

6) A lot of the UFO sightings involved craft that was perceived to be too small to endure the vast distances required to fly from their planet to planet Earth. Some craft sightings appeared to disobey normal flight characteristics and were deemed to display flight patterns too erratic to be verified as being under intelligent control.

7) The ramification with the Cheshire cat analogy is that the UFOs can display flight characteristics impossible for man to duplicate. Their ability to disappear into the atmosphere, often in a matter of seconds, befuddles scientist's attempts to understand how this can happen.

J. ALLEN HYNEK

Citation:
Dr. J. Allen Hynek: extracted from: Allen Hynek Jacques Vallee1. jpeg - United States Government/Public Domain

Hynek concluded that there is sufficient plausibility and motivation to study, argue, and learn from research on all the arguments against UFOs and extraterrestrials except for number 5 listed above. The epistemology of understanding all the parameters of interstellar space travel is too insurmountable to offer validity to the extraterrestrial hypothesis or ETH.

An accomplished American physicist named James E. McDonald, a physics professor at the University of Arizona in Tucson, was at one time a senior physicist at the Institute for Atmospheric Physics. One of his most famous and powerful lectures was presented to the American Association for the Advancement of Science (AAAS) in 1969 titled *Science in Default: Twenty-Two Years of Inadequate UFO Investigations.*

The *Science in Default: Twenty-Two Years of Inadequate UFO Investigations* talk was a critique of the then-recent publication of Dr. Edward U. Condon's book report summary *The Condon Report.* As the principal author of the 1968 piece, Condon summarized the four-year research project results on the topic of UFOs and was sponsored by the U.S. Air Force.

The Condon Report gave a scathing set of conclusions on the UFOs. They dismissed the entire UFO phenomenon by stating, "Nothing has come from the study of UFOs in the past twenty-one years that has added to scientific knowledge. Careful consideration of the record leads us to conclude that further extensive study of UFOs probably cannot be justified in the expectation that science will be advanced thereby."[9]

A curious sidenote is highlighted here about the aspect of Dr. Condon being elected to lead the study and research team that produced *The Condon Report.* Condon's primary career work in the physical sciences was in the discipline of quantum mechanics. He had no experience in other subjects more relevant to ufology such as interstellar travel, atmospheric, and aerospace design engineering and observed propulsion characteristics of suspect craft is based on alternative physics, which is based on what has been gathered thus far on all these subjects until today.

There are at least two significant points that should be noted from this conference and Condon's speech. The first describes McDonald's conclusions about refuting the ETH.

Present evidence surely does not amount to incontrovertible proof of the extraterrestrial hypothesis. What I find scientifically disma-

> *yin* is that, while a large body of UFO evidence
> now seems to point in no other direction than
> the extraterrestrial hypothesis, the profoundly
> important implications of that possibility are
> going unconsidered by the scientific community
> because this entire problem has been imputed to
> be little more than a nonsense matter unworthy
> of serious scientific attention.[10]

The second point is that James McDonald received a memo written by Edward Condon's chief secretary, Dr. Robert Low before, his speech on August 9, 1966. This occurred at the beginning of the research that produced *The Condon Report*. The memo details Robert Low's prognosis and strategy for proceeding with the U.S. Air Force, University of Colorado, and Dr. Condon's UFO investigation. It is a recommendation for assembling an entire team of investigators, each of whom had a reputation for being nonbelievers on the issues. This research team, comprised of scientists within their various disciplines, had their nonbeliever mind-set already established at the start of the investigation to appease the concerns of University of Colorado officials. The project then proceeded as normal for the term, and the science team conducted routine research and interviews.[11]

The conclusions, however, were already assured to reside in the mind-sets of the team as initially assembled. Any final summary reports contained the teams' preconceived fallacious conclusions. Was *The Condon Report* a dedicated, truthful, and accurate analysis of extraterrestrials and UFO hypotheses? This issue will be discussed later in the "Disclosure" chapters.

Consequently, the reported conclusions were not surprising to most of the insiders within the ufology community. The low memo exemplifies some of the above reasons scientists and skeptics have for not affirming the UFO or alien phenomenon.

Some scientists share the same ideology and perspective of the ufology community. Astrophysicist James E. McDonald, a chief whistle-blower of *The Condon Project* office is one of these. Despite the closed-minded attitudes of some scientists, a large number have historically offered their input regarding the subject. Some of their comments are noted in the "What Scientists Are Saying" chapter of this book.

Consider this quote from Professor Hermann Oberth, German rocket scientist, and Werner von Braun's professor in a radio interview in Barcelona, Spain. In May 1965, at a conference of the First Astronomical Week Society, he declared, "We must consider real a fact of which we possess over eight thousand certain sightings. I cannot say if they are or are not interplanetary vehicles, but nobody can doubt anymore their existence."

Dr. Michio Kaku, PhD and famed astrophysicist, cofounder of the physics subject of string theory, responded recently in an interview from *The Observer* article, "Why Aliens May Exist but Aren't Landing on the White House Lawn." The news broadcaster stated that the Kepler Telescope spacecraft might have discovered significant evidence of an existing civilization in a planetary system relatively close to Earth in galactic terms. He started his discussion by stating, "Some people say if there are intelligent life-forms out there, and I think there are…"[12]

Dr. Kaku participated in many discussions about the subject during his many years of scientific research. In 2005, he spoke with ABC's Peter Jennings on his special airing of *UFOs: Seeing Is Believing*, an ABC News special documentary, "You simply cannot dismiss the possibility that some of these UFO sightings are sightings of some object created by an advanced civilization; a civilization far out in space, a civilization perhaps millions of years 'ahead of us' in technology. You simply cannot discount that possibility."[13]

Dr. Jacques Vallee, PhD—astrophysicist, computer scientist, corporate executive, and ufologist—wrote in his 1990 book, *Confrontations*:

> Skeptics, who flatly deny the existence of any unexplained phenomenon in the name of *rationalism*, are among the primary contributors to the rejection of science by the public. People are not stupid, and they know very well when they

> have seen something out of the ordinary. When
> a so-called expert tells them the object must have
> been the moon or a mirage, he is teaching the
> public that science is impotent or unwilling to
> pursue the study of the unknown.[14]

A repetitive cause of jumping to conclusions within ufology and the science refutation debates involves the practice of the analysis of facts and evidence that include multiple fields of science. A scientist is trained explicitly in his particular field of discipline. His training is significantly unique to that area of endeavor. The analysis and final explanation of any sighting or a sighting with more evidence often come without any investigation of factors that involve the other sciences or by that one particular scientist who is trained in his discipline but not in any others.

An example of this is in the July 1952 Washington, DC wave of UFO sightings. Evidence that included sightings from hundreds, if not thousands of eyewitnesses, is presented later in the "What People Are Seeing Especially Trained Public" chapter. Additional evidence in the case involved a lot of radar tracking returns, a lack of sound detection by any witness at any time throughout, and unusual flight characteristics exhibited by the craft, among others. After the wave of that last weekend in July, a possible attempt at a dedicated investigation by Edward J. Ruppelt of Project Blue Book was extinguished when the Air Force informed him that he would not be offered any support when he arrived in Washington, DC.

With characteristics from different science fields involved in a case like this, after literally no investigation except for a couple of interviews with officers on duty from a couple of the encounter sites, the final conclusion of "temperature inversion" was made by Major General John Samford on July 29 of that year, not forty-eight hours after the end of the last episode within the encounter. No study of any factors was undertaken, and while General Samford was only the messenger who made the press announcement, any thoughtful scientific study and analysis was left entirely out of the investigation.

Another scientist who provided an erroneous excuse for the Washington, DC wave came from respected and credible scientist Donald Menzel, an astrophysicist at Harvard University in the 1950s and 1960s. The entire Washington, DC sighting wave began on July 14 and continued over more than two weeks culminating with the Samford press conference of July 29, 1952.

An airline flight crew in flight that night had a clear and extensive view of multiple objects located in the direction of the Virginia coast. Without conducting or participating in an investigative study, Menzel dismissed the incident as anyof the following: meteors, searchlights on the clouds (the night was cloudless, which enabled the eyewitnesses'obseravtion in the first place), or cockpit glass reflections and fireflies trapped on or within the windshield. The reader was left to select which excuse(s) he preferred. Which one was it, Dr. Menzel?

This example is typical of the more significant recurring problem of opinions and excuses offered by the noted experts without any study, total lack of knowledge of the facts, or inappropriate use of the responder's scientific area of expertise when offering a response. It is a crux of the scientist versus ufologist debate and one of the biggest complaints the ufology field has had in this area.

An example of extraneous factors interfering with a factual investigation of a UFO encounter was the Michigan Incident of March 1966. This case ended with the swamp gas excuse provided by Dr. J. Allen Hynek, employed by the Air Force at that time. This extraneous factor originates from the circumstance in which Hynek was just starting to recover from a broken jaw at the time when he received the assignment. He probably did not feel entirely energized to participate in the physical activity of such an investigation.

The facts, in this case, were that the event theater was spread over two counties in southern Michigan, in addition to more radar evidence from a nearby Air Force installation. There were hundreds, if not thousands, of eyewitnesses who had information and data. These facts demanded a dedicated and thorough investigation. Giving consideration to the body of evidence involving the study of multiple science disciplines, Hynek still deemed it appropriate to conveniently excuse the incident wave as being caused by swamp gas.

A few problems ufology researchers have faced for a long time indicate a lack of agreement, synergy, and precision with the terminology used in the field and the presence of some of the very same fraternal human factor problems that science communities confront.

The first point about the lack of language agreement, synergy, and preciseness is significant in two areas. In one aspect, the ufology municipality has no control over the long history of mass media contributing to the problem. When the modern age of ufology began in the late 1940s, the defining term was *flying saucers* and was created solely from the press broadcasts of Kenneth Arnold's description of "saucers skipping on the water," a description of his encounter in June 1947. The term unidentified flying objects did not enter our consciousness until the 1950s and gradually replaced the use of the name "flying saucers." The mass media saw fantastic marketing and graphic opportunities to shorten this to UFOs, and soon after, the term became commonplace in society.

In addition, the ufology community was unable to prevent the acknowledgment of the UFO as a negative term with a negative connotation in society. Only after a couple of decades was the term unidentified aerial phenomena or UAP adopted by ufology. There, however, arose a problem of a lack of precision with the term. Many researchers continued to use UFO, as is frequently done in this book to demonstrate this issue. The problem of usage was that as time went on, there were more and more reports involving the underwater UFO otherwise known as unidentified submerged objects or USO. There are many other examples of the language ambiguity in the ufology discipline.

There are, however, some in the other science disciplines, such as the new chemical elements element 115. This compound was recently verified by the world's chemistry community, the International Union of Pure and Applied Chemistry (IUPAC), which oversaw development of the name for that new element Moscovium or "Mc," the symbol that is used for the element. Hence, all the chemical science communities will refer to element 115 as Moscovium from now on.

On the subject of industry rules and regulations, we can expand this notion to a macro perspective. In this view, it is apparent there is no governing body regulating matters within the ufology industry. All the sciences have their individual international governing organizations. What they bring to their fields is more efficiency and transparency, which leads to more knowledge that is accessible to community members and the public directly. These benefits are provided and enjoyed more quickly when new technological communication improvements merge into this system. Even when two nations who are politically opposed to other matters of sociopolitics are involved, the resident scientists of those nations do not share this opposition. That situation is part of the professional municipal that helps the discipline to act as one. The best example of this notion would be to look at the continent of Antarctica. Antarctica has no political structure or body of communal laws as in other countries. Their reason for existence is only for the free professional development and research discovery of the scientific community. If you wish to live in a politically free society, the only requirement for living in Antarctica is that you are a scientist, infrastructure worker, or part of the support personnel.

The lack of an overall ufology-governing body may have some implications toward the difficulties the industry has ineffective communication with the public. Most of the time, the mass media is in the middle of an information dissemination process. When a ufologist has the knowledge to report to the public, often, the media, if present, blocks effective communication—linguistics and lack of cohesive use of language interfere with its success. A media reporter will use a lot of linguistic license in his or her stories, and this hurts the entire process. Also the media works with its industry and internal organization agenda with its own objectives, which is the source of another continuing problem that the ufology field has faced since the beginning of the modern age of ufology. But this one is not with the science disciplines.

The second problem indicates the continuation of what can be referred to as the human factors problem within ufology. It is the same landscape of lack of cooperation and unity within ufology that hinders the progress of the sciences. It cannot be avoided in any endeavor where many humans are involved; this is obvious. But the sciences do have the *peer review* research mechanism that aids in this area.

The peer review is hardly perfect. The fraternal aspects of the human factors of science discourse, as would be if there was a ufology peer review, do interfere with this mechanism; but the extremely large size of this science review helps buffer the negative ramifications the fraternal aspects introduce. Scientists from all over the world who, by geographical separation may be at least partially insulated from potential favoritism a peer review can often produce, can still offer efficient and thoughtful peer review input. There will always be the science great debates as noted in many literary works. If ufology utilized peer reviews similar to those of the science disciplines, the level of understanding between the sciences and ufology would improve.

References

[1] Ref 1: Ruppelt, Edward J. 1956. *The Report on Unidentified Flying Objects*. Pg. 256. Copyright 1956 by Edward J. Ruppelt. Published by Ace Books in conjunction with Doubleday and Co., Inc.

[2] Yin Yang Ref 2: Unknown author. RRRGroup, "Kenneth Arnold and the pelicans" (Wednesday, April 4, 2007); URL accessed July 29, 2016.

[3] Yin Yang Ref 3: Hellman, Hal. 1998. *Great Feuds in Science: Ten of the Liveliest Disputes Ever*. Copyright 1998 by Hal Hellman. Published 1998 by John Wiley and Sons, Inc.

[4] Yin Yang Ref 4: Mooney, Chris. 2011. "Is Reasoning Built for Winning Arguments, Rather Than Finding Truth?" *Discover Magazine*. April 25, 2011. Retrieved July 29, 2016, from: http://discovermagazine.com/intersection/2011/04/25/is-reasoning-built-for-winning-arguments-rather-than-finding-truth/#.V73tDI-cG9o

[5] Sinclair, Upton. *I, Candidate for Governor: And How I Got Licked*. (1935). University of California Press. 1994. Pg. 109.

[6] NASA Administrator. Warp Drive, When? FAQ of: Is Warp Drive Real? March 10, 2015, as updated November 4, 2015. www. NASA.gov/centers/glenn/technology/warp/warpfaq_prt.htm.

[7] Ref Kuhn: Kuhn, Thomas S. 1962. *The Structure of Scientific Revolutions*. Copyright 1962 by The University of Chicago. Published 1962 by The University of Chicago Press.

8   Hynek, J. Allen. 1983. The Case Against ET. In Walter J. Andrus Jr., and Dennis W. Stacy (eds). MUFON UFO Symposium.

9   Condon Ref 9: Condon, Edward U. 1968. *Final Report of the Scientific Study of Unidentified Flying Objects*, Edward U. Condon, Scientific Director, Daniel S. Gillmor, Editor. Paperback edition copyright 1968. Published by Bantam Books.

10 McDonald, James E. *Science in Default: Twenty-Two Years of Inadequate UFO Investigations*. December 27, 1969. American Association for the Advancement of Science. 134th Meeting General Symposium, Unidentified Flying Objects.

11 Low, Robert. Robert Low "Trick Would Be" Memo and Transcription. August 9, 1966. Memo to E. James Archer and Thurston C. Manning. Reproduced by NICAP. Nicap.org/docs/660809lowmemmo.htm

12 The Observer.com. Dr. Michio Kaku on Why Aliens May Exist, But Aren't Landing on the White House Lawn. November 4, 2015. Observer Media as Publishers.

13 ABC News. USOs: Seeing Is Believing Broadcast. Host Peter Jennings. UFOs: Seeing Is Believing. 01hr18–01hr22. February 24, 2005. Publishers: American Broadcasting Company Television News.

14 Vallee, J. Confrontations. 1990. New York: Ballantine Books. Robert Low "Trick Would Be" Memo and Transcription (below memo jpegs)

CHAPTER 6

Who Has Seen Them?

I feel that the Air Force has not been giving out all the
available information on the unidentified flying objects.
You cannot disregard so many unimpeachable sources.

—The Honorable John W. McCormack
U.S. Speaker of the House of Representatives
January 1965[1]

Key words: vagary, orographic clouds

How many people claim to have observed an unidentified flying object? It is impossible to say. It would seem that since the beginning of the modern ufology era, some may conclude that many thousands, maybe millions, of reports have been made. Logically speaking, the most recent records have not had a chance to get lost in oblivion as is the fate of older archives. Also development of more efficient media technologies is evidence of the notion that more people are observing or encountering UFO and extraterrestrial episodes than ever before.

An additional factor that contributes to the enormity of the problem is the perceived small percentage of eyewitnesses who actually report a sighting in the public domain. A good guesstimate is the probability of this is less than ten percent.

The latter is the bigger reason that even today's modern age of ufology cannot start to reconcile this issue. Today, there are more outlets and options for one to report a sighting and more high profile outlets than ever before. Only the most dedicated of coordination projects could even think of combining all the data from active UFO reporting agencies to tabulate just a small percentage of the multitude of sightings that occur; real-event UFO sightings.

It is generally accepted that the modern era of ufology originated around 1947 when a wave of episodes in June and July of that year motivated society to take slightly more notice of these events. Two of this wave of events were the Kenneth Arnold sighting in Washington State on June 24 and the Roswell, New Mexico, crash on July 3. More of this topic will be addressed later, as they were catalysts for how many segments of society today, including the government and the military, eventually adopted explanation protocols using concepts that are still practiced nowadays. The term *ufology* was also coined as a result of the events of 1947 and developed and became more formal after the June Arnold and Roswell events.

There have been untold and incalculable numbers of sighting and encounter events before the 1947 modern era of ufology began. There were fewer communication technologies, media, record keeping, and historical archiving. So as the archaeology documentation lessen farther back in history, it does not necessarily mean that people throughout history saw fewer UFOs. They were just not documented then due to a lack of accessibility to durable archiving technologies. Also the survivability of archived recording devices lessens farther back in time. So we know and will discover later that through history, there have been untold quantities of UFO sighting events. The archaeological record is not as helpful as we would like it to be. But discoveries are constantly made that will add to this record.

There are still plenty of material reports of UFOs and extraterrestrial entities, extrabiological entities (EBEs), alien entities, and the like that could fill volumes. As noted, fewer than ten percent of episodes ever get reported to the public.

No element of society is exempt from being known to have had an observation episode. Famous people with reliable reputations are included among the eyewitnesses. It will be helpful if we document some reports from famous people who have also observed UFOs. This becomes significant for the eyewitness. These individuals would reason that if they told the public about a sighting, such a report would subject them to potential public ridicule from the high degree of uncertainty and anxiousness apparent with a lack of understanding about the subject. Hence, that person would recognize how much he could potentially lose in income and professional and personal reputation, ridicule, etc. This is a significant factor in ultimate disclosure.

Whatever extraordinary or ordinary circumstances made him or her famous the circumstance of experiencing a UFO event can happen to anybody and become deleterious due to this lack of societal understanding.

This section chronicles numerous lists of famous people who have had one or more UFO sighting events. There are many more examples than those revealed below. The first list contains names of many international historical, political, and social leaders who, through their credibility and notoriety, have been able to contribute societal initiatives and exhibit a higher risk of negative ramifications for their disclosure.

Governor Fyfe Symington

This survey starts with the most recent encounters. Fyfe Symington was an Air Force officer pilot before serving as the governor of Arizona. On March 13, 1997, he became part of a mass UFO sighting named "The Phoenix Lights." The incident is highlighted here in the chapter, "What People Are Observing: The Public."

He was then acting governor in 1997 and used some of his official channels to investigate what happened that night. He was unsuccessful, and no one from the Air Force or other governmental offices was becoming involved. After a poorly thought-out publicity stunt, a couple of months later, which greatly upset the public and especially the eyewitnesses to the Phoenix Lights, he dropped the matter until his term in office was finished.

Symington finally reported his experience years later. In 2007, he went on an informal "lecture circuit." Here are the statements from those interviews:

> I witnessed a delta-shaped craft silently navigate over Squaw Peak, this dramatically large,

very distinctive leading edge with enormous lights. As a pilot and former Air Force officer, I can definitively say that this craft did not resemble any man-made object; it was certainly not high-altitude flares because flares don't fly in formation.[2]

Symington continued,

The gigantic triangular and unidentified V-shaped objects as gliding slowly and silently. The eerie, lighted vehicles were bigger than many football fields up to a mile long. These were clearly solid, technological flying machines that blocked out the stars.[3]

I'm a pilot and I know just about every machine that flies. It was bigger than anything that I've ever seen. Other people saw it, responsible people. I don't know why people would ridicule it.[4]

And finally,

We want the government to stop putting out stories that perpetuate the myth that all UFOs can be explained away in down-to-earth conventional terms. Incidents like these are not going to go away. About a year ago (2006), Chicago's O'Hare International Airport experienced a UFO event. When it comes to events, we deserve more openness in government, especially our own.[5]

President George H. W. Bush

The Senior President Bush served as Ronald Reagan's vice president before his term as president. While President Bush has not revealed any personal sighting of his own, his work history also included service that culminated with being named as the director of the CIA during the Carter Administration.

His 1988 presidential campaign produced guarded remarks and quotes concerning UFO and extraterrestrial matters. On March 7, 1988, while in Rogers, Arkansas, a UFO researcher named Charles Huffer had this exchange with Bush:

> "Mr. Bush, Mr. President, will you tell the people the truth about UFOs?"
>
> Bush replied, "Yeah. If we can find it, what it is. We are really interested."
>
> Huffer said, "Okay, you're going to get it."
>
> Bush replied, "Why don't you send me some information about it?"
>
> Huffer responded, "Naw, you're a CIA man. You know all that stuff."
>
> Bush ended the conversation by saying, "I know some, I know a fair amount."[6]

Vice President Bush handled President Reagan's Daily Briefings and other intelligence publication from the CIA during those terms. The CIA regarded him so much that the organization named their headquarters the George Bush Center for Intelligence. Given his history and life experiences, could Bush be a latent treasure trove of knowledge about what the government knows about alleged UFO and extraterrestrial matters?[7]

Presidential Candidate Dennis Kucinich

Even though he conducted a presidential campaign in 2008, Dennis Kucinich maintained his privacy for over twenty-five years pertaining to the incident that took place at his friend, actress Shirley MacLaine's home in Washington State for over twenty-five years. In September 1982, Kucinich was staying there with MacLaine's long-time assistant, Paul Costanzo, and his friend who has maintained her anonymity.

Here is a recount of the events. At dinner, Kucinich spotted a light in the distance. Moments later, after observing it through a telescope, Kucinich called both Costanzo and his girlfriend outside to watch it. "It was a hovering light, which soon divided into two and then three—three charcoal-gray, triangular craft, flying in a tight wedge, having red and green lights with a laser-like red light at the tail. Each triangle was roughly the size of a large van. The craft approached within two hundred yards suspended over the field and emitted a quiet, throbbing sound. The craft held steady in midair for perhaps a minute then sped away." The next day, the group observed military helicopters surveying the valley around MacLaine's home for the morning.[8]

President Ronald Reagan

President Reagan has had multiple experiences of his own. Naturally, a U.S. president has an exceptional opportunity to be involved in ufology due to the ongoing dilemma with the concerns of government knowledge and possible secrecy and disclosure about the subject. Included here are two separate sighting encounters and example of such involvement. First are the sightings.

Both of these occurred when he was governor of California. The first one was on the night the governor and Mrs. Nancy Reagan were to attend a party hosted by movie actor William Holden. They were late in arriving to the party. When they did arrive, the governor and the First Lady sought out old friends, actress Lucille Ball and comedian Steve Allen, to tell them of their experience. The governor and First Lady were driving down the California Coast Highway and stopped, along with other witnesses from the general public, to observe the episode.[9]

The second came at the end of his last term as governor in late 1974. Air Force Colonel Bill Paynter, the pilot of Governor Reagan's personal airplane, was witness to this encounter.

In a statement he made to Norman Miller, Washington Bureau Chief for the *Wall Street Journal*, the governor's plane was making an approach to land in Bakersfield California. During the descent, Reagan noticed a strange light behind the plane. "We followed it for several minutes. It was a bright white light. We followed it to Bakersfield, and all of a sudden, to our utter amazement, it went straight up into the heavens."

Colonel Paynter stated, "It appeared to be several hundred yards away, a fairly steady light until it begun to accelerate. Then it appeared to elongate. Then the light took off. It went up a for-ty-five-degree angle at a high rate of speed. Everyone on the plane was surprised." The UFO went from normal speed cruise to a fantastic speed instantly.[10]

President Reagan's involvement event came on June 27, 1981, at the screening of the then-new Steven Spielberg movie *ET: The Extraterrestrial*. In attendance were such dignitaries as Supreme Court Justice Sandra Day O'Connor; George H.W. Bush, Sr. (future president and former director of the Central Intelligence Agency in the mid-1970s); NASA administrator James Beggs, former deputy director of the CIA Vernon Walters; Chief of Staff James Baker; British ambassador to the United States Sir John Nicolas Henderson; Chairman of the 1984 Olympics Peter Uberroth; Undersecretary of State for Security James Buckley; and Spielberg, among many others.

Spielberg later recounted, "And he said, 'I wanted to thank you for bringing ET to the White House. We really enjoyed your movie. And there are some people in this room who know that everything on that screen is absolutely true.' And he said it without smiling!"[11]

President Jimmy Carter

Two years before he was to become governor of Georgia, President Carter was preparing a speech on January 6, 1969, at a Leary, Georgia Lions Club. Sometime after 7:00 p.m., another guest called Carter's attention to visible lights to the west of their position. President Carter, who served for over seven years in the Navy, recalled watching the lights "for over ten minutes" as they appeared over thirty degrees over the horizon. In that time, the lights of the object changed colors gradually from blue to red then back again in succession. President Carter felt the lights were "not solid in nature but of a self-illuminating type."[12,13]

It should be noted that there is and will be a continued practice of the U.S. presidents reconfirming their experiences and not back-tracking from their original disclosure reports. Presidents Reagan, Carter, and others continued their quest for more dissemination of knowledge whether through efforts at government disclosure or con-tinued publicity of their encounters.

Sir Eric Gairy, Grenada's Prime Minister

Motivated primarily by his own alleged encounter Prime Minister Gairy launched a movement to elevate the field of ufology to a higher global stage. Starting in 1974, he worked to assemble a team of scientists, politicians, and military leaders who would peti-tion the United Nations General Assembly to officially recognize and place a permanent existing status on the UFO and extraterrestrial phenomena. This legislation could then allow the UN to prepare a more substantial and comprehensive study protocol.

Gairy met with such heads of state as President Jimmy Carter on September 9, 1977, then UN Secretary General Kurt Waldheim of Austria and various British military and political leaders. He assem-bled a science team that included ufology pioneer J. Allen Hynek, whose experiences are described later; Jacques Vallee; Dr. David Saunders; once member of the Edward U. Condon Project Blue Book team; U.S. Apollo astronaut Gordon Cooper; and various ufol-ogists including Leonard Stringfield, Claude Poher from the France UFO study organization GEPAN, and journalist Lee Speigel, whose career would be jump-started by participation in this venture.[14,15]

Gairy would continue his efforts to call for a dedicated and continual scientific investigation paradigm that would advance the study of UFOs and extraterrestrials, as well as seek full disclosure of knowledge by all the world governments. Here are some excerpts from Gairy's speech:

I come now to a matter of great concern to Grenada. I refer to the subject of unidentified flying objects (UFOs). Sightings of UFOs are not restricted to just one or two parts of our planet. Reports come from all over the world today. The question is now being increasingly asked, why should man be precluded from information on UFOs, a matter of great interest and importance to man, while at the same time he is fed so many trivialities which can contribute nothing to his enrichment or to the advancement of mankind.

Gairy continued by stating,

Irrespective of theory, however, the major research groups are dedicated foremost to achieving the following main objectives in their endeavors: to study all significant reports on unidentified flying objects; to disseminate the substantive results of such study to the public and the news media; to work cooperatively with the United Nations to help establish a communications system through which important data can be rapidly exchanged internationally.[16]

President Gerald R. Ford

President Ford's office enabled him to be involved with the ufology subject, though he has not claimed to have had a sighting or any other type of encounter. In fact, his most prolific moments came while he was U.S. Minority House Leader in March and April 1966. He was positioning to influence a dedicated investigation of the UFO phenomenon as a result of the wave over Southern Michigan and Northern Ohio. In response to the many concerned eyewitnesses to the events that began March 13, Congressman Ford called for Congress to persuade formation of such investigative units. The Michigan flap is better known as the "swamp gas incident," and more details are provided elsewhere in this book.

Although he personally could do nothing else later in this area, President Ford's efforts did not go unrewarded. Congressional hearings in April 1966 brought the subject more publicity. This led ultimately to a later 1966 announcement by the Air Force that a special investigative research study was being formed in conjunction with the University of Colorado. This came to be known as *The Condon Report* of the book title published in 1969.[17]

President Richard M. Nixon

A synopsis of *Project Blue Book*, which was authored by Dr. Edward Condon, is supplied in the "Disclosure" chapter of this text. The book was released to the public in 1969 under pressure from the Air Force to expedite the publication. This was because President Nixon had just been sworn in as president and had held a long-running feud with Condon. For years, President Nixon had consistently challenged Condon's activities and security clearances within the Project Blue Book along with other professional duties and projects within the Air Force and the federal government. The Air Force had maintained a program, since the 1950s and evidenced by their many actions and announcements over the years, to enact a strategy of minimizing and debunking all UFO-related experiences. Their objective was to ultimately end the reporting of all such incidents from the public domain. They did not want to have a new president interfere with their efforts at such closure. This was part of their secrecy and disclosure campaign, and they saw Nixon as a threat to that end.

While President Nixon had never disclosed a UFO sighting of his own, a situation occurred in February 1973 that contributed to his legacy of having more than cursory knowledge of extraterrestrial matters. Whenever he traveled to Florida, Nixon flew into Homestead Air Force Base, which was nearby his southern presidential retreat in Key Biscayne.

On this trip, he met with old friend television actor Jackie Gleason, who maintained a long and keen interest in ufology. At midnight of February 19, President Nixon allegedly escaped from the watchful security of the Secret Service long enough to escort Gleason through the front gates of Homestead and took him back to the hangar's entrance, and through the many uniformed security personnel to where pieces of wreckage of one of their craft, and later bodies of deceased extraterrestrial biological entities (EBE's) were stored. The station guards recognized the president and just retreated when his car approached.

According to Gleason,

> There were a number of labs we passed through first. Nixon pointed out what he said was wreckage from a flying saucer. Next, we went into an inner chamber, and there were six of eight of what looked like glass-topped Coke freezers. Inside them were the mangled remains of what I took to be children. I forget whether he said they had three or four fingers on each hand, but they definitely were not human.[18,19]

U.S. Senator Richard B. Russell

Russell served in the U.S. Congress from 1932 until shortly before his death in 1971. He also served as chairman of the Senate Armed Services Committee from 1951–1953 and 1955–1969.[20] Russell's UFO observation came while in Russia on October 4, 1955.

A little after 7:00 p.m., while on a train traveling through the Transcaucasia region, between Atjaty and Adzhijabul and en route to Prague, Czechoslovakia, Russell sighted,

> Two round and circular unconventional aircraft resembling flying discs or flying saucers seen taking off almost vertically one minute apart. Disc aircraft ascended near dusk with outer sur-

face revolving slowly to right with two lights stationary on top near middle part. Sparks or flame
seen coming from aircraft. No protrusions seen
on aircraft. Both craft ascended vertically to six
thousand feet then speed increased sharply in
horizontal flight on northeast heading, like a
discus in flight, after sighting Soviet trainmen
because excited and lowered curtains and refused
permission to look out windows.[21,22]

The first official and top secret report was prepared and sealed
by U.S. Attache Lt. Col. Thomas Ryan. In addition to Russell, three
other U.S. witnesses were confirmed to have reported the sighting:
Ruben Efron, Russell's interpreter; Col. E.U. Hathaway, Russell's
aide; and an unidentified fourth witness. Russell saw the first craft
take off and called Efron and Hathaway over to the window where
they all observed the second ship rise vertically like the first, then
level off at six thousand feet stationary for a few seconds, then vector
northeast very quickly on a level flight path.

Colonel Ryan summarized in his report: "An eyewitness account
of the ascent and flight of an unconventional craft by three highly
reliable United States observers." The CIA and the FBI prepared
their own case history reports. The CIA detailed the sighting of the
fourth eyewitness, but the identity of this person was never reported.
Another report by John Foster Dulles, then U.S. secretary of state,
was briefed on October 18,1955, by the FBI. Their report, prepared
and sealed on November 4, 1955, quoted Dulles as saying, "Lt. Col.
Hathaway's testimony would support existence of a flying saucer."
The Los Angeles, California, *Examiner*, as reported by Tom Towers
in his column *Aviation News*, printed the contents of a letter from
Russell in response to requests for disclosing information about the
encounter. Senator Russell responded, "Permit me to acknowledge
your letters. I have discussed this matter with the affected agencies
of the government, and they are of the opinion that it is not wise to
publicize this matter at this time."[23]

President Dwight D. Eisenhower

President Eisenhower was perhaps the best qualified example of a U.S. president that could offer insights into the UFO phenomenon. His career history in the military then politics and also the time in the 1950s which he served as president were all significant factors which offered very much credibility and validity into these insights. He also had his own sightings.

This personal sighting came during his tenure as supreme commander of the North Atlantic Treaty Organization (NATO) while aboard the aircraft carrier USS Franklin D. Roosevelt on September 20 of that year. The events of that day and following week were part of *Operation Mainbrace* exercises. The crowd which was with him in September 1952 to witness this wave of widespread events totaled over eighty thousand military personnel within the NATO scientific, military, civilian service, and peripheral personnel. All were highly trained service people.

President Eisenhower was also central to many alleged situations through his years involving summit encounters with representatives of extraterrestrial nations. While a consensus of most of these encounters is to discourage any attribution of accuracy and reality to them, one is at least substantiated by multiple eyewitnesses who were with him that day in February 1954. While in Palm Springs, California, the president excused himself from dinner with a complaint of a dental problem. Many other people in his entourage fielded the onslaught from the press over his health issues while other witnesses confirmed him being at Edwards Air Force Base (then known as Muroc Field) that Saturday night and Sunday morning.[24]

President Eisenhower was also intimately known to have collaborated with Winston Churchill over disclosure and secrecy issues and strategies about UFO sightings and encounters by both American and British military and civilian witnesses in the 1950s.[25]

His great-granddaughter, Laura Magdalene Eisenhower, has taken up the path of some of the alleged involvement of President Eisenhower before and during his terms of both military and public office. Among her publications is the presentation she gave in

April 2014 at San Marino Citizen Disclosure Hearings. This speech highlighted some details of the history of President Eisenhower's unproven participation in ufology events and of the more lengthy history of the U.S. government's participation in secrecy and disclosure issues.[26]

Sir Winston Churchill

While the British World War II prime minister did not disclose his own eyewitness, UFO event Churchill's obvious involvement and proximity to matters of the ufology influence has been coming to disclosure for some time now. Their Ministry of Defense is the issuing originator of catalogs of previously top-secret documents that disclose Churchill's knowledge and involvement in many UFO encounters by military service personnel.

The former prime minister, regarding one example, allegedly ordered the unexplained RAF incident over the east coast of England should be kept secret for fifty years because it would "provoke mass panic." Mr. Churchill declared that the incident "should be immediately classified for at least fifty years and its status reviewed by a future prime minister."[27,28,29]

President Harry S. Truman

At the dawn of the modern era of ufology and UFOs, no one had a harder path to navigate than President Truman. It appears logical that if extraterrestrial entities exist and have visited Earth that the main motivation might have been man's crossing into the Nuclear Age. If ETs had the means with which to get to Earth, they had the means to observe human activity such as the explosions in 1945 that ended World War II and the many years of carnage that precipitated those events.

President Truman had the experience of being involved in such an enormous undertaking. His experience undoubtedly assisted him through the postwar years when sightings of unexplained phenomena of ufology interest exploded exponentially in both basic numbers and in numbers that were being made public. While he did not himself report his own sighting experience in any public domain, President Truman oversaw the times when many UFO and alien events occurred. From the Foo Fighters of the World War II battle theater to the alleged Roswell crash in July 1947, President Truman's diplomacy skills faced many problematic situations. Other events included the UFO wave over Washington, DC in July 1952 and even in his last days as President in January 1953 with the Robertson Panel.

Perhaps, the most enduring and nuanced problematic situations came in 1947 in which the Roswell incident was intricately engaged. Within weeks after Roswell, *Project Sign* was created by President Truman to serve as an investigation unit of UFO sightings and reports from the public. Also allegedly, a new office of investigation, *Majestic 12*, also known in ufology as MJ-12 or MAJIC 12, was organized to deal with the more covert aspects of the ufology phenomena in September of that year.[30]

Kenneth Arnold

While the Roswell incident christened the generally-accepted modern ufology movement, a truer establishment of the movement of the modern era of ufology resulted from this earlier incident on June 24, 1947. If not for this sighting, Kenneth Arnold, as untold thousands of eyewitnesses would also lay eventuality to, would have lived his life a lot more anonymously.

Arnold had been a private pilot for over ten years. He had accumulated over a thousand hours of flight time. The afternoon in the skies around Mt. Rainier in Washington State were in mild, sunny weather conditions with a slight wind aloft. While inspecting the terrain for the remains of a U.S. Marines C-46 cargo plane which had crashed before his attention was diverted by the "first flash of light." He grew to notice nine of what he thought were reflections and, after performing some tests with regarding sighting parameters, became fixated on "nine boomerang-shaped" craft flying in a long-chain formation. Sometimes, Arnold could see them on edge, but they alternately turned flat on to where they became harder to see. They were flying at incredible speeds and weaving up and down slightly in altitude even though the wind conditions were not affecting his plane like that.

Arnold would later describe eight of them as convex-shaped and the ninth as "crescent-shaped" he reported the movement like "saucers skipping on water." The resulting newspaper stories created the term *flying saucers*. These saucers flew behind a lower mountain peak in the Rainier chain but returned to visual observation. The velocity was later calculated by Arnold to be more than 1,200 mph. No manmade craft had even yet broken the sound barrier (about 767 mph) as officially recognized, until October 14, 1947, by U.S. test pilot Chuck Yaeger.

Among the support to Arnold's credibility was the first interview he held with the *East Oregonian* newspaper. The story summarized his credibility as follows:

> Arnold had the makings of a reliable witness. He was a respected businessman and experienced pilot and seemed to be neither exaggerating what he had seen nor adding sensational details to his report. He also gave the impression of a careful observer.[31]

Among the refutation debunking explanations offered were a collection of three different explanations spread over eighteen years by Harvard University astronomer and alleged confidant for the Majic-12 investigative unit Donald Menzel. In 1953, Menzel suggested that Arnold had seen "clouds of snow blown from the mountains south of Mt. Rainier." In 1963, Menzel explained that Arnold had seen orographic clouds. *Orographic lift* occurs when an air mass is forced from a low elevation to a higher elevation as it moves over rising terrain. As the air mass gains altitude, it quickly cools down adiabatically, which can raise the relative humidity to one hundred percent and create clouds and, under the right conditions, precipitation.[32]

The third Menzel refutation attempt came in 1971 when it was explained that Arnold might have merely seen spots of water on his planes' windows. This was maintained by Menzel, with no explanation as to why he did not offer one in either 1953 or 1961, even though Arnold had also reported on the event. He specifically reported to the *East Oregonian* and other publications that he had rolled down his windows at the beginning of the sighting as one of his tests to eliminate such factors and attributions.[33]

President Franklin Delano Roosevelt

It was never rumored that President Roosevelt had been an eyewitness to a UFO sighting. His office, though, is rumored to have been a principle participant in an alleged agreement between the American government and a race of "Grey extraterrestrials" in 1934. On July 11, the so-called Grenada Treaty was signed. In her presentation at the World Symposium on UFOs and Related Phenomena and Extraterrestrials and World Politics' Citizen Disclosure Hearings, Laura M. Eisenhower, granddaughter of President Dwight D. Eisenhower, made these statements:

> According to a whistleblower and ex-operative of M-16 (Britain's Secret Intelligence Service), the first treaty between the Grey extraterrestrials and the American Government under Roosevelt was signed in July 1934. The treaty stated that in return for the Greys providing high technology, the United States would allow the Greys to proceed unhindered with human abductions, ongoing genetic experiments.[34]

President Thomas Jefferson

An article from an 1802 publication of *Transactions of the American Philosophical Society* textualizes the report of an extraordinary encounter by then Vice President Jefferson.

A phenomenon was seen to pass Baton Rouge, April 5, 1800. It was first seen in the South West and moved so rapidly as to disappear in the North East in about a quarter of a minute. It appeared to be the size of a large house, seventy or eighty feet long, two hundred yards above the surface, wholly luminous, of a colour which may be called crimson red. A considerable degree of heat was felt but no electrical sensation.

Via a discussion of the article by author Bill Thayer of the University of Chicago, an analysis of the facts was made regarding this event. An explanation of the object being a meteorite was dismissed in the original article for the following reasons:

An object of this size would have probably have left a much larger trace of itself. It was traveling at no more than 2,200 km an hour. He speaks of it, at any rate, as on a more or less level trajectory. Further, Thayer obtained results of scientific experimental measurements of the geophysics of the parameters of the event.

The model yields a seismic effect somewhat less than that reported by the witnesses and a fifty-five-decibel sound level, similarly less than what was reported. Also there should have been a sonic boom if it was a house-sized object coming in at a meteoric speed, it would have been a huge event, with no survivors for miles, flattened trees, etc.[35]

There were no reports of casualties or such residual characteristics as would be demonstrated by a meteoric catastrophe, including the observations of the spectrum of colors observed as the object passed overhead.

President George Washington

Prior to being elected the first president of the United States, George Washington had a more nuanced yet exotic encounter while still commanding the U.S. armed forces during the American Revolutionary War. In early 1778, General Washington had what today may be classified as an "encounter of the third kind." The archives at Valley Forge, Pennsylvania, and the U.S. Library of Congress have filed copies of the article from the *National Tribune* from 1880 that reproduces the reported account of a soldier present during the day of Washington's vision.

According to the article, General Washington experienced a midday encounter from "a singularly beautiful female." He was induced by this entity to experience numerous graphical representations depicting future wars of the republic of America. The first was of the Revolutionary War and its outcome. The second was the American Civil War and its outcome. The third was a war that has yet to be fought. The encounter ended with this passage:

> While the stars remain and the heavens send down dew upon the earth, so long shall the union last. Son of the Republic, what you have seen is thus interpreted: three great perils will come to the Republic. The most fearful is the third (World War II), but in this greatest conflict, the whole world united shall not prevail against her. Let every child of the Republic learn to live for his God, his land, and the union.[36]

James Everell

America's first UFO was sighted and documented over the Charles River in Boston. Governor John Winthrop described the incident in one of his volumes in *Winthrop's Journal*. Winthrop's experience occurred in 1639.

In this year, one James Everell—a sober, discreet man—and two others saw a great light in the night at Muddy River. When it stood still, it flamed up and was about three yards square; when it ran, it was contracted into the figure of a swine: it ran as swift as an arrow towards Charlton (Charlestown today) and so up and down (for) about two or three hours. They were come down in their lighter about a mile, and when it was over, they found themselves carried quite back against the tide to the place they came from. Divers other credible persons saw the same light, after, about the same place.[37]

Governor Winthrop, a renowned public servant of Colonial Massachusetts in the 1600s, held in high repute and credibility, also documented the "First USO (Underwater) UFO Sighting" in America. This occurred on the North End of Boston in January 1644. In addition, in 1647, a highly respected minister, Reverend Cotton Mather, reported in his journal *Magnalia Christi Americana* a significant sighting in New Haven, Connecticut.

Reverend Cotton Mather had his own sightings after the 1647 New Haven events. One of these came on November 26, 1668, when he and several New Englanders observed "bright starlike points" transit across the moon. These comments were documented in NASA's *Chronological Catalog of Reported Lunar Events – NASA Technical Report* R-277-July 1968.[38]

Christopher Columbus

Even Christopher Columbus, as was documented, saw a UFO. While patrolling the deck of the Santa Maria at about 10:00 p.m. on October 11, 1492, the following was chronicled.

After sunset, he called Pero Gutierrez, the steward of the king's dais, and told him that there seemed to be a light and for him to look, and thus, he did and saw it. He also told Rodrigo Sanchez de Segovia who saw nothing because he was not in a place where he could see it. It was seen once or twice, and it was like a small wax candle that rose and lifted up, which too few seemed to be an indication of land. After a short time, it vanished, only to reappear several times during the night, each time dancing up and down in sudden and passing gleams. The light, first seen four hours before land was sighted, was never explained.

Columbus's son, Ferdinand, also verified the sighting as looking like a candle that intermidently rose and fell through the air.[39,40,41]

Among the various interpretations of what this sighting could have been, including meteors, bioluminescence of the marine plankton reflecting into the atmosphere and the natives of Watlings Island throwing fireballs into the sky. The sightings occurred while Columbus's fleet was still over thirty-five miles out to sea. The trajectory of the lights carried the objects into the atmosphere as, from their vantage point, the display was still too far from the horizon to have been caused by human activity without the technology to launch the fireballs with anything except their physical strength.

Philip III Duke of Burgundy

On November 1, 1461, a strange object shaped like a ship, from which fire was seen flowing, passed over the town of Arras in France.[42]

A direct quote from Jacques Duclerc's, a chronicler, and counselor to Duke Philip the Good, *Memoirs of a Freeman of Arras*:

A fiery thing like an iron rod on good length and as large as one half of the moon was seen in the sky for a little less than a quarter of an hour.[43]

The bright object "spirals upward, spins around, rolls over 'like a loose watch' and disappears."[44]

William of Newburgh's Chronicle

An observation made by W. Newburgh described in 1290 at Byland Abbey, Yorkshire, "a large silvery disk flying slowly while the abbot and monks were in the refectorium, a flat, round, shining, silvery object (discus) flew over the abbey and caused the utmost terror."[45,46]

This second list is somewhat more focused in the crosscut of society that it represents. These participants include astronauts who have had UFO sighting encounters.

Cosmonaut Victor Afanasyev

One of many astronauts having their own sightings is Russian cosmonaut, Victor Afanasyev. In April 1979, Afanasyev lifted off from Star City on a mission to dock with the Soviet Salyut 6 space station. While still en route to the station, he saw "an unidentified object turn toward the craft and begin tailing it" through space.

"It followed us during half of our orbit," Afanaysev began and,

> We observed it on the light side, and when we entered the shadow side, it disappeared completely. It was an engineering structure, made from some form of metal, approximately forty meters long with inner hulls. The object was narrow her and wider here, and inside, there were openings. Some places had projections like small wings. The object stayed very close to us. We photographed it, and our photos showed it to be twenty-three to twenty-eight meters away.[47]

He provided drawings and additional statements in later reports to Soviet government agencies of "an elongated flying triangle UFO seen flying in formation. I think we are not alone, something of extraterrestrial origin as visited Earth. The craft turned toward ours, followed us, and flew formation twenty-five to twenty-nine meters away. We photographed the metallic structure that was and forty meters long." The film was later confiscated.[48,49]

Astronaut Major Gordon Cooper

To his acceptance, Gordon Cooper was a multiple eyewitness. In 1951, he had his first sighting while piloting an Air Force F-86 Sabrejet over Germany. "I did have occasion in 1951 to have two days of observation of man flights of them, of different sizes, flying in fighter formation, generally from east to west over Europe."[50]

His second sighting occurred when Cooper was an astronaut on a NASA *Mercury* orbital space mission. On May 15, 1963, while in orbit, Cooper had an audio interview with the Muchea (Perth), Australia, tracking station of a sighting of a "glowing greenish object" coming directly from his "twelve o'clock" directly in front of him and approaching swiftly. The Muchea radar confirmed the episode. Various national television networks, including the National Broadcasting Company, reported on air of the event. Later, NASA placed a complete news blackout on the episode and made no further statements.

Astronaut Donald Slayton

He is popularly remembered as another U.S. Apollo astronaut. Donald "Deke" Slayton also had an episode in 1951 while with the Air Force. He was flying a P-51 fighter over Minneapolis, Minnesota.

> I was testing a P-51. I was at about ten thousand feet on a nice, bright, sunny afternoon. As soon as I got behind the darn thing, it looked like a saucer, a disk. It then suddenly (started) going away from me, and there I was, running at about three hundred miles per hour. All of a sud-

den, the damn thing just took off. It pulled about a forty-five-degree climbing turn and accelerated and just flat disappeared.[51]

Slayton reported this incident to his commanding officer a couple of days later. The colonel told Slayton to, "Get your ass over to Intelligence in the morning and give them a briefing." Slayton interviewed the next morning. The Intelligence officers confided with Slayton this additional information:

> The day you saw this object a local company was flying-high-altitude research balloons. They had a light airplane tracking it and a station wagon on the ground. Both observers were watching this balloon and had seen this object come up beside the balloon. The object appeared to hover, then it took off like hell. The guys on the ground tracked it with a theodolite (a surveying instrument with a rotating telescope for measuring horizontal and vertical angles), and they computed its speed at four thousand miles per hour. I guess they were trying to tell me I wasn't crazy.[52]

Pilot Joseph A. Walker

Walker was a NASA test pilot given the duty of detecting UFOs during some of his flights aboard the X-15 rocket. On May 11, 1962, he had his second sighting incident in one month. In April, he had filmed a half-dozen UFOs on another flight. Walker disclosed this information at a conference on the peaceful uses of space research in Seattle, Washington, after his X-15 project was terminated. At that lecture, he stated, "I don't feel like speculating about them. All I know is what appeared in the film which was developed after the flight."[53]

There has not been a release of any of the photographs from NASA into the public domain.

Pilot Robert White

Major Robert White, USAF, had his sighting experience, not in a space capsule but in an X-15 rocket about sixty miles airborne. This event occurred just after Joseph Walker's sighting, on July 17, 1962. White reported, "I have no idea what it could be. It was greyish in color and about thirty to fifty feet away. There *are* things out there! There absolutely is!"[54]

James McDivitt (Brigadier General, USAF, Ret) and Edward White, Lt. Colonel:

McDivitt, command pilot of the Gemini 4 space mission, had his UFO sighting on June 4, 1965. On the second day of their orbital mission over Hawaii while White was asleep, McDivitt happened to see an unidentified flying object. He described it as looking "like a beer can or a pop can and with a little thing like maybe like a pencil or something sticking out of it." He got a camera and took a few photographs but did not have time to set exposure or focus properly. He believes that since it was visible to him, it must have been in an orbit close to that of his spacecraft.[55]

NASA has never released these photographs to the public.

Astronauts Captain James Lovell and Colonel Frank Borman

Navy captain Lovell and USAF colonel Borman also experienced sightings during the Gemini space mission project. On December 4, 1965, at the end of their second orbit, Gemini 7 had this communication with Gemini Control at Cape Kennedy, Florida.

Lovell:	Boge at ten o'clock high.
Capcom:	This is Houston. Say again seven.
Lovell:	Said we have a bogey at ten o'clock high.
Capcom:	Roger.

(At this point, the live broadcast of the conversation is interrupted by Capcom.)

Capcom: Gemini 7, is that the booster or is
that an actual sighting?
Lovell: We have several actual sighting.
Capcom: Estimated distance or size?
Lovell: We also have the booster in sight.[56]

Many years later, there was a rumor circulating that Frank
Borman would rescind any comments he made about the sighting.
It was Lovell who communicated the incident from the official taped
records of the Gemini 7 flight.

Astronauts Neil Armstrong and Edwin Aldrin

Armstrong, a U.S. Navy officer, and Aldrin, a USAF officer,
may have had perhaps the most curious and profound experience
of all the human astronauts to date. During their Apollo 11 moon
landing, both Armstrong and Aldrin allegedly sighted more than two
different craft sitting at the edge of a crater not far from the landing
site of the Lunar Module on the surface. Aldrin also filmed some of
the actions from both within and outside the module.

A few years later at a NASA symposium, Armstrong confirmed
that the story was true. According to Maurice Chatelain, former
chief of NASA Communications Systems, Armstrong had actually
reported to mission control having witnessed the landing and return
observation of two UFOs and their occupants at the Apollo landing
site. Chatelain acknowledged that most of the Gemini and Apollo
missions, the last two space projects until the initial development of
the orbital space station technology, were observed by unidentified
and unexplained objects and reported by the astronauts involved.
Chatelain stated, "I think that Walter Schirra aboard Mercury 8 (the
space mission project immediately preceding the Gemini project) was
the first of the astronauts to use the code name 'Santa Claus' to indi-

cate the presence of flying saucers next to space capsules. However, his announcements were barely noticed by the general public."[57,58,59]

The next categorical listing contains identification of some of the many scientists who have reported their own incidents in some fashion or another. Many of them reported their encounters through one of the national reporting agencies such as National Investigations Committee on Aerial Phenomena (NICAP), Mutual UFO Network (MUFON), and the National UFO Reporting Center (NUFORC).

Here is a brief list of some of the names of scientists who have reported sightings to NICAP:

Physicist Carl A. Mitchell
Astronomer Clyde Tombaugh
Syemour L. Hess
Geochemist J.D. Laudermilk
Astronomer Walter N. Webb
Astronomer W. Gordon Graham
Biologist Dr. Charles H. Otis
Astronomer Dr. James C. Bartlett, Jr.
Chemist Wells Alan Webb (two sightings)
Astronomer H. Percy Wilkins
Meteorologist Dr. Marcos Guerci
Meteorologist R.H. Kleyweg
Astronomer Frank Halstead
Physicist Prof. Henry Carlock
Astronomer Jacques Chapuis
Zoologist Ivan T. Sanderson
Chemist T.C. Shafer
Biochemist Lee Ball
Meteorologist R.J. Villelia
Physicist Prof. C.A. Maney (and six others)
Geologist John Zimmerman[60]

Here is this third "scientist sighting event" list.

Cosmologist Stephen Hawking

Although he has not publicized his own sighting experience, the world's most celebrated active cosmologist has become more active in his allegiance to and participation in a current movement for a new investigative search for intelligent extraterrestrial life. In July 2015, he assisted in launching a most expensive intensive study named "Breakthrough Listen." A Russian technology businessman, Yuri Milner, helped announce the details at the Royal Society of London.

The effort, which is being funded by Milner, is for a purported $135 million over a ten-year period. Hawking stated, "I am here today because I believe the breakthrough initiatives are incredibly important. In an infinite universe, there must be other occurrences of life. Somewhere in the cosmos, perhaps, intelligent life may be watching. It's time to commit to finding the answer, to search for life beyond Earth. We must know."

Yuri Milner added this to the presentation: "We are launching the most comprehensive search program ever. Breakthrough Listen takes the search for intelligent life in the universe to a completely new level."

The project will use some of the largest telescopes on Earth, searching far deeper into the universe than before for radio spectrum and laser signals. Australia will play a crucial role in the project with the Parkes Observatory in New South Wales signing a multi-million-dollar contract to scan radio waves for life in the cosmos. Breakthrough Listen will only search and listen for activity; it will not broadcast any communicative messages. Following from this origin, there will be a separate initiative called the "Breakthrough Message" that will open a strategic competition to the public for input as to a future program of such messaging broadcasts.

Hawking has had a long history of concern for such search for contact. Hawking voiced his fears at the Breakthrough event, saying, "We don't know much about aliens, but we know about humans. If you look at history, contact between humans and less intelligent organisms have often been disastrous from their point of view, and encounters between civilizations with advanced versus primitive technologies have gone badly for the less advanced."

Ann Druyan, cofounder and CEO of Cosmos Studios and widow of the late cosmologist Carl Sagan, was part of the announcement panel and will work on the Breakthrough Message initiative. She seemed much more hopeful about the nature of an advanced alien civilization and the future of humanity. Her comments were as follows:

> "We may get to a period in our future where we outgrow our evolutionary baggage and evolve to become less violent and shortsighted," Druyan said at the media event. "My hope is that extraterrestrial civilizations are not only more technologically proficient than we are but more aware of the rarity and preciousness of life in the cosmos."[61,62]

Physicist Paul R. Hill

A highly regarded and revered aerodynamicist, physicist, and aerospace engineer, Hill spent most of his working career at NASA and NACA (NASA predecessor) from 1939–1970. The Paul Hill Special Collection, housed at the Archives of American Aerospace, contains many artifacts and documentation of his work including many scientific research papers he authored or coauthored and archives of some of his work with ramjet technology, planning for the Apollo Lunar Landing project and later space station study.

Hill had two separate sightings. The first sighting came on July 16, 1952, at the beginning of the well-documented Washington, DC government buildings flap of that summer. Hill and another man sighted "several craft flying in a level vector through" Hampton Roads, Virginia. Afterward, they made the following comment:

> At 9:00 p.m., we were standing near the ocean looking south when we saw two

amber-colored lights (that were) much too large to be aircraft lights, silently traveling north. Just before the lights got abreast of us, they made a 180-degree turn and started back toward the spot Hill, and his companion had first seen them. As they turned, the two lights seemed to jockey for position in the formation. A third light came out of the west and joined the first two, then several more lights joined them as they moved south.[63]

The weather was clear that evening, but thunderstorms further west were situated, which came to be entered as a cause of evidence controversy. An explanation given by a fleet of Air Force B-26 bombers flying over Hampton Roads by Project Blue Book, as the sighting was reported by Hill the next day, was that it was not possible. According to the pilots, "None of them were over Hampton Roads. In fact, all them had generally stayed well south of Norfolk until about 10:30 p.m. because of thunderstorm activity northwest of Langley." Additional factors that directed determination of an unidentified fleet of airborne craft were: the craft were silent in their maneuvers, and the quantity (one single light) and color (amber) were not characteristic of display by any man-made aircraft. In addition, Hill reported if it was a man-made craft, it would have been much bigger than any man-made craft in existence. Lastly, Hills's credibility, reputation, and fame as a scientist from the National Advisory Committee for Aeronautics and NACA confirm the validity of this sighting.[64]

Hill went on to devote years researching the physics properties of UFOs and obtaining knowledge about the unexplained existence of the craft. His 1995 book, *Unconventional Flying Objects: A Scientific Analysis*, helped the reader understand the many technical aspects involved with interstellar space travel. *Unconventional Flying Objects: A Scientific Analysis* effectively translates presents these concepts in a clear and comprehensible manner.[65]

Astronomer Clyde Tombaugh

One many professional astronomers, Tombaugh, had multiple sighting incidents in the late 1940s and 1950s, all in New Mexico, after the conclusion of World War II. The first one occurred on August 20, 1949, near his home in Las Cruces, New Mexico. He saw a group of four rectangular objects with white lights suspended for a few seconds before disappearing. Tombaugh's wife, Patricia, was with him and observed the same incident.

His second sighting was at the White Sands observatory about a year after his first. This object was "many times brighter than Venus at its brightest, going from the zenith to the southern horizon in about three seconds." Tombaugh described this craft like his earlier sighting; it exhibited the same flight maneuvers in that the explanation of these being meteors was dismissed. While he was quick to explain the first one possibly being some type of temperature inversion, and that Tombaugh was not confident about even that categorization, and the second sighting offered no known natural explanations.[66]

He later reported during 1948 through the early 1950s he had witnessed at least three of the mysterious "green fireballs," which were the focus of an investigation by the U.S. Air Force in a study known as "Project Twinkle."

His later statements included the following:

> I have seen three objects in the last seven years which defied any explanation of known phenomena, such as Venus, atmospheric optic, meteors, or planes. I am a professional, highly skilled astronomer. Also I have seen three green fireballs. Several reputable scientists are being unscientific in refusing to entertain the possibility of extraterrestrial origin and nature.[67]

Later, in 1957, Tombaugh offered further statements in an Associated Press article that appeared in the *Alamogordo Daily News* titled *Celestial Visitors May Be Invading Earth's Atmosphere*:

> Although our own solar system is believed to support no other life, other stars in the galaxy may have hundreds of thousands of habitable worlds. Races on these worlds may have been able to utilize the tremendous amounts of power to bridge the space between the stars. These things, which do not appear to be directed, are unlike any other phenomena I ever observed. Their apparent lack of obedience to the ordinary laws of celestial motion gives credence.[68]

Tombaugh had a fourth unexplainable incident while observing Mars in August 1941. The bright flashes he observed were recollected when he was witness to many U.S. atomic explosion experiments in New Mexico. This attribution was confirmed in a letter from White Sands Missile Range then-commander Robert McLaughlin to astrophysicist, James Van Allen (the discoverer of the Van Allen Belt of electromagnetic radiation that surrounds Earth).[69]

NICAP has published file lists of "UFO Sightings by Scientists." Clyde Tombaugh, along with other accomplished and renowned scientists, made reports of their encounters to the NICAP for the inclusion to these lists.[70]

Physicist Wernher von Braun

The name Wernher von Braun can be added to the list of those who had accomplished careers in professions that attracted their involvement in the ufology, if only vicariously in the public domain. Von Braun's name has been attached to many UFO and extraterrestrial encounters through his professional life. This started early in his professional years while still in occupied Nazi Germany prior to World War II.

For example, there is a detailed exhibit at the International UFO Museum and Research Center in Roswell, New Mexico, describing the alleged "Alien Saucer Crash in 1937 Nazi Germany" in which von Braun's name is mentioned. At that time, he was a scientist at the Peenemunde rocket facility. Nazi Germany was the subject of a rich assortment of involvement in many UFO incidents. It is no secret that one of Adolf Hitler's obsessions was that of paranormal phenomena, including the possibility of extraterrestrial life.

Von Braun would step into another bevy of UFO involvements in his postwar life in the United States. He was the top supervisor overseeing NASA's Juno rocket-testing experimental program. On December 12, 1958, the first Juno II test was launched. While that test was not entirely successful in meeting its objectives, the reentry vehicle was tracked reentering Earth's atmosphere accompanied by six other objects. In later months, von Braun released statements to multiple media press outlets in the United States and West Germany when interviewed about the Juno II incident. Statements such as the following:

> We find ourselves faced by powers which are far stronger than we had hitherto assumed and whose base is at present unknown to us. More I cannot say at present. We are now engaged in entering into a closer contact with those powers, and in six or nine months' time, it may be possible to speak with more precision on the matter.[71,72,73]

Astronomer Dr. J Allen Hynek

One of the founding forefathers of the modern era of ufology, Hynek was a *vagary*. His early career was spent in the duty of the U.S. Air Force and government as an expert astronomer and point person for explanations to the public of UFO sightings. He was also a professor at Ohio State University then in the same capacity and head of the astronomy department at Northwestern University in Chicago while he worked for the government.

Hynek also had a sighting and photography session of his own.

> Allen was aboard an airliner when he suddenly noticed a white object at his altitude, seemingly flying at the same speed as the plane. He made sure it wasn't a reflection, and he convinced himself it must be some faraway cloud with an unusual shape. He pulled out his camera to see how fast he could snap pictures. In all, he took two pairs of stereoscopic photographs and gave it no more thought.

The photographs appeared in a book authored by Hynek and Vallee in 1975, *The Edge of Reality*. They may or may not be of a flying saucer, but they are certainly not clouds. The importance of stereoscopic photographs cannot be overemphasized. Such a camera is of outstanding evidentiary value. Hynek, in effect, had captured a possible Holy Grail on film.[74,75]

Physicist Herman Oberth

Oberth along with American rocketeer Robert Goddard and Russian rocket scientist Konstantin Tsiolkovsky are generally considered the three grandfathers of modern-day rocketry. Tsiolkovsky began his work in the 1800s. Oberth and Goddard added much to this foundation in the early 1900s.

Oberth was a major contributor to the rocket research effort of Nazi Germany along with Wernher von Braun. In 1929, he published his expanded spaceflight dissertation, *Wege zur Raumschiffahrt* (*Ways to Spaceflight*). It was this vision that motivated and encouraged Wernher von Braun to strive for what he accomplished at NASA later.

He never reported his sighting, but Oberth was naturally fascinated by the UFO subject throughout his adult life. He had many opinions regarding UFOs. In the *American Weekly Magazine*, Oberth stated, "It is my thesis that flying saucers are real and that they are space ships from another solar system. I think that they possibly are manned by intelligent observers who are members of a race that may have been investigating our earth for centuries."[76]

In *The Flying Saucer Review*, Oberth wrote an article titled "They Come from Outer Space." In it, he discussed the reports of "strange luminous objects in the sky" and the "Shining Shields" documented by Pliny, an elder in ancient Rome.

> Having weighed all the pros and cons, I find the explanation of flying discs from outer space the most likely one. I call this the "Uraniden" (Greek, 'Uranos) hypothesis because from our viewpoint the hypothetical beings appear to come from the sky.[77]

He also talks about the ideas of space theology and further discusses the Uraniden hypothesis in his 1966 book, *Katechismus der Uraniden*.

Astronomer Edmund Halley

As the discoverer of the famous comet that bears his name, astronomer Halley certainly had an untold awareness of the universe because he observed it for many decades in his life. He was witness to two such encounters. The first one occurred in March 1676. Halley watched a "vast body apparently bigger than the moon. It made a noise like the rattling of a great cart over stones." From his calculations, Halley concluded the object moved "in a matter of minutes" at a velocity greater than nine thousand miles per hour.[78]

There are other accomplished scientists, government and military leaders, and reputable people from many walks of life who have reported their own sightings of unexplainable aerial objects. One more is briefly listed below:

> Air Marshal Mohammed Abdul Azim Daudpota, a decorated officer in the Pakistan Air

Force at the time and future chief of air staff of the Air Force of Zimbabwe, had his sighting in July 1985. According to the Air Marshal, "This was no ordinary UFO. Scores of people saw it. It was no illusion, no deception, and no imagination."[79]

The lists are quite lengthy, but they could have been much longer. Here, the main points are the sightings/encounters phenomena are not a new or unique subject nowadays; and what emerges is that there has been an abundance of similar characteristics recorded throughout history. The sample size chosen for this text only includes a limited number of famous people throughout history. Additionally, not all prominent individual's events have been logged. Furthermore, there has been no entry for any non-famous events, past or present.

Here, no intent to eliminate any individual events due to a conclusion as that of a natural or explainable occurrence has been analyzed. The provocations offered by many scientists is the use of such belittling conclusions as "swamp gas," "that every single sighting ever recorded can be explained by man-made or natural phenomena," "psychological abnormalities of the witness," or "witnesses shouldn't use to open a mind so that their brains leak out." In reality, each is senseless and follows no scientific regimen. In fact, it is the debunker's menu of a la carte options of lower resort when the tools of science and logic elude them.

Again, these famous citizens in history and of the modern era witnessed an event or events that, if broadcast could, first of all, have had some adverse effects on their future lives. Secondly, no one intended to profit from them.

The catalog of non-famous people as witnesses, if compiled, would contain many times the corresponding list of famous people mentioned here. Some non-famous witnesses will be discussed in later chapters, and some of their stories will then become more famous to you.

So here we have assembled a sample of witnesses from diverse backgrounds on a lengthy timescale. Military, religious, and business leaders, presidents, scientists, heads of state, and entertainers, and the list goes on and on believe the sightings are real. Some of them unequivocally did not receive explanations from the knowledge and natural sciences.

While a UFO sighting is an uncommon daily occurrence, it barely scratches the surface of one of the spheres of high strangeness. A different level (or levels) of the sphere are the alien sightings, contacts, abduction, and/or communication events. The other lists identify some famous alien witnesses and contactee/experiencers. Again, there are no non-famous people listed here. The point is, there have been far too many witnesses, so to even try to dismiss all them is a useless, and unsuccessful conclusion will become part of your conclusion tree regarding hypotheses I and II.

Here is a more complete biography on J. Allen (Josef Allen) Hynek (1910–1986). He was a United States astronomer, professor, and science advisor for over twenty-two years who also served in the U.S. Air. Dr. Hynek was also the author of the now infamous "swamp gas" quote that a series of UFO sightings over Michigan on March 20 and 21, 1966, were explained by him as being caused by "swamp gas." While at that time serving in capacity as a U.S. Air Force science advisor, he was compelled by the government to offer that conclusion to the Dexter and Hillsdale, Michigan, authorities and residents. The reaction by affected Michigan residents to Hynek's press conference summary was unanimous and negative.

Dr. Hynek had a change of philosophy in later years and eventually became a strong advocate for the plausibility of the UFO and extraterrestrial hypotheses. It has been suggested that his greatest single quest as a scientist was to convince the world that far more rational and rigorous investigation of these phenomena are needed.

Early evidence of the shift in Hynek's opinions appeared in 1953 when he wrote an article for the April 1953 issue of the *Journal of the Optical Society of America* titled, "Unusual Aerial Phenomena," which contained what would become perhaps Hynek's best-known statement:

> Ridicule is not part of the scientific method,
> and people should not be taught that it is. The

steady flow of reports, often made in concert by
reliable observers, raises questions of scientific
obligation and responsibility. Is there *any* residue that is worthy of scientific attention? Or if
there isn't, does not an obligation exist to say so
to the public—not in words of open ridicule but
seriously, to keep faith with the trust the public
places in science and scientists?[80]

Hynek also created the first functional and widely used
unidentified flying object and extraterrestrial encounter classification system. There are many different levels in this system, each
comprising what is called "incremental spheres of high strangeness." This organization, the UFO and alien sightings/encounters
system, was called "the close encounter classification system" or CE
"x." His original chart is provided below. It should be noted that
modifications have been made to this original system as part of a
continual refinement effort after Hynek's original scale was published in 1972.[81]

I.   Distant encounters:
    DE-1
    DE-2
    DE-3

Nocturnal lights or (DE-1): These are sightings of well-defined
lights in the night sky whose appearance and/or motion are not
explainable when regarding conventional light sources. The lights
appear most often as red, blue, orange, or white. They form the largest group of UFO reports.[82]

Daylight disk or (DE-2): Daytime sightings are generally of oval
or disc-shaped, metallic-appearing objects. They can appear high in
the sky or close to the ground, and they are often reported to hover.
They can seem to disappear with astounding speed.[83]

Radar-visual cases or (DE-3): Of particular significance are
unidentified "blips" on radar screens that coincide with and confirm

simultaneous visual sightings by the same or other witnesses. These cases are infrequent.[84]

II. Close encounters:
CE-1
CE-2
CE-3

Close encounters of the first kind (CE-1): Though the witness observes a UFO nearby, there appears to be no interaction with either the witness or the environment.[85]

Close encounters of the second kind (CE-2): These encounters include details of interaction between the UFO and the environment that may vary from interference with car ignition systems and electronic gear to imprints or burns on the ground and physical effects on plants, animals, and humans.[86]

Close encounters of the third kind (CE-3): In this category, occupants of a UFO, entities that are humanlike ("humanoid") or not humanlike in appearance, have been reported. There is usually no direct contact or communication with the witness. However, in recent years, reports of incidents involving very close contact, even detainment of witnesses, have increased.[87]

Dr. Hynek compiled an evidence chart defining different levels that may be encountered in an investigation. These levels are:

1. Physical Trace
2. Medical Records
3. Radar Scope Photos
4. Photographs

In addition to eyewitness reports, scientific evidence for the presence of something very unusual falls within these categories:

1. *Physical Traces.* Compressed and dehydrated vegetation, broken tree branches, and imprints in the ground have all been reported. Sometimes, a soil sample taken from

an area where a UFO had been seen close to the ground will be determined, through laboratory analysis, to have undergone heating or other chemical changes not true of the control sample.

2. *Medical Records.* Medical verification of burns, eye inflammations, temporary blindness, and other physiological effects attributed to encounters with UFOs—even the healing of previous conditions—can also constitute evidence, especially when the medical examiner can determine no other cause for the effect.

3. *Radarscope Photos.* A tape of traces from a radar screen on which a "blip" of a UFO is appearing is a powerful adjunct to a visual sighting because it can be studied at leisure instead of during the heat of the moment of the actual sighting.

4. *Photographs.* While it might seem that photographs would be the best evidence for UFOs, this has not been the case. Hoaxes can be exposed very easily. But even those photos that pass the test of instrumented analysis and computer enhancement often show nothing more than an object of unknown nature, usually some distance from the camera, and very often out of focus. For proper analysis of a photo, the negative must be available, and the photographer, witnesses, and circumstances must be known. In a few exceptional cases, photos do exist that have been thoroughly examined and appear to show a structured craft.

Hynek did not formally introduce these attributes into any permanent classification chart or system form. These are merely examples of types of evidence that could be observed or acquired and crosscut all the different types of distant or close encounters. This cross-sectional configuration attribution, according to Hynek, would make it most logical not to attempt to structure it as any subsystem as a "stand alone" but to acknowledge them as possible occurrences among any of the distant or close encounters types.[88]

Since 1972, four other deeper levels have been added to his close encounters classification system:

CE-4

CE-5

CE-6

CE-7

CE-4: CE-4 should be described as "cases when witnesses experienced a transformation of their sense of reality," thus including non-abduction cases where absurd, hallucinatory, or dreamlike events are associated with UFO encounters.[89,90]

CE-5: A report of deliberate human behavior that was followed by an obvious response from an unidentified object and/or humanoid. The response of the craft or being included effects suggesting its response was not merely coincidental.[91,92]

CE-6: Death of a human or animal associated with a UFO sighting, although some might consider this as a more severe example of a second-kind encounter (Hynek Classification Ref 11, 12).

CE-7: The creation of a human/alien hybrid, either by sexual reproduction or by artificial scientific methods.[93,94]

From this discussion, it is evident that many famous and renowned individuals have experienced different event levels of Dr. Hynek's close encounter classification system.

We know there is an extensive encyclopedia inventory of UFO sightings and extraterrestrial experiences in the historical record. While a UFO may not always contain a manned pilot of some sort, the ones that do are observed to be far more numerous. So to the "pilots" of those UFOs, if they are not human, then they are "?".

Do UFOs exist?

Does ET exist?

References

[1] Quote Ref 1: McCormack, John W. 1965. *True* magazine editorial review. January 1965, Volume 46, Number 332. Published by Fawcett Publications Inc.

[2] Symington Ref 2: Symington, Fyfe. 2007. Symington: I saw a UFO in the Arizona sky. Special to CNN. November 9, 2007. Copyright 2016 by Cable News Network. Retrieved September 29, 2016, from: http://www.cnn.com/2007/TECH/science/11/09/simington.ufocommentary/index.html?_s=PM:TECH.

[3] Symington Ref 3: Shanks, Jon. 2007. Former Arizona Gov. Admits UFO Sighting on Night of Phoenix Lights. National Ledger, March 18, 2007. Copyright 2004–2016 by The National Ledger LLC. Retrieved September 29, 2016, from: http://www.nationalledger.com/artman/publish/article_272612175.shtml.

[4] Symington Ref 4: Associated Press. 2007. Former Ariz. governor boosts UFO claims. NBC News, March 23, 2007. Copyright 2013 by The Associated Press. Retrieved September 29, 2016 from: http://www.nbcnews.com/id/17761943/ns/technology_and_science-space/t/former-ariz-governor-boosts-ufo-claims/.

[5] Symington Ref 5: Symington, Fyfe. 2007. Symington: I saw a UFO in the Arizona sky. Special to CNN. November 9, 2007. Copyright 2016 by Cable News Network. Retrieved September 29, 2016, from: http://www.cnn.com/2007/TECH/science/11/09/simington.ufocommentary/index.html?_s=PM:TECH.

[6] George HW Bush Ref 6: Cameron, Grant. 2012. George Bush Forty-First President; January 21, 1989–January 20, 1993. Retrieved October 20, 2016, from: http://www.presidentialufo.com/old_site/bush_ufo_story_htm.

[7] George HW Bush Ref 7: Courson, Paul. Cable News Network. (26 April 1999). "Former President Bush honored at emotional ceremony renaming CIA headquarters." Copyright 199–2016 by CNN.

[8] Kucinich Ref 8: Phillips, Michael M. 2008. What Kucinich Saw: Witnesses Describe His Close Encounter. The Wall Street Journal, January 2, 2008. Copyright 2008 by The Wall Street Journal, Inc. Retrieved September 29, 2016, from: http://www.wsj.com/articles/SB119923872081461417.

[9] Reagan Ref 9: Brochu, James. 1990. Lucy in the Afternoon: An intimate Memoir of Lucille Ball, p. 125. April 1990. Copyright 1990 by Reed Business Information, Inc.

[10] Reagan Ref 10: Cameron, Grant. 2009. Ronald Reagan's UFO Sightings. August 13, 2009. Retrieved September 29, 2016, from: http://www.presidentialufo.com/ronald-reagan/204-ronald-reagans-ufo-sightings

[11] Reagan Ref 11: Cameron, Grant. 2012. The Day You Have Been Waiting For: Disclosure. October 15, 2012. Retrieved September 29, 2016, from: http://www.presidentialufo.com/ronald-reagan/463-the-day-you-have-been-waiting-for-disclosure.

[12] Carter Ref 12: Carter, President James. 2015. *A Full Life: Reflections at Ninety*, p. 196–197. Copyright 2015 by Jimmy Carter. Published by Simon & Schuster.

[13] Carter Ref 13: Carter, Jimmy. 1969. Report to the International UFO Bureau in Oklahoma City. Retrieved September 29, 2016, from: http://www.nicap.org/waves/CarterSightingRptOct1969.pdf.

[14] Gairy Ref 14: Cameron, Grant. Presidential UFO Pictures – Jimmy Carter. Copyright Grant Cameron. Retrieved October 20, 2016, from: http://www.presidentialufo.com/old_site/carter2.htm

[15] Gairy Ref 15: Gross, Patrick. Dr. J. Allen Hynek Speaking at the United Nations November 27, 1978. Copyright Patrick Gross. Retrieved October 20, 2016, from: http://ufologie.patrickgross.org/htm/hynekun.htm.

[16] Gairy 16: Gairy Speech – Address by Sir Eric M. gairy, Prime Minister and Minister for External Affairs of Grenada, United Nations, General Assembly, Thirty-Third Session, Thirty-Second Plenary Meeting, Thursday, October 12, 1978. Retrieved October 20, 2016, from: http://www.thegrenadarevolutiononline.com/gairy12oct1978.html.

[17] Ford Ref 17: Barrow, Robert, Argosy UFO, Cameron, Grant. How Presidents Have Handled the Topic of UFOs. UFO Evidence: Scientific Study of the UFO Phenomenon and the Search for Extraterrestrial Life. Winter 1977–1978. Copyright 2011 by ufoevidence.org. Retrieved September 29, 2016, from: http://www.ufoevidence.org/documents/doc1797.htm.

[18] Nixon Ref 18: Cameron, Grant. 2009. Richard M. Nixon. Presidential UFO. Retrieved October 20, 2016, from: http://www.presidentialufo.com/richard-nixon/84-richard-m-nixon.

[19] Nixon Ref 19: Baltimore Post Examiner. 2015. The Presidents and UFOs: A Secret History from FDR to Obama. Retrieved October 20, 2016, from: http://baltimorepostexaminer.com/presidents-and-ufos/2015/10/19.

[20] Russell Ref 20: United States Senate. United States Senate Committee on Armed Services. Retrieve October 20, 2016, from: http://www.armed-services.senate.gov/.

[21] Russell Ref 21: U.S. Air Force Intelligence Information Report. Prepared by U.S. Air Attache Lt. Col. Thomas S. Ryan October 13, 1955. Retrieved October 20, 2016, from: http://www.nicap.org/reports/551004russia_report_swords38C.pdf.

[22] Russell Ref 22: London Daily Express. Senior U.S. senator's report of seeing "two UFOs" was "covered up" secret documents show. October 3, 2016. Retrieved October 20, 2016, from: http://www.express.co.uk/news/weird/578055/Senior-US-senator-s-report-TWO-UFOs-covered-up.

[23] Russell Ref 23: Towers, Tom. 1957. Aviation News. Los Angeles Examiner, January 20, 1957. Copyright Los Angeles Examiner.

[24] Eisenhower Ref 24: Cameron, Grant. 2012. Eisenhower and His Alien Contacts: Part I. Presidential UFO website, November 21, 2012.

Retrieved September 29, 2016, from: http://www.presidentialufo.com/dwight-d-eisenhower/472-eisenhower-and-his-alien-contacts-part-1.

25 Churchill Ref 1

26 Eisenhower Ref 25: Eisenhower, Laura M. 2014. Full ET Disclosure April 2014, The Global Alien Treaty, Your True Human History. Presented March 30, 2014, at the Twenty-Second World Symposium on UFOs and Related Phenomena; Extraterrestrials and World Politics. Copyright 2014 by Laura Magdalene Eisenhower. Retrieved September 29, 2016, from: http://50kview.blogspot.com/p/full-et-disclosure-2014-global.html.

27 Churchill Ref 26: Daily Mail Reporter. 2010. Churchill and Eisenhower agreed to cover up RAF plane's UFO encounter during WWII. The Daily Mail, August 5, 2010. Retrieved September 29, 2016, from: http://www.dailymail.co.uk/sciencetech/article-1299994/Churchill-Eisenhower-agreed-cover-UFO-encounter-WWII.html.

28 Churchill Ref 27: British Broadcasting Company. 2010. Churchill ordered UFO Cover-up National Archives show. August 5, 2010. Copyright 2010 by the BBC. Retrieved September 29, 2016, from: http://www.bbc.com/news/uk-10853905.

29 Churchill Ref 28: Clarke, David. 2012. *The UFO Files: The Inside Story of Real-Life Sightings, Second Edition.* Published 2012 by Bloomsbury.

30 Truman Ref 29: Cameron, Grant. 2009. President Harry S. Truman. Presidential UFO website, July 31, 2009. Retrieved September 29, 2016, from: http://www.presidentialufo.com/harry-s-truman/64-president-harry-s-truman.

31 Arnold Ref 30: Dash, Mike. 2000. *Borderlands: The Ultimate Exploration of the Unknown.* Published 2000 by Overland Press. ISBN 0-87951-724-7.

32 Arnold Ref 31: Wikipedia: Orographic Lift. Wikimedia Foundation, Inc. Retrieved October 20, 2016, from: https://en.wikipedia.org/wiki/Orographic_lift.

33 Arnold Ref 32: Clark, Jerome. 2005. *The UFO Encyclopedia: The Phenomenon from the Beginning*, A-K. Published 2005 (and also earlier version 1998) by Detroit: Omnigraphics. ISBN 0-7808-0097-4.

34 FDR Ref 33: Eisenhower, Laura M. 2014. Full ET DISCLOSURE APRIL 2014 - The Global Alien Treaty, Your True Human History, March 30, 2014. Copyright 2014 by Laura M. Eisenhower. Retrieved September 29, 2016, from: http://50kview.blogspot.com/p/full-et-disclosure-2014-global.html.

35 Jefferson Ref 1: Thayer, Bill. 2015. *Transactions of the American Philosophical Society, Volume 6* (1802), p. 25; as reproduced in; Description of a Singular Phenomenon seen at Baton Rouge, by William Dunbar, Esq. communicated by Thomas Jefferson, President A.P.S. Copyright 2015 by Bill Thayer. Retrieved September 29, 2016, from: http://penelope.uchicago.edu/Thayer/E/Journals/TAPS/6/Baton_Rouge_Phenomenon*.html.

36 Washington Ref 35: Sherman, Anthony. 1778. Washington's Vision. As appeared in the National Tribune, 1880. Catalogued at Library of Congress

and Valley Forge. Copyright 1998–2011 by Independence Hall Association. Retrieved September 29, 2016, from: http://www.ushistory.org/valley forge/washington/vision.html.

[37] Muddy River Ref 36: Celebrate Boston. First UFO Sighting in America Muddy River, 1639. Copyright 2015 by CelebrateBoston.com. Retrieved October 20, 2016, from: http://www.celebrateboston.com/ufo/first-ufo-sighting.htm

[38] Muddy River Ref 37: *Chronological Catalog of Reported Lunar Events – NASA Technical Report R-277-July 1968.* Retrieved October 20, 2016, from: http://www.astrosurf.com/luxorion/ltp-NASA_R-277-1500-1799s.htm

[39] Columbus Ref 38: Blum, Ralph and Judy. 1974. *Beyond Earth: Man's Contact with UFOs.* Published November 1974 by Phillips Publishing Company, which is referring to "The Life and Voyages of Christopher Columbus" (1850).

[40] Columbus Ref 38: Bourne, E.G. as editor. 1906. "Christopher Columbus, Journal (1492)." The Northmen, Columbus, and Cabot (New York, 1906). Swarthmore College. Retrieved September 29, 2016, from: http://www.swarthmore.edu/SocSci/bdorsey1/41docs/01-col.html.

[41] Columbus Ref 40: Sir Clement Robert Markham. The journal of Christopher Columbus: (during his first voyage, 1492–1493), Ayer Publishing, 1972, p. 36.

[42] Philip III Ref 1: Wilkins, H.T. 1954. Flying Saucers on the Attack. Published 1954 by Citadel Press.

[43] Philip III Ref 41: Wilkins, H.T. 1954. Flying Saucers on the Attack. Published 1954 by Citadel Press.

[44] Philip III Ref 42: Vallee, Jacques. 1965. *UFOs in Space: Anatomy of a Phenomenon,* p. 11. Copyright 1965 by Henry Regnery Company.

[45] Philip III Ref 43: Schemm, Paul. Associated Press. 2010. Sleuths study ancient UFOs. NBC News – Cosmic Log. Wednesday, October 27, 2010. Retrieved October 20, 2016, from: http://cosmiclog.nbcnews.com/_news/2010/10/27/5361166-sleuths-study-ancient-ufos.

[46] William Newburgh Ref 44: Vallee, Jacques. 1965. *UFOs in Space: Anatomy of a Phenomenon,* p. 11. Copyright 1965 by Henry Regnery Company.

[47] William Newburgh Ref 45: Wilkins, Harold T. 1954. *Flying Saucers on the Attack,* p. 177. Copyright 1954 by Harold T. Wilkins. Published by Ace Books, Inc., New York, NY.

[48] Afanasyev Ref 46: Soviet Cosmonaut shadowed by UFO in Earth orbit. Retrieved October 20, 2016, from: http://www.ufoevidence.org/cases/case392.htm.

[49] Afanasyev Ref 47: Filer, George A. 2002. Cosmonauts Saw UFO – Believe ET Is Here. Mutual UFO Network Eastern. July 24, 2002. Filers Files #30-7-24-2. Retrieved October 20, 2016, from: http://www.rense.com/general27/filers72402.htm..

[50] Afanasyev Ref 48: Drawing of Afanasyev UFO available at: http://www.filers-files.com/news/images.php?id=69

[51] Astronaut Ref 49: Good, Timothy. 1999. *Above Top Secret: The Worldwide UFO Cover-up*. Published May 31, 1999, by Diane Publishing Company.

[52] Astronaut Ref 50: Good, Timothy. 1999. *Above Top Secret: The Worldwide UFO Cover-up*. Published May 31, 1999, by Diane Publishing Company.

[53] Astronaut Ref 51: Ridge, Fran. 2000. Astronaut Deke Slayton's UFO Encounter. National Investigations Committee on Aerial Phenomena. May 19, 2000. Retrieved September 29, 2016, from: http://www.nicap.org/reports/511212hastings_report2.htm.

[54] Astronaut Ref 52: Retrieved September 29, 2016, from: http://paul.rutgers.edu/~mcgrew/ufo/astronauts.

[55] Astronaut Ref 53: Retrieved September 29, 2016, from: http://paul.rutgers.edu/~mcgrew/ufo/astronauts.

[56] Astronaut Ref 54: McDivitt, James (June 29, 1999). Oral History Transcript (PDF). PDF with Doug Ward. Elk Lake, Michigan. Retrieved June 25, 2016.

[57] Astronaut Ref 56: UFO Sightings by Astronauts. Syti.net. Retrieved September 29, 2016, from: www.syti.net/UFOsightings.html.

[58] Astronaut Ref 57: Neil Armstrong, Michael Collins, Buzz Aldrin, Apollo 11 July 1969. Retrieved September 29, 2016, from: http://ronrecord.com/astronauts/armstrong-collins-aldrin.html.

[59] Astronaut Ref 58: Welsh, Jennifer. 2014. Buzz Aldrin Describes His UFO Encounter during Apollo 11. Business Insider – Science, July 8, 2014. Copyright 2016 by Business Insider Inc. Retrieved September 29, 2016, from: http://www.businessinsider.com/buzz-aldrins-apollo-11-ufo-encounter-2014-7.

[60] Scientist Sightings Ref 59: National Investigation Committee on Aerial Phenomena (NICAP). UFO Sightings by Scientists. Retrieved October 20, 2016, from: http://www.nicap.org/ufoe/ufoe49_50.htm.

[61] Hawking Ref 60: Cofield, Calla. 2015. *Stephen Hawking: Intelligent Aliens Could Destroy Humanity, But Let's Search Anyway*. Space.com, July 21, 2015. Copyright 2016 by Space.com Purch. Retrieved September 29, 2016, from: http://www.space.com/29999-stephen-hawking-intelligent-alien-life-danger.html.

[62] Hawking Ref 61: Brennan, Bridget. 2015. Stephen Hawking launches biggest search ever for alien life. ABC News-Australia, July 20, 2015. Copyright 2016 by ABC News. Retrieved September 29, 2016, from: http://www.abc.net.au/news/2015-07-21/hawking-launches-biggest-ever-search-for-alien-life/6635296.

[63] Hill Ref 62: Ruppelt, Edward J. 1956. Report on Unidentified Flying Objects, p. 208. Copyright 1956 by Edward Ruppelt. Published 1956 by Doubleday, New York.

[64] Hill Ref 63: Ruppelt 1956, p. 209.

[65] Hill Ref 64: Hill, Paul R. 1995. *Unconventional Flying Objects: A Scientific Analysis*. Copyright 1995 by Julie M. Hill. Published 1995 by Hampton Roads Publishing Company.

[66] Tombaugh Ref 65: Steiger, Brad. 1976. Project Blue Book, p. 280. Published by Ballantine Books. ISBN 0-345-34525-8.

[67] Tombaugh Ref 66: Ledger, Don. 2004. UFO UpDates. Published September 20, 2004. Retrieved September 29, 2016.

[68] Tombaugh Ref 67: Clark, Jerry. 1997. UFO Encyclopedia. Volume 2, p. 896.

[69] Tombaugh Ref 68: McLaughin, Robert. 1949. Letter from Robert McLaughlin to James Van Allen. May 12, 1949. Roswell Proof. Retrieved September 29, 2016.

[70] Tombaugh Ref 5: NICAP. 1962. UFO Sightings by Scientists. Retrieved September 29, 2016, from: http://www.nicap.org/ufoe/ufoe49_50.htm.

[71] Von Braun Ref 70: Creighton, Gordon W. 1961. Unidentified Satellites. National Investigative Committee on Aerial Phenomena (NICAP). Retrieved September 29, 2016, from: http://www.nicap.org/docs/usats.htm.

[72] Von Braun Ref 71: Flying Saucer Review. 1961. Volume 7, No. 1 – Jan/Feb 1961, pp. 3–6.

[73] Von Braun Ref 72: Good, Timothy. 1989. Above Top Secret. Published September 1989 by Quill.

[74] Hynek Ref 73: Hynek, J. Allen, and Vallee, J. 1975. *The Edge of Reality: A Progress Report on the Unidentified Flying Objects*, with Jacques Vallée. Published December 15, 1975, by CreateSpace Independent Publishing Platform. ISBN 978-0-8092-8150-3.

[75] Hynek Ref 2: Clark, Jerome (1998). *The UFO Book: Encyclopedia of the Extraterrestrial*. Visible Ink. p. 305. Published June 1, 1998, as two sub edition by Omnigraphics, Inc. ISBN 1-57859-029-9.

[76] Oberth Ref 75: Schuessler, John L., "Statements about Flying Saucers and Extraterrestrial Life Made by Prof. Hermann Oberth, German Rocket Scientist" 2002; for example, the American Weekly article also appeared in the Washington Post and Times Herald, pg. AW4, and Milwaukee Sentinel.

[77] Oberth Ref 76: Hermann Oberth "They Come from Outer Space" Flying Saucer Review Volume 1 Number 2, May–June 1955 pp. 12–14.

[78] Halley Ref 77: Edmund Halley UFO Case, 1676 Estimated Speed of UFO 9,600 MPH. February 2012. Retrieved September 29, 2016, from: http://www.educatinghumanity.com/2012/02/edmund-hallwy-ufo-case-1676-esti-mate.html

[79] Daudpota Ref 78: Birnes, William J. 2004 UFO Magazine; UFO Encyclopedia, p. 110. Copyright 2004 by UFO Magazine. Published by Pocket Books.

[80] Hynek Classification Ref 79: Hynek, J. Allen. Unusual Aerial Phenomena. Journal of the Optical Society of America, April 1953. Retrieved September 29, 2016, from: http://www.ufoevidence.org/documents/doc1985.htm.

[81] Hynek Classification Ref 80: J. Allen Hynek. 1972. The UFO Experience: A Scientific Inquiry; On the Strangeness of UFO Reports, p. 25–35. Copyright 1972 by J. Allen Hynek. Published by Ballantine Books.

[82] Hynek Classification Ref 81: Hynek (1972), p. 42–45.

[83] Hynek Classification Ref 82: Hynek p. 61–65.

[84] Hynek Classification Ref 83: Hynek p. 80–82.

[85] Hynek Classification Ref 84: Hynek, p. 98–100.

[86] Hynek Classification Ref 85: Hynek, p. 126–129.

[87] Hynek Classification Ref 86: Hynek, p. 158–164.

[88] Hynek Classification Ref 87: J. Allen Hynek. 1972. The UFO Experience: A Scientific Inquiry; On the Strangeness of UFO Reports, p. 25–35. Copyright 1972 by J. Allen Hynek. Published by Ballantine Books.

[89] Hynek Classification Ref 88: Vallee, Jacques. "Physical Analysis in Ten Cases of Unexplained Aerial Objects with Material Samples." 1998. *Journal of Scientific Exploration*. Vol. 12, No. 3, pp. 359–375. URL accessed 23 August 2009

[90] Hynek Classification Ref 89: Vallee, Jacques F. 2007. A System of Classification and Reliability Indicators for the Analysis of the Behavior of Unidentified Aerial Phenomena. Confrontations. Copyright 1990 by Jacques F. Vallee. Published 1990 by New York: Ballantine.

[91] Hynek Classification Ref 90: Haines, Richard. 1999. CE-5: Close Encounters of the Fifth Kind. Published 1999 by Sourcebooks, Inc.

[92] Hynek Classification Ref 91: Vallee, Jacques F. 2007. A System of Classification and Reliability Indicators for the Analysis of the Behavior of Unidentified Aerial Phenomena. Confrontations. Copyright 1990 by Jacques F. Vallee. Published 1990 by New York: Ballantine.

[93] Hynek Classification Ref 92: Judith Joyce, "The Weiser Field Guide to the Paranormal: Abductions, Apparitions, ESP, Synchronicity, and More Unexplained Phenomena from Other Realms" 2010, pp. 7.

[94] Hynek Classification Ref 93: Vallee, Jacques F. 2007. A System of Classification and Reliability Indicators for the Analysis of the Behavior of Unidentified Aerial Phenomena. Confrontations. Copyright 1990 by Jacques F. Vallee. Published 1990 by New York: Ballantine.

CHAPTER 7

Man's Technology Explosion

Who are we? We find that we live on an insignificant planet of a humdrum star lost in a galaxy tucked away in some forgotten corner of a universe in which there are far more galaxies than people.

—Carl Sagan[1]

Key words: deoxyribonucleic acid, macromolecule, chromosome, nucleotide, naturally selective

What do painting a picture and computers have in common?

Paint a picture in your mind for a moment. A member of your family has been diagnosed with a serious disease. The long-term prognosis is not promising; chances for survival are less than fifty percent. Current treatments may include regular blood transfusions, physical therapy, surgery, chemotherapy, and/or radiation, maybe even a tissue resection or organ transplant.

This picture could have been visualized from a time around 1980, sometime within the past thirty-five years. The cutting-edge technologies at that time did not offer abundant hope. While these treatments all are still predominately used in today's listing disease treatment's most-used therapies, in only thirty-five years, man has opened the door to a whole new set of technologies that point to a quantum leap of hope and faith about curing once incurable diseases. The word *cure* is now more relevant in the field of medicine than ever before.

No doubt a significant catalyst of this achievement has been the evolutionary path taken by the computer. Computers have enabled inventions, technologies, and possibilities that would not have allowed many of discoveries even to be imaginable. Much of our fast airplane/jet technology would not exist without computers because the physics involved with speed and airborne objects in our atmosphere would make it impossible for a pilot to fly without them. Speaking of our earth's atmosphere, it would be even that much more polluted without the computers used to better regulate the burning of fuels in everything from automobiles and industrial plants to our homes.

Additionally, man would not have been able to leave Earth's atmosphere without the computer revolution; our communication potential would have been hampered immensely, so the "global village" ideology and all its functions would be nowhere near the level they are today. It is doubtful that without them the global village would not be able to approach the maturity level it has currently achieved for a long time into the future.

Computers are the physical tools that allow the progression and function of many real-world machines, processes, and advances in science technology. Those high-speed jets cannot fly without them. Machines of all types around the earth and above it would not exist without them. Robots, medical instrumentation, and more intricate laboratory experiment and diagnosis machines would not exist.

So what does the medical disease picture exercise have to do with computers? Whether man's history of computer evolution was aided with/by discovery/recovery of alien technology as is liberally sprinkled throughout ufology is not the point. The existence and evolutionary path of computers have permitted another technology, indeed nothing less than a collection of entire scientific subjects of study, to be spawned from the evolution of computers and what they can accomplish.

Three capital letters describe the purest summation of one of some "commonalities" that may link planet Earth to the extraterrestrial universe. They are DNA.

What exactly is DNA? It is not an acronym for "does not adapt." Deoxyribonucleic acid is a macromolecule made of the very same chemistry—those carbon, hydrogen, oxygen, nitrogen, and phosphorus atoms—that make up most *everything* in this universe! Huge fact here! DNA is constituted of some of the very same chemical elements that make up all other matter, whether organic or inorganic. It is just that the combinations of chemical elements and compounds are unique. They are not different in the empirical, fundamental sense that they are somehow exotic atoms and elements. The quantities of each element present within and the arrangement of those elements (chemical bonds) in a three-dimensional space may be different from another substance.

Back to DNA in a moment, more importantly, atoms, chemical elements, compounds, and substances that comprise all matter in our universe are the same; so far as we know. Science has tabled and cataloged 115 currently known and confirmed chemical elements. It is called the periodic table of elements. The chemistry profession is working on including nine more elements.

The majority of these are produced or, as a chemist would say, reacted to occur naturally. About one-third of them have to be induced to react in a laboratory setting. Here is a three-letter word that will provide the answer to the question, "How the remaining two-thirds which occur naturally are actually produced?" The answer is that the sun does this. Our sun and all the stars which were ever present in the universe since the Big Bang have made all the chemical elements which are the basis of all matter you see today.

Our sun has made and continues to make many of the basic chemical substances and quantities of such that exist today and will exist in the future. Our Earth has helped the sun to do this since its creation. We have also helped via our activities. This synthesis includes the elements that, when assembled in their own unique way, makes up our DNA. The sun, in its life cycle, takes hydrogen and manufactures helium, carbon, oxygen, nitrogen, and some other basic elements from it over time. Supernovae—star explosions that can be seen in an astronomical photo catalog—produce additional complex elements such as gold and silver. Also chemical reactions not made within the sun's furnace but anywhere else in the universe such as via a nuclear explosion as one of many types of reactions can produce different chemical substances. Remember, elements (atoms) bonded together make all the compounds and substances that exist including DNA.

An amazing thing, this chemistry! The same all throughout the universe, including Earth, Mars, Zeta Reticuli, the Milky Way, and Andromeda. DNA (just think of the initials of deoxyribonucleic acid from now on) is comprised of chemical substances. Different numbers of atoms of these elements bond in such a unique way to comprise the DNA molecule. DNA is the blueprint of life as we know it.

All living things on Earth consist of DNA molecules. This includes humans, animals, plants, and so on. Visualize DNA as a skeleton, a "backbone." A backbone is a substrate or a platform on which another substance is attached to it. DNA is primarily comprised of the elements nitrogen, phosphorus, carbon, hydrogen, and oxygen. A collection of these elements bonds to each other to make chromosomes and genes. Chromosomes and genes are a unique chemical molecule set. DNA is made up of a series of chromosomes laid in what could be visualized as a long string of pearls. Chromosomes are composed of a collection of genes. Finally, genes consist of a collection of another molecule set called genetic bases or nucleotides.

Proceeding from the largest structures to the smallest substances, DNA is made up of a collection of chromosomes. Each chromosome consists of a collection of genes; each gene is composed of a group of bases. Bases are also known as nucleotides, and all DNA is comprised mostly of the chemical elements nitrogen, oxygen, carbon, hydrogen, and phosphorus.

Visualization appears in this manner: if unwound fully, a DNA sequence is a string of pearls measuring approximately six feet long. The next smallest subunits within that six-foot length are the forty-six chromosomes each human cell contains. Other living organisms living on Earth have more or less. Yes, some have more than humans. Each chromosome is made up of many genes. Within each gene, a collection of the most basic substance of DNA is packaged, which is the base or nucleotide. Every base molecule is aligned sequentially, like the very long string of pearls. Each group of these bases forms a single gene; a group of genes together and sequentially placed forms a single chromosome. Humans have forty-six chromosomes in their

DNA. The entire sequence just mentioned comprises the DNA sequence string. One of these DNA sequences exists in every single cell of our bodies. Nature has endowed us with a pretty simple yet exceedingly exquisite biological constitution in that there are only four distinct base nucleotides; however, when strung together like a nearly infinite sequence of "beads on a necklace," they extend to total over twenty-four thousand genes and forty-six chromosomes. The entire human DNA sequence, even though an entire sequence is packed into every cell in our body, when unwrapped and straightened measures about six feet in length. Exquisite biology indeed!

Now get ready for an awesome announcement about how all this chemistry is built. Base nucleotides make up genes, and genes make up chromosomes. Together, all these elements make up a DNA sequence. A gene can be built of anywhere from a few dozen to many thousands of individual bases. A DNA genome is comprised of a few hundred to tens of thousands of genes in all of the life-forms present on Earth!

Human DNA contains somewhere from twenty-two thousand to thirty-five thousand genes. We do not know the exact number at present. It will be slightly different (off by a few) from one person to the next. This is because every human's DNA is unique. This is even more so than our fingerprints. The DNA and cell cycle of creation is continuous because this is what keeps any organism alive. Occasionally, a person is either missing a gene or has an extra one. The entire process is a superbly complex one.

Broken down, human DNA is built from roughly 3+ billion of these nitrogenous chemical bases. Each base in a DNA molecule has the chemical element nitrogen as its core element. All other organisms contain more or less of these number sets.

DNA is the platform that contains the chemical blueprint of an organism's composition. This means that what every organism is, what is seen to be in real life, and what it may pass onto its offspring is attributable to its DNA platform and unique sequence of these chromosomes, genes, and base nucleotides.

All organisms of a particular species appear to be different and are different because of its sequence pattern, the collection of DNA—which equals the total of its chromosomes, genes, and bases—is unique from all others. The reason for this uniqueness lies within the blueprint structure of the chromosomes, genes, and bases.

A better and more accurate visualization is this: the DNA molecule is a string of pearls that you have tied to something. As noted previously, it measures about six feet with the pearls, which are the base nucleotides on the string. Now take one entire six-foot length of this series and wind it around another six-foot segment to make a thicker string. These two strings are attached to each other at each end and along every pearl from end to end. Each pearl base nucleotide connects with another pearl base nucleotide consistently in a pattern on the other string. This is the DNA platform and is known as a "double helix." The entire structure looks like an extension ladder that is twisted around itself along its length.

The entire DNA sequence string can be imagined as a string of pearls about six feet long. The string starts with the first series of nucleotides attached to each other in a meaningful way. This first set of nucleotides (bases) in sum make up the first gene. This could be imagined as the first pearl. In the next part of the chain, another series of pearl base nucleotides is counted to make another gene (pearl). The next gene starts in succession and so on.

This process continues for all the genes that are related until the end of one of these gene groups is reached. All that gene group has a unique relationship with each other to perform vital biological functions for that organism. The term for this group of genes is known as a chromosome. When all the chromosomes have been sequenced, we arrive at the other end of the entire genetic DNA genome sequence. Essentially, we arrive at the end of the "road" of the string of pearls. There are many pearls in an image of a genome.

Let's go back to the very beginning of the DNA string to show this in another way. At the very beginning lies the first base nucleotide. It is chemically bonded to another of the bases, which are bonded to a third base and so on. Every base nucleotide is also attached to its complement on the other string across from it. This configuration is visualized as a "rung on a ladder." This sequence of base arrangement, where any of the four bases can attach next to any other and to its complement across from it, repeats until it has made an entire gene. It could take many thousands of bases as noted before to sequence just one gene.

After the first gene is sequenced, the base nucleotides continue down the entire DNA molecule to make genes number 2, 3, 4, and so on until an entire chromosome is sequenced. This pattern repeats until all twenty-three chromosomes are sequenced. This now constitutes an entire DNA molecule sequence. When attached to its complementary string, as mentioned, the whole double helix structure thus contains the forty-six chromosomes and the twenty-four thousand to thirty-five thousand individual genes in addition to the billions of individual bases of each DNA molecule that is wrapped up inside of every cell in our body. The double helix structure looks like a ladder that is twisted around itself. Wow!

It is just awesome that a particular sequence of just four chemical base nucleotides can make up an entire DNA sequence molecule. Additionally, this series is wrapped and coiled up to fit inside of an individual cell. This sequence is what is known as a genome. Every living organism has such a genome. Different species possess a different length of genome. This is logical because each species has different genetic requirements that determine its way of existence. All plants, bacteria, viruses, animals, and humans—every living organism in our universe—have a genome according to our definition of a living organism.

Just within the past thirty-five years or so, man has begun to make very significant headway toward the understanding of the subject disciplines that make up the science of DNA. This is thanks in a large part to the evolution of computer technology. Without such an evolution, we could not begin to acknowledge, much less understand, what DNA is all about. One other precursor creation needs an explanation before we proceed with the sciences of DNA.

This precursor creation of the study of DNA is known as the science discipline of genetics. It originated in the 1850s, 1860s, and 1870s and was the product of work by a Franciscan monk named Gregor Mendel. In years of work breeding pea plants and an uncommon dedication to the meticulous observation, cataloging, and mathematical analysis of scientific study, Mendel himself was the genius needed to create this science of genetics.

Mendel utilized a lot of the process known as the scientific method in his hypothesis formulation, experimentation, observations, analysis, and conclusions drawn from his work. This is a particularly good study of the process of practicing the scientific method without any of the human factors biases that are introduced into many scientific research study projects. The reason for this is Mendel worked alone. It is characteristic of the accomplishments of a lot of geniuses such as Einstein, Newton, Alfred Wegener, and Galileo. In a science history context and an example of how science communities, also known as paradigms, operate today, Mendel's work and accomplishments were even shunned and ignored by all the science communities for over thirty-five years. These science critics ridiculed and condemned Mendel's work as lacking meaning and was critically disclaimed as pseudoscience. Mendel was not part of the very real and historically perpetual fraternal atmosphere and discourse of the science communities that existed then and continues today. It was only around the early 1900s that science took note of his accomplishments.

It is a shame Mendel was not "accepted as a scientist." His work received no acceptance in the science field until the twentieth century after his death. Now the concept of breeding, in which the study of genetics finds some of its roots, was in use for a long period before Mendel's time. Some plants and some animal species were crossbred to produce other new mixes of offspring breeds, but what was achieved was only a very basic understanding of the process of genetic evolution. Nothing was known about the process's details until Mendel arrived. Mendel discovered and created the term *gene*. He originally used the term traits of inheritance[2] to describe what is now known as a gene. Thus, he single-handedly invented the science of genetics. Mendel also jump-started a few other sciences including bioinformatics, biogenetics, and gene therapy, to name a few.

A lot of knowledge about genetics was acquired in the decades following Mendel. In 1953, an American biologist, James Watson, and a British physicist, Francis Crick, discovered the DNA double helix shape that is now recognized as the twisted ladder of the DNA molecule sequence. Following that came the arrival of computer technology ignited the investigative sciences and offshoots of the biotechnology disciplines.

Once enough computer power was generated to facilitate processing the many gigabytes and terabytes of data required to conduct a meaningful genetic research project, the task of identifying the human DNA blueprint could transpire.

In the 1990s, biotechnology science exploded with such achievements. Entire gene sequences were capable of being identified and studied. This started a discovery of how unique human character traits were made and how they transform into reality. These are called phenotypic traits. The image of the human being that we see is called the phenotype. A comprehensive study that culminated with a mapping of the entire human genetic sequence, called the Human Genome Project, was initiated by different groups of science teams. Each of these communities had their own strategy for accomplishing this project. It was a good idea in that results of such an enormous project could be combined so that the time to completion was shortened considerably.

It took about ten years, but the finish line in creating and mapping the first entire human genome was crossed on April 14, 2003. What it constructed was the human genome that has within its DNA molecule forty-six chromosomes, over twenty-four thousand genes and over three billion chemical base nucleotides. A copy of the genetic sequence resides in every cell in our bodies.

The Human Genome Project took ten years to complete and cost millions of dollars to accomplish. Remarkably, since 2003, we have advanced science to the point where it is now possible for anyone to have his or her DNA genome identified and sequenced. In addition, anyone can have his or her entire genome recorded onto microchip slides that fit into the palm of a hand at a cost of one thousand dollars!

Genomes have been sequenced and archived for many different life species. The National Institutes of Health and GenBank are two of a large number of participating research organizations working in the field.

These accomplishments have opened many possibilities in medicine, biotechnology, agriculture, pharmacology, law, and environmental sciences. We are near the apex of curing and preventing a large number of diseases, increasing food supply by something far more mathematically exponential than linear, creating many ways to help "clean up" our planet making this a truly better place to live in and to live it!

Here are some of the most exciting specific things we are now doing with the knowledge gained from these burgeoning fields of study. The potential now exists for millions of diabetics to be treated with pure insulin that can be mass-produced from yeast or bacteria and pancreas beta cells that originated from them. In this process, pancreas beta cells from their own bodies are introduced to either the yeast or bacteria media via a process known as recombinant DNA, a transforming technology. The process manufactures insulin products identical to that of their own bodies' production. Mass quantities can be manufactured far quicker and cheaper than the human body can.

We are on the precipice of finding cures for a lot of families of diseases through genetic treatments, repairs, and gene therapies. Diseases such as hemophilia, sickle cell anemia, phenylketonuria (PKU), leukodystrophy diseases, and many other afflictions caused by genetic inheritance of what are called recessive genes are simultaneously being studied by many different science communities. It is probably just a matter of time when these disease families will be cured along a short timeline.

Genetic technology has already been used extensively in the legal and forensics fields to free innocent prisoners and absolve them from unjust convictions and also to protect the innocent, not yet convicted.

Biotechnology in agriculture is allowing nontoxic methods of increasing food production in a mathematically exponential fashion. They are also reducing the incidences of loss of production due to disease and/or infestation. Animal husbandry is also benefiting from the same breakthroughs of science knowledge.

In pharmacology, when a disease has not yet been curable, gene therapies are being used as effective treatments. For example, many drugs besides insulin are designed to combat as treatments the side effects of viruses and hurtful disease bacteria that become resistant to older or currently in-use drug treatments.

What we can summarize from this, thus far, is the discovery of DNA with the assistance of computers has created an exponential explosion of knowledge about us, our earth, and our universe. Now knowing that every living organism in our universe contains DNA and needs it to live, we know that living extraterrestrials have it too. Another term for extraterrestrials used in ufology research is "extra-biological entity" or EBE. Remember that all matter in the universe—our pets, plants, the animal kingdom, bacteria, and extraterrestrials—are comprised of the same chemical elements and substances bonded together in the same way that make up their DNA, cells, tissues, organs, and organ systems. Our uniqueness as a species and as living (organic) matter originates from the amount of each chemical element that is used to make the blueprint DNA sequences of our physical architecture. Sequences of the DNA genomes determine how we appear to the outside. This is known as an organism's phenotype, as noted.

So DNA is universal among humankind and all living things in the same way as it would be with extraterrestrials. There is a circular rhythm to DNA. There is also a connection and a circular rhythm to man, aliens, and DNA.

Which brings us to a fascinating subject, one of the very same subject variables for numerous readings throughout this book. The DNA discussion was presented for many reasons. First, it is hoped that some things about how humans are built, albeit at one of the most basic levels, have been beneficial and informative. DNA is specifically a magnificently complex creation of nature. The intensity and dynamic of the evolutionary process that involves DNA is genuinely awe-inspiring.

Second, we as a species have begun, but have only begun, to tap into an exceedingly elaborate creation of nature that governs all life as we know and have defined life to be. This is quite an accomplishment for our species. Subsequently, we should begin to recognize that we have only begun to acquire an understanding of DNA and its entire role in life and our universe.

Medical science, pharmacology, and related "treatment and cure" life sciences are all feeling the effects of the complexities of the study of the genetic and related sciences and their contributions; however, progress is slow. A small step forward seems to beget a step or two back. In its totality, DNA and genetics are supremely intricate and complicated subjects to master and to make progress in their understanding of knowledge. The effects of genetic treatment to cure disease and/or treat bodily injury may require a long time frame to evaluate the pathway of the intervention.

Look how long Mother Nature needs to enact genetic alterations to a species. Study the Homo sapien lineage back on the anthropological timeline and see how long it took for the proto-homo sapien species to morph into the next species. Life-forms adapt as being naturally selective toward their DNA and their environment. These are two of the mechanisms that produce change to a specie's lineage.

Third, knowing some, if not all, currently known medical interventions involving DNA/gene therapy of any kind takes time to produce results. If an intervention involves genetic therapy on the parental dimension in the end point of childbearing, this procedure would take nine months before the first step of intervention is complete.

In the next chapters, there is a discussion of possibilities for alleged extraterrestrial beings or EBEs to wish to visit Earth. Some reasoning scenarios will be explored. According to ufology literature, there exists a recurring theme as to how the extraordinarily complex subject of DNA could also spark the interest of EBEs. Humans as a species have learned some things about this subject. This makes us intelligent, sort of, but it gives another really good reason for other intelligent life-forms to wish to get to know us more.

According to the literature, a reason EBEs have allegedly visited Earth is to explore the subject of our DNA in some contextual manner. In the next chapters I start to explore this notion and the many other notions present in the UFO and extraterrestrial hypotheses investigation.

References

[1] Sagan, Carl. 1980. Cosmos, p. 193. Random House.

[2] Genetics Ref 2 (Mendel): Noble D (September 2008). "Genes and Causation" (Free full text). Philosophical Transactions. Series A, Mathematical, Physical, and Engineering Sciences. **366** (1878): 3001–3015. Bibcode:2008RSPTA.366.3001N. doi:10.1098/rsta.2008.0086. PMID 18559318

CHAPTER 8

Why Here? Why Not Somewhere Else?

Now the effect of that is to produce some kind of new species to bring us together to produce a hybrid species which will populate the earth or will be there to carry evolution forward, after the human race has completed what it is now doing, namely the destruction of the earth as a living system. It's an awkward coming together of a less embodied species than we are, and us, for this evolutionary purpose.

—John E. Mack, MD[1]

Key words: ad hoc, post hoc, pseudoscience, anthropocentric, natural selection

Why here? Why not somewhere else? If they exist, why would they want to or need to come to our planet specifically? These questions must be contemplated in order to apply evidence for the corroboration of any possibility for extraterrestrials to be able to come here. There have been far too many alien event encounters researched and reported (and unreported) by far too many human beings from all walks of life for too long to not give proper context and attention to this question. The science of the "how's" of this subject are dealt with in other chapters.

If EBEs have already been here, then why did they come? For this we ponder, "Why here?" If extraterrestrials were not/have not/will not come here, then the subject would not have/is not/will not be written about as extensively as it has/is/will. There would not be any events that have taken or will take place, no evidence, no afteref-fects of and on the millions of experiencers, contactees, observers in history, now and into the future.

We could try to discover insight into this question by using a highly effective technique that humans have used in all trains of thought, including the sciences, for as long as their documented exis-tence. This technique is best explained by first providing an example then describing the process man uses to move toward a conclusion to this problem.

The example is this: to answer the question, "Why did/does ET wish to come here to Earth?" a technique that provides a background source of knowledge of experience and of context use asks the ques-tion, "Why does man wish to go somewhere else?" In this context, this means man wishes or needs to go to other places, i.e. explore other places. This is asked contextually about the many explorers in history that voyaged around the Earth (and to the moon and the planets) to other lands.

An example like this is used to look at the problem from a dif-ferent perspective, a different lens. This is how most of humankind's new knowledge is discovered. A researcher investigates the question and hypothetical statement from a different and dedicated viewpoint and not from a fallacious *ad hoc* (thought of, done, or performed with a particular purpose and/or end point scenario in mind with-out allowing for alternative outcomes) or a *post hoc* (a Latin term for "after this" or falsely assuming that a later event is caused by an earlier event) viewpoint. What could be used as a better perspective than one that we have some familiarity with? Why has man wanted or needed to explore and trek all around the earth and beyond?

This method draws on past experiences and perspectives that are analogous to the hypothesis under study. This then serves as a benchmark and starting point to help move the process toward answering the question, proving the theory, or to infer a conclu-sion(s). Here, the evidence is being obtained and then observed through a different lens. That which is being observed is taken from relevant documented experiences that happened in a situational past. Most of humankind's questions are studied in this manner. They use past experience as a study parameter instead of trying to utilize only unknowns as the cause-and-effect factors. Is it possible to come to correct conclusions when the only evidence you use comes from entirely unknown parameters? The term *pseudoscience* (when a collection of knowledge, facts that may or may not be equivalent to the level of a discovery, has not been deemed discovered according to the rules of the scientific method) is applied precisely when one uses the alternative of obtaining knowledge from entirely unknown parameters exclusively and without gaining evidence from using any other sources, including any known template.

The known template in this instance involves humankind's propensity to explore other places. This is then used as a bridge to cross over in applying those facts to the question that needs to be answered. So to help answer the question, "Why would ET wish to come here?" we could start by drawing on our prior knowledge base regarding what we have experienced about the subject of humankind's exploration of other places. Then by injecting humankind into the "why come to Earth" scenario, make observations and move toward inferring some output or end point.

Similarities exist in using some of the premises of the scientific method. We draw on our prior knowledge about the subject, make observations, and move into an analysis phase and then toward some conclusion, whether by making inferences or something else. The technique of inference is mostly relied upon in our determinations as to whether the subject involves science, business, or just about anything in our day-to-day lives. This process is how humans have worked through these questions since the times of Aristotle and his contemporaries in ancient Greece.

So using a benchmark of past human experiences on the subject of exploring, we first insert, "Why would *mankind* want to come here to our earth?" Again, "coming to Earth" correlates with the technique of observing the situation through the alternative lens of "humankind exploring other places on Earth (and the moon and planets)."

Here is a list of why throughout the centuries humankind has wanted to explore other places on Earth and why we wish to explore more places both on and outside of Earth's realm. Each item in the list is presented first, then a brief explanation follows.

1. Being curious about knowing there are other *places*, physical pieces of land yet to be explored. These are not just for visiting the places themselves which would be tantamount to sightseeing. This is meant to be a purposeful reason for the voyage. A primary motivation for this draws from knowing that man needed (and needs) to address his survival needs by locating and obtaining resources like food and other materials and things to bring back home and that meet his day-to-day needs. Going to and observing what these places have to offer helps ensure there are at least opportunities for man's survival. This object of curiosity neatly flows to item 2.

2. Being curious about knowing there are others in other places, or just other places yet to be explored containing sustaining necessities that just happen to have other beings there also. Throughout man's history, one main motivation for exploration was the explorers knew about the existence of others in those faraway places. The possibility of obtaining more food and goods and equipment in this scenario grow exponentially. They may also present many more problems or hurdles to overcome in order to achieve the explorer's objectives.

3. The need to explore new lands to possibly move there in the future. Maybe the explorer's homeland population is too crowded. To paraphrase, "The grass may be greener."

4. The idea of moving a criminal subset of the explorer's home population to a new and far-off land to ensure the carrying out of the criminal sentences of the convicted. The historical timelines in Australia and a lot of nations in South America over the last few hundred years are evidential examples of this idea.

5. Acquisition of new food and equipment that logically infer from 1 and 2.

6. To possibly set a foundation for colonization to escape a collective persecution of some kind back in their home-land. This derives from examples from the events and motivations of the early Pilgrims of New England and Virginia in the 1600s. A review of that history will provide an appropriate context with which to explore this notion.

7. Provision as achievement of some sort of evolution, whether technological, social, or else. An example of technologi-cal evolution is extracted from our history of manned air flight and followed further into genesis and evolution of a manned space program to date. In just over one hundred years, we achieved air flight (the Wright Brothers), com-mercial flight, and then developed rockets and spacecraft that are making initial explorations of the rest of our solar system. Stop and think for a moment about all the techno-logical inventions, products, and spin-offs from the history of manned air flight that have found a place in our every-day lives!

One example of social evolution involves the growing and con-tinuing cooperation of every member nation that now work together in that very same manned flight and space program's development just noted. For a moment, stop and think about the relationships among many of the people of our earth who have evolved from noth-ing. Look at the International Space Station, which involves the use of many nations' resources. Why is it called "International?" This could be a reason to want to go there.

Another example of social evolution resides in Antarctica, which is a continent that is "owned" not by any nation but by the people of Earth as a whole, working together for common and productive purposes. Thanks to science, Antarctica is not a nation in the politi-cal sense but in the human sense. This could serve as a reason (in this case it was a reason) to want to go there.

An example of the "else" factor of evolution achievement could come from an expanded knowledge base in any area of human existence—medical, economic, cultural, etc., an interaction of explorers with natives.

8. To expand the explorer's knowledge, in general, and then to pass on that knowledge to his homeland. Even knowing that elements of this reason can be combined with any of the others man possesses a certain curiosity to acquire knowledge, "just because it is there to acquire" if for no other reason.

9. An attempt to spread their culture, way of life, or religion through contact with other people in other lands.

10. Establish contacts with other people to acquire allies for protection against other external threats. This is an example of cooperation that is a primary study focus in biology and anthropology sciences.

11. As opposed to item 6, maybe a detrimental long-term environmental or climate change in their homeland or other natural circumstance forces them to search for a new home.

When the list begins to explain why man may wish to explore places away from the earth, it becomes much less feasible to consider most of the previous items in proper context, at least currently. Our species is just not ready or able to want to explore the moon or other planets to discover new foods, cultures, colonies, and habitats (at least not for now).

12. So the list here narrows considerably to include the reasoning for item 8 in the long term and item 3 in the far long term. Also item 11 could become a reality if you are a believer that current global climate change as a reason to explore will inevitably force us to need to move elsewhere as a long, long term threat. Hopefully, you understand this reasoning.

13. Man also wishes to explore the moon and the planets in order to help understand the history, current state of, and much of the future of our planet Earth better. More under-

standing knowledge of how and what causes the earth to exist as it does would help to solve and or prevent problems present or anticipated or inferred in the future.

Of note is a survey of additional reasons for humankind to wish to trek here that I did not include. They include the following:

Continuing with our alternative lens perspective, it is possible man would want to travel to Earth and "check up" on our civilization to ensure that we are on an existential path that is benevolent to both ourselves and to the environment around us that includes planet Earth and maybe beyond. In reality, we have possessed the technology to manipulate the atomic structure of chemical compounds for over seventy years. We have found both efficient and productive technologies to help our civilization grow and prosper, including nuclear electrical power for both civilian and military purposes. In the very same vein, this technology has produced crucially destructive and malevolent end points. Nuclear power plant "meltdowns," explosions, and other warlike manifestations litter our recent history.

If we were the explorers who traveled to Earth, we would probably be more advanced technologically, civically, civilly, and spiritually than the natives; in this example, it is us. The checking-up process is to ensure that the natives are not an imminent threat to the environment around planet Earth. If the natives have the technology to inflict severe, irreparable destruction on their planet and to planets nearby, of which we currently have the capability, then there would be a marked interest to monitor activities to protect Earth and its surroundings. Additionally, the process may be to ensure we do not harm ourselves as a civilization for the same reasons.

In their book, *Intelligent Life in the Universe*, I.S. Shklovskii and Carl Sagan exemplify this way of thinking: "The rate of technical advance of our civilization is very great. An extraterrestrial society might want to contact an emerging technical civilization as soon as possible, perhaps to head off a nuclear annihilation—one possible consequence of intensive technological development—or perhaps for other reasons."[2]

Now let's take this template and write on it the supposition as it relates to extraterrestrials. Let's obtain and observe through a new lens of perspective. If one can put the speculation into its simplest form, the question then becomes, "Why would extraterrestrials want to come here?"

Looking at the previous list:

Item 1 would be feasible in the presumption that extraterrestrials are seeking food and material resources to import back to their home planet or the means and knowledge to produce these foods and materials instead of the food and materials themselves. Remember that if this is an actual reason for such an exploration by EBEs, it may be more feasible and easier just to have the means and knowledge instead of the burden of having to transport the actual physical goods back to their home.

For item 2, the possibilities exist but are blocked by one big problem thus far. ETs may already have discovered we exist through their technology conduit that allows them to literally travel to our planet and previous to that, their detection of our existence by other means such as telescopes, like what we are currently endeavoring to accomplish. Possibly, they do exist but have not yet shown this to our satisfaction.

Item 3 suggests that a nation of ETs may have concluded Earth is a viable place to move to maybe to escape their dying home, or to provide an extension of item 6, to escape some internal persecution back home.

For item 4, this could be possible, though, in the early stages from the ET perspective. Certainly, a place to move any criminal element to in another solar system is as distant enough as to be an effective sentence for the accused. What crimes would ETs commit?

Item 7 seems less tangible, but if an advanced nation is visiting Earth, they still have an enormous opportunity to learn new things despite their higher level of advanced technology relative to ours. I do not think there is much correlation between variables defined as visitors with advanced technology and a lack of opportunity or desire or both to learn and explore new lands. A logical extension to this would be to include item 8 at this point.

There could be a recurring problem when item 9 is mentioned. The problem is we are still awaiting extensive contact from ET nation representatives that a lot of humans hold as a prerequisite to their acknowledgment of such. Item 10 embodies that same problem. There has also not been any large scale occupation of "refugee" ETs on Earth of which we are aware.

Could a global danger have occurred or come to exist on ETs home planet? Man has taken quite a few minute steps recently to add to our exploration knowledge of other planets and the cosmos and appears to be moving toward a conclusion that there may be *lots* and *lots* of exoplanets that exist under so many physical environments. Most all these seem to us to be uninhabitable. Perhaps, the answer to a global climate or environmental habitability crisis would be a "most certainly!" Our current level of understanding about the natural sciences as applied to Earth proves it is more than feasible.

From the initial list above, item 13 can also most certainly apply to an ET culture. Even if they are advanced enough to be capable of travel to our world, is the universe so homogeneous as to prevent them from learning new things about their home planet from observing ours?

Now let's address another reason set for possible motivations for EBEs to travel to Earth. This set could offer new items of motivation for them to want to endure exploration missions. Therefore, the list is expounded upon I should point out that in a scenario where many EBE civilizations exist. This factor may influence the occurrence of multiple reasons for their exploration activities. This is not only because some of these EBEs would be motivated to explore and make contact for reasons different from the others but also because any particular EBE society would make contact for many reasons.

That said, below is a list of possible items, followed by explanation and insight.

1. Earth's unique resources
2. Man's unique resources
3. Revenue
4. To enslave us

5. To prevent mass extinction of other species we are killing off
6. Medical DNA – to help save us from extinction
7. For food
8. To kill us off
9. Earth is "in the way" literally
10. Amusement – maybe not golf, but…
11. Weapons testing
12. How about any of the reason topics connected with man's propensity to explore?
13. To check on the status of the technological advancement of our civilization
14. To prevent mass extinction of ourselves, as the only native species on Earth capable of doing so
15. Medical DNA to help save them from extinction

Let's advance through this list. Item 1 would accurately dispel many rebuttals from skeptics. If an EBE explorer group was only interested in some geophysical or chemical resources unique to Earth, then they would be acting exactly as we would expect them to. There would be absolutely no need for them to want to make contact with us. Most times during Earth's history, a contact (or detection) scenario would be a function of the human population at that time. We do know that global population has been increasing only very incrementally since writing and symbolism were first used by humans. Only in the past hundreds of years has there been a population explosion. Any explorers capable of transporting through interstellar space would have the capability of conducting mining operations without any assistance, unless that contact would be for purposes of eliciting a sort of employment by what would then be "humans as miners."

This notion of humans as miners became a big part of the career work of author Zecharia Sitchin. From the 1950s, Sitchin's theses included the deciphering of ancient Mesopotamia cuneiform texts discovered by various archaeologists before the 1800s. They describe the exploration voyages of an extraterrestrial civilization called the Anunnaki. Contact was made allegedly in a time before the appearance of the earliest Neolithic civilizations in Mesopotamia. Earth humans became miners of the minerals, especially gold, that was needed by the Anunnaki to save their home planet Nibiru.

Except for this alleged illustration, an advanced race of beings faced with this problem would have planned to bring at least some of the equipment necessary to perform the mining operations themselves without any help on board their craft. Earth may have been more feasible to them despite the presence of humans as a distraction.

Item 2 could be an attractive idea to and a learning opportunity for a knowledgeable extraterrestrial. Many people, including many learned scientists, have the notion that because an EBE can travel to here from there through space, EBEs are exponentially and supremely superior in every format of existence. There is a sort of anthropocentrism in their rationale. Because scientists themselves cannot explain EBEs existence that given what critics say, they will dismiss their existence entirely. Alternately, those scientists will attribute the EBEs to being millions of years more advanced than we are. They conclude that EBEs have not visited Earth because they are superior beings. So since they are believed to be millions of years more advanced than we are, we have "nothing to show them."

We may have some unique or set of unique traits, characteristics, skills, or other special attributes that are attractive or useful for some reason. For example, our DNA, the instructional blueprint for genetic sequencing of the makeup and existence for biological life, is unique to every one of us. DNA identification is humankind's best single attribute of uniqueness among each individual. This holds, so far as we know, for all life. Item 15 will provide more discussion on this topic.

But more importantly, we may have other traits of benefit to them that offer an EBE traveler help in a problem-solving nature that of which we could be entirely unaware. Such a talent would not initially be able to be duplicated by the ET explorers and, therefore, useful to study. This opportunity would mean contact. In their book, *Intelligent Life in the Universe*, authors Shklovskii and Sagan provide a possible criterion: "While any organism of the earth could be duplicated by an advanced extraterrestrial society, the original and duplicate are still different. 'The American psychologist Ruth Ellen Galper has pointed out that we carefully distinguish between natural and *cultured pearl.*'[3]

Could item 3 be a possible scenario? With an open mind, yes, it could. To be clear, I am referring to revenue as some sort of currency of trade or exchange. For us, the concept of revenue means when we as individuals work, we earn revenue via a paycheck. When a business sells a product or service, it is traded revenue for that product or service. In legal terms, both scenarios refer to contracts.

If an EBE party was interested in obtaining some product or service and willing to pay for it, such a contract is either implicit, without any written document, or explicit with one. Here, a train of thought references the notion of EBE's superior advancement just noted. Would such a civilization that is significantly far more advanced in evolution than we even practice such a concept? If so, does this mean at present or at some time in their past? This would have to be answered individually by contemplating whether all intelligent species use and cycle through the notions of trade, currency, and revenue in their evolutionary life cycle.

Item 4 could be subject to some reasonable situational refutation. If ETs explored Earth, they may already have the technical knowledge to have mechanized or artificial intelligent robots or androids to serve a slave's purpose. The ufology literature is full of reported encounters with aliens who are concluded to appear, act, and behave like such an example of artificially intelligent androids. Indeed, with an abundant amount of documented encounters with alleged extraterrestrials, if we were to explore some of their characteristics and that of their activities, our impression could be one of the scout ship occupants being exactly this type of servant.

If anyone is a proponent of there being an existing synergy among being a slave and living in a conquered society, then there need not be another item for which to describe this on this list.

Item 5 takes into consideration the idea that other life-forms exist on Earth. If an EBE party has only recently learned of our existence, the group might not know enough to offer a contact scenario with us. If they have enough knowledge to understand that our decimation of other life-forms is happening and they deem it bad enough to initiate communication us, they surely would. This also means we have been under some level of surveillance by them.

Perhaps, the other life-form we are decimating is more significant to the success of some other process we are not aware of but that they are. If such a scenario was transformed into a reality, it is anybody's guess as to when the EBE party would initiate contact.

Now go back to item 6 and review the DNA passage. Also recall the chapter "Man's Technology Explosion." There will be an abundance of discussion attributed to this notion and reasoning. But this is probably not about saving us from extinction unless EBEs are some type of missionary. Item 15 will continue with this thread.

If extraterrestrials have in reality never visited Earth before, then there is a possibility this reason, item 7, could in fact happen. In reality, any of these reasons could become real occurrences if extra-terrestrials have never made contact with Earth before. In item 7, this notion has been raised previously through our broadcasting media, television. Back in the 1960s, an episode of *The Twilight Zone* titled "To Serve Man" was broadcast. It was based on a science-fiction story of the same name from a few years before.

The Kanamits, a race of EBEs, made the first contact with Earth. The representative met with the United Nations and told them, via telepathy, their motive was to aid humanity. Then the representative left a book; a sort of "Bible" for them to use and to learn about the Kanamit. The book was written in the Kanamit language. The Kanamits invited earthlings to visit their home planet. Everything went well until the book was deciphered. It turned out to be a cookbook and humans were the main ingredient.[4,5]

I would conjecture that if we were, in fact, on the menu of some aspect of an alien harvest from the past, then too many humans would have suddenly turned up missing for everyone not to have noticed.

If any circumstances had happened which would lead to a scenario involving items 8, 9, or both in the past, then any of those have already taken place. If there would have been a reason in the past for an EBE party to "kill us off" and contact was made, they would have done so. Possibly, such an invasion party did so, and we are part of the "phoenix of humanity." There is precedence of this from writings in ancient texts and religious scriptures from all resident civilizations of a gigantic catastrophe and/or flood that destroyed most of the earth.

It is also possible that when we look at ourselves, *anthropocentrically*, we do not visualize ourselves as being like the cast of rabbit refugees in the classic novel, *Watership Down*. To EBEs, though, that maybe how they would view us if item 9 was to have been demonstrated in real life at any time in the past. Besides, if the earth was literally in the way of something that had extraterrestrials as an attribute in the past and contact was made, then steps would have already been taken to cure that problem. Yet we know of no evidence of this ever occurring.

A translation for item 10 is that an amusement is defined as some sort of zoo where life-forms are gathered for recreational purposes. The notion of amusement as would be defined within the confines of a zoo has probably no basis in reality. Again, this discussion is based on what may have happened in the past.

Humanity so far has not been restrained from doing its own exploring of our home planet and our planetary neighbors. We admittedly have not physically ventured past our own moon, but our ships have penetrated the environment of all the planets now. There have been no outward signs of prevention for us to explore freely, thus far. There have been some reports in the research literature of events that occurred on the moon during the American Apollo lunar landings that may have some implications on the moon's dark side and extraterrestrials. These are, as of now, unsubstantiated.

Some people could visualize what the term amusement means. There are many attributes to the notion of studying that arise from this. At a zoo, some being is studied by the observer. There could be some plausibility in this scenario. A zoo implies (infers) there is a level of more intimate contact that would reveal the identity of the EBEs. This would appear as if you were the zoo exhibit. Studying, on the other hand, does not have this restriction.

Studying could be done from afar. Ask any astronomer or cosmologist about this notion. When an astronomer becomes physically able to travel to the stars and does so, does his title remain as "astronomer," or does it change to "astrogeologist" or "exogeologist?" It does depend upon what level of studying is considered. If the study was of individuals, more intimate contact would be required. If the studying was of our overall population trend, the population trend of other earthen life-forms or monitoring our climate and geology, then EBEs could keep their distance.

Item 11, like all the other items under this investigation, does not mean our weapons. It means theirs. This does not make any sense as a requirement for them to make contact with us. If EBEs wished to test their own weapons, they would do so on some barren, lifeless satellite body, unless they were trying to gauge the effectiveness of their artillery by including life-forms in their experiment. In this scenario, they may not even bother to tell us about the imminent testing procedure. They have the means to conduct their tests on any of a large number of planetary or satellite bodies.

When man was conducting the very same type of experiments with weapons, he did not have this luxury. Back in the 1950s, during the atomic weapons testing element of the Cold War, the United States had to warn the residents of the Indonesian islands of the Bikini Atoll to evacuate due to the impending explosion of a thermonuclear bomb (a hydrogen bomb). Unfortunately, the testing went amok, and many thousands more of native residents had to be evacuated from their homes due to all the aftereffects of the hydrogen bombs that had exploded on March 1, 1954. This fiasco was known as Castle Bravo.

A curious quote regarding the possibility of weapons testing activities in ancient history came from a reputable nuclear physicist, J. Robert Oppenheimer. In 1952, during a speech at the University of Rochester, he said, "Ancient cities, whose brick and stone walls have literally been vitrified—that is, fused together—can be found in India, Ireland, Scotland, France, Turkey, and other places. There is no logical explanation for the vitrification of stone forts and cities, except from an atomic blast."[6]

I noted in item 10 that there has been some level of research conducted over claims of possible extraterrestrial contact on the moon. Item 12 allows for us to take the step, however small, of physically being the contactor, at least in part. This could happen deliberately or by accident as it may seem to appear with the Apollo lunar landing episodes.

Many reliable and responsible persons—including those directly connected with NASA and the various space programs, from astronauts to control, research, and support officers—have come forward with descriptions of incidents that occurred during the lunar landings and other Apollo and non-Apollo space missions. There are also live astronaut mission recordings in the public domain of some of the more compelling observation details of anomalous objects while on the lunar surface and also while approaching and exiting. Other incidents of high peculiarity were observed by different Apollo Command Module flight paths on both the near and far side of the lunar surface. Additional incidents occurred and were recorded during other space missions, such as Mercury and Gemini. Photographical imaging of at least some of these reported incidents are known to exist.

Item 13 would suggest a program of contact already occurring if it was and that would have been ongoing for a long time. EBEs may practice some sort of "prime directive" fashioned similar to our conceptualization and context. The longer a time period of involvement for a possible scenario, the more susceptible to detection extraterrestrials would be subject to. Questions also arise as to the EBE's level of interest and motivations for the same. How vested is their interest in watching us evolve as a species? Was there ever a time when the ancestors of the current living EBE representative researchers broke the prime directive and interfered with our development? All these alternatives suggest a possibility for contact somewhere along the dimension.

Item 14 suggests some level of benevolence on the part of EBEs. The ufology literature is full of alleged abduction encounters explaining that the human contactees had been shown visual scenarios. These montages often showed clairvoyant visualizations of a final destruction of humankind. It was expressed to the contactees that we are the cause of what will happen if we do not alter our life course.

This item, 15, would appear to be the most compelling and most complex of possibilities for why extraterrestrials would want to "come" here and to come "here," which could possibly be somewhat of a necessity. In other words, there would possibly be a need to "come" and a need to come "here." The subject of DNA is the most complex and unique on this list. The options for reasons of travel and contact to obtain DNA are fewer than, say, available options for coming to mine natural resources or to check on the progress of our technological advancement. There are options other than Earth that the local universe made accessible for and the availability of obtaining natural resources. Extraterrestrials may not need Earth for this purpose as there are plenty of other planets, satellites, and asteroids nearby and accessible. Checking on our advancement does not trigger thoughts of an urgent need for something so much as wondering what special connection they have with us on that attribute that compels them to come here.

Overall, DNA represents a most basic survival attribute if there was a need or want for it. Because of its nature, structure, and purpose, having or lacking the right type of or exact specification of DNA is a requirement for survival or some other most vital life function and purpose. The exact specification of DNA required for the contact scenario cannot just be plucked off of any other planetary body. While many levels of life-forms may exist on a high percentage of planets and satellites, it is a special kind of DNA that may be needed.

Many intelligent people have presumed and conjectured that EBEs, if they do exist, would not need to travel here because there is nothing on this planet they cannot produce or manufacture themselves. They are millions of years more advanced than we are. According to them, EBEs must be superior because their advanced technologies would have allowed them to discover a way to travel through interstellar space to reach us if they had come here in the first place. But because they are so advanced, they have no need to, so they have not and will not according to them.

EBEs do not exist. Maybe some of the EBEs are getting too much credit from those people. What came first in an evolution of a society's technology—the warp drive transport through inter-stellar space, or the ability to successfully alter life through genetic engineering?

Humans know a tiny bit about warp drive and a little about DNA; however, there is a lot we also do not know about them. I am not at all convinced this is a million-year chasm, but this is not the point now. Here are a couple of facts about DNA that make it exceedingly complex, therefore, not too easy to manipulate by any intelligent species.

Warp drive is part of a system of physics, engineering, and mechanics where the environment that it must work in is more stable and less fluid than that of DNA. The fabric of interstellar space lacks fluidity in some respect because it does not change in its operation or manifestation. We can plan and design an engine to apply warp drive and have confidence that it will work throughout interstellar space without having to change the engine design.

DNA is fluid and constantly changing. The first reason for this is based on the entire ideology of *natural selection*. It is the theory whereby organisms better adapted (or changed) to their environment, tend to survive and produce more offspring, is dependent (caused by) the DNA function. These changes are basic and contin-ual and are cause-and-effect correlated with many variables such as environment, the process of reproduction, cell cycle, and mutations plus many more; considerably more than the warp drive state.

To put this in another context, as stated in the last chapter, "DNA," which is the blueprint responsible for the reproduction of every living cell in our bodies in exactly the right way at precisely the right time each day (the typical cell cycle length) contains forty-six chromosomes, over twenty-four thousand genes and over three bil-lion base nucleotides. All our cells and the DNA contained within each are built from a variety of proteins and enzymes, and both of these are vital to life as we know it.

A very important step in the function of our DNA is to copy or replicate the entire DNA genome sequence as the first step of a multistep process of cellular reproduction. After the copying step, the DNA is transcribed and translated. Only after the completion of these three steps can the cell have the ability to divide and create a copy of itself with the new DNA. The role of proteins and enzymes in the cellular and DNA reproduction cycle is irreplaceable.

For example, the main takeaway point is due to the real and natural construction of the human body with cells and DNA structured the way they are and the need for protein synthesis in which it is processed in our bodies, it is possible that the number of proteins that could exist is on the order of 20e20. This is not to say any of us, either as individuals or as a species, has ever used that many different proteins before in history; but the indicated process makes this end point possible.

Remember that the DNA process is fluid and dynamic in itself and, additionally, in an ever-changing environment. Furthermore, it takes years to finish the project and see the result of genetic reproduction, which is the same thing as producing offspring. This is comparable to parenting a child and seeing the child grow into adulthood.

With this information, it now becomes much more apparent that while extraterrestrials would presumably be capable of DNA manipulation, even they would be laden with the complexities contained within a reproduction or manipulation process which require a pathway, taking years to accomplish.

Consequently, one cannot just take DNA and impose instantaneous "magic" and produce the desired results in one half-hour period. The process requires nature to enable the biochemical pathway of reactions according to nature's timetable. This is the crux of the vitality and complexity the DNA manipulation process entails. EBEs also have limits. Those limits are dictated by nature's rules and how the universe works. Sometimes, the rules can be broken, as with warp drive or wormholes. But a new living and fully functioning organism, on the order of a human or an extraterrestrial, cannot be created immediately.

Interstellar space travel would not be a trivial accomplishment for any species. The challenges of interstellar space travel can be conquered. Yet the challenges of manipulating a DNA sequence have not due to its ever-changing landscape. So many changes occur in DNA that translates to different organisms. Naturally, changes to DNA can come from mutations, environment, reproduction, and "crossing over" of sections of DNA. Artificial changes are also obtained from gene splicing, gene repair, and gene therapy, in addition to the others mentioned previously. So an extraterrestrial may need our DNA for some type of these modifications. One of those ways could include using our DNA (or that of another species inherent to Earth) to repair theirs in an attempt to save them and their reproductive potential to save their species. Another scenario might include perpetuation of their species by creating one that is hybrid and involves DNA from Earth species. A third scenario might involve DNA manipulation to allow the EBE party to physically exist on Earth. This may also be an answer to the critics' refutation that "they have never been seen on Earth so they do not exist." It could also be a reason that there are more reports in the literature of alleged encounters with "greys" that act more like androids than any other except for the mere sighting of UFOs. Maybe it would be androids or artificially intelligent entities that make contact on Earth because the earth's atmosphere and landscape is somehow toxic to the real EBEs. Subsequently, they may not be able to enter the earth's airspace because of this.

References

[1] Mack, John E., MD. 1996. Interview with John Mack Psychiatrist, Harvard University. NOVA Television episode. Public Broadcasting System. February 27, 1996. http://www.pbs.org/wgbh/nova/aliens/johnmack.html.

[2] Shklovskii, I.S. and Sagan, Carl. 1966. *Intelligent Life in the Universe*. Pg. 462. Copyright 1966 by Holden-Day, Inc.

[3] Shklovskii, I.S. and Sagan, Carl. 1966. *Intelligent Life in the Universe*. Pg. 463. Copyright 1966 by Holden-Day, Inc.

[4] The Twilight Zone. 1962. C.B.S. Television. Episode Number 89. Broadcast March 2, 1962. Copyright and Published by Columbia Broadcast System.

[5] Knight, Damon. 1950. To Serve Man. Copyright 1950 by Damon Knight. Published November 1950 in *Galaxy Science Fiction* magazine, Volume 1, Number 2. Galaxy Publication Corporation.

[6] Strutt, Stephen N. 2014. *Out of the Bottomless Pit*. Pg. 136. Copyright and Published 2014 by Stephen N. Strutt. Partner Publisher Paragon Publishers, Rothersthorpe.

CHAPTER 9

How to Move My UFO
I Have a Long Way to Go and a Short Time to Get There (Or to Go; How Do I Get Here From There?)

To raise new questions, new possibilities, to regard old problems from a new angle, requires creative imagination and marks real advance in science.

—Albert Einstein[1]

Key words: gravity assist, angular momentum, ion propulsion, beam-powered rockets, ramjets, external sails, external laser-beamed energy, antimatter, torsion spin propulsion, subatomic particle matter energy, alcubierre drive, thrust-mass ratio, energy density, specific impulse, exhaust velocity, neural pathway, spin, rotation and charge (in chemistry), inference, chirality (in chemistry)

Science is currently unable, without much difficulty, to address issues of nature or inquiry beyond those which occur in three dimensions. As I have discussed elsewhere, answers to questions concerning matters involving the fourth and additional dimensions, those Einstein and other theoretical physicists and cosmologists have been hard at work researching for centuries, are very elusive and create even more questions for each one that is answered. A large part of this example and more generally any such matter of scientific inquiry demands the requirement for inventing and perfecting new measuring instruments and other technologies to be able to address the questions being asked. In the theoretical physics field, the CERN Hadron Collider in Switzerland was needed before any such research could be conducted.

Another example from the past became the most significant claim to accomplishment for Danish astronomer and engineer Tyge Ottesen Brahe, popularly known today as Tycho Brahe. Brahe was obsessed with the astronomical universe, but his genius was equally skilled at improving the instrumentation and processes that drove his successes and to those of in addition to such later astronomers as Galileo and Johannes Kepler.

Whenever a discussion about the existence of UFOs and aliens arises, an argument over the feasibility for them "getting here from there and back" quickly enters the discussion. The discourse then creates a massive amount of confusion and doubt. Next, a misguided and mistaken general debasing of the entire argument often moves the conversation toward a premature restructuring or a conclusion. It is out of ignorance that this discourse plays out.

When it comes to the idea of vehicular propulsion, we, as a species, are not entirely ignorant. We know more about land-based propulsion than air-based and especially out-of-atmosphere propulsion. Regarding space propulsion, here are some of those things in the physics and engineering technology with which we have acquired knowledge. We have achieved certainty about the scale of measurement involved with measurements of interstellar distances. Regarding spacecraft propulsion, the out-of-atmosphere type, here is what we know. We have been only using chemical propellants as a species since humankind ventured into space. This was only in 1957. We are starting to utilize a basic ion fuel propulsion technology in ships for some of the longer space missions. Mostly, though, our spaceships use the gravity assist from a planet to increase the ships' velocity. Our ships cannot carry enough fuel to continually burn it during the flights. The gravity assist is accomplished by using mathematics to calculate where, when, and at what angle the craft hits the other planets' orbit. This represents the entire state of our technology as of today.

Humankind has begun to accomplish an initial discovery and conduct a basic analysis of many other worlds which exist outside of our solar system. There have been over five thousand confirmed extraplanetary bodies orbiting their own host stars. We do possess some very useful tools which help in deciphering some of the many mysteries that reside outside the earth's atmosphere. The mathematical knowledge is also there for us to begin to understand the bigness of the things that come with describing cosmic distances and of any possibility of moving some tangible objects and ships out into space.

Almost by a significant probability of default humankind has much more to add to his toolbox in these areas. It is a problem of the "scaling up" of the perceived problems of cosmic distance and adding to our toolbox the knowledge of technology and accomplishment to the conquering of those issues of distance. For example, it has been generally postulated and verified by many bits of incremental evidence that the species Homo sapien originated in equatorial Africa.

When prehistoric man solved the immediate and most basic tangible survival problems, he could then begin to solve some of the difficulties involved with moving to other faraway lands. The explorer man gained knowledge, bit by small bit, about the existence of those other lands. The distances involved were scaled-up land distances in numbers they were not used to. Often, a changing climate was the primary driver for the migration. Early Homo sapien succeeded, however in making these movements attributable to whatever inventions and technologies of the time he knew of or invented.

Eventually, some new problems of distance appeared when early man first confronted big bodies of water. The discovery of oceans and seas created new problems to solve. Man did just that. By the times of early archival man, which included the civilizations around the Mediterranean Sea, Australia and Africa and with undeniable proof of their existence from their archival written records, artifacts, glyphs, and structures there were still, as was known to those societies, other people existing in faraway lands of present-day Northern Europe, Southern Africa, and the Asian continent. Scaled up problems of distance again were presented as formidable barriers in reaching those faraway destinations and societies. Now often, immense bodies of ocean had to be successfully traversed. Mostly, they were much larger than the land routes their ancestors were used to traveling.

Cutting-edge technologies and inventions of that time were eventually accepted and put to use to conquer that problem of the scaled up of distances. This paradigm of Homo sapiens' migration and exploration exploits were repeated time and time again and included a history of the discovery of the western world of Earth, the Southern continents including Antarctica, and finally the moon. Still to come very soon will be Mars and beyond. All these later migrations involved distances of successively greater length; greater than that of a successive linear increase.

In taking yet another greater than linear leap how to do Mars and beyond? It was and will be the same way humankind "did" the Mediterranean, Northern Europe, Southern Africa, the Western Hemisphere, and the Southern continents including Antarctica. The greater distances, with each iteration example through history and into the future, have and will be scaled up. With this, the problems of "how to get there" will also be scaled up. This process will continue into the next iterations of examples; the ones that are still to be discovered, solved, and accepted.

Earlier societies experienced the same things we do today when describing the process of solving distance problems to reach new lands, wherever they may be. They saw anomalies, whether they were of other foreign people, their homes, their tools, their means of transport, or their foods. Man's curiosity and creativity worked on the problems, making new inventions to come closer and closer to attaining whatever goals they sought in reaching those goals. Most of the time, those goals included traveling to the new lands and probably trading with those peoples or assimilating into those lands in one way or another.

Our problems in these areas today are no different. In the current paradigm, we are seeking answers to the possible existence of other extraterrestrial life and maybe civilizations created by intelligent creatures. This exploration is here to help you make the most informed and correct reasoned and thoughtful decision you can make for yourself. In the most basic ways, it is a survey in which you are being prepared to answer just two questions; one about UFOs and one about extraterrestrial life.

One of the huge issues is embedded within the mystery of intuitiveness that "they" exist. How did and would they get here?

The scaling up of the current iteration of the distance phenomenon seems initially daunting. We are not talking a scaling up of, as problematic of the ancients, a distance of only a few thousand miles. The scale-up is much more enormous than that. The knowledge needed to accomplish these ends (not if but when) is also, by necessity, scaled up exponentially.

Often, historical man created new knowledge by accident and lucky discovery. These could be termed *eureka moments* when the new knowledge was significant enough to create new technologies or ways of doing things. But just as important historical and current man created new knowledge additionally and exactly by taking then current knowledge and, in very small increments, building it into more evolved versions of the same technologies and eventually into entirely new technologies that accomplish the same things but only better. For example, man walked on land before he started floating on water using unmanufactured pieces of wood. He eventually fashioned and transferred enough knowledge from other intuitiveness into manufacturing a boat then later better and faster boats. The scaled-up distance problem then was solved. Then further knowledge transfer of intuitiveness was compiled to then eventually invent the airplane. Many small incremental developments were made that led to faster aircraft and jet planes. This was followed by the development of chemical rocket technology and hence better rockets. Currently, an ion propulsion technology is a reality and is being used by space agencies around the world for a new generation of spacecraft. As knowledge about this technology grows, they will become more commonplace and perform better. The takeaway point is that the process is only incremental unless some sort of knowledge explosion happens.

Here is what is important to take from this. Because technology is a direct function of a knowledge set that synergizes in a back-and-forth relationship with *all* the natural sciences and that in themselves also relate to and embody the current "STEM" educational opportunity, this existing knowledge from all these relevant subjects are the building blocks used to obtain new knowledge. Knowing details about relevant contributions to the problem from all the science areas helps in discovering the new knowledge. STEM is short for "science, technology, engineering, and mathematics."

Additionally important is a recognition and remembrance that people, the humans who make up our societies and civilizations of Earth, are the most important links that bind science, technology, and us together. One cannot function without the others. When investigating any relationships that exist among science, technology, and society, a visualization that could be represented is as a weight lifter competing at the Olympics. He or she could be a member of any one or all nations, lifting the barbell over his or her head. Science represents one set of weight slabs, technology represents the other set of weight slabs, and the human is the driving force causing both to function together. The barbell represents both *communication* among all the entities and as a notion of *rationality* that helps all them to communicate and discover new knowledge and truth rationally, efficiently, and reliably.

When it comes back to a discussion about the possibility of "how did/would other beings traverse the vast interstellar distances to get here and back," it is a worthy conclusion that such a combination of knowledge from many of the natural sciences is requisite for a successful analysis. For this topic, there is no shortage of proposals offered by many science and technology communities.

Often, these examples are discussed from the anthropocentric viewpoint of "man being the one to get there and back" and not from an alternative viewpoint of "extraterrestrials getting here and then back to their home." Man has thought in this perspective since ancient times. This is one ramification of our continued use of the ancient Greek philosophy and science ideologies, which have been pedagogically practiced for over 2,500 years.

Below is an inclusive (somewhat) chart showing numerous types of propulsion design systems that humans have had the creativity to imagine up as possible answers to questions of "getting there from here." This next iterative leap of the exponential distance problem has many presentations to offer; many of them are only in a speculative design stage at a schematic demonstration level. The instruments needed to be invented to create these new technologies can be found on many drawing boards around the world.

The Propulsion Theories

Chemical Rockets
        External Sails
Ion Propulsion Rockets
        Nuclear Fission and Fusion
Beam-Powered Rockets
        Interstellar Ramjets
External Laser-Beamed Energy
        Antimatter
Beamed Propulsion
        Wormholes
External Energy
        Subatomic Particle Matter Energy
Artificial Gravity
        Alcubierre Drive
Torsion Spin Propulsion

At the outset, the discussion here is all about propulsion. There have been many conversations, held in the media, on the potential for the practical use of heavy chemical elements for the operation of onboard spacecraft systems. For example, the New Horizons space probe, which NASA successfully missioned to Pluto in 2015, uses plutonium pellets to power the ships' onboard electrical systems.

The means of propelling the craft, however, is governed by two facets: first by ejection of pressurized chemical gases, such as hydrazine in the New Horizons mission; and additionally by a smart mathematical and physics formula strategy known as *gravity assist*. In the gravity-assist mechanism, the mathematics and engineering formulas are solved to allow the craft to gain velocity without the use of any fuel or propellant by "whipsawing" around a planetary object massive enough to allow the craft to utilize its *angular momentum* to sufficiently "velocity up." *Merriam-Webster* defines angular momentum as "the quantity of rotation of a body, which is the product of its moment of inertia and its angular velocity."[2]

Credit for the propulsion theories list should be given to astro-sciences researchers and their discoveries as far back as the 1800s. Konstantin Tsiolkovsky was the nineteenth-century Russian rocket pioneer and researcher who originally envisioned some of the propulsion hypotheses, which became the theories we recognize today. The function and length of this development timeline happens most often in the science disciplines. From concept-to-practical functionality, it can take over one hundred years for an idea in science to come to practical application. Like the scale-up of the distance dilemma, many incremental improvements to both science and engineering technology design must be achieved for the technology to become successful, useful, and buildable in all practical aspects.

It should also be noted that our civilization is already capable of introducing into practical use other propulsion systems from the above list. For instance, given that the financial resources would be available, we could successfully prove the theory, design, build, and functionally operate a system using some controlled nuclear fission propulsion. However, in addition to the fiscal constraints, many global political barriers exist that are preventing the use of nuclear-operated programs of this nature in near outer space or any space from Earth's orbit and beyond for that matter.

Let us take a glance at what these theories could embody if their study could successfully advance from hypothesis to pseudoscience then to testing, observation, analysis, conclusion, and positive introduction into the realm of mainstream science, functionality, and the reality of the existence of the technology. This is an example of the process of the scientific method as it is practiced today.

One kind of categorization in the whole propulsion theories list and discussion of each theory within the list may assist in how to understand that more than one of these propulsion systems share common characteristics and criteria. From this categorization, a more detailed discussion of each will help better understand what these systems are and how they are adapted to space travel.

We could categorize a handful of technologies that share a common characteristic of the action of "shooting something through a nozzle" or out the back of the craft. This notion was eventually applied from the Sir Issac Newton discovery of the third physics property of motion or law of motion in the late 1600s, which provided a foundation for those technologies. The third law states that, "For every action of force placed on a body of matter there is an equal and opposite reaction upon that same body."[3,4]

The nozzle in this type of rocket provides a physical capability for that reaction mechanism to happen. The nozzle mechanism is part of an example of the first category, which includes chemical or ion-propulsion rockets, ramjets, and nuclear fission and fusion rockets. When something is forced out the back end of a nozzle, the application of Newton's third law of motion explains how the object attached to that nozzle is propelled in the other direction. This happens both within our atmosphere and outside in the vacuum of space.

We are most familiar with these characteristics as they have been part of our experience for almost a century. The presence of the nozzle technology is part of not only chemical rockets but also of ion-propulsion missiles, nuclear fission and fusion, ramjets, beam-powered rockets, and some ship designs that are proposed to use antimatter systems.

Ships with nozzles can only propel a craft forward. If the ship needs to go back, it would require other nozzles strategically placed to accomplish this. The Apollo Command Service Modules and NASA's Space Shuttle both had this mechanism and capability. However, it only worked in the confines of space outside of Earth's atmosphere.

Of course, not every object needs a nozzle attached to it to give it thrust and obtain a forward propulsion. For example, one could "push" directly on an object to make it "go forward" and therefore not need a nozzle. In the vacuum of space, the forward reaction of this kind would propel the object long distances. All the remainder propulsion system types in the propulsion theories list are currently in design or speculation stage.

The next category of propulsion types also uses a pushing mechanism to achieve propulsion but not with the use of a nozzle. All kinds of external sails, external laser-beamed energy, and other beamed propulsion systems have been postulated for as long as chemical rockets. But in total, the immediate prospects of obtaining lasting success for this category of technologies as applied to interstellar travel would, from an intuitive perspective, currently appear spurious.

There are six candidates remaining in the master propulsion theories list. These do not use any physical action of thrust or of pushing or pulling to achieve propulsion. Instead, there would appear to be some "manipulation of matter or something going on" within these systems. Truly exotic would be the best way to describe the remaining technologies. This category includes antimatter, antigravity, wormholes, external energy, torsion spin propulsion, and the alcubierre drive as, at least thought-of proposals.

We cannot change the laws of nature, but we can learn to manipulate them and will need to be able to do so if we wish to explore other interstellar planetary and orbital extraterrestrial destinations we are now discovering at a faster than linear rate. We also know about the distances involved and the scale-up requirements from the mathematics of "going the measured distance from here to there." Is all this possible not only for humans but any other intelligent extraterrestrial beings with the same aspirations?

Some of you have heard of the New Horizons mission to Pluto and the Voyager missions. New Horizons passed by Pluto in the summer of 2015 and is now headed to some other dwarf planets farther out. The two Voyager craft were launched in the 1970s. They are now entering a very far outer region of our solar system. The craft blasted off using chemical propellant. After burning through the fuel, they used the gravity-assist technology I have described. Now they just coast using no other technology with which to accelerate or slow down. Voyagers 1 and 2 have been traveling for about forty years to get to where they are today, which is about twelve billion miles away from Earth.

These craft have covered a distance of only a few billion miles. The nearest known solar system with planetary candidates that could support any type of organic life, in our knowledge, are trillions of miles away; hundreds of thousands of times farther than even where Pluto and the Voyager aircraft are at this time.

With this in mind, we can safely conclude that chemical propulsion will not be practical if even generously granted as a possible solution to travel successfully to the closest star system. So toward our discussion of "how to move my UFO," let us quickly move away from the technologies of chemical propulsion.

The discussion thus far has continued from an anthropocentric perspective. In order to analyze our two main hypotheses about the possible existence of UFOs and extraterrestrial intelligent beings, we have to reconcile the notion that if humanity can achieve any of these propulsion technology objectives or at least dream them up and experiment with the schematics. Then other civilizations, if they exist, can also do this. It is such an anthropocentric perspective that prevents one from allowing for any opportunity to perceive and maybe learn from different viewpoints. Having an open mind is a goal of all research activities in these and all scientific endeavors.

For a discussion about any propulsion system that utilizes a nozzle design or like-kind procedure, some concepts such as *thrust-mass ratios, energy density, specific impulse*, and *exhaust velocity* are necessary baseline measures of potential success of obtaining the fastest speeds both within that design parameter and in comparison with other designs.

Thrust mass describes a concept of the relationship between physical force and the mass it is acting upon. It is an inverse relationship that can be thought of both intuitively and mathematically. Here, mass is different from weight, especially to a physics scientist. Weight is embedded with a characteristic of the forces of gravity while a measurement of mass is not.

Simply put, the more thrust per unit mass the propellant can produce, the faster the velocity a rocketship can attain.[5] We cannot use the term *weight* here; in physics, *mass* and *weight* are two entirely different concepts; this is a no-no! The comparison of all rocket type thrust-producing propulsion systems may be useful in a visualization. The propellant (fuel) with the smallest thrust-mass ratios is chemical rocket technologies. For our discussion, liquid chemical rocket systems produce the smallest thrust-mass ratios. All we need to know here is that the smallest thrust-mass ratios are equivalent to the slowest ships. Hence, these are the most undesirable ships to use for interstellar travel.

Liquid chemical rocket propulsion produces the smallest of these thrust-mass ratios and the slowest velocities. Solid chemical propellant rocket thrust-mass ratios are slightly higher than liquid chemical rocket technology. Ion propulsion systems deliver more thrust-mass ratios. On a scale-up hierarchy, ion propulsion systems slot in next. NASA is currently using this system in operating the Dawn Mission that is flying to the Near-Earth asteroid Vesta. These ion propulsion technologies have a higher thrust-mass ratio and, therefore, higher velocity than chemical rockets. Ion propulsion is the next attainable technology in the very immediate future of interplanetary (not interstellar) space travel.

A diagram of a nuclear propulsion system would move the thrust-mass-versus-velocity trend upward and to the right in a graphical image using the thrust-mass ratio on an x-axis and velocity on a y-axis. Two main types of nuclear reaction can occur: fission or fusion. In nature, atomic fusion enables a higher thrust-mass/higher velocity than does nuclear fission. Also here, there are propulsion possibilities for gaining more efficiencies that mean still more velocity potential than the others discussed. This is due to some future study analyses that allow for the introduction of onboard fuel requirement alternatives. What this means is some types of nuclear fuel could be obtained or manufactured when the rocket ship is in transit on its mission journey. The advantage is the ship does not have to carry all the fuel when it begins its journey, which means a lighter/smaller mass for the ship and a higher thrust-mass ratio. Consequently, the "need for speed" is satisfied!

Moving still further upward and to the right of our thrust-mass-versus-velocity graph are some future study conjectures for possible uses of antimatter technology and an introduction of some non-rocket/non-nozzle technologies into this scenario. We can think of antimatter as being allowed to fit into the nozzle category of spacecraft propulsion protocols mainly because its components consist of matter, as we know them to be. This matter has a chemistry to it. This means there would be a distinct chemical reaction occurring when chemical matter propellant reacts with something else to produce the propulsion of the aircraft. This reaction occurs with all rocket-type systems, including antimatter. As with all propellants that utilize nozzle technology, there will be an exhaust velocity and specific impulse reaction that together produce the forward thrust of the ship.

Loosely defined, *thrust* is the force that moves a rocket through the air. Thrust is generated by the rocket engine through the reaction of accelerating a mass of gas. The gas is accelerated to the rear, and the rocket is accelerated in the opposite direction; and in order to accelerate the gas, a propulsion system is required.

Why are we interested in specific impulse? First, it gives us a quick way to determine the thrust of a rocket, if we know the weight flow rate through the nozzle. Second, it is an indication of engine efficiency.[6]

Specific impulse and exhaust velocity are directly related in a mechanistic sense. Increasing specific impulse in a propellant reaction and injection increases the exhaust velocity of the burned fuel. This conjoined mechanism creates faster rocket velocity which is the main objective.[7]

Antimatter is a concept which has been popularized in many science-fiction movies. In its most basic form, a particle of matter crashes into another particle of antimatter. The matter-antimatter reaction takes place causing the annihilation of the two opposed particles of matter. They literally "bump into" each other. This reaction produces far more efficient (higher) thrust-mass ratio than other rocket nozzle-type systems. Consequently, the energy density is higher than the other propulsion systems discussed so far.

Energy density is the amount of energy that can be stored in the given mass of a substance or system. The higher the energy density of a system or material, the greater the amount of energy stored in its mass. Energy can be stored in many different types of substances and systems.

A material can release energy in four types of reactions. These reactions are nuclear, chemical, electrochemical, and electrical. When calculating the amount of energy in a system, most often, only useful or extractable energy is measured.[8]

The end point is that the velocity of the spacecraft under a matter-antimatter propulsion system design can theoretically approach (or exceed) significantly large fractions of light-speed velocity. Indeed, the Starfleet governing body of *Star Trek* television fame embraced antimatter as its propulsion technology.

Two key concepts within the discussion of matter versus antimatter are the spin of the opposed pair of matter and antimatter particles that interact with each other and the electromagnetic charge of the couple. These are important concepts to remember for discussions on propulsion theories. All of us perceive and experience this notion of spin every day in our lives, though only in the back of our consciousness. Spin happens everywhere. It occurs with every atom in the universe, though we cannot see this occurring. Spin occurs every day when the earth rotates on its axis. All planets and stars do the same.

First, everything in this universe has a physical property of spin. The spin property of matter is a foundational concept of what causes reactions in chemistry of all kinds including the kind in the matter-antimatter example just described. While we cannot directly visualize the spin part of the chemical reaction between atoms and molecules, we can only directly visualize the results. For example, the northern lights are caused by chemical reactions in the earth's upper atmosphere. We cannot see each atom interacting with its specific spin properties because atoms spin too fast for us to see them and, obviously, they are too small. But we can see the beautiful colors the reaction produces. The spin property helps to produce this, and it is the same with all chemical reactions.

Additionally, the reaction of matter with antimatter involves electromagnetic charge. When each pair of matter/antimatter particles collide, both atoms are annihilated. The energy produced from this annihilation is also caused by the opposite electromagnetic charges the two atoms have. An atom of matter comes packaged with a proton that has a positive electromagnetic charge of 1, as defined in the chemistry literature, and an electron that has a negative charge of 1.[9] There are also neutrons in the atoms, but they have zero charges; just a mass. Conversely, an atom of antimatter has the charge of the respective constituents reversed. So the antimatter proton has a negative *charge* of 1, and the electron has a *positive charge* of 1. Other than that, the two atoms are identical in constitution and nature.

When this pair of atoms collide, they produce an enormous amount of energy that is much higher than any other chemical reaction, even with atomic fission and fusion reactions. Matter-versus-antimatter propulsion technology is placed still further up and to the right on the "thrust-mass-to-velocity" graph curve. This is the most efficient type of chemical reaction possible that has practical application possibilities. The problematic situation is being able to manufacture sufficient amounts of usable antimatter, which is a first major hurdle to overcome for any evolution into a practical development application of the technology.

So does this mean that we will not see human-built spaceships with matter-antimatter propulsion systems? Maybe not anytime soon. Theoretically, we do know antimatter exists and can currently manufacture single atoms of it. On a mass measurement scale, though, all that we have been able to produce amounts to a tiny fraction of a microgram over a production time frame of months and years. Forget the cost-of-current-manufacturing factor, there is no practical arrangement yet in existence to sufficiently ramp-up this technology. Scientifically, when a new pseudoscience evolves to being considered as mainstream science, the antimatter propulsion theory is still not even a theory; it is always primarily pseudoscience.

Remember that a field of study or discipline is considered by the science communities to be pseudoscience when its epistemological structure cannot be successfully proven by a sufficient number of the science communities. The title is just a prelude attribution to, when the theories and knowledge become so much accepted and proven by those communities of peers that the pseudoscience morphs into then being classified as mainstream science. There is no set "body" of science communities or peers who oversee this protocol. All science professionals "shoot from the hip" and use the term as they so desire.

For example, the field of genetics was proclaimed as pseudoscience by the science communities and peers back when Gregor Mendel was experimenting with his pea plants in the second half of the 1800s. Fast-forward to today and look now at what we have learned about genetics, DNA, and what entire science disciplines he single-handedly created and contributed to and the monumental knowledge he left to us to further explore! Albert Einstein was another genius pioneer in the same dilemma whose theories about space-time and relativity were scoffed at as pseudoscience for years before their acceptance into mainstream science.

A third example of the progression of pseudoscience theory to that of mainstream science was the work of Alfred Wegener, the geophysicist and meteorologist of the early 1900s. He discovered the theory of continental drift, which explains that all of Earth's continental land masses floated on top of Earth's mantle layer. He also created new fields of science associated with geology and geophysics but was also disparaged and rebuked by the science experts of the day, such as Mendel and Einstein.

Wegener's theoretical arguments and evidence were not enough conclusive proof at the time for the scientific communities to absolutely confirm the drifting of the continents. That definitive proof came in the form of the discovery of the concept of mantle convection of the earth's mantle layer later in the 1960s. Unfortunately, because the time-consuming and slow methodical nature of science discourse and paradigm activity and revolution prevent discoveries moving swiftly from pseudoscience to mainstream, Wegener did not live to see his theories confirmed. Another analogy to describe the nature of the situation would be to congratulate geniuses like Mendel, Einstein, and Wegener for the vision they had in discovering and inventing the things they did even though mainstream science could not imagine these things for themselves. Nowadays, if we ask any biologist, physicist, or geologist who should be in the biological sciences, theoretical physics, and the geological "halls of fame," it would be all three.

The next couple designs still use some mechanistic elements of the "nozzle" propulsion technology but attack the problem of creating increasing thrust-mass ratio and hence increased specific impulse, exhaust velocity, and final ship velocity from another direction. The first of these two designs is the operation of a system that uses a type of atom collector and ramjet engine apparatus. A physicist named Robert W. Bussard, a fellow at the International Academy of Astronautics, hypothesized the atomic fusion ramjet concept in 1960. The spacecraft would carry such a collector, envisioned by Bussard as a kind of "scoop" along with the ramjet engine. This type of apparatus would provide a substantially reduced mass of the entire craft and make the resulting thrust-mass ratio much higher, thus creating increased velocity.[10]

The atom collector would accumulate hydrogen atoms collected from along its course vector on the interstellar space highway and react with the atoms by a hydrogen proton-proton fusion reaction. The resulting nuclear fusion reaction would expel the energy through its energy density gradient and out the back nozzle of the craft. This type of nuclear reaction is the exact mechanism that our sun and stars use to burn the same atomic hydrogen, helium, and other matter elements to produce energy. Thus, the craft would not have to carry all its fuel on board making the craft lighter. The reactor itself would be a lighter apparatus than those used in chemical burning rockets. This is another way to increase the thrust-mass ratio in our favor for use in matter propulsion systems of any kind.[11]

This design has not attained even a higher level of proof of concept for mainstream testing along its evolution of idea-to-practical use. It is because of design considerations that require reconciliation with both the most effective collector apparatus and the nuclear power plant itself. The general concept is perceived by many to be presently doable as an investigation of the mathematical foundations has proved its viability. The mathematical foundation of a proposal is the first design considered for any of this type of hypothesis even before a mock-up proof-of-concept consideration can be undertaken. From this, we can see the design-time-to-practical usage and fiscal constraints as problematic.[12]

The next type of propulsion apparatus also uses the design consideration of making the craft lighter to reduce the mass portion of the thrust-mass ratio equation. Thrust and mass numbers work inversely with each other in creating a velocity gradient. An energy beam propulsion concept accomplishes this by eliminating the weight of a nuclear reactor and, instead, sets it up and uses it on the ground (Earth). The propellant is still on board the craft. The nuclear reactor beams lasers to the ship and ignites the propellant, which then propels the energy out the rear of the craft. Velocity and acceleration are achieved and maintained continuously throughout the journey. Other systems for braking and maneuverability would be designed into the craft with presumably a similar level of mass considerations. An offshoot design consideration of the beamed propulsion would allow for the ship to use an atomic collector apparatus similar to the preceding ramjet proposal, ultimately relieving the need to carry the total amount of fuel on board the craft the entire journey.[13,14]

Interstellar distances may potentially be as problematic simply because since the laser beams are originating on Earth, how is a level of efficiency realized as the ship moves farther away? A laser generator system portable enough to be carried on board, thus increasing the mass of the entire ship could be proposed into the schematics.

Other proposals use the general concept of using a beam to propel the craft without pushing energy out the rear through a nozzle. Some of the proposals include variations of external sails, including solar and magnetic sails, or even electric propulsion of a similar imagining as that used in electric cars on our terrestrial highways today. These sail systems must take both sides of the thrust-mass ratio into consideration. Thus, any design for future potential as a deep-space vehicle is still in process for if and when the plan can proceed to the next level of proof of concept.

Here is a good time to provide a bit more knowledge about the concept of thrust. In physics, thrust is the mechanism that propels a combusted propellant in one direction. The thrust gradient enables the resultant motion of the object or ship in the other direction. This is exactly Newton's third law of motion. Thrust, in the physical sciences, can include power transmission from an apparatus that does not burn or react matter of any fluid or solid fuel. The process of electricity and even the physical force of pushing a mass can produce thrust. We are most familiar with references to "thrust" as what we see when a gasoline- or diesel-powered car, jet plane, or even a space shuttle is enabled to move.

More or less, in all these modes of transportation lighter weight (the mass of an object as described when within Earth's atmosphere) is a principal objective. Where, in the equation "force = thrust/weight (mass)," a smaller/lighter weight (mass) creates more force. In these designs, a more concerted effort is needed to amplify the thrust component than with other designs. This is because the power requirement is coming from our sun, photons of light from other sources, or other sources of electrons. Those are not as rich a source of electrons as nuclear reaction or matter-antimatter.

There are other designs not mentioned which utilize other propulsion design systems and which are not of the exotic variety. We have seen the vast number of less exotic technologies that, while positioned at different points along a lengthy timeline of achievability, could be moved faster because we currently have the knowledge to establish and engage with projects to test the next step of a proof of concept for most of these plans. There are metascience and engineering impediments with some of the progress such as political, legislative, and legal hurdles (nuclear power in outer space-1/27/1967 Outer Space Treaty) and fiscal considerations that are problematic to this timeline. The main point is we possess some useful knowledge about most of the design concepts discussed so far.

As we look at the next list of more exotic forms of propulsion, we can reconnect with this book's two main hypotheses to the possibilities presented in this chapter. We also turn away from the purely anthropocentric viewpoint surveyed so far because, from our perspective, these exotic designs are items that are associated with the most speculative theories of propulsion. They fit into the scientific pseudoscience category by the amassed knowledge of mainstream science communities.

Some of the more modestly exotic notions of propulsion that do not use a rocket design involve a redistribution of or imbalance of gravity influence around the volume area of the spacecraft to induce thrust. What this means is explained below.

Picture an airplane flying around a neighborhood. In Earth's atmosphere, there is an air pressure that envelopes our entire planet. When that airplane taxies down the runway to lift off, its engines are creating an air pressure imbalance from what exists underneath its wings to what exists above its wings. When that air pressure imbalance becomes large enough as the airplane increases its velocity down the runway, its wings and that air pressure imbalance creates lift, and the plane takes off into the sky. The continued engine operation, either a propeller or jet engine system, sustain that lift mechanism, and the plane can be seen flying through neighborhoods.

The force of gravity in the universe behaves like that air pressure blanketing the earth. Though gravity is not air pressure, it is the blanket of force that influences all planetary bodies in outer space. Wherever there are two bodies, planetary or lunar near each other, or objects like spacecraft, the influence of gravity is present. Whether nature creates gravity or is artificially created, such as what would occur if one of our spacecraft were designed to allow for it to spin or rotate enough to create such a stronger gravity that affects whatever is inside of the craft, gravity is present. In fact, it occurs whenever two or more bodies (even spaceships) have a gravity attraction on each other. Gravity is part of the layers of space.

Just like the example of the airplane creating lift in our atmosphere by an air pressure imbalance thus allowing it to fly an imbalance in the force of gravity and the amount of its subsequent influence on the spacecraft could allow for the ship to be propelled in a specific direction. This mechanism is the starting point for applications of a mechanism to allow for spacecraft to be powered through space at tremendous velocities. So the creation of a gravity gradient imbalance in the three-dimensional volume of space around the aircraft could allow for it to move much faster than would be allowed by chemical reaction forces alone. If such a craft were to be observed in operation, a chemical reaction of combusted propellant being forced out a nozzle or from the rear of the aircraft would not be visible. This application could hold if the craft was traveling through outer space or within an atmosphere such as Earth's own atmosphere.

The perception of the imbalance within a physical system can also apply to the concept of wormholes and propulsion. Take the notion of space-time. This discovery was proven by Albert Einstein. With an application of time travel via a wormhole mechanism, Einstein thought the fabric of space would be folded over somehow toward another edge of itself. This then allows the spacecraft to take a shorter path across these two folds. The result of this event would be that, instead of the craft moving in an entirely straight line toward its destination the entire trip, it could cut across those two folds and arrive at its destination in a fraction of the time it would take for the straight-line journey, at least from the perspective of the persons aboard that spacecraft. The craft would travel through an altered three-dimensional space and use time, as the *fourth dimension*, to cruise that wormhole through to its destination. This would be back in the three-dimensional space we observe.

The following are two examples of this process. The first one works like this. Take a sheet of paper to symbolize the fabric of space. Now fold one end of it over toward the other end. The first or starting end that will be folded over represents the starting point of the spacecraft. The end that is folded upon is the destination point. Folding one end over toward the other shortens the distance that we observe in three dimensions between the two points of the beginning and the destination points. In outer space, it is conjectured that a wormhole will be created via this process. The craft then travels through this wormhole toward its destination point in a lot shorter time than it would take to go the traditional straight-line path through three-dimensional space if the paper was not folded over.

A second visualization involves how the human brain is illustrated in any book in conjunction with the process that is taken when a thought of information or memory travels through it. We can refer to a photograph or drawing of a human brain in a magazine, textbook, or video. Specifically, recall the many folds the fabric of the brain has as it is tucked inside of the human skull or that of any animal for that matter. Human brain matter is like the folded sheet of paper used in the last example except it has many folds. But all the folds can assist in visualizing the many folds that could occur in the fabric of space and allow for wormholes to be created and shortcuts to then exist for a ship to travel through in greatly expedited time.

Suppose now that we are the captains of a microscopically tiny craft just like the one portrayed in the 1966 movie *Fantastic Voyage*. The ship in this example will transport a cargo of information or memory through our brains to its destination in another part of it for processing. Here, brain folds represent the three-dimensional fabric of space that we observe. It is just that this fabric already happens to be folded over such as our image of the human or animal brain.

Our ship can now take two distinct paths of varying length to get from point A to point B. Our craft can take that cargo of thought and move along only among and within the folds of our brains to arrive at its destination, and they would zigzag back and forth down the three dimensions of just each fold. This is similar to the path that is taken in our brains when we create a thought or are exposed to some sensory stimuli that cause them to create that thought. The *neural pathway* is that set of sequential individual neurons that fire one at a time to stimulate the next neuron in the path just like if we ran an electrical charge through it from one end or like a necklace of metallic pearls. A nervous reaction such as this only takes a minuscule fraction of a second!

Our ship can use an alternate route. It can travel across the folds instead of down them at some point, using this as a shortcut to our destination. The time it takes for the transport of that cargo of information through to its destination via this method will be shorter from our perspective as a captain of the spacecraft and also from that of an outside observer.

As long as there is a pathway that would enable the ship to travel through or along a shortcut, this is how a wormhole would theoretically operate. According to our understanding of the science involved and what Einstein proclaimed in his theories of relativity stating, wormholes involve a mathematical construct using a type of negative mass with which we know very little.

The overall notion of an imbalance of forces, whether of a gravity gradient or space-time fabric or something else, is a driving mechanism of many of the more exotic forms of spacecraft propulsion. This is to say that to make the technology go would require some manipulation of nature to create an imbalance. The remaining items on the propulsion theories list not yet discussed can now be thought of as involving a mechanism whereby a spacecraft exploits an imbalance of some forces in nature. They do not necessarily include the chemical burning or other reactions associated with such reactions to propel a craft.

For example, the Alcubierre drive, also known as the Alcubierre metric, proposes that such an imbalance could be created in the space-time fabric that could propel a craft to faster-than-light velocities. The mechanism for this notion involves creating the imbalance by contracting space in front of the craft and expanding space behind the ship to propel it in the desired direction. The Alcubierre drive uses Einstein's existing theories of relativity to allow for the use of an energy density that is less than that of the energy contained in a vacuum.

The successful Alcubierre drive mechanism would be able to contract or shrink the fabric of space in front of the spacecraft in the direction it is heading. It would simultaneously need to expand the fabric of space behind the craft. This means an energy density field with *negative mass* must be created and harnessed to facilitate the mechanism to work. If it could, the speculation allows for faster-than-light propulsion. A similar observation that would occur in an Alcubierre drive device, as well as a wormhole mechanism, is there would be no combusted or burned chemical matter streaming from the craft.[15,16]

ALCUBIERRE DRIVE

Citation:
Alcubierre Drive
Digital art by Les Bossinas (Cortez III Service Corp.), 1998

A more recent expansion of proposals for the foundational aspects of exotic propulsion systems trace back to 1900s research on the nature of how the atomic particles that makeup matter interact in our universe. More specifically, the knowledge that we have accumulated for hundreds of years about the chemistry and physics of the behaviors of matter is being used to explore applications that include exotic propulsion systems. These research communities are taking some of the very most fundamental behaviors and properties of matter and applying those behaviors to move toward hypothetical statements about new ways to propel bodies (ships) through space.

Here is where we can connect with the previous discussion about the natural properties of spin and charge regarding atomic particles. Remember that all visible matter in the universe is comprised of atoms and molecules, a group of atoms. We will put aside those types of invisible material such as dark matter or dark energy for now and focus on matter in the visible spectrum. Every visible atom and molecule is composed of subatomic particles: the electrons, protons, neutrons, and other nucleic particles that makeup these substances.

Each atomic structure exhibits the universal physical properties of spin and charge. Relating to the atomic property characteristic of spin, all science and nature spin means rotation. It is precisely equivalent to say then that all visible matter is composed of molecules, atoms, and subatomic particles in which all demonstrate the natural physical properties of rotation spin and charge.

We cannot, obviously, observe every atomic or subatomic particle spin or rotatation. But we can conclude the existence of these properties by the logic and reasoning of *inference*. The scientific method, the construct of science practice in any context ever practiced by man, includes inference as an absolute requirement. We can infer that rotation spin and charge exist from known characteristics of the nature and properties of matter in our universe.

Spin is "an intrinsic form of angular momentum carried by elementary particles, composite particles (hadrons), and atomic nuclei."[17,18] A good practical analogy is watching a spinning top. The earth does "spin"; this is known as "rotation around its axis." One spin takes twenty-four hours to complete. So in practical terms, the earth rotates. The characteristic of a spin is that one complete rotation of the spinning object takes place in fractions of a second. The concept is normally left to the domain of quantum physics and quantum mechanics. But this is relevant to a discussion of exotic forms of interstellar propulsion.

Individual atoms are neutral; they contain the same number of protons as electrons. Remembering that a proton carries an electromagnetic charge of $+1$ and an electron carries an electromagnetic charge of -1, the total charge of an atom of this type will be zero. A charged particle or atom will have an uneven total of protons and electrons to cause an imbalance. By definition, an ion is an electrically charged particle produced by either removing electrons from a neutral atom to give a positive ion or adding electrons to a neutral atom to provide a negative ion. When an ion is formed, the number of protons does not change.[19]

We can also see that our universe is full of examples of rotation, spin, and charge. For instance, we can readily observe a tornado's or hurricane's winds rotate around its central vortex. We can see a top spinning when we twirl on its top cap. In numerous graphic animations, video producers show a planet or star rotating around another star at an advanced speed. In real time, our visualization perceives it to be much slower, but you have to take into account the size and mass of whatever example you are observing.

To gain a better understanding of spin, rotation, and their effects on gravity is by locating and watching a centrifuge in operation. All science laboratories use them as do most doctors' offices. NASA, all other partner companies in the aerospace industry, and all the world's astronauts use them in their training.

The centrifuge operates by the rotation and spin motion of the apparatus itself and/or an appendage attached to it. In the case of the astronauts' centrifuge, the apparatus spins and rotates around a pivot mechanism. The astronaut is enclosed inside the apparatus. When the centrifuge is on, the rotation creates a gravity imbalance. The astronaut feels increased g-forces or a stronger gravity force upon his or her body. The faster the centrifuge rotates, the more of these gravity forces are felt. The human subject would be able to tolerate about ten g's of force before any additional g-forces would begin to impede his bodily functions. We would experience a less obvious version of this artificial creation of gravity imbalance if we were to ride a roller coaster or other amusement park ride or on an airplane.

A centrifuge that is used by science, medical research, and in doctor's offices can rotate and spin much faster than the astronaut's model. Used for separating biological substances, these apparatus can turn to over one hundred thousand g-forces. These forces are needed for the isolation and study of biomedical sample substances for research experimentation and drug treatment research. Centrifuges have been in existence for practical and extensive use since 1875. English engineer Benjamin Robins first envisioned the concept in the mid-1700s.[20]

In chemistry class, students are taught that molecules, atoms, and subatomic particles are all rotating or spinning. They do this much too quickly for any human to be able to observe. So we have to logically infer these phenomena from peripheral characteristics of how atoms behave in nature; often from what end points these characteristics produce or manufacture in nature.

The main point here is rotation and spin are among the most basic characteristics to the nature of how our physical and chemical universe operates. These represent some of the laws of Mother Nature that have to be followed or are available to be manipulated in some way to achieve our desired effects and goals. Those goals, in this discussion, would be to find alternative means of interstellar propulsion.

Being that subatomic particles, atoms, and molecules are all the basic building blocks of matter, and they all have rotation and spin, and often ionic charge, it must be concluded or inferred that all matter must contain these characteristics to exist. In all the natural science disciplines, these laws are indisputable. The study of chemistry and quantum mechanics investigate these phenomena at the smallest known levels—those of the subatomic particles, atoms, and molecules. Quantum physics also involves particles such as *quarks* and *leptons*. It is these chemistry building blocks that help to define the biological processes in nature that life engages. Our bodies—from our bones to our muscles, organs, and brains, to our DNA—all exist because of chemistry or the chemistry of life. Similarly, the geological sciences constituency of rocks and minerals and their life cycles are derived from what display the chemistry lays blueprints for them. All other physics disciplines act accordingly. The studies of astronomy, engineering, and all other applied sciences exist because of matter in which chemistry is among the most fundamental natural building blocks.

Everything in nature has rotation and spin associated with it. When thinking about and observing events in nature of this characteristic and speculating about how the mechanism of interstellar propulsion would be feasible, could there be an open possibility that the properties of rotation and spin be included as factors?

Let us not also forget about the notion of atomic charge in the chemistry of the natural sciences. *Charge* here means the plus or negative charge, or the + or -, of chemistry. The plus and negative charges of chemistry are characteristics of attractive or repulsive forces in nature that are the building blocks of chemical elements and molecule, as those we have just introduced. Without these charges, atoms and molecules of matter, as we have come to know them and will continue to know in the future, would not exist; it would not be real in the sense that we would recognize.

A quick summary of the notion of chemical charge is that most of the chemical elements, the ones that appear on the periodic table of elements, have an unnatural number of electrons rotating or spinning around their respective nuclei. This means, naturally speaking, atoms of most chemical elements either seek to obtain or seek to share or give up at least one electron. The atoms that do this wish to be of neutral or zero charge. When an atom does not have a neutral charge, it then desires to join with another atom of exactly the opposite numeric charge in order to again become of neutral charge. This is a major motivation for the process of chemical bonding to happen in nature.

Atoms that exist are all rotating or spinning as defined from that which is happening inside their nuclei and also from the electrons in orbit around each atom. This can be viewed exactly how we picture our earth and the other planets orbiting around the sun. It is the same process.

The way a proton acts within the nucleus of an atom compels it to hold a plus charge or a "+" charge. The way an electron acts when in orbit around its nucleus compels it to hold a negative charge or a "-" charge. The most frequently occurring version of a chemical element has the same number of protons as electrons. Thus, the charges of the subatomic particles of that atom element equate to zero. So there is no electromagnetic charge to that atom.

When we read in a chemistry text about a molecule having a "1+" charge, it means there is one more proton present within that molecule than electrons. Or take a molecule that has a "2-" charge to it. This means there are two more electrons than protons within that molecule. That molecule is unbalanced and takes on the charge status and a number of the more frequent charged particles within that atom or molecule.

Atoms or molecules with a nonzero charge have a propensity to be attracted to other atoms or molecules with an opposite charge. This process of joining is the very basis of chemical bonding and that which forms all the different compounds of matter we have currently.

So altogether in nature, even the tiniest of atomic and sub-atomic particles contain within them these properties or characteristics of rotation spin and charge. The union of these particle types in nature creates larger and more substantial structures of matter that include the largest objects in our universe, namely planets and stars in the cosmos. All the matter objects within also contain the same characteristic properties of rotation spin and charge.

Let us now try to relate these scientific laws to those of propulsion and interstellar travel. The study of these notions of rotation and spin began extensively in the mid and late 1900s to construct theories about what has come to be known as the torsion field. Given that the scientific laws of rotation and spin hold for every elementary particle of matter and of every object in the universe, the theory of the existence of a force field, which is the torsion field, is concluded thus far to influence all those particles of matter and of all the objects in the universe.

Torsion is a phenomenon in nature partially defined as a sort of "twisting" effect. This twisting mechanism affects all matter and space. It has been thought to have influenced time itself. Torsion is stored in all objects and all matter. When other forces influence the object, such as temperature and pressure changes, for example, these cause the torsion force to be instantaneously transferred from the source object to a destination object. Among the effects of this cause-and-effect event include changes in gravity and time dilation, among others. Claude Swanson, PhD wrote extensively on this subject. His book is titled *The Synchronized Universe*[21] and a seperate research paper titled "The Torsion Field and the Aura."[22] These contextualizes the entire study of the torsion field. Swanson's resume includes postgraduate work at MIT, Princeton, and Cornell Universities and formed and headed his own applied physics company performing relevant research for organizations such as the U.S. Army and Navy, the C.I.A. Dupont, and United Technologies.

The information within Dr. Swanson's publications describes a variety of contextualized applications of the study of torsion fields. The basic learning concepts are taken from the notions of rotation spin and the electrical charge properties of matter. Applications from these notions infer possibilities of propulsion which include interstellar travel.

Torsion fields result from the interaction of matter in certain capacities. The chemistry concept of *chirality* applies to a study of torsion fields. *Chirality* means "handedness and asymmetry." Here is how its creator, Sir William Lord Kelvin, defined it: "I call any geometrical figure, or group of points, 'chiral,' and say that it has chirality if its image in a plane mirror, ideally realized, cannot be brought to coincide with itself."[23]

This handedness, vital to the study of organic chemistry, includes an asymmetric construct of matter. A good way to visualize the concept of chirality would be to put either hand up to a mirror as the image Lord Kelvin described in his definition. Neither of them is superimposable on the other. This is exactly a visualization of organic compounds (ones that contain carbon) that are not symmetric or superimposable on each other. This is a crucial concept behind the existence of many of the pharmaceuticals in existence at the present time.

Handedness and asymmetry apply to the interaction of matter with some outside force(s) acting on it. This outside force could be any property of nature such as a change in temperature, pressure, or gravity for example, or any combination of the three. These are only a few properties among others. When any of these forces interact with matter, the state of handedness affects the activity. The torsion field energy radiates away from the impacted matter and toward some destination. This is achieved instantaneously.

The implications of the torsion field include changes in mass and energy density and time dilation. All these phenomena have been replicated and proven in experimental laboratories over the entire world since the 1970s. It is the application and ramp-up of the theory of the torsion field forces to be functioned to alter the energy density of matter or gravity, among other forces, it creates the possibilities for interstellar space travel at light speed or faster.[24,25]

The torsion field forces could be incorporated to work with fuel sources that would either be stored on board a spacecraft or, as noted earlier, collected on route to its destination. The output of the torsion field forces would be utilized to produce even more superlative thrust-mass ratios, and thus, even faster propulsion velocities for space vehicles than any of the other theories are speculated to produce.

As explained in Swanson's article "The Torsion Field and the Aura," replicable laboratory experiments have offered reliable support for the scientific foundations of the theory. Conclusions which can be drawn from this body of research indicate that knowledge is proceeding on the right path toward some potential future acceptance and practicality of the torsion field theory. This paradigm is an example of what defines a fundamental epistemology of scientific discovery; the possibilities that exist from following the scientific method toward discovery and acceptance (at least non-refutation) of the theory.

Are any of these technologies possible in reality? Can any civilization, as seen from the perspective of the other observer, take any of these technologies and design interstellar spaceships which are able to attain close to, as fast as, or faster than light speed velocity? We know the mathematics already indicate that it is possible for these technologies to accomplish this end point.

We know ion propulsion is possible and practical; we are already on the threshold of use as we speak. We know nuclear fission propulsion is possible. Other extraneous factors, such as the political and economic climates surrounding these communities, are current impediments that block further efforts toward practicality. Nuclear fusion propulsion technologies are significantly behind in the race toward practicality. Humanity, though, already understands and is attempting to translate the mathematics of beamed propulsion, ramjets, and matter-antimatter technology into experimentation. Humankind does seem to be significantly primordial in the defined landscape of these speculative propulsion technologies. We have been able to use automobiles and airplanes in a practical sense for only slightly over one hundred years and have already amassed the body of knowledge to at least conceptually understand most of these other futuristic propulsion technologies. From our perspective, motivations, and how we channel these futuristic visions, who knows how long it will take for us to be able to drive our speculations into invention and use for interstellar travel?

References

[1] Einstein, Dr. Albert. 1938. *Evolution of Physics: The Growth of Ideas from Early Concepts to Relativity and Quanta*, pg. 92. Copyright 1938 by Albert Einstein and Leopold Infeld. Published 1938 by Cambridge University Press.

[2] *Merriam-Webster*, 2016. Angular Momentum definition. Copyright 2016 by Merriam-Webster, Incorporated.

[3] Newton, Issac. 1728. *Philosophiæ Naturalis Principia Mathematica*. Published and translated 1728 by Benjamin Motte.

[4] Newton, Issac. 1729. *Mathematical Principles of Natural Philosophy*, p. 20 under, "Axioms or Laws of Motion." Translated by Andrew Motte. Retrieved July 29, 2016, from: https://books.google.com/books?id=Tm0FAAAAQAAJ&pg=PA19#v=onepage&q&f=false.

[5] Propulsion Theory Ref 5: NASA. 2015 General Thrust Equation. Last updated May 5, 2015, by Glenn Research Center. Copyright 2015 by NASA. Retrieved July 29, 2016, from: https://www.grc.nasa.gov/www/k-12/airplane/thrsteq.html.

[6] Propulsion Theory Ref 6: NASA. 2015. Specific Impulse. Last updated May 5, 2015, by Glenn Research Center. Copyright 2015 by NASA. Retrieved July 29, 2016, from: https://www.grc.nasa.gov/www/k-12/airplane/specimp.html.

[7] Propulsion Theory Ref 7: Braeunig, Robert A. 2012. Rocket Technology. Rocket and Space Technology. Compiled, edited, and written in part by Robert A. Braeunig, 1997, 2005, 2007, 2009, 2012. Copyright 2012 by Robert A. Braeunig. Retrieved July 29, 2016, from: http://www.braeunig.us/space/propuls.htm.

[8] Propulsion Theory Ref 8: Hanania, Jordan; Heffernan, Braden; Jenden, James; et al. 2016. Energy Density; a discussion. Copyright 2016 by University of Calgary and Jason Donev. Retrieved July 29, 2016, from: http://energyeducation.ca/encyclopedia/Energy_density.

[9] Propulsion Ref 9: Definitions. American Chemistry Society. Copyright 2016 from American Chemical Society. Retrieved July 29, 2016, from: https://www.acs.org/content/acs/en.html.

10 Propulsion Ref 10: Bussard, Robert W. 1990. Fusion as Electric Propulsion, p. 567–574. Journal of Propulsion and Power, Volume 6, number 5. September–October 1990.

11 Propulsion Ref 11: Bussard, Robert W. 1960. Galactic Matter and Interstellar Flight. Acta Astronatica, Volume VI, p. 179–195, 1960.

12 Propulsion Ref 8: Andrews, Dana G. 1994. Cost Considerations for Interstellar Missions. Acta Astronautica, Volume 34, p. 357–365. Retrieved June 26, 2016, from: http://adsabs.harvard.edu/abs/1994AcAau..34..357A.

13 Propulsion Ref 13: Michaelis, M.M. and Forbes, A. 2006. Laser Propulsion: A Review. South African Journal of Science, Volume 102, numbers 7 and 8. Retrieved June 26, 2016, from: http://researchspace.csir.co.za/dspace/bitstream/10204/1014/1/Michaelis1_2006.pdf.

14 Propulsion Ref 14: Kantrowitz, A. 1988. Proceedings of the International Conference on Lasers '87, F.J. Duarte, Ed. (STS Press, Mc Lean, VA, 1988).

15 Propulsion Ref 15: Jones, Andrew Z. 2016. Alcubierre Drive. Copyright 2016 by About Inc. Retrieved June 26, 2016, from: http://physics.about.com/od/physicsatod/g/Alcubierredrive.htm.

16 Propulsion Ref 12: Alcubierre, Miguel (1994). "The warp drive: hyper-fast travel within general relativity." Classical and Quantum Gravity. Volume 11, number 5. L73–L77. arXiv:gr-qc/0009013.

17 Propulsion Ref 17: Griffiths, David (2005). Introduction to Quantum Mechanics (2nd ed.). pp. 183–184.

18 Propulsion Ref 14: Thompson, William J. (1994). *Angular Momentum: An Illustrated Guide to Rotational Symmetries for Physical Systems.* Wiley. ISBN 0-471-55264-X.

19 Propulsion Ref 19: Atoms vs/ Ions. 2016. Copyright 2016 from Purdue University. Retrieved June 26, 2016, from: http://chemed.chem.purdue.edu/genchem/topicreview/bp/ch2/atom_ion.html.

20 How to Move Ref 20: Chisholm, Hugh, ed. (1911). "Robins, Benjamin." Encyclopædia Britannica23 (11th ed.). Cambridge University Press. p. 422.

21 Swanson, Claude, PhD. 2003, 2010. *The Synchronized Universe.* Copyright by Claude Swanson, PhD. Poseidia Press. Also available at www.SynchronizedUniverse.com.

22 Swanson, Claude, PhD. 2008. "The Torsion Field and the Aura." Subtle Energies and Energy Medicine. 2008, Volume 19, Number 3, Pgs. 43–89.

23 Chirality Ref 23: Sir William Thomson Lord Kelvin (1894). "The Molecular Tactics of a Crystal." Clarendon Press

24 Swanson, Claude, PhD. 2003, 2010. The Synchronized Universe. Copyright by Claude Swanson, PhD. Poseidia Press. Also available at www.SynchronizedUniverse.com.

25 Swanson, Claude, PhD. 2008. "The Torsion Field and the Aura." Subtle Energies and Energy Medicine. 2008, Volume 19, Number 3, Pgs. 43–89.

CHAPTER 10

Ancient Texts of High Strangeness

A priestly and noble man, considered respectable and reliable, once looked up and saw bright lights zigzagging across the sky. As he watched the lights turned into wheels and wings that emitted shining sparks, finally a weird object evolved and as it landed, four creatures, descended from the craft and approached the lone observer.

—Otto Billig, *Flying Saucers: Magic in the Skies*[1]

Key words: concordance, translation bias, parma

From lions and bulls and horses and cattle and whales to Enki, Marduk, and Ishtar, all Sumeria; Ra, Hathor, Vishnu, Lakshmi, all India; Hu, Shu, Nu-Kua, all China; Kukulcan, Gukumatz, Hunab Ku, all Maya; Viracocha, Inti, Mamaquilla, all Inca; and Osiris, Ra, Isis, Amun, all Egypt. From animals as deities to supreme gods that supplanted them. Through history, humankind has left much knowledge and communication to their descendants in many forms. Textual scripture is one popular format.

When the conversation is about the temporal issues among ancient texts and their archives, one important function should be considered but often isn't. When temporal factors arise, such as in situations when the text is found much later by archaeologists or other researchers, the creators of the text would need to ensure that the medium and/or language used will survive through time. In other words, will my descendants be able to decipher and read my writing?

Think about this in your situation in the twenty-first century. Our civilization uses computers as a main creation tool for recording our writings. The hundreds of languages spoken on Earth are the vital main tools of communication the computers are recording. The storage devices—such as flash drives, hard drives, DVDs, etc.—are analogous to the tablets the ancient Sumerians, Egyptians, Chinese, Indians, and Mayans used to permanently record their thoughts and knowledge. The question I ask you now is, "Do you think flash drives, hard drives, DVDs, and the computers with which to create our writings will survive so our descendants of, say, five hundred years from now will be able to read them?" Throughout history, do you think anyone has even thought about this dilemma?

Ancient art and symbolic artifacts predate any other form of communication of archival value. Archaeology is still writing the optimum timeline, but petroglyphs dating to over forty thousand years ago have been cataloged.

Art, artifacts, sculptures, and relics told stories and revealed details about the ancient's way of life. Hence, one relic could have spoken thousands of words. As writing and language text technologies were created about seven thousand years ago, the communication needed to impart an idea or story used symbols as words. As history progressed, words (as in Arabic, Greek, Roman) replaced the ancient relic dialogue. Eventually, languages were established to replace the symbolism and stories proliferated in the earliest known relic language.

Arts, petroglyphs, and symbolic objects and carvings conveyed information about humans' lives in ancient times; their annals, so to speak. These diaries passed on a historical record of the source's experiences of his most important life events. All the ancient chronicles archived what humans saw and experienced in their lives.

The medium through which the communication occurs—whether by petroglyphs, symbols, sculpture, or relics—text is just the instrumentality of the "transference of information." The content of the communication is the most important and relevant constituent to ancient humans and modern humans.

For example, when NASA was designing the Pioneer 10 and 11 spacecraft, they had the idea of carrying a message from our civilization on board the vessel. They enlisted Carl Sagan, astrophysicist Frank Drake, and Carl Sagan's first wife Linda Salzman Sagan to create a design for the project. They crafted the gold-anodized pictorial plaque that was finally used.

The plaque design is a pretty simple pictorial—a rudimentary version of the real things. The visualization of our solar system and of the man and woman is a rough-cut version of reality without much elegance. A curious thought would be to imagine if the plaque was not discovered for five hundred years or one hundred for that matter. The representatives who make the discovery may be from an advanced civilization themselves. How would the discoverers of the plaque(s) interpret such a find? Could and would they interpret it as an ancient artifact from a creator who was less evolved and lacked the sophistication of the modern day?

Written texts and alphabets were invented as a much younger form of communication among humans after symbolic artifacts and petroglyphs before that. As time went on, civilizations experienced more and different events in their lives. They required a better and more accurate description of the syntax of communication by language.

This evolution was working toward the eventual appearance of first the cuneiform texts of ancient Sumeria of six thousand years ago. Ancient Egyptian texts appeared as early as 4,600 years ago. The earliest Hebrew texts, which evolved as an offshoot of Egyptian (both commonly known as Afro-Asiatic languages), arrived around 3,200 years ago. Arabic languages emerged only a few hundred years after the earliest Hebrew texts.

More than three thousand years ago, most of the great ancient civilizations, including Greece and India, had established a written text. Curiously, one great ancient civilization, the Inca, had no written language per se. They had a writing system and a word communication system that did not use words or symbols but functioned by using a complex system of rope cords, knots, and colored ropes called Quipu.[2,3]

The Quipu originated about 4,600 years ago. The Quipu is an example of a "high wonderment" instead of "high strangeness." How could a civilization perpetuate such a language with all-inclusive functionality for words, numbers, and mathematics? Quipu communicated and archived every aspect of ancient Incan society!

The vast majority of ancient texts still available today depict, as mentioned, events in the lives of the earliest history documentarians. They recorded what they saw or experienced using terminology the best they could according to the date and their dictionary. We do the same thing today. There is nothing different in the procedures just described from then until the present.

Ancient written texts often morphed into archives of religious significance. Many of the events communicated were of special times or situations on those days that the events happened. Designated people often wrote texts within that particular society. Just as it is today, who wants to record an archive of an "everyday diary" unless it holds some very special significance?

Such unique texts were often elevated to the status of religious significance as the best way to communicate events of "high strangeness." These strange events were of situations and/or describing objects observed or objectivities that were real and that occurred. There was no terminology to describe a lot of these events. Just as man uses written language today, when he comes across a concept, event, or observation that he has not seen before, he must create a new term to describe that newly discovered concept, event, or object. Just as man is ignorant today to new events, observations, and ideas about how the natural world works, so ancient civilizations were just as ignorant in their day as to how the natural world worked in a lot of ways. These events were significant enough to drastically change the course of the evolution of the societies and the lives of the citizens of them in special ways. The combination of these factors indicated that the translation thus often took on a divine significance.

Hence, a large number of ancient texts describing these events and divine origins subsequently became attributes of these events. Today, we enact the very same process in our dealings with the same subject. When an observer or reporter of today documents and archives special events for posterity, she may not use divinity as an attribute as much as the ancients did because our knowledge of the natural world is greater than theirs was. When an extraterrestrial (provided they exist) encounter happens, we do not treat them as gods per se. But today, we are not as technologically behind them as humans were in those ancient times. So it works, sort of.

Ancient texts of the type we are interested in are not literature in an attempt to entertain the reader. They never were. This notion gets tragically interjected, weaved, and excused into the discourse by skeptics, debunkers, and naysayers. Literature or written texts did not even enter an earthly civilization's archives until the Greeks, Romans, Chinese, and East Indians created the foundations for the studies of philosophy only about 2,700 years ago. It was just then that any study, by the ancients themselves, of the natural world begot a genre of storytelling of any entertainment depiction. Indeed, no written texts were able to be mass-produced for use as entertainment pieces for its peoples back then. Ancient texts were annals of real events appropriate for serious and dedicated study in a scientific realm and not as literary flights of fancy. A lot of those written texts survive today thankfully for our analysis and knowledge.

Throughout our history, humankind has only been able to communicate through imagery our thoughts and ideas by way of analogy. This is evident in every written recorded text in all formats such as prose, poetry, and song. Communication by analogy is inherent in other forms of communication such as symbolic artifacts and art. The notion of comparison is basic to this notion of analogy. Also categorization and classification are complementary notions of analogy. This is partially a hardwired subroutine of how our brains function and have evolved from ancient times.

Representations of nature are thought of and communicated differently from person to person. It is this contextualization that makes us unique. Additionally, this communication discourse operates in all aspects of society. Dialogue like this has been utilized by man for all recorded history. As being a hardwired morphology of our brains, this has not changed over time. So communication of and learning of concepts, ideas, and events through media such as texts, artifacts, and oral lore remain a constant epistemology by biological default.

There are two aspects necessary to promote any kind of effective and truthful communication. The first aspect is a literal one. The second is an interpretative aspect. These two aspects are not mutually exclusive in any situation. They are present in any analysis of ancient texts, artifacts, structures, art, and so on. When we view a recorded archived depiction of an event from ancient times or observe a real-time encounter, we see a literal aspect and are then left to interpret this encounterin a contextual way.

In *Intelligent Life in the Universe*, author Guy Consolmagno, astronomer of the Vatican Observatory, Vatican State, describes how these notions have influenced the study of ancient texts in this manner:

> A videotape is not science, nor history. Only a human interpretation of the events constitutes human intelligent understanding of those events. And different historians can and will have different interpretations.

He continues,

> We must recognize that the Bible is divine science, a work about God. It does not intend to be physical science.[4]

There is a literal and an interpretative aspect to ideas and representations of nature that must be adhered to in order to correctly visualize the environment around us. It is the sum of both the literal and interpretative aspects that encompass the process of explanations of events in nature and science. It is this "nature of knowledge" that defines what we know. Consolmagno continues with this thought: "In understanding or interpreting what we know, be it in science or the Bible, in the final analysis, all language and all explanation, whether biblical or scientific, is by analogy." He then continues, "No equation (in science) is ever a perfect description, but rather an approximation of reality that is always simplified and incomplete. Science never gives us an absolute, complete knowledge of the truth."[5]

To better visualize this notion, what Consolmagno is referring to is a delineation of what constitutes the study of philosophy and the study of science. Philosophy is the study of the fundamental nature of knowledge, reality, and existence. Science, as is discussed in great detail elsewhere in this book, takes the epistemological tools of the literal and interpretative aspects, and via its organization of the scientific method, *applies* this knowledge and the study of the event to help explain the nature and reality of the event. Philosophy is the first step of sorts, and science is the logical progression as the next step. These two aspects are required for any analysis, including any study of a scientific nature.

This awareness, therefore, also applies to any study of ancient texts and also to old artifacts, art, and structures all as forms of the communication of events and thoughts. Both literal and interpretative elements are involved as tools to strive for achievement of a correct and enduring theory and law in nature.

We know the ancients recorded events based on what they saw, heard, and experienced descriptive of their best attempts with the language at hand. Ancient texts, in some ways, are manifestations of the wonderment of the event the observer experienced. The events themselves were real. The written descriptions utilized the writer's best knowledge and explanation to pass down to its future generations. Again, we do this today in the same manner so our descendants will read about our events and experiences in the same way.

In their book *Intelligent Life in the Universe*, authors Iosif S. Shklovskii and Carl Sagan recognize the importance of societies archiving their special events. Examples abound in history of societies recording these occurences even when the community did not have a written language, as noted of the Inca, or written recording mechanism, as noted of the Tlingit society of the northeastern coast of North America in the late eighteenth century. This did not impede successful documentation of first contacts with alleged extraterrestrial explorers.

It is easier for the native receivers of the visit to record such encounters if they had a written language. Those who did not, including civilizations precursor to even the Sumerians, resorted to drawings and other symbolic artifacts.

Shklovskii and Sagan used the example of the Tlingit and first contact with French explorer La Perouse in 1786 to drive their conclusion that such first contacts can be recorded:

> In a reconstructible manner, the encounter between La Perouse and the Tlingit suggest contact with an alien civilization will be recorded in a reconstructible manner and be greatly aided if (1) the account is committed to written record soon after the event; (2) a major change is effected in the contacted society; and (3) no attempt is made by the contacting civilization to disguise its exogenous nature.[6]

Two websites are available to aid in expanding our knowledge of the history and cataloging of ancient texts from all over Earth's civilizations. The first website is www.ancienttexts.org. The second site is www.sacredtexts.com. They both reproduce early texts from Greece, Rome, China, India, Mesopotamia, Egypt, and many others.

Every ancient civilization that had a recorded written history has conducted this archiving to our advantage. However, it is tragically unfortunate that, even with the presence of these websites and all other catalogs of ancient texts, multitudes of immeasurably valuable history have also been lost to future humankind (to our descendants and us). There are many causes of this state of affairs.

For example, the ancient Library of Alexandria was an agency division of the museum of Alexandria. This museum complex was built in the third century BCE. It was by far the largest repository of ancient knowledge from across the globe at that time. All the famous scholars of the ancient world studied at the library and museum. The great library suffered multiple fires over the next two hundred years. It is generally agreed by many nalysts of today that these series of fires were caused by various factions and sympathizers of the Roman Empire.

These fires caused the eternal destruction of much of the knowledge of the ancients. During the degradation of the main library, an adjunct library was built. It too was eventually destroyed in AD 391 by Coptic Pope Theophilus of Alexandria. Therefore, vital ancient knowledge was lost to our availability never to be recoverable.

A second great tragedy was the almost entire destruction of all the written knowledge archives the file the ancient Maya civilization had ever recorded. These were known as codices. Spanish conquistador Hernan Cortez de Monroy y Pizarro commanded a three-year conquest of the current nation of Mexico, the ancient homeland of the Maya civilization, in 1521–1523. The Aztec super culture that included the Mayan culture was ravaged by the Cortez military war effort. Among the high losses to historical oblivion were the vast majority of these codices, Cortez's armies burned the written historical archives of the Mayan culture. The loss of knowledge accumulated over two thousand years' worth of the Mayan contributions to Earth's and man's history.

A third example of the useless wanton destruction of man's heritage took place in South America. Around the same time in the 1500s, another Spanish conquistador, Francisco Pizarro, was permitted by his homeland to conquer the Inca nation in Peru. In 1532, a war was waged between Pizarro's army and the Incas; by 1572, the last of the Inca territories was annexed.

The Spanish Empire's doctrine of religious conversion of all conquered peoples to Roman Catholicism was anathema to the natives' cultural text archives. The result was the destruction of the Inca Quipu libraries of historical archives on the scale as large as that of the codices archives of the Mayans and Aztecs.

Generally, humans of military status hid innumerable treasures of written Earth and man's history. Immeasurable quantities of man's pedigree can, thus, never be recovered. Current and future researchers are left to piece together the remains with even fewer pieces than what is contained in a jigsaw puzzle.

We are left with a somewhat permanently incomplete historical record of ancient man and Earth. Answers to lots of documented significant events of the ancient Egyptian, Mediterranean, Aztec, Mayan, and Inca, among others are gone. A fundamental idea holds that incomplete records of the subject lead to more speculations introduced into analysis about that subject. More speculations lead to more occurrences of the logical fallacy known as jumping to conclusions in the research and analysis. Proponents and opponents of the extraterrestrial hypothesis both fall victim to this conclusion.

I will briefly summarize some of the human ancient texts that describe events where and extraterrestrial motivation or causation has attributes. An elaborate description of any of these events that caused the texts to be written is left to the future volume of this book series. You will learn of the events, and I will point you in the direction of access to such more elaborate descriptions as we continue. The point of the short introductions to these texts is that there are a lot of them existing in the various historical records of ancient cultures.

Sumeria

If we consider a chronology of ancient texts, we need to start with the oldest known archive, precisely, the documentation of ancient Sumeria. The first communities that have been attributed to the national designation of Samaria as its own nation appeared about five thousand years ago. For four thousand or more years before that, the Middle East region was considered a "center of humanity." The first evidence of the agricultural evolution from this time indicates evolution pointing away from the techniques of nomadic survival to that of a basic communal concept. Small villages eventually grew into small city-states. Because of the geographical proximity advantages these villages' residents (a new term never before presumed to have had meeting at that time) communicated with each other. The first stages of steady commodity bartering that simultaneously appeared then allowed for the evolution of the communnities.

These city-states were mostly autonomous.[7] Separate districts of commerce and temples were built in the centers of every region. The presence of the temple concept indicated another evolution toward a communal way of existence for humanity. Here, the city-states developed a profound devotion to symbols of their continued survival. These symbols were documented in early pictograms of animals: lions, bulls, and horses. It was the belief that this concept required to create new terminology—religion—to the human language.

Before five thousand years ago, the common threads in animals as religious icons changed drastically. For any measure of absolute certainty, no one has been able to answer this dilemma, not in science, literature, or folklore.

An early text's account of events was written in Sumeria about five thousand years ago. The Sumerian story of Enuma Elis was committed to seven cuneiform clay tablets containing over one hundred lines on each tablet. They were recovered intact for the most part by Austen Henry Layard of Ashurbomipal. The original archaeological site is Kouyunjik, part of ancient Nineveh or modern day Iraq. George Smith, a British archaeologist, translated the cuneiform tablets and published some of them in an enduring documentary titled "The Chaldaean Account of Genesis." This was part of the longer written work that has been transcribed and known as *The Epic of Gilgamesh*.

This Sumerian work details the events of the Creation of Man, the Fall of Man, and the Great Flood, the Tower of Babel, in addition to the patriarchs and fables and legends of the gods. It is evident from this collection that a concerted cultural and symbolic transformation away from the worship of animals as deities and to a more human fixation.

The Sumerian version of this cultural evolution identified the arrival and colonization of the Annunaki to Mesopotamia. The Annunaki were supposedly an alien race who were seeking to obtain gold and minerals for rescuing their home civilization from destruction. Humans were employed as slaves to mine the desired treasures for the Annunaki, and the Annunaki overseers maintained the city-states of that time.

Many written works followed, including the Christian Bible and Hebrew Torah, which are widely believed to have origins from, at the very least, an approximation to the Enuma Elis.[8] An excellent source of information on this topic can be found in the book *The Chaldean Account of Genesis*. This manuscript was originally written in 1876 by George Smith.

While the Ubaid and later Sumerian cultures were the first civilizations documented to utilize the convention of communal living, nomadic groups traversed all the continents prior to seven thousand years ago. This means that India, China, Japan, Australia, Europe, Africa, North and South America, and Polynesia were inhabited. All the races were hunter-gatherers who formed their own version of communal living. These times of evolution varied by hundreds or thousands of years.

India

Scientific study is continuously enriching India's civilization and written records of its creation and history of events relating to high strangeness. This is true of all ancient human civilizations as archaeology and the sciences uncover more pieces of the largest jigsaw puzzle that we know.[9] An excellent detailed source of reference on this topic is at the website, www.ancient.eu/timeline/India.

India habitation dates back over ten thousand years when Hindu nomads and traders navigated forth and back to the Middle East and the Far East. Far East expeditions were made mainly to China and Japan. The Indus Valley was regional to the origin of the Tantra Shastra texts. To this day, written accounts of events of this time to eight thousand years BCE are poorly translated and understood. What is more important is a lot of proper names of such entities as gods, i.e. Shiva and Kali also appear in the Hebrew Torah texts. Some of the earliest Torah texts extensively describe the migrations of Abraham and his descendants by trade passage roots that reached the Indus Valley. Many occurrences of similarly described relationships of the events involving the gods, symbolism, descriptions of events, beliefs, studies, exercises, etc., are cross-related in both bodies of written texts. Dated archaeological finds to seven thousand years ago our evidence of the origins of widespread religious activities. Evidence also of contact and communication with Sumeria dates back to around four thousand years ago. The Indus Valley was referred to as Meluhha to the Sumerians.[10]

By 4,500 years ago, any remaining usage of Tantra was left behind to history or as yet been discovered. But the common threads of multicultural assimilation of and creation of new languages continued in the Indus Valley region. The first of the earliest identified Vedic text languages were the Vedic, proto-Indio-Iranian, and proto-Indio-Aryan with definite ancient Arabic undertones. This was very soon followed by Vedic Sanskrit also which very soon transformed into just Sanskrit.

Vedic text record currently holds India's earliest records of their historical events, including reference to some which may have been influenced by extraterrestrial visits. There are four Veda text volumes: the Rig Veda, Sana Veda, Yajur Veda, and Atharva Veda. The four volumes, via a hymnal written format, describe the creation of the universe (the big bang), our solar system, and so on. Also depicted and noted is that the gods of that period were now picto-portrayed or humanely and no longer as animals. Animals place in the deity paradigm are scaled back to more of a quasi-divine nature only; still vital but not genuinely Godlike.

Descended next are the Puranas and more liturgical Hindu epic texts from the Ramayanas and the Mahabharata. It was from these bodies of translations that we learn of great battles fought by extraterrestrials using flying machines called "vimanas" and weapons akin with atomic power weapons with similar destructive forces.

Here are examples of text passages from the Ramayana: "An aerial chariot, the Pushpaka, conveys many people to the capital of Ayodhya. The sky is full of stupendous flying machines, dark as night but picked out by lights with a yellowish glare;"[11] and, "Now Vata's chariot's greatness! Breaking goes it and thunderous is its noise. To heaven it touches makes light lurid (a red fiery glare) and whirls dust upon the earth," as referenced from the Rig-Veda. Vata is the Aryan god of the wind.

From the works of the Mahabharata, "The cruel Salva had come mounted on the Sanbha chariot that can go anywhere, and from it, he killed many Valeant Vrishni youths and evilly devastated all the city parks."[12]

Egypt

Ancient Egypt had, as its version of the Genesis or beginning, writings and glyphs about Zep Tepi, meaning "first time." In what here is also recorded of pre-ancient Egypt's knowledge of "waters of the abyss," the kings of this time were known as the Neteru, meaning "neters or Gods."

The waters of the abyss, often referenced in the Egyptian records, come from times before the rise of Egypt as a powerful civilization. This era would not commence until after the Great Flood documented in the Christian and Hebrew scriptures around the time of Noah and the Ark.

This timeline is conjectured to position pre-ancient Egypt much further back in time than the Great Flood. The first of the Egyptian pyramids are presumed to have been built a few hundred years later. The waters of the abyss gave rise to the recorded name of Nun (or Nu) and was referred to in the earliest known Egyptian writings as the "father of the god." This god was later known as Ra, the sun god. Logically, this makes sense as to any native of that time the emergence of a weather/climate pattern that included sunlit days after flooding rains of whatever proportion would allow for the formation of such a thinking process.

Nun, the father of the gods, was named for the watery abyss that surrounded a "globe of life" that would later emerge. The parallels to Judeo-Christian and Indian scriptures are just one of the many examples of different societies documenting their events in their own fashion and format. Nun was the only civilization that existed on Earth before land reappeared after the Great Flood.

Ancient Egyptian texts, known as the Pyramid Texts, are widely considered to be among the oldest known existing writings of earthly antiquity. Written into the walls and sarcophagi of the pyramids at Saqqara, the religious scriptures are estimated to date back to around 2400 BCE. They describe the death and resurrection of Osiris. Many ancient astronaut theorists also attribute this event to that of extra-terrestrial influence. Osiris was murdered by Set, his brother, who usurped Osiris's throne. Isis, Osiris's wife, brought him back to life and later bore another son to Osiris.

Of note are a couple of facts regarding ancient Egypt and documented texts and cultural and cosmic influences. One involves the recurring thematic of the astronomical and societal references to the star constellation Orion in the northern night skies. It is speculated that Osiris was born on a mother ship in a region around the Orion nebula.

The ancient Egyptian texts, including the Pyramid Texts, are the oldest known written documents of scriptures known today. Additionally, ancient Egypt civilization is the standing longest unbroken or unconquered human civilization in our history. In their time, they were the most knowledgeable society in a lot of subjects, including astronomy, mathematics, agriculture, law, and engineering. A second fact implies that any of the ancient texts offer abundant reference to the Orion star constellation. Indeed, the great pyramids, as well noted as other pyramid structures around the world, depict a precise mathematical model and replica of the schematic of the Orion belt. One can see other parallel thematic analyses in ancient Maya, Southeast Asia, and the Middle East.

An ancient text, known as the Tulli Papyrus, depicted a 1480 BCE encounter of two mass sightings of unidentified flying craft by Pharaoh Thutmosis III and his court from the palace city that bore his namesake. The first of the two events were chronicled as depicted thusly:

> In the year 22, third month of winter, sixth hour of the day, the scribes of the House of Life found that it was a circle of fire that was coming in the sky. It had no head, the breadth of its mouth had a foul odour. Its body (was) one rod long (approx. 150 feet) and one rod large, it had no voice. A few days later another mass sighting report came from the Palace of Pharaoh

Thutmosis III: "Now after some days had passed over these things, Lo! They were more numerous than anything. They were shining in the sky more than the sun to the limits of the four supports of heaven. Powerful was the position of the fiery disks. Circles of fire hovered over the palace while fishes, winged creatures, and other objects rained down from the sky."[13]

Ancient China

Written documentation of activities of Earth in ancient recorded times includes a survey of civilizations all around the globe that seem to parallel their achievements and activities. To say that any one ancient culture was the leader in being the most advanced earthly society is not exactly fair to the others. We know that images of many great earthly events were also experienced and recorded by all those ancient societies.

The Mayans were expert astronomers, but so were the Egyptians. The Chinese were experts at medicine, but the others were also in their own ways. We know from the Silk Road in Asia to the spread of the influences of the extended family of Abraham in the Old Testaments that the peoples of ancient Earth were more capable of such contact than what they were given credit for even a couple of generations ago. Similarly, there is even a growing speculation of feasibility of crossing of the Pacific Ocean to and from ancient North and South American nations by Polynesian explorers that attests to this.

The earliest known connections made to ancient China and the feasibility of extraterrestrial influence come around 2700 BCE. The yellow emperor, Huang-Di, lived from 2697–2598 BCE. He is considered the ancestor of all Chinese. While it is unanimously agreed to that he did live, whether he was partly mythical has been debated among most Chinese scholars. It is the reconciliation of this dimension that then allows for an extraterrestrial framework.

Huang-Di's birth was marked by a "radiance from the great star Chi in the constellation of the Big Dipper (aka Ursa Major)."

Huang-Di was attributed the credit for the ultimate unification of China. He brought advanced medical technology and knowledge to their society. He also introduced China to the manufacturing process of silk and taught students on the cultural influences that eventually became the study of Taoism. His reign on earthly China lasted over one hundred years. He was acknowledged to have been alive both before and after his time on Earth. His return to the skies was written to have been facilitated by a metallic dragon, which descended from the sky taking Huang-Di away.

In ancient China, as well as in other cultures, their descriptions of concepts and instrumentalities had to fit their personal vocabulary as noted elsewhere in *The Humaniverse Guide to Better Reasoning & Decision-Making*. The Chinese depicted dragons as immense and powerful machines and/or entities as living beings and creatures. A large number of them could project fire from their mouths or bellies. It was believed some of Huang-Di's teachings introduced the parallel of the concept of the dragon, applying many events or experiences to the members of the ancient China nation. These are all conjectured, in the literature, to have meaning and attribution to extraterrestrials.

In the Sanxingdui Museum located in Guanghan Province, Southwest China resides statues and artifacts recovered from various archaeological expeditions. These qualify as artifacts of high strangeness. Among them, a nine-foot-tall statue representing a king of the Shu Kingdom, more popularly known as Sichuan in those ancient times. Among the more than ten thousand artifacts are finds of gold, jade, bronze, marble, pottery figurines and statues of animals, fish, and humans. Current dating research has placed the age of the Sichuan collection to over five thousand years old. The collection has stimulated speculation of unanswered questions about the creator's advanced technologies and engineering of such treasures.

The Christian Bible

Perhaps, the single most significant iconic record of ancient text known to the Western World is the Christian Bible. The Bible has two testament volumes: an Old and a New Testament. The Old

Testament was assembled by ancient Hebrew prophets and affirms a permanent connection between ancient Judaism and the later Christianity following.

As noted, there are many correlations of the Bible with the Torah, the Koran, the Indian Tantra Shastra texts, records of ancient Egypt, and of many other cultures and continents of documentation of comparative mythology. For example, the Great Flood depicted in chapters 6 and 7 of Genesis was also documented in the Hindu Puranas (Vimanas), the Mesopotamian Flood cuneiform tablets, of which the Epic of Gilgamesh is part of, the ancient Greek Deucalion mythology, the K'iche and Maya texts of Popol Vuh in Mesoamerica. Also included are the Muisca, Canari, and Inca Confederations in South America and the Lac Courte Oreilles Ojibwa tribes of ancient North America. Chinese and Russian archaic texts document the same events.

It should be noted that these comparative mythological accounts are culturally contextualized so physical descriptions of an artifact or prop in related stories may contain this contextualization. All the texts describe the same events using available terminology within their individual languages.

This is an on-point example of the usage of comparative theme reporting among the different societies of those times to provide a posterity for their descendants. Each of them describes similar occurrences of great actual events and, in turn, were verified by a widely geographically dispersed body of witnesses.

The Christian Bible, as is also disclosed in the Hebrew Torah, contains an abundance of examples of accounts whereas people of the Middle East experienced situations with entities of a middle level of existence between those of man and of God (Yahweh, as defined from ancient Hebrew texts). The books of Genesis, Exodus, Ezekiel, Kings, Isaiah, Psalms, Job, Jeremiah, Zechariah, Jude, and the book of Enoch (only a small portion of which is translated in certain versions of the Bible) are just some of the original portrayals of encounters with gods and their missions and instrumentalities.

For example, in the book of Kings II 2:2, 11–12: "When the Lord was about to take Elijah up to heaven in a whirlwind. And it came to pass, as they still went on, and talked, that behold, there appeared a chariot of fire, and parted them (Elijah and Elisha) both asunder; and Elijah went up by a whirlwind into Heaven. Elisha saw this and cried out, 'My father! My father! The chariots and horsemen of Israel!'" Or consider Psalms 18:10: "And he rode upon a cherub (generally associated with the wheel-discs encountered by Ezekiel) and did fly." He did fly upon the wings of the wind. Or in Isaiah 60:8: "Who are these that fly as a cloud?"

A most significant depiction of something anomalous was the scripture from the book of Ezekiel 1:1–28: "Now it came to pass in the thirtieth year. The heavens were opened and I saw visions of God. Behold, a stormy wind came out of the north, a great cloud, with brightness around it, and fire flashing forth continuously, and in the midst of the fire, as it were gleaming metal. And out of the midst thereof came the likeness of four living creatures. They had a human likeness. Their legs were straight, and the soles of their feet were like the soles of a calf's foot. And they sparkled like burnished bronze. Now I beheld the living creatures, behold, one wheel upon the earth. The appearance of the wheels and their work was like unto a beryl; and they four had one likeness; and their work was as it were a wheel within a wheel. And I saw as it were glowing metal, as the appearance of fire within it round about, from the appearance of his loins and upwards."

The book of Genesis was seminal in documenting encounters of monumental importance in ancient days: the destruction of the cities of Sodom and Gomorrah in chapter 19 or in Exodus chapter 18 when the manna from heaven was ministered to the Israeli refugees in the desert to sustain them until they reached their promised land. The book of Exodus chapters 33 and 34 tell of Moses's encounters with God and the eventual creation of the tablets which contained the Ten Commandments. Or in Mark chapter 13:26: "And then shall they see the Son of Man coming in the clouds, with great power and glory."[14]

The New Testament continued this thematic throughout. The book of John often referenced other races in his writings. In John 1:1: "Who was present in the beginning"; John 10:18: "Coming to lay down His life and take it up again not only as the Son of Man but also as a Child of other races?" Further, in John 10:14–16: "I am the Good Shepherd. I know My own and My own know me, just as the Father knows Me and I know the Father. And I lay down My life for the sheep. I have other sheep that do not belong to this fold. I must bring them also, and they will listen to My voice. So there will be one flock, one shepherd."

Ancient Hebrew

When reading Scripture, whether Hebrew, Christian, or that of other texts, we could question whether some words which were used to describe objects or concepts in that time seem to hold the same meanings of today's language. In every historical time period, the natives had their own vocabulary of words to describe concepts or objects. The point here is that we today cannot inject our current vocabulary into interpretation of ancient texts without fatally altering their true native meaning.

The discourse here focuses on an area of linguistics that relates to translation bias. Indeed, if you traveled in a time machine back to an ancient civilization and were given the responsibility of writing scripture of that day, you could only use vernacular of that society of that time and day. If you tried to use the language you knew existed in the future time from where you just came from, no one would then be able to understand your alternative vernacular. Creating words is fine for a culture that would be dedicated to use those words regularly. But ancient writers could not anticipate or invent appropriate language to use in their writings that societies hundreds or thousands of years later would understand at least without much difficulty. Therefore, those ancient text writers would be confined to using linguistics, vernacular, and syntax of their day.

One of the most far-reaching and lucrative aspects of studying the Hebrew Torah, a substantial body of which encompasses the writings of the Christian Old Testament books of their Holy Bible, is the liberal propensity and opportunity to insert many interpretations from the many different versions of either of those great works.

Thousands of years have passed since the original authors recorded the events of their societies. The literature and societal discourse today are rife with different interpretations of the occurences that shaped ancient civilizations across the globe.

Those interpretations have made use, either by design or circumstance, of linguistic construct, verbiage, and syntax that is more conducive to today's society. Thus, the reader is now introduced with a version of translation bias that on one hand may help him acknowledge the events about which he was reading. But on the other hand, as summed exquisitely in this adage, describing events of the ancients, "something is lost in the translation."

The Hebrew Torah and the Christian Bible provide a nearly identical description, when these levels of linguistic bias are exposed, of society and events of the ancient Middle East in the millennium before the alleged arrival of Jesus Christ. The Hebrews do not acknowledge Jesus's life; the Christians do. So the lives of Abraham, Moses, Ezekiel, and the other great people of that time are documented similarly on the most basic levels of reality and substance.

When it comes to understanding how and why the ancient writers wrote what they did, there are significant cultural, societal, historical, and life experience biases that beset all them in the same quantity as any writer of today. What makes it more difficult to determine what the ancient texts, including all them of all the antiquated civilizations of Earth, really meant is based more on additional interpretations given by descendant writers up through today. There have been over four thousand years to enact this discourse.

Remember that the ancient original writers—even though they had their own cultural, societal, historical, and life experience biases—were writing from a firsthand or at least secondhand perspective. That could never be said about such translations and reinterpretations written over the thousands of years later.

When it comes to Torah or Old Testament Scripture, even though they had to use vernacular of their day, they were documenting events that happened to their society during their time. When the dictionary of ancient Israel, for instance, was used to produce Scripture, they did not have the vernacular we may enjoy millennia later. Their descriptions were from firsthand and secondhand accounts using the tools available to them.

Remember also that in the millennia that followed the time of the original writings the Scripture's translators had perspectives about the universe which were different from those of today. For example, at that time, the translators thought the earth was at the center of the universe. This bias, along with others, are examples of many more translator biases that influenced their written works. There also existed many cultural biases which influenced translator's writings. Another example of a source for translator bias is explained by exploring the writing device known as a concordance.

A concordance is an alphabetical index of terms used in a text. With all the biblical texts and translations written, there exists many different concordances attributing to the same biblical topics. There are different terms found in these concordances. The translators have introduced translational bias into these writings using their own interpretations of the original scripture. While this is necessary and common for arguing perspectives, the bias has often distorted these original scriptures. When the temporal factor is introduced, where many translators from different periods in history are reinterpreting the same biblical texts and scripture, the bias eventually destroys the original meaning and intent of the source material.

Let's go through a few examples of word translation and contextual accuracy as they relate to any biases, misusage, and misinterpretation and as they exist in the Hebrew Torah and Christian Bible. These examples are all with respect to our main theme of UFO and extraterrestrial existence.

The first term to be discussed is *heaven*. In the Hebrew Old Testament, *shamayim* is the term used to represent heaven. When we hear the term *heaven* today, we think of the afterlife. The term *shamayim* was used by the ancient Hebrews not to describe heaven but as is the only correct English translation means the "sky, outer space, cosmos, or universe." Somehow, the ancient translators moved away from such valid translations for whatever reason(s).

When the word *shamayim* is inserted correctly into verse, even the syntax and meaning of the verse becomes more logical and apparent. For instance, in Genesis 1:8: "And God called the expanse shamayim." The translators substituted heaven into their versions. *Shamayim* means the "sky, outer space, or a description of the cosmos; it is not heaven." Even today, astronomers use a colloquial pluralization of the heavens to describe the cosmos.

In Isaiah 13:10, "For the stars of shamayim and their constellations," correctly describes certain stars of outer space. The term *heaven* is incorrectly used in place of shamayim in many translations. In Genesis 15:5, "And he took him outside and said, 'Now look toward the shamayim and count the stars, if you are able to count them.'" The term *heaven* is often transposed into many translations. Again, this is incorrect and affects reliability. Most of the translation errors in this example appear in the biblical manuscripts.

Separately, here are some examples of the correct usage of shamayim and its translations to accurate synonym usage. In Psalms 11:4, "The Lord is in his Holy Temple; the Lord's throne is in Outer Space." In Psalms 89:5–6, "And the universe will praise thy wonders, O Lord; Thy faithfulness also in the assembly of the holy ones. For who in outer space is comparable to the Lord?"

From the second book of Kings, it was said, "The spinning chariots of fire that took the prophet Elijah off into *shamayim*" (2 Kings: 2:11). The term *heaven* was used and its improper translation changes the whole image of what was meant in both a literal syntax and substantive context. In Job 22:12, "Is not God in the height of outer space? Look at the distant stars, how high they are!" If the term *heaven* was used, one can see how the usage does not fit into the syntax, substantiveness, or validity of what was being communicated in the verse.

A second example of translation bias concerns the use of the word *God*. Various biblical dictionaries, including *The Hebrew and Chaldee Dictionary* and *Strong's Exhaustive Concordance of the Bible*, continue the doctrine of biblical scholarship by overusing the term to apply it as a universal description of any God written about in the Bible or Torah. The translation bias has blurred proper usage to the point where one cannot differentiate between God, as Yahweh, defined in Exodus by Moses at Mount Sinai and any other god.

The term *Yahweh*, used in the Hebrew Bible and as described by Moses in the book of Exodus, has itself been mistranslated over the millennia to where now the term *Jehovah* is being used by certain religious sects. Other misuses of the term *God* or *Yahweh* include *Lord God*, *Sovereign Lord*, or simply *Lord*. If we compare a collection of biblical scholarship dictionaries, there would be significant divergence and incongruity among usage of the same terms and concepts. This is what makes accurate interpretations more difficult.

The original Hebrew writers of the ancient texts had a name for gods other than Yahweh. It was Elohim. *Elohim* is the general term assigned to the named gods, all of a lesser presence than Yahweh, to unnamed gods in both Hebrew and Christian Scriptures, pagan gods, and carved idols that were worshipped then. Remember that Hebrew and Christian doctrine introduced a permanent notion of a "one God" universe only within the last epoch before the Christian coming of Christ. Up to this era, there were many worshipped gods written in the texts among all the civilizations of Earth. Thus, a successful evolution toward one-god worship was an arduous and painstaking task for humankind in those societies. The various nations of Earth adopted this notion in general around the same epoch.

Thus, Yahweh or sometimes seen as Lord Yahweh is the accurate name for the Supreme Almighty God. The Elohim is the categorization given to all the lesser gods of the time and also contains use of the terms *God* and the *mighty ones*. So in Exodus 20:2–3: "I am Yahweh of the mighty ones who brought you out of Egypt, out of the house of slavery. You shall (place) no other mighty ones before me."

When translation bias is removed from the evolution of terms such as *heaven, shamayim, God, Yahweh*, and *Elohim*, credit is restored to the original writers of the Hebrew and Christian Scriptures. Similarly, full substantiveness is restored for a more correct meaning of the events they were trying to communicate.

In the book of Deuteronomy 10:17: "For Yahweh of the mighty ones (read Elohim) is the mightiest of the mighty ones and the master of masters; the great, the powerful, the awesome mighty ones." In Psalms 97:9: "For you are Yahweh, highest over all the earth; you are exalted far above all the mighty ones (Elohim)." And in Genesis

1:1: "In the beginning the mighty ones (originally read, God) created the cosmos and the earth." Later in Psalms 115:3: "But our mighty one (Elohim) is in shamayim." And further on in the book of Psalms 103:19–22: "Yahweh has set his throne in shamayim and has soverign rule over the universe. Bless Yahweh all you his mighty ones (Elohim), mighty in strength, acting at his behest, obedient to his command. Bless Yahweh all you his hosts, his ministers who carry out his will. Bless Yahweh, all his creatures, everywhere in his domain. Bless Yahweh, my soul!"

What must be accomplished and this would be no easy task is to recognize that the original writers of the biblical texts had a specific concordance index at their disposal. When they wrote their scriptures, they knew nothing about our concordances of today. If this was recognized, then a reversion back to what concordance was, theirs would enable a more real and accurate version of the scriptures to be finally published. Hence, the confusion that is translation bias would be eliminated.[15]

Ancient Rome

Ancient texts from the Roman perspective move away from a Scripture-based framework of other civilizations but also provide no shortage of data. Around 241 BCE, when Rome defeated Carthage in the First Punic War, the platform was built for Rome to become a superpower.

As noted previously, written scholarship of ancient Roman history has more contextual influence from philosophy than from religious thought. Therefore, there were more military and secular influences and overtones and laic undertones in the writings with fewer mythological say, for example, the Indian Veda and Rama Bharata texts.

Dedicated scholars in ancient Rome at that time included Cassius Dio, Livy, Julius Obsequens, and Pliny. Lucius Cassius Dio (AD 155–235) was a Roman consul and historian who wrote eighty volumes about Roman history, curiously in Greek penmanship only until we learn that he was of Greek origin. Julius Obsequens was a Roman writer who is believed to have lived in the middle of the fourth century AD. The only work associated with his name is the *Prodigiorum Libellus* (Book of Prodigies), extracted from an epitome or abridgment written by Livy. *De prodigiis* was constructed as an account of the wonders and portents that occurred in Rome between 249 BC–12 BC.

Titus Livius (64 BCE [approx]–AD 17), commonly known as Livy, wrote far more than Obsequens and about as much as Dio about Roman history. His central work *Ab Urbe Condita Libri* spanned the recognized Genesis of Roman history approximately 753 BCE until his own lifetime during the reign of Emperor Augustus. Livy was the truest of philosophers of the four authors mentioned. Gaius Plinias Secundus (AD 23–AD 79), commonly known as Pliny the Elder, was also a philosopher more than a historian but maintained a life-long friendship with Emperor Vespasian beginning in his years in the military service in the Roman Navy and Army.

The most beneficial format with which to portray ancient Roman text episodes of extraterrestrial influence is to list them chronologically. Here is a compilation of those passages:

223 BCE *from Cassius Dio; Roman History Book I*: "At Ariminium a bright light like the day blazed out at night. In many portions of Italy three moons became visible in the nighttime."

222 BCE *from Pliny, Natural History, Book II, Chapter XXXII*: "Also three moons have appeared at once, for instance, in the consul-ship of Gnaeus Domitius and Gaius Fannius."

218 BCE *from Livy, History, Book XXII, Chapter I*: "Glowing lamps were seen in the sky at Praeneste, a shield was observed at Apri and in the Amiterno district, the sky was all on fire, and men in white garments appear." Additionally, *from Livy, Book XXI, Chapter LXII*: "A spectacle of ships gleamed in the sky."

217 BCE *also from Livy, History, Book XXII, Chapter I*: "At Faleri the sky had seemed to be rent as it were with a great fissure and through the opening a bright light had shone." Additionally, *from Livy, Book XXII, Chapter IX* and *from Orosius, Chapter IV*: "At Apri shields (parmas) were seen in the sky." A *parma* is a shield made of hardened metals, usually iron, bronze, or others.

216 BCE *from Julius Obsequens, Prodigiorum Libellus, Chapter LXVI*: "Things like ships were seen in the sky over Italy. At Arpi (180 Roman miles, east of Rome, in Apulia) a round shield was seen in the sky. At Capua, the sky was all on fire, and one saw figures like ships."

214 BCE *from Julius Obsequens, Prodigiorum Libellus, Chapter LXVI*: "At Hadria an altar was seen in the sky and about it the forms of men in white clothes."

212 BCE *from Livy, History Book XXV, Chapter VII*: "At Reate a huge stone (saxum) was seen flying about."

It should be noted that the 218, 217, 216, 214, and 212 BCE encounters were witnessed by Roman militia during the Second Punic War, which lasted from 218 until 201 BCE.

173 BCE *from Livy, History Book XXV, Chapter VII*: "At Lanuvium a spectacle of a great fleet was said to have been seen in the sky."

163 BCE *from Obsequens, Prodigiorum Libellus, Chapter CXIV*: "In the courtship of Tiberius Gracchus and Manius Juventus at Capua the sun was seen by night. At Formice two suns were seen by day. The sky was afire. In Cephalienia a trumpet seemed to sound from the sky. There was a rain of earth. A windstorm demolished houses and laid crops flat in the field. By night an apparent sun shone at Pisaurum."

122 BCE *from Obsequens, Prodigiorum Libellus, Chapter CXIV*: "In Gaul three suns and three moons were seen."

99 BCE *from Obsequens, Prodigiorum Libellus, Chapter CXIV*: "When C. Murius and L. Valerius were consuls, in Tarquinia, there fell in different places, a thing like a flaming torch, and it came suddenly from the sky. Toward sunset, a round object like a globe or round or circular shield took its path in the sky, from west to east."

90 BCE *from Obsequens, Prodigiorum Libellus, Chapter CXIV*: "In the territory of Spoletium (sixty-five Roman miles north of Rome, in Umbria) a globe of fire, of golden colour, appeared burning in the north with a terrific noise in the sky, then fell to the earth, gyrating. It then seemed to increase in size, rose from the earth, and ascended into the sky, where it obscured the disc of the sun, with its brilliance. It revolved toward the eastern quadrant of the sky."

85 BCE *from Pliny, Natural History, Book II, Chapter XXXIV:* "In the consulship of Lucius Valerius and Caius Marius a burning shield scattering sparks ran across the sky."

81 BCE *from Julius Obsequens, Prodigiorum Libellus, Chapter CXIV:* "Near Spoletium a gold-colored fireball rolled down to the ground, increased in size; seemed to move off the ground toward the east and was big enough to blot out the sun."

73 BCE *from Julius Obsequens, Prodigiorum Libellus, Chapter CXIV:* "While Roman legions were engaged in battle near the Black Sea against King Mithridates a huge flaming object fell down between the two armies. It was said to have a shape like a wine jar and was the color of molten lead. With no apparent change of weather, but all on a sudden, the sky burst asunder, and a huge, flame-like body was seen to fall between the two armies. In shape, it was most like a wine jar (pithoi), and in color, like molten silver. Both sides were astonished at the sight and separated. This marvel, as they say, occurred in Phrygia, at a place called Otryae."

It should be noted that this encounter could qualify as a unique mass-sighting incident equivalent to some of this century's most prolific military encounters, such as Operation Mainbrace, and much more prolific than even The Rendlesham-Bentwaters Incident and the Shag Harbour, Nova Scotia, Canada Incident.

66 BCE *from Pliny, Natural History, Book II, Chapter XXXV:* "In the consulship of Gnaeus Octavius and Gaius Suetonius a spark was seen to fall from a star and increase in size as it approached the earth. After becoming as large as the moon, it diffused a sort of cloudy daylight and then returning to the sky changed into a torch. This is the only record of its occurrence. It was seen by the proconsul Silenus and his suite."

48 BCE *from Cassius Dio, Roman History, Book IV:* "Thunderbolts had fallen upon Pompey's camp. A fire had appeared in the air over Caesar's camp and had fallen upon Pompey's. In Syria, two young men announced the result of the battle in Thessaly and then vanished."

42 BCE *from Julius Obsequens, Prodigiorum Libellus, Chapter CXIV:* "Something like a sort of weapon, or missile, rose with a great noise from the earth and soared into the sky."

12 BCE *from Julius Obsequens, Prodigiorum Libellus, Chapter
CXIV:* "A comet-like object hovered days over Rome for several then
melted into flashes resembling torches."[16,17,18]

In modern UFO reports, it is common to hear about either
glassy fibers or a chalky substance left behind by the UFO, known as
angel hair. Ancient statements also include angel hair.

In AD 196, the historian Cassius Dio wrote: "A fine rain resembling silver descended from a clear sky upon the Forum of Augustus.
I did not, it is true, see it as it was falling, but noticed it after it had
fallen, and by means of it, I plated some bronze coins with silver.
They retained the same appearance for three days, but by the fourth
day, all the substance rubbed on them had disappeared."

Two other "rains of chalk" were reported in Cales 214 BC and
in Rome 98 BC.

Pope Pius I's brother was probably the only witness of this UFO
sighting near Via Campana, Italy, around AD 150, which read, "On
a sunny day, a 'beast' like a piece of pottery (ceramos) about one
hundred feet in size, multicolored on top and shooting out fiery rays,
landed in a dust cloud accompanied by a 'maiden' clad in white."

Stothers concluded that, "This collection of what might be
termed ancient UFO reports has been pulled from a much larger
number of reports of aerial objects, most of whose identifications
with known phenomena are either certain or at least probable.
Embedded in the mass of relatively explicable ancient reports, however, is a small set of unexplained (or at least not wholly explained)
reports from presumably credible witnesses."

"Any viable theory must reckon with the extraordinary persistence and consistency of the phenomena discussed here over many
centuries."

References

1   Billig, Otto. 1982. *Flying Saucers: Magic in the Skies.* Copyright 1982 by
    Schenkman Publishing Company Inc.
2   Quipu. 1996. Quipu – Ancient Scripts.com/quipu.html/. Copyright 1996–
    2012 by Lawrence Lo.

3   Knotty Problems: The Ancient Writing System of the Inca. www.archaeologist.about.com.

4   Consolmagno, Guy. 2005. *Intelligent Life in the Universe*. Catholic belief and the search for extraterrestrial intelligent life. Pgs. 26–27. Copyrighted and published by The Incorporated Catholic Truth Society.

5   Consolmagno, Guy. 2005. *Intelligent Life in the Universe*. Catholic belief and the search for extraterrestrial intelligent life. Pg. 24. Copyrighted and published by The Incorporated Catholic Truth Society.

6   Shklovskii, I.S. and Sagan, Carl. 1966. *Intelligent Life in the Universe*. Pg. 453. Copyright 1966 by Holden-Day, Inc.

7   The History Guide. Lecture 2: "Ancient Western Asia and the Civilization of Mesopotamia." Copyright 2000 by Steven Kreis. Last revised July 19, 2014.

8   Smith, George. 1876. *The Chaldaean Account of Genesis*. Published by Londons. Low, Marston, Searle and Rivington. Original copyright 1876 by George Smith. Currently not in copyright.

9   www.Ancient.eu/timeline/India. Some rights reserved. (2009–2015). Ancient History Encyclopedia limited UK.

10   www.Ancient.eu/timeline/India. Some rights reserved. (2009–2015). Ancient History Encyclopedia limited UK.

11   Mahavira of Bhavabhuti (a Jain text of the eighth century sourced from older texts).

12   Alien Races and Descriptions 3. www.Bibliotecapleyades.net/vida_alien/alien_race05.htm/.

13   Ref Egypt 13: Tully Papyrus, as translated by Prince Boris de Rachewiltz and R. Cedric Leonard. Retrieved from: http://www.ancient-code.com/arrival-sky-gods-ancient-egyptian-papyrus-details-mass-ufo-sighting/.

14   Jessup. Morris K. 1956. UFO and the Bible, First Edition. Copyright 1956 by The Citadel Press.

15   Ref Hebrew 15: Graham, W.L. 2000. On the Wings of Angels: The Extraterrestrial Theology of the Hebrew Record. Copyright 2000 (revised 2009) by W.L. Graham. Retrieved from: http://www.biblerealitycheck.com/WingsTreatise.htm.

16   Rome Ref 16: Unknown author. Earliest Known Records of UFO and Alien Sightings. Educating Humanity. Retrieved June 4, 2016, from: http://www.educatinghumanity.com/2011/08/ufo-and-aliens-earliest-records-of-ufos.html.

17   Rome Ref 17: Retrieved June 4, 2016, from Frances Fontaine netseeker2@sprint.ca. January 26, 2001.

18   Rome Ref 18: Excerpted from *Reader's Digest* "Mysteries of the Unexplained" pages 207–209. Retrieved June 4, 2016, from: http://rense.com/general7/ages.htm.

CHAPTER 11

Ancient Structures of High Strangeness Part 1

It is mere rubbish thinking at present of the origin of life;
one might as well think of the origin of matter.

—Charles Darwin, 1859[1]

Key words: megalith, monolith, obelisk, oopart, ziggurat, holistic, cartography, cargo cults, henge, pastoralism, trilithon, sacbeob, cenote, stele

Any subject investigation that requires a study of characteristic parameters along some timeline is extremely difficult to process. To put this thought another way, if the investigation necessitates that we look at how the study items may have different influences according to the time in history, they occurred the extreme difficulty of obtaining accurate, valid, reliable, and meaningful conclusions become obvious. This problem arises when some "thing" from two different times requires a comparison.

Part of this extreme difficulty originates from what lens the researcher is looking through at the present state of conditions. This means a scientist or anybody else will have unique perspectives and viewpoints regarding how they construct the various parts of the perception of what they are visualizing today and in today's times. These perspectives are shaped uniquely for every single person from their total life experiences. These life experiences are shaped by the observers' cultural background, the entire library of everything he has learned in his lifetime and how he learned it, in addition to the influences of other people in his lifelong learning process and the historical background of the individual and his life environment. In other words, his life experiences are the total collection of his experiences in their environmental context. It is through this person's unique lens that creates his involvement in the process. This is only one notion in the broader format of notions that together sum the total visualization of how some "thing" is perceived by all.

Another concept to introduce into the study is how the characteristic parameter is portrayed or documented on this stage of history. In other words, when comparing characteristic parameters—whether they are tangible objects, processes of how a thing was performed, built, or accomplished—a comparison between this characteristic in history versus how it is thought about today is a difficult thing to reconcile.

When some "thing" is investigated from a time in history and is compared with the "thing" as it is today, the researcher's unique lens of the today item, as well as the contextual lenses of the people who built the presentation of the historical thing, become two factors in this dilemma.

Here is an example of these notions. In keeping with our themes, there are many difficulties reconciling the presence and existence of the ancient Egyptian pyramids. These structures are thousands of years old yet still intact despite the processes and amount of "real" geologic time that passed since their construction, which is thousands of years! Some of the characteristic parameters present in an investigation of the pyramids include the following: first, they are magnificently massive structures; second, they were built thousands of years ago and still intact even today; third, we conjecture that people from that era did not have the "tools" we have today to build those *megaliths*; fourth, the "how and why's" of their construction are additional parameters and some of the most important.

When we start to investigate these parameters, we are influenced by insight from our unique life experiences, the contextual lenses of all the literature written by various scholars on the subject, as well as the notions related to how we would "do that thing today" with what we know about doing it. We would then attempt to inter-ject that way of thinking into historical time.

So in trying to reconcile how the ancient Egyptian pyramids were built, we would, thus, introduce a form of *bias* or *biases* into the process. After all this, we could conclude, "Those people could never have constructed them because they would have needed the technol-ogy of today to do so." or, "They needed help from somewhere." or that, "They could certainly have done so." But they would have had to overcome what would seem to be insurmountable technical odds.

Well, the pyramids are real because we can see and touch them in this case, so they do exist and, thus, we cannot conclude they were not built. This example of people's biases in their way of think-ing about the existence of the pyramids continues by explaining that they have a difficult time reconciling how and why and sometimes when they were built and by whom.

People today live in a world where a construction crew takes any combination of manufactured steel, wood, other metals, cement, brick, and whatever else we use to make a structure. The construc-tion crew goes to a job site five or more days per week to work and earn a monetary living that enables them to provide food and shelter for their families. In our society today, we use money as the transfer medium for this process.

The ancient Egyptians were doing the same when the pyra-mids were being constructed. However, we have a difficult time fathoming this. Citizens of ancient Egypt, along with many other nations of that era, lived in a different sociopolitical and socioeco-nomic structure than we do today. The work paradigm remains the same. The construction crews went to work at the site each day and performed their duties, so their pharaoh would supply them with the accoutrements necessary for their livelihoods and their family's survival. To build a structure today such as a skyscraper, an athletic stadium, or anything built on a large scale takes thousands of con-struction workers, engineers, toolmakers, work-in-process manu-facturers, and all the other support personnel needed to have the materials required to complete the job. They all earn their living today in this manner.

So did the pyramid builders. The same community of workers from a variety of occupations who built the Cheops, Giza, Khufu, and others in Egypt and around the world are doing the same things today. There is no difference. The visualization of the infrastructure for that type of project today, though, would be different than that of an ancient time for these reasons. It certainly would appear strange to us at first glance.

The materials and techniques used were different, as they will be 2,000+ years from now when our descendants perform the same duties as we do today and as they did 2,000+ years ago. It is remarkable in any survey of ancient structures that even thousands of years later, through a slice of modified geologic time, many still stand intact today and in good shape! Why the ancients did not use wood for construction but stone instead, which scientifically stands up to time best of all, is puzzling. Their use of brick and similar earthly compounds additionally were treated sufficiently then to be proven to exist today, yet we have problems maintaining a brick structure for even over one hundred years or a road to drive on for more than a few years. Today, we think it much more improbable, if not impossible, that the ancients could accomplish structures such as these with "so much less" than we have today. It is disappointing that we as of yet cannot travel back to that era and observe firsthand for ourselves and then accurately bring the knowledge back to our world. I conclude this would be a paradigm changer.

So it would assist the researcher immensely to recognize, to account for, and to eliminate and mitigate the effects of any biases that could exist to break down their analysis. Part of this involves having and maintaining an open mind when doing an investigation. It is two separate things to recognize this and to call out to everybody that they are trained to do this. It is another third and most important thing to practice it.

About an investigation of ancient structures use the term *oopart*, which describes some of the critical thinking that applies. Earth is fairly and some may say profusely populated with very old and magnificent structures, some of which are undiscovered. Our thoughts grapple with painting a picture depicting how we would build that structure today and how they were able to or how we were not capable of building it back then. But again, they did as they do exist.

In continuing with an example, the statement "so what" if it did take one hundred thousand Egyptian citizens over twenty years to complete the Cheops Pyramid may come to mind. So what if some of the workers were slaves, as is alleged, and forced to do their work. Today, we use workers from the state prison systems to maintain some of our highways. For others, construction of the pyramid was an occupational project that permitted them to support their families. Whatever political structure was in place then is not the point. Nowadays, it is our construction workers' occupation that permits them to support their families. It is additionally laudable and fantastic that the Egyptian civilization was able to allocate this amount of human resources to these and that they did not have to use as many human resources for other occupations and pastimes over the centuries, such as war and conquest. While no human civilization in history is free from the label of having engaged in war, the Egyptians maintained an excellent legacy of primarily refraining from such activities for over three thousand years!

Ancient structures were built and maintained by member civilizations all over Earth for thousands of years after the "technology explosion" of about twelve thousand years ago when the invention of agricultural systems caused a creation of the concepts of civilization and communal existence in village living arrangements. The intentions and practice of any investigator of ancient structures within a context of uncertainty and questions about the possibilities (oopart) should consistently include considering questions of "the how's, why's, and sometimes the when's and who's" of the civilization that built them. Remaining conscious of and holding on to these questions will help to sustain a practice of pragmatism that will assist in producing effective and valid conclusions for these issues that could easily become lost on a tangent of untruths and some line of conclusion different from that of rationality. There is no philosophical aid here. The objective is to get to the reality and truth of it all and not to let any biases interfere with the process or stray from the salient points.

So many things that are important or necessary to our lives are factored in from the sky in one way or another. If we are thinking about growing food, weather, or heat, we look up to the sun. Stargazers look to the night sky for their recreation. Items of a religious tone also cause thoughts to be applied upward. Indeed, places of worship are adorned with spires, domes, stupas, or crosses and have been since well the ancient times. Looking up to the sky is an important physical activity today as it was for the ancients.

Because just like today, when we do things like constructing buildings or other structures, the sky becomes part of that conversation, as it did back then. When talking about ancient structures, because the notion of "looking up into the sky" is so universal today as it was back then, an appropriate question to consider is, "What was or is our motivation for looking up into the sky?"

The sky was as important to the ancients as it is to us today. When the Neolithic Revolution was borne, villagers became more dependent on successful growing seasons. So watching the sky became vital to crop sowers. They were less equipped with knowledge about the subject of mass agricultural proliferation than we are today. This deficit naturally caused a heightened sense of anxiety among the villages when the weather and climate turned unfavorable for production. As a compelling course of nature, the ancients tried what they knew to produce a favorable food-growing objective. That often meant watching the sky or inducing behaviors they thought might compel or help the sky to produce the favorable weather needed. They could not control the sky, but they could ask it for help.

The early Neolithic Age archives are symbolic of Earth's civilizations worshipping various parts of the sky as in what is called a religious behavior. Hence, there were "sun gods," "moon gods," "star gods," "rain gods," and many animals as gods worshipped as at least "demigods." The symbolism that these gods represented were cause for the ancients to partake in, to obtain more food, to survive, and to raise families. It was these mechanisms controlled by the sun,

weather, climate, plants, and animals that worked together to produce more food.

When the Neolithic Age evolved into more of a city-state era,
the ancients needed more structures to accommodate their new
global habit. Their devoted attention to the sky evolved with their
migration, and the structures accommodated all this. As time went
on up until about 3000 BCE, the devotion to the various "sky gods"
and "animal and plant gods" maintained their allure for all our
civilizations.

The buildings increased in size and became more sophisticated
in a relative sense. Villages grew over the millennia and became small
cities. Eventually, when the worship congregations of urban communities merged became a sizeable one, a need for buildings to accommodate this worship developed. So the first multistory and especial-
ly-adorned structures came into existence.

However, this evolution to about 3000 BCE was very slow.
Most of the ancient cultures, at least for our current discoveries, did
not leave anything more than one-floor structures for our archaeologists to discover. Additionally, most of these ancient societies used
either wood, uncut stone, or soil substances of some type to serve as
adhesion materials.

The first known structures utilizing a pyramidal architecture
have been identified as having existed in ancient Mesopotamia.
The Ubaid society formed in around 4000 BCE, built and maintained an early version of what later was called *ziggurats*. The Ubaid
shaped their structural forms roughly like what we know as classical
Egyptian pyramids. They were step-tiered, made of rounded bricks,
and—according to archaeologists today—built with no mortar
materials. The mainstream ziggurats populated the Mesopotamia
region, around what is known as present-day Iraq and Iran following
the reign of the Ubaid toward the end of the Early Dynastic period
approximately 3000 BCE. The Egyptian early pyramid history did
not take any known existence for a few hundred years.

The speculated primary use of those basic Ubaid pyramids
and later the ziggurats was for general ceremonial purposes. The
archaeological record on the Ubaid pyramids is scant due to the

materials used, age, and lack of knowledge about preservation techniques at that time. There is also no reference in archived history that indicates any use of special spatial, physical orientation as to their construction. This means the early Ubaid and later ziggurats did not incorporate any astronomical, cosmological, or spiritual aspects into them.

There are even more ancient examples of ancient structures which have recently been discovered. Archaeologists' unearthing of Gobekli Tepe in present-day Turkey and Baalbek, in present-day Lebanon, are known to have existed in service as living structures. Gobekli Tepe is not a freestanding structure; rather an extraordinary living structure carved into rock and earth. The original Baalbek was more freestanding and purposed as a type of temple ceremonial symbolism that is dated back to over seven thousand years. Thus, the most general concept of a four-or-more-walled structure with a roof was being used but nothing else more elaborate.

As noted, the ancient Egyptians started their building renaissance at the end of the Early Dynastic Age, following 3000 BCE. After the onset of the Bronze Age but still millennia later, the ancient Incas, Mayans, Asians, and Europeans created their own concept of large structures. Most of the ancient civilizations on Earth by about 500 BCE had archived the existence of these ancient structures, at least those we are aware of. It is likely much more exist and have yet to be discovered.

An example of this dynamic concerning cutting-edge knowledge of new pyramid technology and our lack of it involves current investigations of prospective pyramidal structures in Indonesia and Bosnia. Gunung Padang in West Java is conjectured to have dated origins from 7000 BCE, while the Bosnia investigation theorizes that three or more pyramid structures are situated around the ancient Balkan city of Visoko. The dating for these alleged pyramids is estimated to be thousands of years old.

With the preceding perspective defined, there emerges a desire for some study of what was happening around the time the Egyptians were well into their pyramid-building era and takes the timeline to about the 1000's BCE.

It is imperative that the structures under this experiment involve the pyramids only. These are "like" structures and can provide some direct comparison and analysis. By around 700 BCE, pyramids were rising in many places on Earth. The era of pyramid construction on Earth lasted approximately two thousand years, as concluded from the latest archaeological estimates. Instead of looking solely at "How could these people have built the pyramids?" another lens for investigation could include, "How do they compare?"

The pyramids and another type of structure known as the *obelisk* became visible at the end of the Late Neolithic Era. Obelisks are categorized in two groups: *megaliths* and *monoliths*. A megalith was a term created to describe a collection of very large structures concentrated in close proximity. A monolith is a singular version of a very large structure. The main building substrates of both were of stone materials. A monolith has an additional characteristic of being constructed from only one stone block. These single block monoliths are confirmed to weigh up to a thousand tons or more!

Obelisks, the main type of monolith, were often situated at the flanks of a pyramid structure or as a form of entrance threshold into a series of other more massive megalithic structures. A lot of the ancient monoliths stood and stand today many meters high, some of them over 200 feet high (60+ meters). Most of them were carved from one huge stone block.

At the very end of the Neolithic period in anthropological study, the phenomenon of building enormous structures became evident throughout most of the European continent. Stonehenge is an example. The very sudden appearance of a lot of these structures across a large percentage of Earth's land mass gave cause for the eventual creation of a period with which to classify people and civilizations of the time. Thus, the Megalithic Period is the name given based on this period. In the future, more of these structural discoveries will be made. The megalithic period will accommodate these new additions to the historical record.

An initial comparison summary concerns the geometrical and architectural aspects of pyramid designs. An atypical and currently unexplainable similarity among them is that a lot of extremely sophisticated engineering and mathematics design resulted in their successful construction. The societies that produced most of these structures include Egypt, Meso-America (ancient Mexico and Central America), Inca (Peru and Bolivia), India, China, Southeast Asia and Indonesia, and possibly Bosnia. More will be discovered in the future.

All them built pyramids that contain elements of unconventional or out-of-place oopart characteristics. One example of oopart takes in the notion of "looking up to the sky." This notion involves knowing that the building of pyramids incorporated, to a great extent, some design involvement with the star constellations of the night sky. The notion appears to have had its genesis only after the end of the Early Dynastic period of 3000 BCE in Mesopotamia.

Often, as in the Giza and Teotihuacan megaliths, the blueprints on many basic structures duplicated the spatial representations of Orion's Belt and the Sirius star systems. The builders accomplished a measured and highly complex, even by today's standards, geometrical characteristic into the engineering of the pyramids. Their design expertise in these areas was extraordinary and more accurate than what could even be accomplished by today's engineers to the admission of many of them. This is another oopart about pyramid construction unable to be realistically answered by modern science.

From our construction of the first of the Mayan pyramids was started sometime during the last millennium BCE. The Giza pyramids were presumed to have been started around 2600 BCE. Current conjecture indicates some similar structures in India and China date to around this time, if not earlier. Anomalous to this is that, regarding the geometrical aspects, the formal creation of the discipline of geometry mathematics was invented by the Greek philosopher Pythagoras around 530 BCE.

What were these geographically different societies thinking when they were conceiving their pyramid megaliths? Could they have been in contact with each other? It is certainly possible that, with a land bridge existing between Egypt and Asia, they could have traveled on their own. This opens up a possibility that each society from as far away as Egypt, China, and India could have made contact. Our knowledge presumes that the first Egyptian pyramids were started around 2500 BCE.

An inconsistency with the travel notion is that the ancient Chinese Silk Road trade route, a most prolific economic travel engine and excuse for land travel, came into existence only around the time Pythagoras invented geometry mathematics during 500 BCE. When this travel notion is extended to other civilizations such as those in the Western Hemisphere, no land bridge connected with their civilizations with either Egypt or China and India to enable them to go there.

Additionally, there are no ancient writings depicting actual contact between civilizations such as these or any other far-off lands. It appears the ancient Mayan, Meso-American, and Inca cultures were successful at least without the help of Egyptian or Asian technologies. It also seems the Egyptians left no writings of physical contact with any Chinese, Indian, or Southeast Asian counterparts, at least before 500 BCE. There was no written evidence of Asian people having contact with or knowledge of Egyptian pyramid technology in the same time frame. However, another anomaly is known today in that the Mayan's written alphabet and some early Indian language tools have striking visual similarities to the Egyptian hieroglyphics.

When determining possible connections between faraway lands toward a possible answer to the pyramid question, "Could any faraway culture have had enough contact with any one to take cultural, engineering, and technical knowledge back and use them in their own pyramid technology?" the only logical explanation so far is to say that Egypt was close enough to Mesopotamia so a land travel route was less hampered by geographical obstacles and, therefore, within a realistic framework of possibility.

Design characteristics of the pyramids of Egypt, Asia, and the Western Hemisphere share similarities. Many designs and descriptive characteristics of other types of ancient artifacts and texts—including those found in ancient Egypt, Asia, and the Western Hemisphere—anomalously compare also. These will be introduced later in this book.

The mathematical and engineering design of this class of structures is a compelling convergence of thought and execution among such widely diverse and separate cultures. The blueprints of many of these pyramids were conceived centuries before the invention of the mathematical disciplines of arithmetic and geometry. While it is possible that the ancients could have had an intuitive awareness of this mathematics and be able to use them in their construction projects for megaliths of this type, size, and placement, it appears more than incredible that this was the case without assistance of some kind.

Another oopart aspect is the ancients needed awareness from other disciplines like engineering, physics, geophysics, geology, chemistry, and meteorology to be successful. Additionally, it is strange to reconcile when thousands of these megaliths also have tunnels, shafts, and passageways running deep within and through them. All this knowledge was necessary to have been able to finish these projects. Some of them took twenty or more years and used thousands of workers. Furthermore, these structures have lasted thousands of years! Examples of these exceptional structures exist all over the planet Earth. Astonishingly, their structural form has remained intact, not destroyed by nature or otherwise. This is a feat which has not been accomplished by humankind since. There is no current justifiable and irrefutable explanation which the sciences and historians have been able to offer for these out-of-place artifacts.

Some cosmetic differences exist between the pyramids of the different ancient societies. At the outset, it is noted that some ancient cultures built their pyramids with differing angular slopes. Some Egyptian pyramids were less acutely angular, often eight-sided (which is only detectable even today by specific sunlight and shadow play on certain days of the year at elevation) and came to a point at their apex. In contrast, megaliths in India and Southeast Asia were more vertically angled. The walls of some Indian, Central American, and Sudanese, more specifically the Kush pyramids, rose higher vertically than did the Egyptian pyramids. These cultures also built pyramids of differing angular measures. Many of the Mexican and Central American pyramids had rounded features as they approached their apex, but some were also vertically sloped. In Cambodia and Asia, the same pattern existed. The great Borobudur Temple pyramid of

Indonesia is unique in that, in addition to the different angular slope, seventy-two stupas sit on the top platforms of the megalith. These stupas were designed as shrines which contain statues of Buddah monks. Counting these, the total number of monks found within the entire pyramid megalith is a total over five hundred. Many Asian and Meso-American pyramids shared similar adornments.

The surface features of the stones themselves were different among cultures primarily due to the native geologic inventory available to them. Stones of ancient Peruvian and Bolivian pyramids were smoother, though another anomaly existed, which was the distance the stone blocks had to be transported from quarry to building site. The Egyptian pyramid stones used limestone and some granite. The Mayan and later the Aztecs of Meso-America also used limestone and some igneous volcanic rock types. The Columbian, Moche, and Peruvian Inca used more adobe and sedimentary rock types. All cultures had their separate formulas for concrete adhesives, though other entirely different variances arise over a discussion of the milling process and physical joining of the stone blocks workers used.

Some of the Meso-American pyramid structures were used for living quarters for the "kings in power" or burial tombs and many as ceremonial sites similar to our churches today. Additionally, some were designated as ritualistic sacrificial sites. More pyramids were built in Meso-America than the rest of the known inventory in the world combined.

Evidence obtained by archaeologists and other scientists points to artistic adornments and archived documentation that existed as some of the most magnificent, beautifully colorful, and ornate surface embellishments of these pyramids in man's history.

Much of the same complementary conclusions can be drawn for the other pyramid inventories from around the world. Egypt used a lot of gold and colored adornments as evidenced from their ancient artifacts, carvings, and texts. Quality natural limestone has a very bright white exterior. This combined with the ornate and brilliantly colorful visual effect we experience from a real-time observation of many of the Egyptian pyramids. Unless these limestone facades were sufficiently treated, though, their sheen and color hues would erode over a short geologic time. They were not, however, and they have remained intact for thousands of years.

The South American pyramid culture was not as ornate as elsewhere in the ancient world. This is a more conjectured analysis because of the lack of dedicated written and otherwise undocumented, historical archive collection that we assume exists.

A large part of the lack of adornment is due to many factors. First, the cultures of ancient Andean-South America were not as tightly integrated as those of Meso-America or elsewhere. The ideology was not the same as the other cultures. While future discoveries may change this notion, there is not much help to be gained from written text archives because these cultures did not possess the necessary language tools to do so. Second, the ancient Incas used a language that, in and of itself, is unlike how today's written language and alphabet appears. This language is known as the *Quipu* and is described in other chapters. The Quipu does not use an alphabet or numbers as we know them. Thirdly, when the Spanish Conquistadors plundered South America in the 1500s, they destroyed many valuable written sources of Inca historical documentation along the way. These historical records are known as *codexes*.

In Asia, the pyramid panorama was lensed a little bit differently than elsewhere. The depiction of some sizeable symbols on exterior facades is slightly altered from similar mind-set depiction translations in Egypt, Meso-America, and South America. The Borobudur Pyramid of Central Java, Indonesia, is a classic example of using large adornments on exterior structures. As noted before, the stupas are dome-shaped shrines that house Buddhist artifacts and relics and also serve as the tombs of Buddhist monks. The Borobudur Pyramid contains seventy-two stupas on its exterior complex. Also the physical symbolisms include 504 Buddhas on the three main platform levels encompassing the massive complex. The stone material is primarily andesitic rock which is representative of the region where volcanic activity shapesits geology.

The exterior facades of pyramids from other cultures were adorned with petroglyph carvings directly into the stones as were those in, for example, Asia, Indonesia, and Egypt. The only apparent difference is the regular adornment of large symbolic stone statues, such as the Borobudur Buddhas and the Terracotta Army statues in ancient China that guarded the entombed emperor from invasion in the afterlife.

Moving away from the superficial nuances that give unique flavor to the cultures that created them and back into the similarities, a description of some of the purposes and motivations of the pyramid panoramas are in order. For example, the Mayan Teotihuacan and Egyptian Giza Pyramid complexes share the same anatomy on many levels. At Teotihuacan, Mexico, the Plaza Pyramids included the Pyramid of the Sun, Pyramid of the Moon, and the Avenue of the Dead. The Egyptian Giza Pyramids of Menkaure, Khafre, and Cheops symbolized the same things to those people. Motivational drivers of these collections included cosmological, spiritual, astronomical, religious, climatic, and economic aspects.

Pyramids were often uniquely positioned to act like calendars. Their spatial platforms frequently allowed for certain visualizations for specific dates in time, such as at the equinoxes or solstices. This notion took on multiple contexts. For instance, the Mayan Temple of K'uk'ulkan Pyramid at Chichen-Itza displayed and still does today an ascension or descending of the great feathered serpent god K'uk'ulkan to and from Earth on the equinoxes. These appear in a form that portrays how unique light shadows snake up and down the steps of the pyramid. This symbolized the annual "energizing of the earth." Elsewhere, for example, in Egypt, many pyramids were aligned with the equinoxes to serve as calendar points of reference for when to plant or harvest crops.

It also enabled the pyramids to incorporate spiritual and religious contexts, as well as agricultural and economic design features, for their use. It also included obtaining knowledge about seasonal weather patterns within their calendar cycle for guidance onwhat crops to plant, when to harvest them, or for animal herding and the like. This was achieved through the engineering and construction of these pyramids in certain physical formations that not only coordinated a calendar with astronomical phenomena such as the Orion constellation but contained other spatial and geographical aides such as timetable that translated to these spiritual, economic, and agricultural-related activities.

It is evident here that the pyramids had more applications in their design strategy than just those that related to the entombment of the rulers who ordered their construction. In fact, the majority of all known pyramids were not built for entombment purposes. Very few of even the most heralded pyramids did not contain actual remains of intended rulers. Overall, there is much less doubt that there was some level of ceremonial significance to many of the pyramids that were built.

Perhaps, this notion of multimodal design strategies produced an overall application attribution. For example, the Mayan Temple of K'uk'ulkan Pyramid at Chichen-Itza displayed, as noted, the ascension or descention of the great feathered serpent god K'uk'ulkanto and from Earth on the equinoxes each year. This attribution aligned with all other attributions preserved within the complex. This equates with the perspective that those civilizations, not only of Meso-America but also over the entire ancient world, being maintained by a certain philosophy.

This would be the life perspective of a *holistic* philosophy. A holistic culture is relating to or concerned with wholes or complete systems, rather than with the analysis of, treatment of, or dissection into parts (holistic medicine attempts to treat both the mind and the body; holistic ecology views humans and the environment as a single system). It is characterized by a comprehension of the parts of something as intimately interconnected and explicable only by reference to the whole.

An exceedingly meaningful, portable, and profound visual takeaway from the principle of holism is to place a map of the world in front of us, a globe or wall map will do. After looking at these props, imagine an airplane that is flying upward into the sky or a rocket ship. As we peer out the window, the border delineations that separate one country from another or a large body of water, like the Great Lakes, with a similar border delineation line crossing through it are potentially evident.

Here is an example of holism. The borders are entirely human-made contrivances. We recognize and understand the process behind these contrivances. Nature and reality do not care about, nor do border delineations exist in nature. In fact, at that time, humankind practiced the most basic form of *cartography*. The ancients made maps but these were only of the stars in the galaxy. It was only after the Neolithic Period was well underway that maps of only a village-sized grid were completed and later discovered. When the evolution of communal living developed sufficiently, the first maps of a city and then a city-state appeared. As the size of the community grew through history, the map depiction grew accordingly. The segmentation process slowly changed humankind's philosophy of holism. It can be said the process of holism, in this regard, eroded gradually to where it stands today.[2,3]

Another example of an attribute being part of an ancient holistic philosophy of design strategy is the Great Sphinx. The Sphinx is situated at the same complex as the Giza Pyramid complex. This megalith certainly qualifies as one of the most magnificent and enigmatic structures of humankind's history. Its length is almost 250 feet, and it is over sixty feet tall. It is unique as the only superstructure of its kind in existence. The colossal jigsaw puzzle that is the Great Pyramid contains the answers to many unsolved questions. But there is general agreement that it's planning, construction, and site location was not an isolated strategy; rather, it was meant to be a significant element of this holistic philosophy ancient Egypt, like all cultures, perpetuated.

Even if there is no question or disagreement that the Great Sphinx represented a part of a more holistic philosophy, there are still many disagreements as to the "who or what does it represent and for what purpose (specific)" it portrays nowadays. The "how" is surprisingly easy. It was carved out of the limestone foundation where it stands; it was carved out of bedrock.

There are a variety of hypotheses pertaining to why it was built. One conjecture is that the Sphinx is another element in the construct of the ancient night sky. The spatial positions of the pyramids, the Nile River, and the Sphinx represent the same positioning of the Belt of Orion, the Milky Way Galaxy, and the constellation of Leo the Lion respectively.[4,5]

A second hypothesis is it was the adornment of the Pharaoh Khafra. Mainstream archaeology aligns that the face (human) may be evidenced to be that of Khafra who ruled Egypt from 2558–2532 BCE.[6]

As for the controversies which persist as to what purpose(s) it may have had to its builders and their descendants, maybe these will be answered once a lasting agreement can be reached as to when it was built. There is currently more disagreement in this aspect. The historically-accepted conclusion was that if the Pharaoh Khafre did order its construction, it would have been done in the twenty-fifth century BCE. But there are no ancient texts, scripts, or other attribution that confirm this. Other more current hypotheses relate back to the holistic philosophy that is attributed to the Sphinx in complement as part of the Great Pyramid complex. When studied in its entirety (holistically), the landscapes that are the pyramid fields are actually spatial representations of the night sky and our Milky Way Galaxy at a point in history. If this is the case, those researchers who estimate the construction of the Sphinx at 7000 and back to 10000 BCE are indeed correct.[7,8]

The best way to summarize the state of agreement that exists today is to reflect on a quote made by archaeologist Mark Lehner of Harvard University in *Smithsonian Magazine*: "Some parts of the Giza complex, the Sphinx included, make up a vast sacred machine designed to harness the power of the sun to sustain the earthly and divine order."[9]

The inventory of Earth's pyramids then incorporated astronomical, cosmological, spiritual, ceremonial, sacrificial human rites, and commercial and agricultural significances into their existence. For such structures that were primarily identical in appearance and function and that evolved in geographically isolated and divergent cultures at about the same historical time, what are the catalyst(s) that could explain how and why they turned out like they did? Another question to ask is how, when up to about 3000 BCE, there were no extraordinary building structures except for a few isolated communal living buildings that were mostly carved into hillside rock and earth and then suddenly a pyramid explosion started?

The Harappan city-state in India thrived in the period of 2500 BCE. Numerous small buildings at the site have been made known to exist. The Harappan disappeared completely by 2000 BCE, and nothing remains of their architecture. Only two thousand years later did any architectural renaissance occur, and this only involved wooden structures. Buddhist temples and stupas did not appear until hundreds of years later.

Chinese architecture included only wooden single-story structures with flat roofs until about AD 500 when evidence of the first brick structures appeared. The ancient Inca and their ancestors are known to have had only single-story structures, also dating back to around 500 BCE. These were made of local fieldstone set into stone foundations. They had access to limestone and granite. These ancients often used belowground level in their buildings. While the ancient Inca and their ancestors built a lot of non-pyramidal structures with out-of-place oopart engineering and structural integrity, such as the structures at Tiwanaku and Puma Punka, their extraordinary pyramid or other large building inventory is sparse.

The ancient Greeks and Romans displayed kindred architectural styles. The first large Greek structures, though not pyramids, appeared around 600 BCE. The pillar-and-step marbleized style was carried over to Rome in approximately 500 BCE when they merged earlier Etruscan influence with that of the Greeks. The Romans adorned their temples with more steps than those of ancient Greece. This seems logical in a context of an innate human motivation to "make or have more of" a certain characteristic or element. This motivation predates the written language of humanity.

The ancient Mayans and their predecessors also had no extraordinary structures to boast of until the first pyramids started to be built. There was nothing in their "Pre-Classic" or "Classic" Eras of before 2500 BCE until the first city-states began to evolve in 700 BCE.

One other comparison element to include here for reference would be that the first castle structures of any kind to appear on Earth did not until around AD 1050.

So until around 3000 BCE, no large megalithic engineered structures were known to exist. Then almost in an instant, a large number of massive pyramids and structures, such as the Sphinx, appeared as if out of nowhere. As a sidenote, there is a growing body of conjecture that the Sphinx was built much earlier than 3000 BCE. This technology explosion leaves unanswered questions such as, "How and why were they built?"; "How and why did they suddenly appear in rapid order?"; and "Who built them and what was the purpose in building them?"

From the ancient hieroglyphs and texts carved into the pyramid face records, it is known that those people knew that the sun, moon, and water were responsible for food crops to grow and for animals to eat and grow themselves. They knew that all these factors and the humans themselves needed water and the sun to survive. The natives knew they were part of a holistic community, not the one of today where we are most motivated by activities and interests only within our microcosm: our paycheck, paying taxes and retirement, and the "art of recreation." Most of their needs were fulfilled by the ruling class and through their ceremonial recreations.

Holism is defined as "the theory that parts of a whole are in intimate interconnection, such that they cannot exist independently of the whole, or cannot be understood without reference to the whole, which is thus regarded as greater than the sum of its parts. Holism is often applied to mental states, language, and ecology."[10]

All the ancient societies had gods to worship and held ceremonies that encouraged their crops and livestock to thrive. But the pyramids were not spatially designed around or pointed to the sun or the moon or to a water source, except in some minor contextual circumstances. The sites around the pyramids were certainly used for such ceremonial purposes as sun or moon or rain worship, among other reasons. However, there was another main reason for their design considerations.

The pyramids were built around and at least thematically directed to stars in the sky! They were engineered and built around notions and thoughts of astronomical devices, such as the constellations in the night sky and their characteristics. Here, the ceremonial events involving the sun, moon, rain, water, and animal gods seem almost peripheral to the purposes of the structures themselves on many levels. If they were built specifically and categorically with those gods in mind, why go to the trouble of precisely calibrating them with the night sky, astronomical and cosmological notions they had?

Remember that these pyramids cannot easily be duplicated by us with even today's technologies over five thousand years later! This also does not bring into a discussion anything about the influences and oopart nature of other ancient structures we have not mentioned yet. Additionally, remember that a lot of these pyramids were built with and contained many internal structural components that curiously align with other elements of a cosmological, astronomical, climate, seasonal, and geophysical nature. For example, tunnels and shafts in the Giza Pyramids are exactly aligned with the sun's seasonal equinoxes and solstices, as are some of the Mayan Pyramids and their peripheral plaza building structures. Some are also aligned with Orion and Sirius and Aldebaran, three more primary oopart points of evidence. The perimeter geometry of the structures themselves maintains the same mind-set.

Besides this reason, for the ancient cultures to "look up to the sky" in designing the pyramids, they also looked up to the sky in recognition of an afterlife for their adorned rulers emperors and kings of their societies. This meant their way of thinking, for whatever motivations or catalysts, directed them to build the pyramids to honor and, under orders from, their rulers; moreover, to design and build them with the heavens of the stars and the heaven of where their rulers would go in their afterlife. The honoring and worshipping of the gods of the sun, moon, rain, crops, and animals were often done on the ceremonial sites around the pyramids but not to serve as the main reason for their existence.

The concept of the pyramid, on these levels, had the most motivational force as being thought of as a machine, albeit with no direct moving parts in the traditional sense. As a device to guide the ruler or pharaoh to his afterlife destination, the pyramids and their accoutrements pointed and guided the ruler's soul and spirituality in the right direction. Often, that direction took them to the Orion constellation.

As noted previously, the implementation of the building of each pyramid merged well with the daily existence, job routines, and outlook of the many thousands of workers who contributed to the building's design and construction. If it took twenty or more years to build one, it did not matter. Ancient societies were structured to accommodate such a holistic economy, even though that concept appears foreign to us today. This is because we view economy and the construct of it today in such a capitalistic form that includes such concepts as supply versus demand, input versus output, profit versus loss, and so on. The ancients lived in a holistic society where their living needs were provided for. This allowed for the more successful undertaking of monumental and lengthy building projects such as the pyramid.

The ancients knew the stars in the night sky were not responsible for their food to grow or for rain or the appropriate short-term weather systems related to these to happen. This is the reason they worshipped separate gods for those things.

So moving beyond this, the ancients noticed and had knowledge of the more magnificent astronomical objects. No doubt when we look at the night sky today, some of the very first things you see are the Belt of Orion, the brightness of Sirius in the Northern Hemisphere, the Southern Cross, the redness of Mars, and the movement of the other planets in our solar system. Regarding these, those who lived during this time also had knowledge of planets as far away as Saturn. This is certainly feasible because they were all naked-eye visible on a regular basis back then, as they are today.

So why the pyramid design considerations for the stars in the night sky? Besides the magnificent astronomical objects, there were many thousands of visible stars and star look-alikes, such as our solar system planets, to align to. It is valid that some of the star targets utilized had more attraction qualities than others. Did this de facto reason become the most favorable for the ancients to then just arbitrarily call these targets "heaven" or "heavens" as in more than one heaven destination for the afterlife of the rulers who directed their pyramid's construction? Why not utilize the Big Dipper or the Southern Cross (for the South American cultures) or the Northern Cross which contains the very bright star Deneb? Additionally, Deneb serves a dual role as a prominent member of the "Summer Triangle," which includes Vega and Altair, two very bright magnitude visible stars. Or were there other motivating reason(s) for the choices they made?

Did rulers each have separate heavens designated for their eventual afterlives? This seems reasonable that the third and second millennia BCE was still an era of polytheism. The notion of a monotheistic one-God religion was at least not found to be documented until a monotheistic construct called Zoroastrianism first appeared about 3,500 years ago in ancient Persia. There, the prophet Zarathushtra preached that there was one God, whom he called Ahura Mazda. *Ahura* means "Lord," and *Mazda* means "wise," so Zoroastrian's God the "Wise Lord." No one knows precisely when Zarathushtra lived, but scholars say it could have been anywhere from 1500 BCE to 600 BCE.[11]

Meanwhile, before that time, many gods were worshipped, and rulers were sent to their own afterlife to be with and serve any one of these gods. So the heavens in a plural sense accommodated many stars as destinations for society's rulers.

Two of the most significant gods and related star symbols were the Egyptian gods, Osiris and his spouse Isis of Orion and Sirius respectively. Hieroglyphs carved directly onto the pyramid texts explicitly state that Osiris was known as Orion himself. He was the supreme god of all Egypt. Isis was also known as Sirius. Isis was the first daughter of Geb, god of the earth, and Nut, goddess of the sky. She married her brother Osiris and together, they conceived Horus. Isis was instrumental in the resurrection of Osiris after he was murdered.

Mesopotamia existed even before Egypt. In 1849, a British Museum archaeologist team led by Austen Henry Layard and Hormuzd Rassam uncovered cuneiform star catalogs from dig sites in present-day Mosul in northern Iraq. These thousands of cuneiform texts documented knowledge of the cosmos from over five thousand years ago before the emergence of the developed Egyptian civilization.

These texts charted and documented the history of activity of astronomical objects and cycles of both galactic and planetary proportions. Gods and goddesses were assigned to many of these objects. These attributes were mimicked by societies all over Europe, Egypt, and Western Asia.

For example, Babylonian cuneiform tablets tell of a "god of justice," known as Ninib. The "lord of wisdom" known as Marduk. The "god of war" named Nergal. The "divine scribe" known as Nebo. Finally, the "goddess of love" was called Ishtar.

The ancient Greeks and Romans then adopted the Babylonian god and goddess references for their own purpose. Ninib, in Greek and Roman, were known as Chronos and Saturn respectively. Marduk was identified as Zeus and Jupiter respectively. Nergal was known as Ares and Mars respectively. Nebo was called Hermes and Mercury respectively. Ishtar was known to the Greeks as Aphrodite and to the Romans as Venus.[12]

The Greek and Roman names are quite well-known. There is no unusual circumstance around this assimilation of such god and goddess adoptions into Greek and Roman culture. Around 540 BCE, King Cyrus of Persia came to power. His cousin, Zaratas, arrived in Babylon around the time of Cyrus's conquest of the city. Zaratas was a teacher of high regard and fame, so much that he welcomed an extended visit from the Greek mathematician, Pythagoras.[13]

King Cyrus was also well-known to the Greek society as Cyrus the Elder. These account for two of many network contacts that provided the network underpinnings which started the process for Greece assimilating some of this knowledge into their history. The Roman culture had similar system connections. Both systems were influenced due to available geographical opportunities many other societies did not enjoy until after ancient times when travel became easy.

Both the Judeo-Christian and the Hebrew Bibles, as noted, documented many encounters in which their various prophets and observers witnessed numerous gods interacting with either the prophets and/or the witnesses themselves. Even though by the times of the history of the Christian and Hebrew Bible Scriptures that the notion of monotheism had substance, there was still a lot of scriptures allocated to other gods and their encounters with mortals.

When gods appeared on Earth, they came from those heavenly stars in the sky. The pyramids were thus engineered around and, taken individually, took a perspective toward a particular star or star group, such as Sirius or the Belt of Orion and sometimes the entire Orion constellation itself. Egypt and the Mayan, for example, made significant use of these star systems. Giza and Teotihuacan Pyramid complexes were designed around the Belt of Orion and the constellation itself. Later, in Greek and Roman times, other systems and planets such as Venus, Mars, Jupiter, and Saturn served as cosmological backdrops for some of their mighty gods. There were many star systems, many gods, and many opportunities for rulers to rise to one of them to continue toward the afterlife.

The structure of the pyramids has provided a multitude of oopart incongruities that, in and among themselves, long for one smoking gun incontrovertible proof that entities from elsewhere in our universe enacted a role of existence and reality in ancient times. For now, the lack of the one overarching smoking gun is substituted by a lot of circumstantial evidence which is symbolized by the presence of many anomalies.

First, there was a technology explosion in just the genesis and development of pyramid structures that occurred in a momentary span of time. Can we compare this explosion with that of today's last century? Take the emergence of successful physical construction of pyramids many times larger than the largest structures of that time using materials and tools of the period and contrast them against the technology explosion of today. Add in a relative essence of proof that they existed up to five thousand years later as another anomaly. Remember, there was no mathematics or science or philosophy over five thousand years ago.

Additionally, societies were holistically organized back then. Life and living accouterments were supplied to citizens in a far different fashion than today. This is why much of this situation is difficult to comprehend and reconcile. Cultures during those times ran with a different day-to-day thematic schema filled with concerns about having enough food, drink, adequate shelter, and good health. Psychologically, these people could be analyzed using Abraham

Maslow's "Need Hierarchy" to determine that they had more concerns over basic physiological need fulfillment than we do today. The satisfaction of these concerns often led to the worship of multiple gods for an extensive amount of time. According to the evidence, such civilizations all over our planet were very successful at these practices.

Compare this with knowing that, just a century and a few years ago, man had only just successfully "flew off the ground for a hundred feet." Today, man is flying machines to beyond Pluto while directing them with computers that did not exist over one-hundred years ago.

Another point of discovery tells of the ancients' visualizations of the astronomical constellations and their relationship with the spirituality of the afterlife for their passed rulers. Before Pythagoras's time in Persia and his later creation of the geometry mathematics during 500 BCE, the ancient people did not even attribute geometric or any other mathematical, scientific, or philosophical thought to any study of the cosmos. They could not. As noted elsewhere in this book, none of the constructs of any of these disciplines even took shape until the time of Pythagoras's immediate ancestors.

The ancient stargazers developed an innate sense of knowledge, spirituality, and purpose by seeing the cosmos as a series of living pictures.[14] Before the time of King Cyrus of Persia, the ancients displayed their knowledge, spirituality, and purpose through the archives that included the hieroglyphics, pyramid texts, and other local cultural texts and the pyramids themselves. The Babylonians, as evidenced by the cuneiforms of the Babylonian Star Catalogs of the last six hundred years BCE, thus created the zodiac and through this construct made the cosmos come alive.

When investigating the seemingly ritualistic format, the ancients took when deifying and worshiping their departed rulers an interesting parallel emerges in relation to present-day phenomena. Here, I will introduce a concept and practice that has and still does occur even today in different parts of the world and introduce a human cognitive and psychological thought process that is analogous from and to similar activities in ancient times.

The practice is called *cargo cults*. There is a lot of investigative discourse which takes the term cargo cult and disseminates it into such subject areas including sociopolitical, socioeconomic, purely political, business, technological, and militaristic realms. Some of these cargo cult investigative issues do not address our main topic which is discussing ancient cultures and the pyramids in correlation to the old human ways of thought with the existence of gods and such.

A most general definition of a cargo cult is a pattern of critical behavior displayed by a large group of people usually in response to some phenomena or series of events that maintain vital and direct influence on some significant aspect of their day-to-day existence. *Merriam-Webster* defines a *cargo cult* as "any of various Melanesian religious groups characterized by the belief that material wealth (as money or manufactured goods) can be obtained through ritual worship."[15]

The Melanesians are an Indonesian society that have experienced behavior since the early twentieth century. Serious study of the cargo cult experiences of many such cultures in Oceania and elsewhere have occurred in the literature since at least the 1940s, though the concept practiced well before that time. The reference to "material wealth" is an example of just one of the many investigative discussions. When the Melanesians are the primary focus of this discourse, it has a metaphorical meaning to all other cargo cults.

Here, the material wealth aspects of the cargo cult are not of interest. Rather, the focus is on a typical example of the proliferation of one that often happened in the 1940s and continues today. I will then compare and contrast this to the same ancient societies' social behaviors. The main takeaway point is to compare how the native peoples reasoned and conducted their rituals which seems to have spanned many millennia.

During World War II, Oceania was front row of the war theater in the Pacific. The Allied and Axis war machinery moved around that entire part of the world and came in contact with many indigenous societies. Most often, the natives had never seen a white person or any of the *accoutrements* these soldiers brought with them. This visualization is one of the military regiments either coming ashore in boats or flying onto land in their airplanes. They brought with them food, shelter, clothing, radios, automobiles, and other equipment. A supply of food, shelter, and clothing was given to the natives on an ongoing basis.

The indigenous people, never having seen any of these "things" or the strange white people themselves before, could not immediately interpret such new things and events. But they knew what food, shelter, and clothing were immediately. They interpreted these things as being good and seemingly abundant. Their way of thinking easily transferred the actual benefits of their new wealth to the other materials and aspects of such a first contact.

Even though they could not understand the multilevel motivations for these strangers to be there, in their world, the natives only knew that they regularly struggled to obtain enough to continue day-to-day survival and to raise families. Their cultures also practiced their ritual routines and worship that were a big part of the communal livelihood, which included the Melanesian community.

Consequently, the white strangers who suddenly appeared with all these supplies became the objects of godlike worship by the aboriginal people. The uniformed soldiers supplied the natives with seemingly unlimited goods.

Then as suddenly as they appeared in the communities, the soldiers left and took all their ships and airplanes and other equipment with them. Naturally, the natives were entirely confused. The various aboriginal communities did not understand the reasons for the god's departure. Their reactions reflected the normal behavior inherent in human beings.

They began participation in a cargo cult phenomenon. This series of activities included building physical replicas of the ships, airplanes, radios, equipment, and shelters in direct and blatant imitation of what they observed and experienced. In the objective that these behaviors would appease the gods and help make them return with more foodstuffs, these objects were built and displayed prominently on the open land and beaches where they had their first contact. The natives also performed ritualistic ceremonial rites in their rationalization that the soldiers would hopefully return with more foodstuffs.

Much discourse has been written and debated over the years about our discussion of the psychological and cognitive facets of such communal behavior. For example, in his 2009 paper "What Happened to Cargo Cults," Ton Otto listed some conclusions demonstrating ways of critical thinking that the natives practiced. "The development of a certain myth dream,[16] a synthesis of various indigenous and foreign narrative elements and religious concepts; the expectation of the help and/or return of the ancestors; the emergence of leaders with special experiences and/or knowledge; and a strong belief in the appearance of an abundance of goods. These characteristics give a fair picture of the prominent aspects of most cargo cults."[17]

In the 1950s, during the genesis of the heyday of cargo cult discourse, researcher Jean Guiart wrote extensively about precepts and motivations for cargo cult behavior. One observation he made was: "Melanesian natives covet the standard of living of the white man with whom they are in contact. By a rather logical method of trial and error, they tried to create the conditions appropriate to the arrival of the long-delayed steamer: organization of camps with military discipline, building of cargo houses, model villages."[18]

Guiart provided an empirical definition of cargo cult behavior as the following: "People in cargo situations behave the way they do to move to certain ends they want to attain. The situation prescribes certain actions if they wish to attain those certain ends. Therefore, they act in the way that we have since discovered. This solves the problem of their behavior."[19]

A most direct psychosocial and psychocultural analytical conclusion was offered by researcher Theodore Schwartz in 1976. He defines cargo cults as "type response, a basic mode of reaction or adjustment to situations of rapid culture change characteristic of an entire area in a specific historical phase."[20] Schwartz concludes that these people experience indicated responses due to factors such as deprivation, stress, unconscious hysterical behavior, and cognitive dissonance.

The takeaway from this analysis of human behavior consists of multiple facets that are empirically comparable to similar documented situations of the ancients, their structures, and their practices. First, this comparison is most interested in human behavioral characteristics. It does not necessarily mean to contrast the building of objects with the construction of pyramids by the ancients. The building of structures is secondary here to the primary aspect of the cognitive thinking patterns and psychology of present-day examples in comparison and contrast to those from ancient times. Here is proof of widespread ritualistic behavior patterns of many cultures living in today's world which worshipped what to them were godlike visitors.

It is all about the immediate actions of the cult to communicate with or obtain things from the gods rather than the sociopolitical or socioeconomic factors or both, often researched that is the end point of the discussion. Cargo cults still happen today. Man's way of thinking nowadays is like that of thousands of years ago. Today, the cults are invoking the presence of the soldiers as gods to bring more food, shelter, and clothing. Back then, the ancients were imploring the presence of their gods to bring them food, shelter, and happiness. In other words, the ancients were asking for their survival. The ritualistic practices manifested from these behaviors took the course that is now our evidence.

According to a thematic of this chapter, present-day thinking may have a problem embracing these examples of such cultures participating in these behaviors. But they exist. As an example of a what-if scenario, consider this. The ancients of all societies experienced many seasons where agriculture and animal husbandry turned unproductive. It occurred primarily because of inclement weather and mini-climatic phases. In response to bad seasonal weather/crop growing patterns, how did they react? Even in the what-if option that they did not invoke worship of gods and goddesses to assist them, it still testifies to the realization of all the hieroglyphs, scripts, and texts revealing that a worship paradigm still happened!

The ancients thought in the same critical reasoning manner that their descendants of up to five thousand years or more are doing today. It seems to be an insult of either credit or ridicule depending on the context we wish to take with this notion. How could such evolved humans of today's world be accused of displaying the same attributes of their distant ancestors? Or how could the ancients be so advanced as to show the same critical thinking patterns that are replicated by their distant descendants today? How can we view this?

Second, the situation signifies a direct correlation to human behavior back then and now. The ancient people had some contact with godlike entities at some point. Those gods and goddesses were reported by the ancient scribes to have performed many deeds such as curing the sick, providing food, knowledge, and wisdom to citizens of those societies. They appeared for a time. Some of them stayed longer. All these events were documented to display some sort of actual existence of genuine experiences the ancients had with vastly superior entities. This documented evidence includes all the hieroglyphs and carvings displayed on pyramids, temples, monoliths, megaliths, *henges*, and other structures to the written texts and as-of-yet undiscovered records.

Then they left. The ancients reacted in the same way humans do today. It is in this relationship of how modern human societies act and respond to such phenomena in the same way ancient societies did that presupposes some possibility that the events of the gods contacting the ancients and then the ancients building pyramids and creating the whole god worship paradigm actually happened as well. There is a correlation between how ancient man and humankind today shapes and forms the same thought processes.

References

[1] Quote Ref: Darwin, Charles. 1863. Letter to J.D. Hooker, March 29, 1863. In F. Burkhardt and S. Smith (eds.), *The Correspondence of Charles Darwin 1863* (1999), Vol. 11, p. 278. Retrieved June 26, 2016.

[2] Holism Ref 2: Bagrow, L., revised by R.A. Skelton (1986). History of Cartography. Transaction Publishers.

[3] Holism Ref 3: Crawford, P.V. (1973). "The perception of graduated squares as cartographic symbols." The Cartographic Journal. **10** (2): 85–88. doi:10.1179/caj.1973.10.2.85.

[4] Sphinx Ref 4: Orser, Charles E. 2003. Race and practice in archaeological interpretation, p. 73. University of Pennsylvania Press. ISBN 978-0-8122-3750-4.

5   Sphinx Ref 5: Hancock, Graham; Bauval, Robert. 1997. *The Message of the Sphinx: A Quest for the Hidden Legacy of Mankind*, p. 271. Published 1997 by Three Rivers Press. ISBN 978-0-517-88852-0.

6   Sphinx Ref 6: Hassan, Selim. 1949. *The Sphinx: Its History in the Light of Recent Excavations*. Published 1949 by Cairo: Government Press.

7   Sphinx Ref 7: Hancock, Graham; Bauval, Robert. 1997. *The Message of the Sphinx: A Quest for the Hidden Legacy of Mankind*, p. 271. Published 1997 by Three Rivers Press. ISBN 978-0-517-88852-0.

8   Sphinx Ref 8: Schoch, Robert M. 1992. Redating the Great Sphinx of Giza. In Circular Times, ed. Collette M. Dowell. Retrieved October 20, 2016, from: http://www.robertschoch.net/Redating%20the%20Great%20Sphinx%20of%20Giza.htm.

9   Sphinx Ref 9: Hadingham, Evan. 2010. Uncovering Secrets of the Sphinx. Smithsonian Magazine, February 2010. Retrieved October 20, 2016, from: http://www.smithsonianmag.com/history/uncovering-secrets-of-the-sphinx-5053442.

10 Holism Ref 10: *Merriam-Webster Collegiate Dictionary*. 2016. *Holistic* definition. Copyright 2016 by Merriam-Webster, Inc.

11 Reference Monotheism 11: http://www.allabouthistory.org/when-did-monotheism-start-faq.htm. Retrieved July 29, 2016.

12 Babylon Star Chart Ref 12: Powell, Robert. 2007. Origins of Star Wisdom. Copyright 2007 by Astrogeographia. Retrieved July 29, 2016, from: www.astrogeographia.org/about_us/origins_of_star_wisdom.

13 Cyrus Ref 13: Guthrie, Kenneth Sylvan and Fideler, David. 1987. *The Pythagorean Sourcebook and Library*. Pg. 125. Copyright 1987 by Phanes Press.

14 Babylon Star Chart Ref 12: Powell, Robert. 2007. Origins of Star Wisdom. Copyright 2007 by Astrogeographia. Retrieved July 29, 2016, from: www.astrogeographia.org/about_us/origins_of_star_wisdom.

15 Cargo Cult Ref 15: *Merriam-Webster Dictionary*. 2016.

16 Cargo Cult Ref 16: Burridge, Kenelm. 1960. *Mambu: A Melanesian Millennium*. Copyright 1960 by Kenelm Burridge. Published by the Routledge, Taylor and Francis group, London, England.

17 Cargo Cult Ref 17: Otto, Ton. 2009. What Happened to the Cargo Cults. Published in Social Analysis. Volume 53; Issue 1. Spring 2009, Pgs. 82–102. Copyright Berghahn Journals. Doi: 10,3167/sa.2009.530106.

18 Cargo Cult Ref 4: Guiart, Jean. 1951. Forerunners of Melanesian Nationalism. Publised in Oceania, Volume XXII; No. 2, Pg. 85. December 1951. Retrieved July 29, 2016, from: http://horizon.documentation.ird.fr/exl-doc/pleins_textes/pleins_textes_5/b_fdi_16-17/22935.pdf.

19 Cargo Cult Ref 19: Jarvie, I.C. 1963. Theories of Cargo Cults: A Critical Analysis. Oceania, Volume XXXIV; No. 1, Pg. 20. September 1963.

20 Cargo Cult Ref 20: Schwartz, Theodore (1976). "The Cargo Cult: A Melanesian Type Response to Change." In DeVos, George A. Responses to Change: Society, Culture, and Personality. New York: Van Nostrand. p. 159, 174. ISBN 0442220944.

CHAPTER 12

Ancient Structures of High Strangeness Part 2

Scientists and engineers may be the most devastated
by the discovery of relatively superior creatures.

—The Brookings Report, 1961

Key words: henge, polytheism, sacbeob, phytomorphic, anthropomorphic, steles, trilithons

There are many other types of structures built in ancient times that defy any kind of rational, credible, and valid accreditation of easy explanation by natural phenomena. When placed in the real context of the ancient world these structures defy humans and considering the resources available to man at that time, the ancients could not have been the sole architects without some oopart involvement.

General discourse is in agreement that humankind's last evolutionary technology explosion occurred about twelve thousand years ago. It is defined as the beginning of the Neolithic Period of man's evolution. This time coincides with the geophysical retreat of glaciers of the last Ice Age. An additional visualization exemplifies that the Great Lakes had fully formed by then. The inventory of known man-made structures that have survived to the present time includes some that have been scientifically dated to over eleven thousand years ago. Reflect on what this means for a moment. The human species has built structures of various kinds that have withstood the geology of time to over eleven millennia!

Adding to this notion is an additional one that the studies of archaeology and paleoanthropology are always discovering new atypical unexplained phenomena. An example of this is an introduction to the topic of ancient earthworks in Kazakhstan.

In October 2015, NASA released a bevy of satellite photographs that reveal the existence of 260 structures on the terrain in the Turgai region of northern Kazakhstan. These structures consist of earthen-raised mounds, trenches, and ramparts sized larger than several football fields. The structures all appear to be constructed to depict specific geometrical shapes. Because of their size, the established patterns cannot be identified except from an aerial perspective. The shapes range from squares, crosses, rings, and diagonal crosses connected by other triangular shapes. Dating analyses conclude the structures were engineered over eight thousand years ago.

Called the "Steppe Geoglyphs" by some scientists, the first set of this collection was discovered by archaeologist Dmitriy Dey in 2007. Continued investigation by Dey prevented the topic from being completely buried in anonymity. Upon the release of the satellite photo catalogue, Compton J. Tucker and Katherine Melocik, NASA senior biospheric scientists responsible for the broadcast, also said that NASA has also assigned the International Space Station crews the project of obtaining a more detailed and nuanced photographic record of the entire region of north central Asia.

The dating conclusion makes sense as there was an indigenous culture, the Mahandzhars, settling in the region around 7000 BCE. Another atypical pattern embedded within the mystery is that the Mahandzhar were known to be a nomadic populous. To have a discovery of such sophisticated structures contradicts the nomadic lifestyle, which itself is also unexplained. Each mound when originally constructed stood over ten feet high and over forty feet in diameter. There are many thousands of these within the Steppe Geoglyphs. Additionally, the thousands of ramparts and trenches adorning the structures point to an indication of the Mahandzhar evolving a culture of settled communal existence.[1]

An overview notion to mention as a prelude to the following discussion concerns the question, "Why did ancient man build structures of stone materials?" Many, though not all, of these nations did have wood and other earthen materials readily available, more accessible, and easier to work with than stone material counterparts. Additionally, the Bronze Age is not currently known to have begun until around 2500 BCE. This is much later than the origin of man's Agricultural Revolution just noted.

In reality, all these materials were used extensively back then as well as today. The extensive use of stone materials represented four ideologies to the ancient cultures. First, they knew living shelters made of stone would last much longer than those of wood or granular earth. They would not need to rebuild, and they would offer better protection from others with malevolent motives. Second, stone monuments to adorn, worship, and guide the gods in their afterlife travels to and from Earth represented a philosophical visual of the eternity of such an afterlife. Third, stone would withstand time so their message and symbolism could be communicated to their descendants. Fourth, stone materials demonstrated a vital and significant contextual connection with ancient societies' holistic ideologies. Stone materials defined a strong connection between the aboriginals and the earth and nature itself. In summary, when one incorporates all the advanced oopart mathematical, astronomical, and cosmological concepts of the way with which the ancient's holistic philosophy used stone materials in their structures, could it be said that the ancients were, in fact, the first philosophers and scientists in humankind's existence?

There are many natural museums around the world which display stories highlighting the existence of ancient structures of many kinds. There are temples, megaliths, monoliths, henges, and even underground cities that have been left for us to admire. Let's start with a survey of henges.

The word *henge* refers to a particular type of earthwork of the late Neolithic period, typically consisting of a roughly circular or oval-shaped bank with an internal ditch surrounding a central flat area of more than twenty meters in diameter. There is typically little, if any, evidence of occupation in a henge, although they may contain ritual structures such as stone circles, timber circles, and coves.[2]

As a precursor to the Iron and Bronze Ages, the Neolithic was primarily a Stone Age. In light of our previous discussion, the ancient cultures of these times also used stone materials because iron and bronze technologies had not been developed then. Ancient cultures of eras after the Neolithic certainly exploited stone technologies for all the reasons presented above. This is another one.

When henges are discussed, thoughts invariably turn to England and to Stonehenge. While most of the henge inventory is situated in Great Britain, they were not known as the first to be built.

That honor verification so far belongs to a place in the Nubian Desert in southern Egypt called Nabta Playa. This region is situated in the middle of the Sahara Desert. Archaeological findings conclude the earliest human activity occurred up to 10000 BCE. At this time, geophysical records indicate the Sahara was a lush savanna with abundant water sources. Natives reportedly used ceramics and used the practice of pastoralism. By around 7000 BCE, entire communities were discovered. Natives used wooden and stone huts built in rows like a present-day residential neighborhood.[3]

By 5000 BCE, the aboriginals left evidence of religious and ritualistic activities that included cattle sacrifice within clay buildings with stone roofs. The evidence also includes a "calendar circle" that has an intricate alignment correspondence with astronomical patterns. This structure was made of stone megaliths standing a few feet high and displayed a working and accurate device of astronomical alignments that are still being studied and expanded upon.

The first findings indicated the Nabta Playa constructed this henge around 4800 BCE. Some of the identified structures from within the site show an accurate replica of the entire Orion constellation, as well as the summer solstice sunrise for that period.[4] More up-to-date investigations ascertain the henge of Nabta Playa also documented alignments with other astronomical objects such as the star systems Sirius, Arcturus, and Alpha Centauri. Researchers J. McKim Malville and Fred Wendorf concluded a recent research paper by stating, "The symbolism embedded in the archaeological record of Nabta Playa in the Fifth Millennium BC is very basic, focused on issues of major practical importance to the nomads: cattle, water, death, earth, sun and stars."[5]

A henge site that does not predate the Nabta Playa but was itself built over a much longer time span and is reasoned to be a predecessor to its neighbor Stonehenge was the Avebury henge. There is no consensus as to accurate dating but suggestions of anywhere from 3000 to 3700 BCE are conjectured. Avebury is also thought of as a henge structure not constructed over the time span of a few hundred years.

The geophysical landscape of Great Britain changed drastically in the later Neolithic Period. What started as dense forested geography of the early Neolithic when the Great Britain landmass was still connected to Continental Europe changed drastically around the time just before the start of the Avebury henge site. The native hunter-gatherers decimated the landscape by a prolonged slash-and-burn land management strategy which converted the region into a kin of the present-day grasslands.

The Avebury project also left continuing evidence of symbolic changes in the philosophy of natives toward their lives and a motivation for acquiring knowledge about the world around them. While some dated artifact symbolism may indicate that the aboriginals may have been global latecomers to notions such as pastoralism and communal agriculture, Avebury represents a leap of knowledge on multiple levels. This is another example of the extraordinary achievements of ancient humankind.

Avebury currently holds the status of being the largest henge complex in the world. The outermost geometric circle is over one thousand meters in circumference. A combination of stone and wood artifacts has been discovered as functional aspects to Avebury which were used at various times during construction. It is assumed the construction took several hundred years as is representative of most of Britain's henge projects. Although the research body remains conjectured as to definitive purposes for the existence of Avebury, some explanations include a ritualistic function to appeal to the powers of nature.[6] An additional major significant purpose and activity set at Avebury includes ancestor worship. This is the recurring theme involving functionality of most of the henges.

Stonehenge is the most noted and explicitly magnificent henge complex, as is exemplified by the ongoing research of other henges which are producing new knowledge as time moves forward. Geographically, Stonehenge is not far from Avebury and a group of other smaller complexes; it is also smaller than Avebury. It is approximately thirty-five percent of its size.

Stonehenge has some peculiarities from the others. In classic henge engineering, the outermost concentric feature is the raised mound bank that circles the structure. Immediately, concentric inward is a ditch that had multiple functions in the complex. Stonehenge's most external concentric piece is actually the characteristic ditch with the bank inside of it. At least this is how the current body of knowledge concludes it to be. Also many of the *trilithons* (two vertically-standing stone pillars supporting a third stone pillar laid perpendicular across the two vertical pillars) were much larger than the other henge examples. Each individual pillar contriving trilithon lintel rings measures twenty-five feet tall and weighs over fifty tons. Each trilithon segment was attached by mortise and ten on joint work. This unique construction feature is not seen around the world in the time of the Stonehenge project, which is currently conjectured to have started sometime before 3500 BCE.[7]

Similar to other British henge projects, it is estimated that Stonehenge took hundreds of years to complete. Started around the same time as Avebury, the landscape must have been prepared by eradicating big tracts of heavily-forested land surface. This was accomplished at the end of the slash-and-burn methodology the ancient communities used at the end of the Neolithic period. Stonehenge also used a lot of wood, stone, and mortar in its construction.

The most prominent construction anomaly which sets Stonehenge apart from the others is the importing and use of thousands of massive stone pillars used for the variety of trilithon segments, altar arrangements, and exo-concentric ring manufacture. One type of pillar, the igneous bluestones, had to be transported from over 150 miles away from present-day Wales in the West. While the bluestones were smaller than the sarsen stones, with gross weight being up to five tons, they were transported (more than sixty total) a lot farther and had to be moved down the mountains from where they were quarried. The sarsen stones, meanwhile, were the 50+ ton slabs and the tallest, those being over 25 feet in length/height when installation was complete. Each trilithon stood 25 feet tall, so the 50+ ton cross members were hoisted to sit atop each pillar pair. These features attract millions of tourists each year.

The Stonehenge construction anomalies help envision one type of out-of-place oopart for the natives of 3500 BCE. No written explanation explaining how this project was completed has been discovered, at least up to now. A best guess by archaeologists and engineers describes a wooden roller mechanism which was pushed the requisite transport distance of anywhere from 40 to over 240 kilometers up and down some mountainous terrain.

Stonehenge has been placed into the general protocol of other henge complexes by many research conclusions in terms of purpose. Also some common functionality is likewise attributed to Stonehenge. However, other functionalities differ markedly and help give it a more prominent place in the public eye. The purpose capacities inferred from study include: ritualistic and ceremonial activities, a burial ground, religious and ancestor worship, and a place of healing.[8]

Other blatant functionality anomalies within Stonehenge present another set of oopart hypotheses. The most comprehensive dedicated and enlightened investigation of Stonehenge was led by astronomer Gerald S. Hawkins in the 1960s. His book *Stonehenge Decoded* explained the discovery of the overall design protocols and motivations of Stonehenge.

Like a lot of ancient structures which contained multiple design purposes, motivations, and arrangements merged into one structure complex, Stonehenge was no exception. Hawkins discovered that Stonehenge was designed as a multifunction astronomical observatory, agricultural, and religious calendar. The layout of the complex is awed for its abilities to produce precise and accurate predictive authority for eclipses, equinoxes, solstices, and other cosmological events that assisted in planting, harvesting, ceremonial dates, and other related functions.[9]

A newly-discovered archaeological find that predates even Stonehenge and Avebury and is about as old as Nabta Playa is the Goseck Circle in the Saxony-Anhalt region of Germany. Discovered only within this century, chief archaeological investigators Francois Bertemes and Peter Biehl of the University of Halle-Wittenberg have begun studying Goseck.

What has been uncovered to date has been artifact radiocarbon dating to about 4800 BCE. There were at least four rings of concentric circles, a raised bank, a ditch, and two palisades. The appearance of the complex, after the researchers reconstructed major portions of it, closely resemble the British henges in many ways. The suspected uses of Goseck also include ritualistic burial ceremonies with the discovery of numerous animal and human bones. But the most significant purpose also contains the astronomical and climate seasonal attributes which duplicate the British henges purposes. The investigation of the Goseck complex, which could accurately be called a henge, is ongoing.

This situation is characteristically the norm with most of the ancient structures, as well noted as the ancient texts, paraphernalia, and such. Most of these complexes have only had a very small percentage of excavation operations performed. For example, the Tiwanaku and Puma Punka complexes have had roughly 2% of their entire sites excavated and explored. In Gobekli Tepi, this figure is about 5%. The Mayan complexes, in general, have had only a fraction more made accessible to thorough investigation. Like any science research study, there is a significant fiscal attribution to being able to conduct their investigations. Many governmental and political impediments also disallow further study of additional ancient sites.

One other significant and correlative anomalous oopart bears mention at this time and will be inspected as we move through our survey. Investigations of almost all these ancient sites have not uncovered any significant tool artifacts which could be attributed to construction activities! From our knowledge of the length and spread of the epoch which defines the Stone Age as attributed to such ancient structures and paraphernalia, we would not expect to find any metallic tool technologies in any remains until after around 2500 BCE.

However, in accordance with our survey of the henges thus far and next the Gobekli Tepe complex and others, we would expect to find stone tools. Not many of such tools nor remnants have yet been found. It would then be a rational supposition that the general question of "How did the ancients build these structures?" is an entirely valid and reliable one.

This notion will be threaded through the entire survey of ancient structures and paraphernalia impedimenta.

In summary, these henge design efficiencies created another set of oopart anomalies on the larger landscape. In yet another part of the ancient world, geographically isolated from other continents, there resides monumental structural artifacts produced by different societies multiple millennia prior to more modern technological ages. The technological tools available to those ancients did not include command of the mathematical and engineering knowledge needed for a successful completion. There is no evidence that can allow one to conclude they were not more advanced than what has been historically displayed by a Stone Age post-Neolithic society. This would include the lack of evidence indicating they did not have knowledge of wheel technology.

The survey of anomalous ancient structures, which extends around the world and spans thousands of years, continues over to Cambodia. Angkor Wat is the largest religious structure in the world. Measuring over four hundred acres, it was built very recently in the twelfth century AD. Angkor Wat was constructed by the Khmer King Suryavarman II regime as a Hindu temple of the Indian god Vishnu. In Hindu mythology, the divas were represented by Mount Meru, and Angkor Wat was designed with this in mind.[10]

An exterior moat and outer wall are over two and one quarter miles long. Early after Angkor Wat began its service, it was gradually transformed from a Hindu center to the current religious significance that of Buddhism. Angkor Wat was entirely a place of worship, ritual, and ceremony. There were no residential aspects in its design.

Construction of Angkor Wat does not offer an unusual oopart attribute issue per se. Yet some features make it a magnificent work of architectural craftsmanship and functionality in any time period. However, when situated back in ancient times, this achievement becomes more fantastic. It incorporates the seemingly typical astronomical fascination elements that humankind has replicated since the last Ice Age. Solar and lunar time and seasonal cycles are a part of the architecture.[11]

Angkor Wat was constructed from over 10 million sandstone blocks each weighing up to 1.5 tons. The outer wall, which is enclosed by the moat, is over 3,300 feet wide and 2,600 feet long. The moat itself is over six hundred feet wide. The stone had to be transported from over twenty-five miles away. It is estimated that it took many thousands of workers to manufacture and install the structure.[12]

The most distinctive feature of Angkor Wat is the textual carving of virtually all the exterior facades that constitute the entire complex. Archived records depict the comprehensive history of the Khmer in ancient times when Hindu, then Buddhist influences dominated Angkor society.

The tool inventories, as defined from the construction period of Angkor Wat, assimilate more sophisticated metallurgic technologies than most of the other ancient structures. These include bronze and iron technologies, as well as stone tools of some fashion. Archaeological finds support this feature. Angkor Wat is one of the few ancient complexes where this phenomenon has occurred.

Gobekli Tepe is another recent discovery attributed to exceptionally ancient origins. Situated on a mountaintop and ridge in the Anatolia Region of Turkey, Gobekli Tepe was first discovered in the 1960s. The initial real significance, though, was not discovered until the last twenty-five years. Gobekli Tepe is estimated to be at least ten thousand and up to twelve thousand years old!

There has been a newly-emerging uncovering of over two hundred T-shaped stone pillars forming twenty concentric circle arrangements. The pillars, which loosely resemble the trilithons of the henges, stand about twenty feet tall. The pillars are set in holes dug out of the bedrock foundation.[13]

The sheer age of Gobekli Tepe is indicative of the vast depths of the complex that need to be excavated in order to access significant new knowledge. With an estimated age of around 10000 BCE, it travels back to the time that may predate man's Agricultural Revolution and the genesis of the Neolithic Age!

There is a good general point to consider when studying an investigation of ancient remains of any kind. The study can be paraphrased by something like this: "The older it is, the deeper one has to dig mostly." The natural processes of geology and geophysics will bury remains and artifacts as big as structures in a very short time. Another visualization example of this fact is that of the Moai of Easter Island. We have mostly seen the images of just the heads and maybe shoulders of the hundreds of relics left on Easter Island. What we have not seen is the earliest paintings of the Moai as they actually appeared. They were head and body built by the indigenous society. The bodies got buried over time, which makes the Easter Island relics all the more oopart.

With Gobekli Tepe, an ample amount of later activity encroached atop the older ruins. Among the oldest discoveries are structures from such times that predate any use of pottery. Many buildings have been unearthed. Ample evidence of rock quarrying has already been discovered. Many monoliths over thirty feet tall have been discovered as have their quarry origins. Remains of some of thet-shaped trilithon-like pillar megaliths have been found still residing in their quarries.

Many other quarries are scattered around the landscape. The ages of last activity in each vary somewhat but are presumed to be among the earliest active sites within the complex's history. Much animal sculpturing was done and has been found but curiously not nearly as much human imagery. Though some of the pillars do show human body parts, such as arms, carved into the T-shaped pillar trilithons. The visualization is of a stylistic representation or the desire to depict images of godlike representations.[14,15]

The sophistication of the carvings are out of place in such a time in that Gobekli Tepe is thought to be archived in a period in which hunting gathering was still the primary food producing activity. The quality of the façade relief of the stone monoliths and relics appear to belie currently accepted notions and conclusions about the attribution possibilities to the natives. Either a rush-to-judgment fallacy is at fault here or some recalibration of the attribution versus timeline scales must be made if it was possible for those humans of over ten thousand years ago to produce such advanced manufacturing and sculpturing.

Due to the scope of necessary excavation currently required, some purposes and motivations for Gobekli Tepe's existence are undergoing more excavation and inspection. Because only a minute percentage of Gobekli Tepe has been unearthed, a lot of investigation remains. One generally agreed upon expected purpose is related to ritualistic and ceremonial activities. Another purpose under verification analysis is that it was an extremely large underground residential city. Due to its age, Gobekli Tepe is thought to have existed in a time of hunter-gatherers that were at least partly nomadic in nature. These features cause Goebkli Tepe to be even more remarkable. Its pillars and monoliths did not need to be transported far relative to other ancient structures of later times. The quarries were within a few miles of the complex. The weights of the various monoliths, though, regularly exceeded twenty tons and were as high as fifty tons. Historically, this predates the wheel, pottery, metallurgy, animal husbandry, and any language of writing identifiable today.[16]

The sheer age of Gobekli Tepe in relation to the many artifacts and structures already discovered and uncovered defies any currently accepted and practiced conclusion techniques used and offered so far. The representations in the carvings, monolithic aspects, ritualistic symbolisms, and structural architecture and manufacturing do not line up with what groups of humans could accomplish in that time. One could not even attribute use of the term *society* to such a people. They had not presumably evolved enough to form permanent village living arrangements yet.

Also because of the relative newness of its original discovery, the Gobekli Tepe complex has yet to yield any tool inventory evidence. As reasoned, if and when any inventory is discovered, it would be expected to consist of Stone Age technology only. As with most of the other complex anomalies of this type, the question of "Then how did they?" continues to stand out.

Next, Derinkuyu is a complex that was built as an underground city in the seventh century BCE. It has been determined that it was a residential urban living environment for up to twenty thousand citizens and all their living necessities. Derinkuyu is the largest of several underground urban environments spread across the Cappadocia region in Turkey. The accoutrements included animal stables, buildings for wine making, and other food manufacturing and processing, storage, schools, and religious palaces.

The mountains the city was excavated from are comprised of igneous rock. Some are very hard and others are soft. This region is highly susceptible to earthquakes and other geological hazards. Amazingly, failure of any of the inner passageways or other buildings would certainly cost many lives. It is therein that lays the anomaly of "How were they engineered the way they were?"

Situated all around Derinkuyu city are doors made of stone blocks which weigh over half a ton. These doors could only be closed from the inside. Only one person was needed to shut or open a door. They were made airtight to prevent anything from getting inside the structure. Purposes for apparatus such as these are unknown. Another fantastic feature of Derinkuyu and the other ancient cities is they were all connected by tunnels that traversed underground for over five miles.

All the inner structures were excavated from the igneous rock mountain to depths of almost three hundred feet. Thirteen separate levels were discovered from initial excavation projects. Results of all investigations to-date explain that the architecture and design were done in this manner to serve as the main living quarters for over twenty thousand residents. An additional feature offered these residents protection from outside invasion. A noted most effective function protected the city from attack during the Byzantine Wars of the late hundreds centuries and later by the Mongolians in the fourteenth century. These are some of the most fantastic features of Derankuyu.[17,18]

Other curious findings about Derinkuyu provide evidence that the city's livestock resided within the massive structure and that there are over fifteen thousand air shafts dug from the surface. Conclusively, they were installed underground to help maintain a continuous survival and long-term day-to-day living arrangement.

Derinkuyu contains no synergies with other common characteristics of other ancient structures such as astronomical, cosmological, or advanced mathematical design conceptualizations nor any extraordinary ritualistic or ceremonial symbolism apparent in other ancient structure complexes. The uniqueness and size and scope of the images we might have of a series of cities that housed thousands of residents would be astonishing. Knowing they were built entirely underground, separated by miles, and connected by tunnels also aids in appreciating that ancient societies should be given more credit than we imagine them to be.

Yet another ancient find, Baalbek, is a current-day town in present-day Lebanon with a population of about ninety-five thousand. The geography that Baalbek resides in has a history dating back as far as nine thousand years or more.[19] Ongoing scientific investigation continues to reveal fresh knowledge which is expected to continue far into the future. The major historical backdrop of extensive knowledge of Baalbek begins with Alexander the Great's conquest in the 330 BCE. For a few hundred years thereafter, Baalbek was known as Heliopolis (meaning "sun city"). Later, the Romans used the name "Heliopolitana" and was then connected to the Roman sky god Jupiter. Later, Baalbek was a site of battles during the initial rise of the Christianity movement.

The Baalbek complex is located within the Heliopolis confines that measure about four miles in circumference. A well-known segment is situated on the western end named the Temple of Jupiter. It is here the Romans performed a major reconstruction project within Baalbek that restored some of the infrastructure and adornment worn away by millennia of use and many earthquake disturbances in the region.

Baalbek's claim to out-of-place oopart anomaly is in the construction methods and the time in history that they were performed.

The Temple of Jupiter was rebuilt atop three monoliths (one of many where Baalbek sits as a gigantic raised platform) called the *three stones* trilithon. These stones are 62 feet long, 14 feet high, and 12 feet wide and weigh about 880 tons each. Further down the site in a place known as the stone of the pregnant woman, another stone weighs over one thousand tons (two million pounds). The largest stone, still further down the complex, weighs over 1,300 tons.[20,21,22] Baalbek contains dozens of these massive foundation limestones once used as support for a variety of megalith and monolith structures within the shell of the complex.

Stone materials used in the overall Baalbek project consisted of limestone, white granite, and marble. Later, the Romans also used indigenous pink granite for some adornments. All stone materials originated in quarries a few miles away in the surrounding mountain ranges.

It is a fact the Romans did not have anything to do with movement of any of these trilithon stones or any other foundational limestone segments. They were originally laid further back in time. The Baalbek stones are the most massive collection of such structural materials in known human history.[23] The archaeological record confirms human activity at the Baalbek site as far back as over 7000 BCE.

There is no unanimous conclusion as to how this anomaly of successful quarry and movement of single rectangular stone segments weighing over 2.6 million pounds could have been transported by humans. As we are aware, the ancients did not have technological assistance from tools such as the wheel or any other that were not believed to be part of human civilization dating as far back as nine thousand years or more.

TEMPLE OF BAALBEK

An historical site located in Western Bolivia in South America, Tiwanaku, is an ancient pre-Colombian locale. A second complex called Pumapunku is situated within the Tiwanaku site. Together, they contain many pre-Inca structural inconsistencies with the dictates of the time and technology capabilities of human societies.

Tiwanaku is thought to have been originally settled close to 1500 BCE. The original purposes and motivations for the Tiwanaku exo-structure remain uncertain due to the lack of a written language of the indigenous societies of this region, including the descendant Incas. Archaeological evidence discovered so far exemplifies typical human activities. This includes peripheral remains of pottery and other food-related life artifacts, as well as common ritualistic symbolisms such as small relics. From the congregation of general types discovered, Tiwanaku society was holistic similar to other global societies. The populations' life needs were accommodated by the ruling elite class of the time.[24]

What makes Tiwanaku and Pumapunka oopart are the advanced design technologies and the extraordinary stone craftsmanship man today could not achieve without tooled machinery and equipment only invented by more recent generations in the twentieth century. The Tiwanaku exo-structure consists of seven regions, including the Akapana, Akapana East, and Pumapunka, among others.

Some of the stones they used for their structures, in which all are indigenous andesite and red sandstone, weighed over 130 tons. They were quarried from around the region near Lake Titicaca, which was situated over six miles away. These had to be moved entirely over hilly terrain. Some of the stones, smaller andesite blocks used in Pumapunku, came from over fifty-five miles away from the other side of Lake Titicaca. These could have been floated across Lake Titicaca, but the remainder of the journey was undoubtedly over hilly terrain.[25] There has been no evidence discovered of use of wheel technology for these ancient projects.

A large pyramid type of ritualistic monument, the Akapana, was once surrounded by a moat. Its base was not squared but rectangular in nature. It measured over eight hundred feet by six hundred feet. The Akapana Pyramid stood only about fifty feet tall. The largest of the construction stones weigh about sixty-six metric tons. The largest stone block at the Tiwanaku and Pumapunku complexes weigh over 130 tons.[26] There is evidence of applied knowledge of astronomical and shaman-type ceremonial activities and human sacrifice rituals as part of their history.

Pumapunku, in keeping with the same thematic, was the truly symbolic and most active ceremonial and religious site in the complex and also the more enigmatic. The Incas maintained the site in an omnipresent religious attribute as they practiced the belief that it was where, "The world was created."[27] It is conjectured that, from a fragmented collection of artifacts discovered thus far, Pumapunku served as the central ceremonial complex at a time after the Akapana and Akapana East activities of the same. This would have taken the construction time frame forward to a few hundred years AD.

The Pumapunku complex is situated about a kilometer away from other parts of Tiwanaku. Current archaeological and geophysical investigation has determined that nearly 98% of the Tiwanaku and Pumapunka complexes are still buried and inaccessible due to present-day governmental concerns about the preservation of the sites. More of Tiwanaku and Pumapunka still lies out of view and needs further excavation.[28,29,30]

A question of two oopart incongruities to the existences of both Tiwanaku and Pumapunku asks, "How were the stones moved from over fifty-five miles away and over water then up and down steep embankments?" An abundant amount of hypotheses about this question are evident. More common theories include using special llama ropes or the use of ramps and inclined planes.[31]

The larger of the oopart anomalies asks about the quarrying, masonry, and precision achieved with the manufacturing and final residence of the stones in design with the structures themselves. These gigantic stones have ornamental, profiling, and cutting engineering so precise and sophisticated that not even a razor blade will fit between any two stones along their many meters' dimensions. The stones have cut features that are so smooth, angularly precise, and flush that machinery we use nowadays could not achieve all the precision that was accomplished with these megaliths and monoliths. Additionally, the use of higher mathematics is seen throughout the entire Pumapunku complex as well as the entire larger Tiwanaku complex itself.[32] Some researchers have offered conclusions that these were accomplished by stone hammers and flat stones and sand. This is inferred because some stone hammers have been excavated in nearby quarries.[33,34]

Pumapunku also exhibits use of I-shaped cramps that substituted for any concrete-related functions used to bind some stones. These cramps were made of a copper-arsenic-nickel bronze compound alloy. The largest stones also used these cramps, but they were made of as yet undetermined chemical composition.[35]

PUMA PUNKU

Citation:
Puma Punku
Brattarb/CC-BY-SA-3.0 (https://creativecommons.
org/licenses/by-sa/3.0/deed.en)

The advanced mathematical and spatial geometry achieved at
Tiwanaku is also inconsistent with the knowledge of the time. A method
of descriptive geometry which conceptualizes an extensive use of proportional sides, triangles, and figures as well as a required standard of
straight measure such as what we use as feet, yards, and meters was contextualized into the entire infrastructure.[36] These were achieved without
any demonstrated use or evidence of the technology of the wheel.

The Tiwanaku and pre-Incas cultural and spiritual practices were
in synch with what was happening in other cultures of the period.
There was prolific application of polytheism with the same attri-
butes of ornate and adorned ritual activities as that of the Egyptians,
Chinese, Indian, Mesopotamian, and Meso-American civilizations.
Pumapunku and Akapana were the core locations of these activities

for the entire world of Peru and Bolivia city-states. Illimani mountain was also connected with Pumapunku and Akapana as, taken together, they represented a sort of gateway between Earth and heaven. The connection of the people to their gods was the driver of Tiwanaku's ongoing existence. As with the other cultures of the time, a what-if scenario that did not include these activities would have meant the end of those societies. One other rite of the Tiwanaku that has been discovered to have been replicated elsewhere was the practice of human sacrifice.

Many have tried to comprehend some of the procedures the ancients demonstrated with use of whatever tools utilized for building these monolithic and megalithic structures, if in fact they did. As with the preceding sample of our survey, very little discovery or recovery of any inventory from the Tiwanaku or Puma Punku complex sites supports other explanations for their construction. None have been offered in the scientific research or refutation/debunking literature stating humans performed all the construction without help from any other source. It is apparent to note the frame of time being discussed in this investigation is thousands of years and dates from around 10000+ BCE until AD 1000+. In those times, humankind began an agricultural revolution, further developed stone tools, and evolved to discover bronze metallurgy and then iron metallurgy.

There is a comprehensive lack of tools which have never been recovered from archaeological sites such as Gobeckli Tepe, Derenkuyu, Mohenjo Daro, Tiwanaku and Puma Punku, Teotihuacan, Chichen Itza, Stonehenge, Avebury, and many other ancient global sites. Discovery of at least some of these tools could help explain and support opinions offered from skeptics who believe humans performed all the design, engineering, construction, and transformation of these geographical sites into the enduring structures we see today without any help from any other sources. The structures themselves have existed for up to twelve thousand years or more, yet the tools purported to have been used are nowhere to be seen.

A public broadcasting system airing of the television show *Nova* episode titled "Secrets of the Lost Empires" in 1998 motivated an interview with archaeologist Julian Richards, a principal investigator of the Stonehenge mystery. When it came to the topic of tools, technology, and potential success for engineering the construction of Stonehenge, the following statements were offered:[37]

Stonehenge: Expert Q&A

- Q: What type of marks, if any, were left on the monoliths as evidence of how they were moved?
- There are no marks on the monoliths that provide evidence of how they were moved.
- Q: In a book I read, it said that they probably put burning branches on a place they wanted to cut then poured cold water on, cracking it. Is this what your experiment showed that they did?
- We didn't really go into the shaping of the stones, but fire is one way of breaking and shaping a stone like sarsen. It obviously carries risks, and having quarried a forty-ton block, it would be unfortunate to crack it in the wrong place. My feeling is that most of the shaping is done by pounding the surface of the stone with mauls ranging in size from footballs to small grapefruit.
- Q: To move the stones, could the ancients have lashed enough logs to the stone to form a cylinder, loop ropes around the complete assembly, and pull on the upper loops to roll the stones to their site?
- This was one of the ideas that Mark and I discussed and then rejected when we were thinking about how we could move the stone. It would certainly work but could be potentially very dangerous when trying to control a forty-ton garden roller going downhill.

Here is a comprehensive experiment to assist in gaining an understanding of the lack of relationship between the tool technology, the presence of the ancient structures, artifacts, and paraphernalia and man's potentials for achieving these manifestations. Think back through your life and try to remember those times you saw things rolling along the ground down a hill or some embankment or any sloping ground. The things can be rolling logs, rocks, or stones, a bottle, or anything else that is the same visualization I have given you. Looking back through our memories, we can seemingly visualize an object such as, for example, a rock, bottle, or ball rolling down a hill or slope.

Now let's take this analogy and transfer it to the exact same set of circumstances except as observed by an ancient native from anywhere in the world. Consider only the natives who lived before the first practically applied use of the wheel as a transportation mechanism. The wheel was actually invented around 3400 BCE, but the only use for the object in this shape was for making pottery. No widespread use of a wheel mechanism engineered for transporting arrived until the ancient Egyptians after 1336 BCE when King Akhenaten used chariot technology.[38]

Next, we can take all the times we observed objects rolling down sloped platforms or ground and transfer this analysis to those people. Consider the sum of all the collective lives and man-years of all the ancients who lived on Earth during that time. This equates to over 2,200 years by another conservative estimate. For all these people who lived during this time, there was no one who could transfer the critical thinking from their observations of objects which rolled along the ground to adapt something, such as the pottery wheel, to this kind of use. It took civilization over 2,200 years to make the adequate connection and then to invent the wheel for this use!

Humans, therefore, with this reality confirmed, could not fathom any other creation, evolution, or use of the wheel even though they had the schematic technology in their possession for millennia before they applied topical use of the device. Yet a lot of intelligent people currently presume and assume that the pyramids, Gobeckli Tepe, Derenkuyu, Mohenjo Daro, Tiwanaku and Puma Punku, Teotihuacan, Chichen Itza, Stonehenge, Avebury, ancient Baalbek, and other superstructures were built by the same people with no sweat.

To summarize so far, the arguments so far and coming up paint a growing portrait of unique and dedicated forms of symbolism embedded within these ancient structures. Many of the ancient cultures all over Earth share these concepts of symbolism in those ancient sites. Their individual megaliths, monoliths, and obelisks share many similarities in design, architecture, construction, purpose, longevity, but an incongruent fit with the times they were constructed.

The ancient Mayan and Aztec structures are other examples of similar purpose and building motivations for other more ancient cultures. Some texture and proliferation idea of Mayan and Aztec cultures was noted in the pyramid's passage. Far more pyramids were built by the indigenous societies in Mexico and Central America than the rest of the world combined.

Also by example of our upcoming discussion of Teotihuacan and Chichen Itza city-states, one cultural commonality shared by the Mayan, Aztec, and pre-Aztec with the ancient pre-Incas and others is the urban planning of their infrastructures. This respective planning was achieved by directly incorporating the pyramids and ceremonial temples into the physical landscape and culture of the city. This means everyone lived within the general confines of the municipal district bounded by these structures, as evidenced by the archaeological remains in acquisition, as well as a measurement of the dimensions of the city. The major city centers or capitals measure in the tens of miles in circumference. Both Teotihuacan and Chichen Itza are estimated to have had populations of over two hundred thousand. Indeed, this is why the Mayans and Aztecs are historically world-renowned as the most advanced ancient urban planners.

Teotihuacan (pronounced teo-tea-wacan) was the earlier built pre-Aztec urban infrastructure. Because at its peak, Teotihuacan's municipality measured over 11.5 square miles, it has presumably earned the description of being called a municipality instead of a complex. The earliest known establishment is believed to have been dated to around 300 BCE. The infrastructure was under construction for another three hundred years after. As noted previously, family life was central to the living space of the Teotihuacan municipality. Residential multifamily buildings continue to stand and are documented by magnificently colorful adorned wall murals narrating life back then. Original tools made of obsidian survived the millennia. Such igneous rocks as obsidian and certain types of jade, such a pyroxene jade, are commonplace due to the volcanic geology present in this region of the world.[39]

There is some disagreement as to the original founders of Teotihuacan. The city-state of its name predates Aztec appearance by about one thousand years.[40] It is most generally agreed to have been a multicultural city center, as evidenced by a diverse socioeconomic history evidenced in wall murals. These murals and other structural adornments take on more importance for the Aztecs and the neighboring Mayans due to the dearth of written scripture and other texts. As noted elsewhere, the Spanish Conquistadors destroyed most of these written records. Their main reason was due to the primarily pagan symbolism documented that was diametrically opposed to the Catholic religious beliefs of their invaders.

Most of the agreement of the original name as attributed to Mayan hieroglyphic texts as Place of Reeds. Spanish invaders were very virulent against the Aztec inhabitants of Teotihuacan. The Main Street was known as the Avenue of the Dead. Two great pyramids, the Pyramid of the Sun and the Pyramid of the Moon, and the Pyramid of the Feathered Serpent adorn the Avenue of the Dead. The Pyramid of the Sun is the third largest pyramid in the world, smaller than the Giza and Cholula structures. A predominant, of many, architectural styles attributed to Teotihuacan were known as talud-tablero. A structure's external side was located beside or atop a rectangular-shaped inner building.

There was never a period of monotheism in ancient Meso-America. Teotihuacan structure and spiritual culture was highly ornate and colorful. Jade and gold were the most popular and valuable metals. The religion of Teotihuacan worshipped eight deities. These deities were known as:

1. The Storm God
2. The Great Goddess
3. The Feathered Serpent
4. The Old God
5. The War Serpent
6. The Netted Jaguar
7. The Pulque God
8. The Fat God[41]

A vast amount of time and effort was dedicated to practicing polytheism in their sociocultural and spiritual world. Recreation and war were contextualized around worship and ritual which included human sacrifice.[42]

CHICHEN ITZA

Citation:
El Castillo in Chichén Itzá
Daniel Schwen/ CC-BY-SA-4.0 International (https://creative-commons.org/licenses/by-sa-4.0/deed.en)

AVENUE OF THE DEAD
at TEOTIHUACAN

Citation:
Avenue of the Dead at Teotihuacan, view from Pirámide de la Luna
MIKHEIL/CC-BY-SA-4.0 International (https://creativecom-mons.org/licenses/by-sa-4.0/deed.en)

As is exemplary of the predominance of ancient structures, it appears requisite that most of them incorporated some multiple levels of sophisticated engineering, mathematical, astronomical, climate-weather, calendar, agricultural, and socioeconomic attributes. Teotihuacan is another example of a polytheistic culture that somehow obtained and used a lot of very advanced architectural techniques to produce structures that defy explanation.

Another Meso-American anomaly is the Mayan Chichen Itza city-state. As with Teotihuacan, the municipality covered an area of two square miles, much smaller than Teotihuacan. Construction began around AD 600. At its peak, Chichen Itza maintained a population of over 125,000. The same municipal urban plan functioned as in Teotihuacan.

The streets were stone paved and called *sacbeob*. Over eighty of these have been discovered so far. One unique infrastructure feature not shared elsewhere was a response to a problem of lack of overland rivers to supply water for Chichen Itza. They solved this problematic situation by planning their infrastructure around sinkholes called cenotes. Even their architectural features incorporated roof drainage systems so rainwater could help replenish these cenotes and provide a year-round water supply for the entire population.[43]

Stone for many structures was imported from Central Mexico hundreds of miles away a large number of cases. The Chichen Itza utilized the same ornate architectural features, extensive hieroglyphic murals, and reliefs on structural facades that the Teotihuacan did. Two unanswered anomalies within these contexts are again, "How was all the stone transported?" and more significantly, "How and where did the Mayan acquire the linguistic hieroglyphics from?"

Some of the more massive structures are the Temple of Kukulkan, the Skull Platform, the Platform of Venus, the Platform of the Eagles, and the Jaguars. The Temple of Kukulkan is also known as El Castillo.[44,45] The steepness of the Chichen Itza pyramids is also unique to Meso-America. Many exterior facades rise at more than a forty-five-degree angle. This is not a feature elsewhere around the ancient world.

A small inventory of stone tools currently resides in various museum sites scattered around Mexico currently. These tools only shed a little knowledge of how small handheld items were fashioned from use of those stone tools. The large more megalithic blocks used for pyramidal and temple construction projects are still absent of the existence, knowledge, and reality of corresponding tool inventory sets.

The Meso-American ancient cultures were shorter in life-span duration than others in ancient times. One reason for this is that geologic climate shifts were frequent. They included extended periods of drought. This is exemplary of the dedicated worship of the sun god, the moon god, and various warrior gods. The extended climate degradations gave rise to more tensions among the natives and more wars resulted. One unique ritual occurred around the Cenote Sagrado. Archaeological discoveries excavated numerous objects from the bottom of the cenote such as gold, carved jade, pottery, skeletons, and a lot of other life artifacts.[46]

Another anomaly that ties into the Chichen Itza pyramids involves the ritualistic architectural feature on El Castillo. Also known as the Pyramid of Kukulkan, the Mayans engineered El Castillo to offer yet another unique worship tribute to the feathered-serpent god, Kukulkan, himself. On both equinoxes each year, available daylight, shadows, and the engineered façade synchronize to perform the act of Kukulkan "rising up the stairs of El Castillo or descending them to Earth." This symbolizes and allows for Kukulkan to be able to descend to Earth to help cure a problem for the native people and to later ascend into heaven again.[47]

In terms of the ritualistic practice of human sacrifice, the Mayans of Chichen Itza left many hieroglyphs depicting such events. Glyphs of decapitations adorn many structural walls around the city. One contribution these glyphs give to our knowledge is numerous depictions of decapitations where "streams of blood in the form of wriggling snakes" is shown.[48] To the Mayan, blood vessels were depicted as snakes. Sometimes, when there is reference to the worship of snakes, the real symbolism goes one level deeper and becomes an attribute to human blood vessels. The Mayan and Aztec and any other language and linguistic protocols of the period did not have terminology that defined blood vessels. The respective society used depictions that generally observed things in nature to describe and define something they did not otherwise have a symbol or term for. This was a very basic part of the human thought process and communication back then.

When discussing language and linguistics, another complexity in drawing accurate conclusions about the practices and capabilities of ancient societies arises. Here, the ancient Meso-Americans left us a treasure trove of carvings and texts depicted as hieroglyphics. These are the exact style and type of communications used in ancient Egypt and the Mesopotamian civilizations before them! All the obelisks, pyramid facades, and tunnel galleries tell the Mayan and Aztec histories with the same linguistics and same style as the Egyptians. How was this evolution portrayed in such geographically dispersed and culturally diverse societies?

We arrive now to an example of an ancient anomalous oopart that was not built by a manufacturing of colossal hard stones of unimaginable weight and subsequent transport of up to hundreds of miles. There is not as much architectural proclivity of advanced mathematics and technical engineering inherent in a visualization of these structures as in other ancient structures. Just an exponentially abundant amount of symbolism that, along with conjectures about spiritual and ritualistic purposes, allows one who has knowledge of such to reflect and compare with the psychology and sociocultural motivations behind the cargo cult phenomenon as noted previously.

The largest single geography devoted to one structural phenomenon in human existence is the Nazca collection. The Nazca Lines are a massive collection of geoglyphs that can be conclusively determined to be the achievement of the native Nazca civilization. Scholars date their creation to as far back as around 500 BCE. They are situated alongside and atop mountain ridges and plateaus between the Peruvian towns of Nazca and Palpa. This region is situated about 250 miles south of Lima, the Peruvian capital.

The Nazca "palette" stretches for over fifty square miles. There are many hundreds of large structural lines, figures, and designs throughout. The dimension of such objects range from a collection of straight lines and geometric figures each in and of itself to zoo-morphic animals, *phytomorphic* plants (having plant-like attributes) and trees, and *anthropomorphic* (having humanlike attributions) castings. The single largest representation measures over two hundred meters across.

The investigative motivation is not directed predominately to an analysis of how they were made but of why. The Nazca Lines were among the easiest ancient design projects to construct. Each figure was made by digging shallow trenches of about six inches deep into the native soil and rock. The trenches were then traced to design specifications. Here, the stone tool sets used are more apparent and supportive of human technology acquisition and realistic usage than most of the others.

This part of Peru has maintained a very arid desert for thousands of years. The native soil is a combination of red-brown iron oxides as the top layer. Underneath the requisite, six inches or so lies the next strata of white clay deposits. When the surface layer of oxides is scraped away, the white layer beneath it is exposed to light. Because the climate around Nazca also produces very little wind, this together with the desert geology allows for a synthesis of designs that would last for many centuries. Thus, the how questions about the building of the Nazca Lines seem pretty easy to explain accurately with the use of stone tools.

The native Nazcas engineered the designs and dug the trenches needing a fraction of the time that other communities needed to construct their own megalithic structures and obelisks. How long the entire Nazca region of geoglyphs took to accomplish this has remained unknown. Some mathematical knowledge would have been necessary, even in an intuitive process of critical thinking to successfully complete the Nazca Lines.

The final structures are renowned for their exactness and symmetry in design at all points where needed. A few years ago, researcher Joe Nickell, a scholar from the University of Kentucky, simulated such a construction. His experiment proved it may have taken only a few weeks for the natives to complete one design. The design parameters of Nickell's experiment used only tools available to the Nazca at the time. Prior archaeological finds that added to the knowledge set included wooden stakes discovered at the ends of a few of the glyphs. Presumably, this was inferred to mean the Nazca utilized this mathematical intuition to develop and complete their designs.[49]

It would have been somewhat difficult for the Nazca to successfully complete all the designs with inherent technology, design tools, and cognitive knowledge. Of all the survey of ancient structures, though, this may have been among the easiest to accomplish. Conclusions by the research communities regarding how Nazca was built seem more plausible than most others of the same endeavor.

Before we move on to the why investigation, a relevant segue warrants another point of brief discussion. Remember that the Nazca designs are each hundreds of feet in one dimension. When one is standing next to a design, no identification is gained from that viewing angle. A person must be, at the very least, on a mountaintop miles away looking down and/or from an airborne viewpoint to gain a proper perspective of what the person sees. Some researchers have inferred there was only a ground perspective for the natives that was purposeful and explained the entire Nazca experience and existence. The Nazca Lines were made to point the native people to nearby mountains as places where their gods could be worshipped.[50] This is shortsighted, fallacious, and probably dismissive of a lot of accuracy and rationality. Undoubtedly, the intent of the final design and construction was to have a significant perspective of analysis and meaning from the air or far enough off its ground zero on higher mountain terrain to make that meaning conceivable.

To begin a summary of the content which is the discipline of ancient structures, there are examples of analyses from research literature that seemingly lack what a proper, accurate, and thorough investigation should undertake in any field of endeavor. A point of discourse between ufology and the sciences is the contention by ufologists that not many scientists are applying the proper and correct methodology to research and a complete lack of thoroughness that has a common end point of such conclusions as a rush to judgment and "my mind is made up" logical and critical thinking fallacies.

A good example of this tract is evident from the debate over the perceived purposes of the Nazca Lines. Many theories have been offered, often resulting from sight unseen and of dubious investigative that omits study of most or all the factors. The following are descriptions of some of these theories.

One published theory links the Nazca Lines to being physically used as giant textile looms to manufacture extensions of yarn materials that were later used in wrapping mummies.[51] The only fact that could correspond with any construct of this study is there was yarn which was used to wrap mummies. A separate theory that describes how the Nazca Lines were made concludes that hot air balloons were used, which declares hot air balloons were the only possible means of flight available.[52] There has never been a discovery of any such materials that, together or separately, could contribute to any construction of a vehicle of this type by any ancient society in man's history. This theory has consequently been met with published rejections from the communities citing a lack of evidence of such balloons.53

Another theory proclaims the designs, specifically including the line images and the animals and plants, were made as fertility symbols and as irrigation systems for water flow. Also theorized is the lines served as a type of astronomical calendar.[54] Yet another example was a theory presented that one of the lines, the large spider geoglyph, is really a pictorial of the Orion constellation and that other lines tracing to the glyph show movement over time of the three Orion belt stars: Alnitak, Alnilam, and Mintaka. This theory was critically refuted with the reasoning that there were other parts of the geoglyph that were not studied and of an omission of consideration of additional evidence made by other people.[55]

LINES OF NAZCA

Citation:

The lack of information and knowledge or the unwillingness to devote proper time to study all of the facts of an investigation have been examples of what the literature has frequently called bad science. Also the discourse over a study topic sometimes involves one human being taking the researcher to task because they have not considered all the evidence offered by everyone else. An example of this references the last noted example of the omission of consideration of other evidence made by other people.

It should be recognized that people will offer different perspectives on the same study. This leads to different conclusions and is a large part of an individual's unique life perspective. Most often, this situation becomes the ultimate catalyst which leads to new inventions and revolutions of the paradigm being practiced. As noted elsewhere, an expert researcher on this topic was the scholar Thomas Kuhn. The manner with which he studied a problem was an innovative one. He was able to do so without disrupting the system of the paradigm while finding new discoveries of knowledge and invention.

The points of refutation regarding the creation and existence of ancient structures and any such connection between native human-kind and possible assistance from others as living entities seem to have recurring themes among the cultures of that period. This period is delineated to reside within the concept of the Late Neolithic and Megalithic Periods, depending on what science discipline you are attributing the term. Numerically, the time period spans from about 5000 BCE to after 1000 CE. This time period cannot be all-inclusive of the structures that fit this study known to exist and have yet to be discovered.

The evolution of ancient structures among individual cultures involves both similarities and unique characteristics. To start with similarities, the pyramids emerged across all cultures. The purposes the indigenous societies contextualized offer many similarities. The most significant include two general themes. The first was the ritu-alistic worship of gods and goddesses and the ceremonial pastimes that defined them. The second was a macro view of astronomy, cos-mology, agriculture, and animal husbandry that were applications of a study of the sky.

The ideologies within the context allowing these practices to be enacted permitted the societal structures that ultimately existed. This is a third similarity among ancient cultures with some localized cus-toms evidently included. What is meant by this is ancient societies were holistic in nature. All of a society's needs for survival and were provided for them by governmental authorities. Analyzing their daily activities, the ancients woke in the early morning, worked their jobs on an extremely communal project, and spent dedicated and signifi-cant amounts of time practicing ritualistic recreations that have been presented. The ancients' whole way of life was structured around an intimacy level with their earth, origins, and future—a future we find nowadays at least foreign and maybe more extreme. Generally, we do not understand much of what it was to be ancient Egyptian or Inca or Chinese. This notion influences research decisions which are made when topics about oopart and out-of-place phenomena arise.

So the ancients conducted their daily lives using much of the same notions about nature, governance, and fulfillment and hope for their future lives. This perception may not be entirely unusual of itself but rather a manifestation of "what it is to be human." This may or may not mean it was possible a hard-wiring origin of the human brain existed or exists. At the least, we are partly a product of our environments. At any reasonable period in time, those environments are different so as to produce an evolution of physical, physiological, and psychological alterations in any living thing and any physical change in a nonliving thing.

A point to be made concerns similarities among ancient peoples that seem less probable and even less possible given the environment and circumstances available to them and on their own. For example, there are numerous similarities in culture, linguistics, architecture, religion, ritual, and appearance between ancient Rome and Greece. Obviously, the geographical barrier that was a basic prerequisite for difficulty of assimilation was not there. So we could observe and conclude it was reasonable for Rome to have appeared like Greece in many ways.

More highly unusual is a similarity between the language and linguistics of ancient Egypt and the Mayan and Aztecs. The hieroglyphics used in both historical text and carvings are the same. This is in addition to the morphology of the pyramids and obelisk monoliths being in lockstep with each other, the general polytheism of worship, and the physical adornments of their structures also reflected the same levels of agreement. If such important manifestations of two cultures separated by over five thousand miles and an ocean and the sea are exactly the same, why then is it that one practiced human ritual and the other did not? There were no wars between Egypt and Meso-America. It would have been necessary for the Egyptians, being the earlier existing civilization, to travel to Central America to impart their influence. There is no evidence of this or of any such contact at any level.[56] One civilization did not know of the other. But all these similarities exist and are peculiarly exact in the amount of correspondence and likeness. How and why?

Similar peculiarities exist among ancient Chinese and Peruvian cultures. There have been abundant correspondences among Chinese pottery, vessels, artifacts, geoglyphs, clothing garments and even carpet designs, and those of ancient American Michica and Chavin cultures.[57]

A peculiarity in a different direction exists among the ancient Indian and Egyptian cultures. There is known substantial interaction between these two cultures in the millennia between 3000 and about 200 BCE. In the earlier times and with the frequent contact among traders and foreign residents of both societies, similarities existed between Egyptian culture and the Indian Harappan people of then northern India. The Harappan built gigantic cities and had the adorned temples, buildings, and paved roadways an observer transported back to those times would also see in Egypt.

Then after about 2000 BCE, the Harappan civilization disappeared. Despite the still frequent contact among the two peoples, the descendants of the Harappan did not construct any buildings again until around 200 BCE. This was the beginning of the Stupa Period of ancient India. The Indians seemingly forgot how to build big stone structures even though the Harappan nation erected some of the most magnificent and functional structures in the ancient world.[58]

In continuing the summary, let us move from the peculiar cultural similarities where there should not be and the equally peculiar non-similarities where there were at one time. There were many examples of peculiarities where structures that should not have been able to have been built were and still exist for our inspection. Let's also move away from the many "easy" explanations from field researchers that dispel any notion of man needing any help to build them.

A list of examples using the obelisk monolith structure concept as related to the possibilities of man successfully building them alone and with the then most advanced technology is included for two reasons. First, the examples are easily transferrable to analyses of similar investigations of the pyramids, henges, temples, and all other structures. Second, the existence of obelisks has not been discussed but is an important out-of-place example of why and how accurate conclusions of their construction solely by humans seem unlikely.

The history of obelisks dates back to ancient dynastic Egyptian times. The oldest still in existence is the Senusret I at Heliopolis and dates back to about 1971 BCE. The oldest known from written and carved records dates back to the Fourth Dynasty with the existence of the Ra-Atum obelisk also in Heliopolis. Most of the early obelisk projects were quarried in this area known to the Egyptians as the Wast. The Wast was divided into two areas, known as the Luxor and the Karnakareas. The most recent archaeological excavations indicate a dated time to around 3200 BCE, around the time of the creation of the then modern Egyptian state when Ra, the sun god, was the supreme being.[59]

Also popularly known as *steles* or stela, the obelisk was quarried on one piece from whatever rock deposit they used. As with the pyramid and temple experiences, most of the time the monoliths were somehow moved miles to their final destination. Steles on record have weighed many hundreds of tons. A few discovered monoliths have been discovered still in quarry that are estimated to weigh well over one thousand tons (two million pounds). Egypt has the greatest number of obelisks on the world scale. The most popular locations are in the Luxor region of the Nile River Valley. A few were used as burial markers, but most were entirely adorned in hieroglyphic carvings and placed in pairs at temple and pyramid entrances and used for other reasons.

Purposes inherent in obelisk ideology are straightforward and reasonably conclude that the linguistic and textual attributes are still directed at the presence of gods, goddesses, astronomy, and cosmology. Obelisks also had attributes related to astronomical earth-sun-moon observations and are related to agriculture and food production. Again, being that most of them were not discovered or seen as burial markers and that nothing else remotely like or as big and heavy as these structures ever existed until the "all of a sudden" time of the late Neolithic or Megalithic periods, how and why did they get there?

Some researchers have conducted simulations in an attempt to answer one question: was it possible for humans to erect such ancient structures on their own. At the outset, it must be mentioned this series of endeavors is one example of only a few dedicated investigations that have been undertaken regarding ancient culture questions and ufology to seriously study such phenomena according to the possibility of humans being able to do these without help. There exists a deficiency in the quantity of, the willingness to, and the motivation for focused and intensive investigations by serious researchers of questions and phenomena related to ufology. Here is a summary of one series of these project simulation experiments.

First, "How can one calculate the weight of an obelisk or any other stone rock monolith?" If one cannot lift it, how can it be weighed? Science, mostly from physics, uses their basic formula for the weight of an object. This equation in known as volume x density = mass. In physics, mass and weight are two concepts with entirely different meanings. In the context of our examples, though, weight is a correct variable concept to use.

The volume variable is complicated by the fact that many objects such as monoliths or pyramid stones are at least partially hidden from view and also that the objects are not exactly cubical or otherwise entirely symmetrical in shape. Its calculation becomes a lot more complicated. The density variable presents more serious problems in that it is the density of the stone that needs to be accurately identified. This requires an exact identification of the type of stone being measured. The indigenous peoples frequently used granite and andesite, the types of stone available to them during that time. Often, there were many types of, say, granite or andesite that were available and then used. These all have different densities. To be precise, calculation is itself complex for the physicist to accurately accomplish.

For archaeological and geophysical experiments, it is a reasonable assumption to allow for an error estimate of up to fifteen percent when calculating weights of heavy objects that cannot be weighed using other methods.[60] Let's investigate a few examples of documented obelisk projects, only some of which were experiments performed by modern research scientists. Some of the experiments were actual events.

The obelisk that stands today at Saint Peter's Square in Rome was erected in 37 CE at the site about three hundred meters away from its current position. Pope Sixtus V served his reign from 1585 until his death in 1590. He wished for the obelisk to be repositioned at the front of Saint Peter's Square. Sixtus V commissioned the project begin immediately.

The obelisk was over 80 feet tall and weighed over 330 tons (660,000 lbs). It was to be moved nine hundred feet per the pope's wishes. The moving project began on April 30, 1586. It took 18 days and 1000 men, 140 carthorses, and 47 cranes to move the obelisk to the erection site.[61] This project was performed in 1586, thousands of years after the period of the majority of obelisk quarrying and building occurred. The weight of it is representative, though not the heaviest or lightest, of the ancient monoliths built in history. The technology of the day in Rome was far more developed than what was possible for the ancients of their time. They had in their possession the wheel, cranes, levers, metallurgy, Leonardo Di Vinci, horsepower, two thousand years of mathematics, engineering, physics, and all the other natural sciences.

This example is an experiment that was conducted many times by the same research team in the 1990s. In late 1999, the team led by research scholar Mark Lehner of the University of Chicago and Roger Hopkins, a Massachusetts stone craftsman, demonstrated this experiment on an episode of PBS Television's *NOVA* series. They were trying to erect a twenty-five-ton obelisk. They were not trying to move it. They had failed in two previous attempts. It took a lot more than 130 people pulling on ropes that constantly broke and an additional dozen workers employing and maneuvering levers to successfully stand the monolith upright. They had to prepare a path for a twenty-foot transport by laying wooden rails into the ground and then preparing a sledge for the obelisk to rest on. They moved it the requisite twenty feet. It took weeks of physical preparation for the event of the short movement of one typical monolith a distance of six meters and one erection procedure.[62]

In February 2012, a commercial company, Stone Valley Materials Quarry in Riverside, California, tried to move a 680,000-pound (340-ton) granite monolith over 100 miles to its final mooring at the Los Angeles County Museum of Art. It took a 1.4 million pound crane (700 tons) and a 44-axle tractor trailer that could output over 2,400 horsepower to even get the obelisk mounted. Steel girders, 208 tires, and two separate power plants were also needed to provide a push-and-pull strategy of locomotion. What remains unknown is how many people were used in the transport and how really long the physical preparations took. The transport project was successful as the apparatus was able to move the object at about five miles per hour. A special road route was required as most available road surfaces were not constructed to bear the weight of such a heavy object. Cranes and other modern machinery typical in the twenty-first century assisted in the project.[63]

PBS *NOVA* Television broadcast another monolith transport experiment in 1997. Roger Hopkins, who teamed with Mark Lehner in a later project, as noted, participated in a series of experiment trials to replicate part of a reproduction of Stonehenge. The only positive results of their numerous attempts was, with over 150 men and the use of dozens of manned levers spaced along the length, they were able to transport a forty-ton stone a short distance.[64]

There are more examples of such experiments and real-life attempts to move massive monoliths of the size and type included in this discussion. Some were successful and most failed. All were performed and documented in times that were only within the past two hundred years. The most advanced technologies of the day were used in all cases except for those specifically conducted as testing experiments. Even in a large number of those modern testing experiments, the researchers had to resort to contemporary technological apparatus to finish the experiment intervention.

These were far more sophisticated than were known available to the ancients. Estimates of the number of men needed to pull or push the obelisk are officially documented. A basic statistical measurement of average and deviation indicates the required numbers of people needed to achieve a successful test exceeded thirty men per ton of mass of movement in many cases. This indicates, if one of the Baalbek *trilithon* blocks of over one thousand tons were to have been transport tested, it would have taken, according to the researcher's calculations, thirty thousand men to pull on the ropes enough to move it.

Success is only possible by the factual prerequisite that none of the equipment used, such as the ropes and manpower, could break or break down during a transport or other construction sequence. Remember that omnipresent in this entire discussion is the immense power of gravity that does not take any time off on Earth. A movement sequence of any kind requires enough force to counteract gravity's force at all times to keep the object from being destroyed. So a three-hundred-ton monolith requires more than three hundred tons of force applied continuously to counteract gravity's force on a flat surface. When the surface is not flat, the forces change. Additional factors such as possession of the wheel or lack of it for locomotion across a land surface will change the gravity dynamic.

The use of such experimental testing and analysis to prove a hypothesis such as this is fraught with so many assumptions and prerequisites it becomes unreliable and invalid as evidence of an adequate explanation. The question of how and if humans of over five or four or three thousand years ago or less or more could have successfully quarried, manufactured, transported, and built the ancient architectural structures has not been answered by the investigative techniques and diligence reported thus far. Moreover, it is a fantastic exclamation point that we can even discuss these issues today with the real evidence still in existence and in remarkably good shape after millennia of weathering, climate, and—because of the long time frame attributed to this subject—geological degradation that has or could have occurred with these structures. Surely, some have been lost to such geological and geophysical processes, but there may be as many or more yet to be discovered. While proving or disproving a hypothesis can be accomplished with one smoking gun or eureka piece of evidence, the vast majority of the time it is the incremental smaller pieces of evidence, discovered by researchers over many studies and experiments, when—taken together in a pattern—provide an analogy to a jigsaw puzzle which results in the proving or disproving of the hypothesis. It happens in science; it happens in the law, in business, in everyday life.

References

[1] Earthworks Ref 1: Blumenthal, Ralph. 2015. NASA Adds to Evidence of Mysterious Ancient Earthworks. New York Times, October 30, 2015. Copyright 2015 by The New York Times Company. Retrieved July 29, 2016, from: http://www.nytimes.com/2015/11/03/science/nasa-adds-to-evidence-of-mysterious-ancient-earthworks.html?_r=0.

[2] Henge Ref 2: Definition of Henge from Wikipedia. Retrieved July 29, 2016.

[3] Henge Ref 3: Wendorf, Fred and Schile, Romuald. 2000. Late Neolithic megalithis structures at Nabta Playa (Sahara), southwestern Egypt. Comparative Archaeology Web, archived from the original. November 26, 2000. Retrieved July 29, 2016, from: https://web.archive.org/web/20110806140123/http://www.comp-archaeology.org/WendorfSAA98.html.

[4] Henge Ref 4: Malville, J. McKim (2015). "Astronomy at Nabta Playa, Egypt." in Ruggles, C.L.N., Handbook of Archaeoastronomy and Ethnoastronomy, 2, New York: Springer Science+Business Media, pp. 1079–1091. ISBN 978-1-4614-6140-1. Retrieved July 29, 2016.

[5] Henge Ref 5: Malville, J McKim; Schild, R.; Wendorf, F.; Brenmer, R (2007). "Astronomy of Nabta Playa." African Skies/Cieux Africains. 11. Bibcode: 2007AfrSk..11....2M. Retrieved July 29, 2016.

[6] Henge Ref 6: Malone, Caroline. 1989. Avebury. London: B.T. Batsford and English Heritage. Pg. 38. ISBN 0-7134-5960-3.

[7] Henge Ref 7: Alexander, Caroline. If the Stones Could Speak: Searching for the Meaning of Stonehenge. 2009. *National Geographic Magazine.* National Geographic Society. Retrieved July 29, 2016.

[8] Henge Ref 8: Kennedy, Maev. 2008. The magic of Stonehenge: new dig finds clues to power of bluestones. The Guardian. September 23, 2008. UK. Retrieved July 29, 2016.

[9] Henge Ref 9: Hawkins, Gerald S. and White, John B. 1965. Stonehenge Decoded. Copyright 1965 by Gerald S. Hawkins. Published January 1, 1965, by Doubleday Books.

[10] Angkor Wat Ref 10: Higham, C. (2014). Early Mainland Southeast Asia. River Books Co., Ltd. pp. 372, 378–379. ISBN 978-616-7389-44-3

[11] Angkor Wat Ref 11: Stencel, Robert, Fred Gifford, and Eleanor Moron. "Astronomy and Cosmology at Angkor Wat." Science 193 (1976): 281–287. (Mannikka, née Moron)

[12] Angkor Wat Ref 12: *Time Life Lost Civilizations Series: Southeast Asia: A Past Regained* (1995). p.67–99.

[13] Gobekli Tepe Ref 13: Curry, Andrew (November 2008). "Gobleki Tepe: The World's First Temple?" Smithsonian.com. Retrieved July 29, 2016.

[14] Gobekli Tepe Ref 14: Klaus Schmidt (2006): *Sie bauten die ersten Temel. Das rätselhafte Heiligtum der Steinzeitjäger. Die archäologische Entdeckung am Göbekli Tepe*. Munich, pp. 83–92.

[15] Gobekli Tepe Ref 15: K. Schmidt, "Göbekli Tepe—the Stone Age Sanctuaries: New results of ongoing excavations with a special focus on sculptures and high reliefs," Documenta Praehistorica XXXVII (2010), 239–256. Retrieved July 29, 2016, from: http://arheologija.ff.uni-lj.si/documenta/authors37/37_21.pdf.

[16] Gobekli Tepe Ref 16: Taracha, Piotr (2009). Religions of second millennium Anatolia. Eisenbrauns, Pg. 12. ISBN 978-3-447-05885-8. Retrieved July 29, 2016.

[17] Derinkuyu Ref 17: Darke, Diana (2011). Eastern Turkey. Bradt Travel Guides, Pgs. 139–140. ISBN 978-1-84162-339-9

[18] Derinkuyu Ref 18: Kinross, Baron Patrick Balfour (1970). Within the Taurus: A Journey in Asiatic Turkey. J. Murray. p. 168. ISBN 978-0-7195-2038-9.

[19] Baalbek Ref 19: Lebanon, Baalbek, Projects, Berlin: German Archaeological Institute. 2004. Archived from the original on October 11, 2004. Retrieved July 29, 2016.

[20] Baalbek Ref 20: Jessup, Samuel (1881), "Ba'albek," Picturesque Palestine, Sinai, and Egypt, Div. II, New York: D. Appleton & Co., illustrated by Henry Fenn & J.D. Woodward, pp. 453–476 .

[21] Baalbek Ref 21: Adam, Jean-Pierre. 1977. About the Baalbek Trilithon: The Transport and Use of the Megaliths. Syria, Volume 54, Numbers 1 and 2, Pgs. 31–63. Doi: 10.3406/syria.

[22] Baalbek Ref 22: Ruprechtsberger, Erwin M. (1999), "Vom Steinbruch zum Jupitertempel von Heliopolis/Baalbek (Libanon) [From the Quarry to the Temple of Jupiter of Heliopolis (Baalbek, Lebanon)]," Linzer Archäologische Forschungen [Linz Archaeological Research], Vol. 30, pp. 7–56 . **(German)**

[23] Baalbek Ref 23: Mark, Joshua J. Baalbek. 2009. Published in Ancient History Encyclopedia September 2, 2009. Retrieved July 29, 2016, from: http://www.ancient.eu/Baalbek/.

[24] Tiwanaku Ref 24: Bahn, Paul G. *Lost Cities*. New York: Welcome Rain, 1999. Retrieved July 29, 2016.

[25] Tiwanaku Ref 25: Vranich, A., 1999, Interpreting the Meaning of Ritual Spaces: The Temple Complex of Pumapunku, Tiwanaku, Bolivia. Doctoral Dissertation, The University of Pennsylvania.

[26] Tiwanaku Ref 26: Ponce Sanginés, C. and G. M. Terrazas, 1970, Acerca De La Procedencia Del Material Lítico De Los Monumentos De Tiwanaku. Publication no. 21.Academia Nacional de Ciencias de Bolivia

[27] Tiwanaku Ref 27: Birx, H. James (2006). *Encyclopedia of Anthropology.* Thousand Oaks, CA: SAGE Publications, Inc.

[28] Tiwanaku Ref 28: Ponce Sanginés, C. and G. M. Terrazas, 1970, Acerca De La Procedencia Del Material Lítico De Los Monumentos De Tiwanaku. Publication no. 21.Academia Nacional de Ciencias de Bolivia

[29] Tiwanaku Ref 29: Ernenweini, E.G., and M. L. Konns, 2007, Subsurface Imaging in Tiwanaku's Monumental Core. Technology and Archaeology Workshop. Dumbarton Oaks Research Library and Collection Washington, DC.

[30] Tiwanaku Ref 30: Williams, P.R., N.C. Couture and D. Blom, 2007. Urban Structure at Tiwanaku: Geophysical Investigations in the Andean Altiplano. In J. Wiseman and F. El-Baz, eds., pp. 423–441. Remote Sensing in Archaeology. Springer, New York.

[31] Tiwanaku Ref 31: Protzen, Jean-Pierre; Stella Nair, 1997. Who Taught the Inca Stonemasons Their Skills? A Comparison of Tiahuanaco and Inca Cut-Stone Masonry: The Journal of the Society of Architectural Historians. vol. 56, no. 2, pp. 146–167.

[32] Tiwanaku Ref 32: Vranich, A., 2006. The Construction and Reconstruction of Ritual Space at Tiwanaku, Bolivia: AD 500–1000. Journal of Field Archaeology 31(2): 121–136.

[33] Tiwanaku Ref 33: Protzen, Jean-Pierre; Stella Nair, 1997. Who Taught the Inca Stonemasons Their Skills? A Comparison of Tiahuanaco and Inca Cut-Stone Masonry: The Journal of the Society of Architectural Historians. vol. 56, no. 2, pp. 146–167.

[34] Tiwanaku Ref 34: Vranich, A., 2006. The Construction and Reconstruction of Ritual Space at Tiwanaku, Bolivia: AD 500–1000. Journal of Field Archaeology 31(2): 121–136.

[35] Tiwanaku Ref 35: Protzen, Jean-Pierre; Stella Nair, 1997. Who Taught the Inca Stonemasons Their Skills? A Comparison of Tiahuanaco and Inca Cut-Stone Masonry: The Journal of the Society of Architectural Historians. vol. 56, no. 2, pp. 146–167.

[36] Tiwanaku Ref 36: Protzen, J.-P., and S.E. Nair, 2000. On Reconstructing Tiwanaku Architecture: The Journal of the Society of Architectural Historians, vol. 59, no., 3, pp. 358–371.

[37] Tools Ref 37: *NOVA* television. 1998. Excerpts from the episode "Secrets of the Lost Empires." Aired May 5, 1998. Retrieved July 27, 2016, from: http://www.pbs.org/wgbh/nova/ancient/stonehenge-questions.html.

[38] Wheel Ref 38: Gambino, Megan. 2009. A Salute to the Wheel. Smithsonian Magazine, June 17, 2009. Retrieved July 27, 2016, from: http://www.smithsonianmag.com/science-nature/a-salute-to-the-wheel-31805121/?no-ist.

[39] Teotihuacan Ref 39: "Teotihuacan." Heilbrunn Timeline of Art History. Department of Arts of Africa, Oceania, and the Americas, The Metropolitan Museum of Art.

40 Teotihuacan Ref 40: Pollard, Elizabeth; Rosenberg, Clifford; Tignor, Robert (2015). Worlds Together Worlds Apart Volume 1 Concise Edition. New York: W.W. Norton & Company. p. 292. ISBN 978-0-393-91847-2.

41 Teotihuacan Ref 41: Miller, Mary; Karl Taube (1993). The Gods and Symbols of Ancient Mexico and the Maya: An Illustrated Dictionary of Mesoamerican Religion, Pgs. 162–163. London: Thames & Hudson. ISBN 0-500-05068-6. OCLC 27667317

42 Teotihuacan Ref 42: Coe, Michael D.; Rex Koontz (1994) [1962]. Mexico: From the Olmecs to the Aztecs. New York: Thames & Hudson. ISBN 0-500-27722-2. OCLC 50131575.

43 Chichen Itza Ref 43: Osorio León, José (2006). "La presencia del Clásico Tardío en Chichen Itza (600-800/830 DC)." In J.P. Laporte, B. Arroyo y H. Mejía. XIX Simposio de Investigaciones Arqueológicas en Guatemala, 2005 (PDF) (in Spanish). Guatemala City, Guatemala: Museo Nacional de Arqueología y Etnología. pp. 455–462. Retrieved 2011-12-15.

44 Chichen Itza Ref 2: Cano, Olga (January–February 2002). "Chichén Itzá, Yucatán (Guía de viajeros)." Arqueología Mexicana (in Spanish). Mexico: Editorial Raíces. IX (53): 80–87. ISSN 0188-8218. OCLC 29789840.

45 Chichen Itza Ref 45: García-Salgado, Tomás (2010). "The Sunlight Effect of the Kukulcán Pyramid or The History of a Line" (PDF). Nexus Network Journal. Retrieved 27 July 2011.

46 Chichen Itza Ref 46: Cano, Olga (January–February 2002). "Chichén Itzá, Yucatán (Guía de viajeros)." Arqueología Mexicana (in Spanish). Mexico: Editorial Raíces. IX (53): 80–87. ISSN 0188-8218. OCLC 29789840.

47 Chichen Itza Ref 47: García-Salgado, Tomás (2010). "The Sunlight Effect of the Kukulcán Pyramid or The History of a Line" (PDF). Nexus Network Journal. Retrieved 27 July 2011.

48 Chichen Itza Ref 48: Piña Chan, Román (1993) [1980]. Chichén Itzá: La ciudad de los brujos del agua (in Spanish). Mexico City: Fondo de Cultura Económica. ISBN 968-16-0289-7. OCLC 7947748.

49 Nazca Ref 49: Nickell, Joe (2005). Unsolved History: Investigating Mysteries of the Past. The University Press of Kentucky ISBN 978-0-8131-9137-9, pp. 13–16

50 Nazca Ref 50: Reinhard, Johan (1996) (6th ed.) The Nazca Lines: A New Perspective on Their Origin and Meaning. Lima: Los Pinos. ISBN 84-89291-17-9.

51 Nazca Ref 51: Stierlin, Henri (1983). La Clé du Mystère. Paris: Albin Michel. ISBN 2-226-01864-6.

52 Nazca Ref 52: "The Theory of Jim Woodman: Science in the Sand." Retrieved July 29, 2016.

53 Nazca Ref 53: Haughton, Brian. (2007). Hidden History: Lost Civilizations, Secret Knowledge, and Ancient Mysteries. Career Press. ISBN 1-56414-897-1.

54 Nazca Ref 54: Brown, Cynthia Stokes (2007). Big History. New York: The New Press. p. 167. ISBN 978-1-59558-196-9.

55 Nazca Ref 55: Aveni, Anthony F. Between the Lines: The Mystery of the Giant Ground Drawings of Ancient Nasca, Peru. Austin, Texas: University of Texas Press. July 1, 2006, ISBN 0-292-70496-8 p.205 [1].

56 Summary Ref 56: Rawlinson, Philip. A Comparison of Ancient Egyptian and Mayan Pyramids. University of Nottingham. Retrieved July 29, 216, from: https://www.academia. edu/16164395/A_comparison_of_Ancient_Egyptian_and_Mayan_pyramids.

57 Summary Ref 57: Miller, Mark. 2015. Scientist explores connection between Shang Dynasty China and ancient Peruvian Cultures. Published January 5, 2015. Retrieved July 29, 2016, from: https://www.ancient-origins.net/news-history-archaeology/scientist-explores-connection-between-china-and-peru-020153.

58 Summary Ref 58: Copyright 2012–2016 Karen Carr, Portland State University. This page last updated Friday, August 19, 2016. http://quatr.us/india/architecture/,

59 Summary Ref 59: Unknown author. Copyright ©1982–2016 Martin Gray. Retrieved July 29, 2016, from: https://sacredsites.com/africa/egypt/obelisk_of_queen_hapshetsut_karnak.html.

60 Summary Ref 60: "Density Variations of Earth Materials." Earthsci.unimelb. edu.au. Retrieved 2010-09-12.

61 Summary Ref 61: "Della trasportatione dell'obelisco vaticano et delle fabriche di nostro signore papa Sisto V fatte dal cavallier Domenico Fontana, architetto di Sva Santita, libro primo. – NYPL Digital Collections." Retrieved 21 August 2015.

62 Summary Ref 62: "Dispatches" (http://www.pbs.org/wgbh/nova/egypt/dispatches/990314.html), NOVA. Retrieved July 29, 2016.

63 Summary Ref 63: Unknown author. Ancient Power Sources of the Gods: Advanced technology and our ancestors. Retrieved July 29, 2016, from: http://www.ancient-code.com/ancient-power-sources-of-the-gods-advanced-technology-and-our-ancestors/.

64 Summary Ref 64: "NOVA/Transcripts/Secrets of Lost Empires/Stonehenge." PBS. 1997-02-11. Retrieved July 29, 2016.

CHAPTER 13

Ancient Paraphernalia of High Strangeness

All knowledge of reality starts from experience.

—Dr. John E. Mack from his book, *Passport to the Cosmos*[1]

Key words: oopart, relative dating method, phenotypic, pareidolia, aophenia, substrate animal husbandry, head shaping, Wondjina, broadsheet board

If we were to collect all the ancient artifacts archaeologists have retrieved dating back to a period of twelve thousand years ago, the alleged time of modern man's first technological revolution, the resulting museum space would fill galleries as large as multiple sports stadiums placed in a row. Even if we were to divide the historical timeline into two periods—a period from 1,000–2,500 years ago and the second one for all the years before that—there would still be enough to fill many stadiums. Most important to this idea is that scientists are finding more artifacts each day. As time goes on, even more treasures will be discovered from each period in man's history, as well as more from time periods prior to twelve thousand years past.

We will be looking at ancient paraphernalia, art, text scriptures, and structures that defy logical explanation and argument of their depositional nature. You may have heard of a term used synonymously with this notion. It is called *high strangeness* and will mean that the artifact(s) under discussion would appear to be sequentially out of place and would not fit in with the civilization, daily lives, or not be very well correlated with the environment, culture, or technology of the time it is attributed. Let us start with ancient artifacts and the term that is used to describe this phenomenon.

That term is *oopart*. What is an oopart, you may still be wondering. An oopart is something like an anachronism except that instead of simply contributing to another time period, an oopart actually existed during that time period.

For example, in 2009, ancient Chinese pottery collections were discovered in the Hunan province. After their excavation and recovery, these artifacts were chemically dated to over eighteen thousand years ago. Pottery is a fire-influenced tool. Fire was a technology known to have existed back then but not for the manufacturing of durable goods. Let's continue with some other strange findings.

In 1935, ancient remains of a culture, called the Hong Shan, were discovered in the Wuerjimulun River Valley of Chifeng, Inner Mongolia.[2] A well-known artifact of Hong Shan Jade called the coiled Dragon Fetus is part of the museum treasure dated to six thousand years ago. The Hong Shan Jade artifacts were manufactured with metal saw blades and drilling instruments. The Hong Shan were also temple and city architects, but their use of such metal tools do not fit with earthly civilizations of over six thousand years ago. Also the Hong Shan Mids temples are pyramidal-shaped and are the oldest discovered yet. Additionally, they contain steps and stairways which afforded the ancients the ability to climb to the top of the megalith. These pyramids predated all Egyptian pyramids by over one thousand years. The Hong Shan used diamond tips in their drills and metal mold casting technology, as is evidenced on the stone cuts.

Recent discoveries claim (but not proven) that a community of geographical neighbors of the Hong Shan, a civilization called the Xinglongina, used many of these technologies over 8,500 years ago. Ancient China is rife with examples of highly advanced metalworking and processing technology that high strangeness does not allow to fit in a time frame dating back to eight thousand years ago.

Ancient jars and vases currently housed in the National Museum of Iraq in Baghdad have been archaeologically dated back to over seven thousand years ago. Adornments on these relics depict intricately detailed portraits of entities which bear little resemblance to a being from planet Earth as we know them to have existed back then or now.[3]

Here are a few ooparts (out of place artifacts) of ancient Egyptian archaeology. From the geographical area around Giza Egyptologist, Sir William Petrie became famous for his lifetime contributions that greatly improved the knowledge of the archaeology discipline. His scholarly accomplishments included the creation of a new method of dating artifacts used in the science of archaeology. This is called the *relative dating method* and was perfected during his career which spanned from the 1870s until his death in 1942.

Shortly after his death, a secret room behind the bookcase wall in his home in Jerusalem was discovered. Found within that vaulted room included two mummified bodies, a collection of ancient mechanical devices that exhibited out-of-place technology, a multi-part mechanical sequence of gold tubing, orbs and cross-like plates all joined, also of out-of-place oopart, as well as linens and other clothing of the period.

The mummies contained skeletons approximately four feet in height. They were *phenotypic*, the visually observable physical characteristics of an organism, characteristic features akin to that of "grey" aliens. They had long heads, huge almond-shaped eye sockets, and spindly arms. Furthermore, additional "oopartifacts" consisted of an apparatus with multiple parts that had a physical mechanistic functionality and were constructed of gold and other metals. In addition, stone tablets which depicted alien-appearing flying craft were removed from the vault room.

All of Dr. Petrie's collection has been archaeologically dated to seven thousand years ago. They are among the very earliest archaeological finds of the ancient Egypt civilization. The Rockefeller Archaeological Museum took possession of some of this collection. They have, to this time, decided not to have their pieces placed on public display. The Petri Museum of Egyptian Archaeology in Malet Place, Gower Street, London, England, has some of the remainder of the collection which is currently on display.

A second "oopartifact" collection dates back to approximately five thousand years ago in the region of Abydos, also in Egypt. Over ten temples and burial fields discovered denoted a very rich civilization which survived for thousands of years. The archaeological recovery projects have been ongoing in these regions of Egypt since the 1840s. The temples include acknowledgments to the lives of Osiris, Ramses II, and SETI I.

In the SETI I Temple of the Nineteenth Dynasty of ancient Egypt the walls are adorned with elaborate inscriptions and hieroglyphs. Many of the figures have a detailed resemblance to modern and current machines of transportation: airplanes, submarines, helicopters, and so on.

For instance, some scientists and psychologists have attributed and excused the interpretation of these artifacts by people to a phenomenon called *pareidolia*. Pareidolia is a perception among humans in which the mind perceives a pattern familiar to themselves where none actually exists. Pareidolia is a sensory form of *aophenia*, where a pattern is perceived from random data. Science attributes it in one way of "seeing things in clouds" to describe the pareidolia episodes. An explanation used in more modern examples is taken from the Michigan 1966 encounter when some 350 residents of Southern Michigan saw the swamp gas phenomenon, as described by J. Allen Hynek. This widely infamous conclusion was used as explanation for those famous sightings to try to help the U.S. Air Force debunk mass UFO sightings.

The reality is that the Egyptians spent a monumental amount of effort producing this historical record which has lasted for millennia. Humans portray what they see and experience and are very imitative by nature. The hieroglyphs and artifacts are real, tangible, and have not decayed to the point of disintegration even over many millennia. This is a further wonder in and of itself! These temples were not art galleries. They exist because of a far more profound nature and cause than of providing only entertainment value to their present-day constituency who lived thousands of years ago.

Lanzhou Stone:

As reported on June 29, 2002, in the Lanzhou Morning News, an amateur archaeologist named Zhilin Wang found a stone while on a field research excursion in the Mazong Mountain area of the Gansu and Xinjiang provinces of Northwest China. The piece measures about 8 × 7 cm, weighs a reported 466 g (just over 1 pound), and is extremely hard according to the results of preliminary laboratory testing performed on the relics.[4]

The "oopartifact" feature is the presence of a screw-threaded metal bar which is completely and tightly embedded within the hard black rock. Laboratory experimental studies were conducted by ten physicists and geologists from around the world.All have concluded thus far that the artifacts' existence dates back to a prehistoric time and civilization. Further, neither the rock nor the metal bar, which contains almost microscopically-machined threads throughout, have shown any contamination or degradation due to environmental, chemical, atmospheric, or organic erosional oxidation factors.

Baghdad Battery:

This collection of artifacts, which consist of ceramic jars of about 14 cm in height, are cylinders made of rolled copper sheets and iron rods. An iron rod fits inside the copper cylinder which then fits inside the ceramic jar. All parts were engineered, and not just naturally formed, to enable a snug fit. This would be not unlike a piston assembly engineered to fit within cylinder housing. Also this would allow a liquid to pass through the membranes created by the functional assembly.

The collection was discovered in 1938 by German archaeologist Wilhelm Konig Khujut Rabu. These artifacts date back to Baghdad, Iraq, the area of origin from around two thousand years ago. One analysis found the pieces contained evidence of the correct engineering, construction, acidic compounds, corrosion, and intact constituency consistent with the modern-day version of today's direct-current battery technology. Additional materials found next to but not directly attached to the jar machines were collections of needles of a few inches in length.[5]

The "oopartifact" technology issue is attributed more toward the actual production of electricity and the knowledge behind the concept, if that was indeed the final objective of the manufacturing of the final units. This truth is unknown. Some controversy exists among scientists and archaeologists as to what uses the Baghdad battery would have had; some potential functionalities could have been for electroplating of jewelry and related symbolic representations of the time or medicinal purposes.

The electroplating functionality could have had very widespread use in the jewelry and adornment industries back then as global civilizations were very prolific in the popularity and potential profitable economic viability.

The batteries may have also had medicinal functions. Two examples of such functionalities can be traced back to the ancient Greek society of before those times. There are artifacts dating back to that time which show extensive use of electrified fish as topical painkillers. Also Chinese civilizations were already using acupuncture techniques during this period. Discovering the needle collections lead more credence to the supposition that the ancients were capable of fashioning equipment for some variety of medicinal uses.

The "oopartifact" issue would involve a highly evolved knowledge set of at least the basic principles of the electron, the flow of those electrons (the electrical current itself) in the visual metamorphic effects of the materials acted upon (the substrate), and the specific ingredients needed to manufacture the electric current. Together, and they must be in this example, an intricate defined mechanism was the final product.

Russian Tooth Wheel:

Oopart paraphernalia shares a global history and widespread heritage. An example from Northern Asia adds to this artifact inventory. In 2012, a resident of Vladivostok named Dimitri took possession of a stock of heating coal for his furnace. As he did so, he noticed an unusual object was embedded in one of the pieces.[6]

Dimitri took the specimen, a geared metal bar, to local scientists for inspection and testing. According to television network Komsomolshaya Pravda, multiple tests were performed by biologist Valery Brier and other geologists from Russian science research organizations. Molecular spectroscopy, chemical agents, and x-ray diffraction experiments concluded that: 1) the bar was composed of 98% aluminum and 2% magnesium, and 2) because of these properties the specimen was very, very light in weight and mass density.[7]

Inferences taken from those analyses conclude the embedded metal bar, which was wholly insulated from environmental and atmospheric exposure except for the single event where it was exposed to the atmosphere for experimentation by breaking the cold block into two pieces, was a sample of a very highly machined and refined technology process. Possible uses for a specimen like this would be in electronic, medical equipment, or other technologies where there was a requirement for extensively engineered devices. This is the first defined "oopartifact" anomaly.

A second oopart anomaly lies in the facts that with only aluminum and magnesium as molecular components of the bar tooth wheel, and because pure aluminum atoms can naturally decay into magnesium atoms, the parent compound could very well have been pure aluminum. Pure aluminum is not a naturally occurring isotope on Earth. Magnesium and hydrogen can be fused to form aluminum via a nuclear fusion technology or via fusion within a superheated source, such as a star like our sun. Understanding that it takes a sun-like fusion reaction of enormously high temperatures to form aluminum concludes that a highly sophisticated and futuristic technological process is used as an alternative to synthesizing pure aluminum. The mechanism does not occur naturally on Earth, nor does the end product which is 100% pure aluminum.

A third oopart anomaly is that because the half-life of the most common isotope of aluminum is 717,000 years. The amount of magnesium present as the by-product of the decay of aluminum atoms in this sample characterizes a potential age of this "oopartifact" would still be at least thousands of years old.

In the initial analysis, the tooth wheel bar was encased in a cold block of Pennsylvanian coal with its origin in the Chernogorodskiy mines north of Mongolia. The term *Pennsylvanian* refers to a geologic period of time dating back to about three hundred million years. Initial popular research studies conjecture that the wheel tooth bar may also be three hundred million years old. This seems unlikely when we consider the evidence involved. Chemistry and other testing analysis indicate, nevertheless, the artifact is still at least thousands of years old. This estimation is assumed because it is much too old for humans of any technological prowess and capability near that time to have made such a specimen from pure aluminum and with technologies in use at the time that did not include the wheel or any metallurgy beyond stone and maybe some bronze technologies. Additionally, no tools have been discovered that could point to positive evidence.

Ubaid Lizardmen:

Between seven thousand and eight thousand years ago, human civilization rose to a unique cultural prominence in prehistoric Mesopotamia. Archaeology and paleoanthropology have named this epoch the Ubaid Period. The evolution anthropology started along a geography of what is now Southern Iraq. The people of the Ubaid developed an advanced proficiency in such societal and cultural achievements as an infrastructure which included paved streets, the use of brick multi-roomed tiered houses, the first temples of nuanced architecture, adorned metal tools, colored pottery, agriculture practices, and animal domestication. This is also known as *animal husbandry*. These were all aspects of a technologically proficient and socially stratified civilization. The Ubaid people spread these cultural influences to civilizations further around the Iraq territory.

The production of clay figurines was extensive among the Ubaid cultures' domains. In 1919, the first inventory of these figurines was discovered at a site called Tell Al' Ubaid by archaeologist Harry Reginald Hall.

The "oopartifact" elements of the Ubaid collection reside in the characteristics of the artifacts themselves. The relicsare depicted with long heads, which was an identified and practiced part of the Ubaid culture called *head shaping*.[8] Unlike the others, these heads are lizard-shaped, with almond-shaped eyes, long attenuated faces, four digits on their appendages, and postures that—according to scientists—do not reflect anything related to ritualism. A suspected link to the ancient Sumerian culture, specifically to the Sumerian God Enki, is apparent in that the lizard symbolized a godlike status in both cultures—the Ubaid and expected immediate ancestor to the Sumerians.

An analytic tool archaeologists and other scientists often use in such investigations is contextual in nature. Here, the scientist takes into account the context of different aspects of the artifacts. The Ubaid artifact analysis uses the other artifacts that were found immediately around the geographic grid in such analysis. Their efforts are to try and construct a contextual landscape and to answer some of the questions such as why, where, when, how, and to what purpose they serve.

The only consistent attributes which can be ascertained so far from this inventory are of a specifically known and dated period. Additionally, there was some profound symbolic meaning or experiential context or the question of "what were they made for," which scientists are presently unable to determine.

A second "oopartifact" idea is not one of a high technology attribution but the consensus of the links of many ancient cultures' adornment and status symbolic recognition of reptilian depictions of deities and the deified. This is another analytical strategy archaeologists, paleontologists, and other scientists use of a contextual nature. They try to link known attributions to similar paraphernalia found in other parts of the world and not just the immediate area around the grid site.

Further head-shaping practices, in addition to the Ubaid depictions in their artifacts, and in reality to an extent, have been discovered in other ancient Earth cultures of that era and in later ones. These examples include: the Huns of Europe, the Mayans, the Inca, Aboriginals of Australia, North American Indians, and Southwest Asia. The earliest known discovery of head shaping, a significant class of the more general title of "cranial deformation" to note other ways in which the skull can be distorted, was found in the Shanidar Cave in Iraq. This artifact dated back to 9000 BCE.[9,10] The general synergy of the context of the elongated heads among these cultures and others include ancient Egypt, the Incas, Africa, the Mayans, and Mesopotamia. Were the ancient Ubaid people symbolizing what they witnessed and experienced?

Antikythera Mechanism:

Toward the end of the year 1900, a group of sponge fishermen discovered the Antikythera mechanism intact within its housing in a shipwreck off the island of Antikythera, Greece. This artifact was recovered in the months after its discovery. Then yearlong scientific testing led to the discovery of the first set of metal gears that fit inside the mechanism.

Investigation of the Antikythera mechanism accelerated when all this shipwreck's impediments—which included statues, glassware, ivory, and points—were moved to the National Museum of Archaeology in Athens. Acatalog full of new discoveries exposed the ultimate complexity of this machine.

An expansive category of "oopartifacts" slowly emerged as a result of the discoveries of the Antikythera mechanism which have led to more and more conclusions and more and more questions. The process of these technologies answering questions about Antikythera and making new ones continues to this day. The apparatus has been most recently dated to approximately 200 BCE and contains at least 30 gears within its housing. A growing number of inscription discoveries have also led to theories about additional purposes of the Antikythera machine being a fully-functioning analog computer.

Numerous researchers have built functioning replicas of Antikythera, and there is a significant consensus that there were contextual combinations of astronomical, calendar, and meteorological aspects to its existence and function. The miniaturization of dozens of moving parts and precision movement of those parts inside the housing only thirteen inches wide by eight inches tall by four inches thick provides an additional "oopartifact" aspect. Such engineering predates a known capability for humans to have been able to produce a crude working copy of this apparatus of this diminutive size by at least 1,800 years.

Even though the ancient Greeks may have had some intuitive notions and knowledge of a few of the underlying conceptual purposes of what the machine was measuring—such as dating, navigation, and astronomical phenomena—the engineering, design, and manufacturing of the technology is out of place with the times and technologies. Other atypical phenomena engaging the Antikythera mechanism include: 1) this was the first Antikythera mechanism discovered; 2) more may be discovered in the future; 3) the various technologies inherent are not fully represented in any other machines of the time and no individual technology or technological diffusion in or responsible for the existence of the Antikythera mechanism did not appear in any historical future record until the last few hundred years, around the 1500s; 4) how were all the variety of technologies from many different science disciplines discovered to have been able to build this computer? 5) were the technologies acquired from elsewhere?

Kozhim, Russia Micro Machine Artifacts:

In 1991, a concentrated deposit of ancient machined metal springs, coils, shafts, plates, and other high-technology artifacts was discovered by geologists in the Kozhim River region of the Komi Republic of Northern Russia. The expansive collections were discovered buried from depths of three to fourteen meters. Additional finds were discovered in the region over the next two years.[11,12]

The Russian Academy of Sciences in Syktyvkar, the Moscow Institute, St. Petersburg Institute, and the Helsinki, Finland Institute each performed micro spectroscopy, radioactive dating, and electron microscope analyses with numerous detailed measurement testing over the next few years.

Findings from that bevy of experimentation determined the artifacts ranged in physical size from about 1-1/4 inches down to less than 1/10000 of an inch! Chemical composition of the paraphernalia collection includes objects made of 100% pure copper, 100% pure tungsten, and 100% pure molybdenum. Those isotopes are practically impossible to find naturally within Earth's terrain or underneath the surface.

Another example of the anomalous features characterized by these samples is that of silver and gold. Nowhere on Earth can one find naturally-occurring 100% pure isotopes of these elements. If gold were 100% pure, it would be extremely soft and malleable.

The majority of pieces are reasoned to have a functional utility in a variety of areas of nanotechnology. The first "oopartifact" aspect comes from the fact that the existence of this nanotechnology is only becoming remotely possible for humans to achieve as of 2016. The collection was found in 1991.

In 1996, Dr. E.W. Matvejeva of the Central Scientific Research Department of Geology and Exploitation of Precious Metals in Moscow reported that the dating analyses performed on the objects was consistent with the geological strata depth location of from ten to over forty feet. These research studies continued until 1999 when one of the principal researchers, Dr. Johannes Fiebag, passed away.[13,14,15]

The second oopart aspect combines findings where within the geology, the found artifacts were positioned, the dating analysis research, the manufacturing requirements and materials used have determined the time frame of the origin of this cache is not of any contemporary period. They date back at least a few thousand years, most likely longer.

Incidentally, metals can be scientifically and empirically dated using chemical analyses. It is far more difficult or impossible to achieve the same purpose with materials made of stone or rock, whether it is sedimentary, metamorphic, or igneous.

These detailed examples of artifacts with unknown origins hardly begin to account for the full inventory of both what unfound artifacts exist and the known inventories which reside in museums. Some professionals in science and academia have decided to dismiss the existence and reality of the artifacts of unknown origin for whatever purpose. Some in the province of extraterrestrial discourse early on relegated their artifacts of interest to a status that of doubtlessly confirming their existence.

In any event, these are a few of the very many cases of high strangeness that exist and in which all or many of the "where, why, when, how, or who" questions have not been answered on any individual and/or collective basis. Now that some discussion of oopart has been presented, we may become inspired to think in another way about man and his anthropology within Earth's history. It appears that the evolution of civilization is not a cut-and-dried assembly of gradual innovations and then applications of new technologies. There exists a very large amount of unexplained situations where such anthropology has left their descendants achievements of their culture which seem out of place with their capabilities at that time.

This part of the survey sampled just a few of the many objects from antiquity that do not fit in any off-the-cuff explanation or neatly compartmentalized critical reasoning which has been offered from varying perspectives. A lot of these views are simply conjectured opinions which have no investigative or concerted research rationality as yet pursued.

This part of *The Humaniverse Guide to Better Reasoning & Decision-Making* discussion is limited to only artifact objects which were small in size. Another less-thought-about aspect of the out-of-place rationale creates more uncertainty when the topic of tools is introduced.

When the possibilities of tool design, usage, and feasibility are attributed to certain of these "oopartifacts," the rationale is reasonably logical. The Baghdad Battery apparatus, for example, were discovered intact and having used either readily available or obtainable raw materials in their final production state and other materials where some manufacturing would have been necessary prior to final assembly. All such materials fit into the manufacturing and environmental capabilities of humans of that time. Hence, is the oopart aspect derived from the "where and how" the intelligent scientific knowledge, schematic design, and engineering capabilities were obtained?

Other ancient artifact impedimenta are not as easily explained on many levels. The Antikythera mechanism or the Hong Shan jade treasures depicted use of sophisticated futuristic tools for their spatial design and construction. The advanced technologies used do not fit within the framework of the capabilities of the natives to create them. Currently, the tools needed for such successful manufacturing have not been discovered in any archaeological recovery projects. There is no record of documentation indicating that these peoples did or could have even invented such instrumentation as the tools needed to build these "oopartifacts" as well as, or more so, invented the final apparata that are the Antikythera computers or the Hong Shan jade treasures. The discussion of tools and the recurring theme I have offered will often arise throughout this entire survey. The notion is extremely anomalous in and of itself!

These ancient "oopartifacts" examples I have presented are just an initial sample of a much larger catalog of evidence of high strangeness which exists on Earth. What other clues about the evolution of humans, their origins, experiences, and the existence history of recovered ancient paraphernalia, visual representations, written textual scripture records, and structures have been discovered?

Our survey now moves in a direction toward more pictorial and written representations of the many experiences and their abnormal features that our ancestors have expressed with regard to UFOs and extraterrestrial entities. These representations have occurred as experienced through the eyes, ears, mind, and ultimately the hand as portrayed by these eyewitnesses and/or as representative agents of the witnesses to these events and experiences.

An abundance of pictorial symbolism is evident as depicted in artistic representations dating from the later Middle Ages back to before the known emergence of the Homo sapien species of over forty-five thousand years ago. As is characteristic of many of the topics associated with this survey, the notion of discovery and recovery of new artifacts of knowledge in all these areas and the fashions of the artifact relic, pictorial visualization, the written word, or the megalithic structure are dynamic activities. They will continue to be insightful and exciting into the foreseeable future as we continue to find many more of them.

Graphic representations with a strange resemblance to the UFO and extraterrestrial entity in the archaeological and paleoanthropological record date back over forty thousand years. The Kimberly Mountain Cave drawings in Western Australia were crafted by the Aborigines of that era. The figures depicted are referred to as *Wondjina* by the aboriginal descendants. These sacred beings have large round heads and gigantic black eyes, similar to the modern-day representations of the "grey's" extraterrestrials. The Wondjina were worshipped by the ancients as godlike in their symbolism.

A similar character, the "Great Martian God," also known as *Jabbaren*, is another glyph painted on the walls of caves in the Tassili Mountains of Sahara Africa. This illustration depicts a group of such "spacemen" painted much larger in the murals than the drawings of the native humans and animals. These glyphs were discovered by French explorer and cave art historian Henry Lhote and documented in his book *The Search for the Tassili Frescoes*.[16,17]

Other petroglyphs fitting within the same category but without much other explanation include the Star-Blower mural drawn during the time of the Hopi Native American tribe in Arizona around 800s–1500s. The murals were symbolic of the existence and travels of the *Kachinas*, a group of watchers or overseers of the Hopi and of the Navajo and Zuni Indians. The Kachinas, known as either Blue Star or Red Star, were supernatural beings known to these Native American's ancients not as gods themselves but representatives of the gods who originated from the same home in the cosmos away from our solar system. The Kachinas had flying machines that helped characterize their status as entities somewhere between that of the Native Americans and the gods.[18,19]

Reversing a temporal direction and going back in time to around 27000 BCE, a cave drawing in Itolo, Tanzania, Africa depicts several disc-shaped objects with domes sitting atop them. In a separate drawing from Kolo, the mural shows four entities surrounding women. Noted is another entity looking down from inside a sort of box or object.[20,21]

Next, moving on to the European continent, the caves of Pech Merle, France, exhibit cave drawings near Le Cabrerets. The murals depict a landscape filled with wild animals presumably indigenous to the area as the natives would have experienced. Toward the left side of the main mural, there is a section portraying an atypical humanoid figure with limbs as well as a tail with its sight directed toward three airborne saucer-shaped objects depicting exhaust trails emitting from their sterns. The illustrations of the animals are all rational representations supported by archaeological cataloging; in fact, they were real and did exist. The humanoid figure and the airborne objects are the anomalies within the mural. The figure was not shown to be interacting with the animals.[22]

This visualization offers an opportunity to investigate how the ancients communicated events, experiences, and stories. Knowing that written language did not appear to exist until around 5000 BCE, the aboriginals used drawings to contain many words or even an entire story within one figure. There is a recurring thematic from peoples of these Epipaleolithic times and before them. Precisely that a mural was a visualization not just of one single event but more of the way an entire society (in a general sense of the word) appeared and was experienced and lived at a point in time. It is most like a modern-day photographer taking many snapshots of different parts of his environment and merging them into one large mural, instead of taking individual snapshots and pasting them into a photo album.

A body of critical forethought was exhibited from the natives regarding the location of the ancient cave art inventory. Why were these pictorials drawn inside of caves? There was an obvious element of permanency reasoned in this practice in that the aboriginals chose indoor venues away from the weather elements to make and display their communications. It may also be true that they did similar projects in the outdoors, but no evidence of such displays from such ancient times has been found maybe because of the same weathering degradation factors. So they did wish to provide a permanent record of their experiences for others after their departure, and intelligent and cogent planning was utilized. When such anomalies appear in glyphs all over the world, it seems more unlikely that these works originate from a disorganized logic.

When a mural such as the one at Pech Merle, France, is encountered, the best way to start an analysis of what one is observing is to recall how the peoples of this culture and time communicated. It was different from the manner with which we connect today, and this is one of the differences we see. So to observe an entirely rational and logical environment of the wildlife in their neighborhood and then to witness an out-of-place depiction like the humanoid figure and the airborne craft in the same mural is extremely unusual at the least.

Also located in France are drawings of additional saucer-shaped airborne objects concluded to have been painted somewhere between 13000 and 10000 BCE in the Paleolithic Era.[23,24]

At Val Camonica, Italy, around 10000 BCE, there are several anomalous drawings of bipedal beings wearing protective clothing suits and very large helmetlike appurtenances. They are holding strange implements that do not resemble any type of weapon the aboriginals were known to have in their tool technology inventory.[25]

In Asia, at around the same time period of the dated Val Camonica cave drawings, there were recent discoveries of multiple paintings near Chhattisgarh, India. These representations show the same type of flying ships that are seen depicted in caves all over the world. Additionally and more significantly, the murals show many humanoid figures with some in various parts of the mural appearing slightly different in some phenotypical characteristics. Some figures had no noses or mouth, some wore large helmets with protective face shields, but all had disproportionally large heads in relation to their bodies. The Indian State Department in Chhattisgarh has recently entered into a project with both NASA and the Indian Space Research Organization to begin a research investigation into the discoveries.[26,27]

Around 5500 BCE, rock cliffs in Segu Canyon, Utah, became the canvass for drawings of over eighty life-sized figures of entities with humanoid characteristics but also some without arms and/or legs. All them had very large heads and are characteristic of many other depictions of this genre. Some scholarly refutations reason that extensive use of shamanistic ritual ceremonies by the natives were responsible for the rock art depictions.[28,29]

Back in Africa, near the Algerian town of Tassili in the Sahara Desert, exists an elaborate collection of over fifteen thousand petroglyphs from local ancient history that dates back over twelve thousand years. Some of the paintings have bizarre depictions of what appear to be spacemen wearing suits, visors, and helmets resembling modern-day astronauts. This takes us to the West African tribe, the Dogon, whose legends say they were guided to the area from another part of Africa that was drying up by fish gods called the Nommo who came in huge ships from the sky. The Dogon nation and culture is still present there today and have their own ties to history of alleged interaction with extraterrestrial entities.[30,31]

From Southern Mexico, Pacal the Great ruled over the Mayan city of Palenque during the seventh century. He was buried inside a pyramid called the Temple of Inscriptions. The intricately carved lid of his sarcophagus has become a classic work of Mayan art and also iconic as alleged evidence for ancient alien theorists. In their view, Pacal is pictured in a spaceship during takeoff with his hand on a control panel, his foot on a pedal, and an oxygen tube in his mouth.[32]

The ancient art phenomenon continues into more modern times. Around dawn on April 14, 1561, residents of Nuremberg, Germany, saw what they described as an aerial battle, followed by the appearance of a large black triangular object and then a large crash outside of the city. According to eyewitnesses, there were hundreds of spheres, cylinders, and other odd-shaped objects that moved erratically overhead. A *broadsheet* news board was published later that month, describing the event in elaborate panoramic detail. The broadsheet, illustrated with a woodcut engraving and text by Hans Glaser, is discussed elsewhere in *The Humaniverse Guide to Better Reasoning & Decision-Making*. The document is archived in the prints and drawings collection at the Zentralbibliothek Zürich in Zurich, Switzerland.[33,34]

Another example of woodcut art and communication exists which references encounter with extraterrestrial entities. On August

7, 1566, in Basel, Switzerland, another aerial battle with which they had no basis for comparison unless any of the residents had heard of the Nuremberg, Germany, experience took place. At dawn of that day, many citizens were reportedly frightened when they went outdoors to witness numerous black spheres involved in a formidable aerial battle for several hours. They reported seeing "many large black balls which moved at high speed in the air toward the sun, then made half-turns, banging one against the others as if they were fighting a battle or in combat with each other, a great number of them became red and igneous, then died out." To date, the only known source that recorded this incident is the 1958 book *A Modern Myth: Things Seen in the Skies* (Ein more modern Mythus Von Dingen, die am Himmel gesehenwerden).[35]

Thousands of other examples of atypical artistic representations have been recorded throughout history. As the frame of time spans over many thousands of years and many anthropological ages, the ancients have used a variety of media to relate their experiences. Additional examples recounted below characterize more recent times, but the depictions remain the same.

In their book *Wonders in the Sky*, authors Jacques Vallee and Chris Aubeck analyze these works of art that display anomalous overtones in addition to hundreds of other such atypical findings. According to Vallee and Aubeck, "You cannot simply say that, because somebody saw something round in the sky in medieval times, it's the same phenomenon that people see today. We're simply describing what people saw and the phenomena associated with it as a contribution to the overall study of the history of the phenomenon."[36] Such pictorial depictions are listed below.

The Annunciation with Saint Emidus (1486) – Artist Carlo Crivelli depicts the Virgin Mary prior to the birth of the baby Jesus. There is a white beam of light directed through the opening in Mary's bedroom wall from an airborne object. This has been explained away as a white dove, a symbol of the Incarnation. There are also many other doves and birds in the sky a distance away from the object which do not look anything like the object portrayed.[37]

The Crucifixion of Christ (1350) – This portrayal, which is housed at the Visoki Decani Monastery in Kosovo, shows multiple airborne flying ships with depictions of pilots at the cockpit controls. Various explanations have dismissed these as the sun and moon with the pilots representing a "man in the moon." A duplicate explanation is the white doves in the *Annunciation with Saint Emidus* were slotted in to also be an explanation toward the attribution of this work.[38]

The Baptism of Christ (1710) – Painted by Dutch artist Arendt de Gelder, this prominently displays a disc-shaped object projecting many beams of light toward the baptismal theater directly below. This artwork resides in the Fitzwilliam Museum in Cambridge, United Kingdom.[39]

The Madonna with Child-Saint Giovannino (1400s) – The Madonna was being observed by some entity inside a flying disc-shaped craft from behind her shoulder. An eyewitness and his dog are also looking up and observing the flying ship. This experience has been excused away by some explanations such as the artist, at this particular time, just felt like painting this object instead of an angel.[40]

MADONNA COL BAMBINO
E SAINT GIOVANNINO

Citation:
Arcangelo di Jacoppo del Sellaio, The Madonna with Saint Giovannino, fifteenth century (https://commons.wikime-dia.org/wiki/Commons:Free_Art_License_1.3)

The Triumph of Summer Tapestry (1538) – This representation was woven in Bruges, Belgium. The tapestry shows several UFOs hovering and maneuvering in the sky above the town and its residents. This tapestry resides in the Bayerisches National Museum in Germany.[41]

The Miracle of the Snow (1428) – Painted by Masolino da Panicale, this portrays the snowfall that showered ancient Rome in August of the fourth century. There is meticulous detail of the residents looking up at an equally meticulously detailed portrayal of Christ Jesus and Mary. Spatially between the parties shows a large group of disc-shaped clouds that were the same size spatially in the painting. A proposed explanation of this scene is that of snow in Rome on an August day with Jesus and Mary watching over the mood.[42]

While giving some artistic license to practitioners in the field, da Panicale was possibly in the mood to draw some clouds in his painting. Meteorologically speaking, if it was snowing that day, most probably, the cloud cover would have been unbroken and not likely to form the precise shapes he portrayed and would, thus, allow da Panicale to give an even more powerful interpretive message. Furthermore, some of the individual clouds show semi-globes sitting atop each of them. Could these be clouds "with cockpits?"

Glorification of The Eucharist (1600s) – Painted by Ventura Salimbeni, this artwork shows a highly strange representation of a sphere with two antennae protruding from its top as the sphere sits between Jesus and the Holy Father. There are two other protuberances that havebeen described as "telescopic eyes" at each bottom corner of the sphere. According to refutations in the literature, the orb is a Sphaera Mundi, a globe-like representation of the universe once common in religious art. The strange lights on the satellite are merely the sun and the moon, and its antennas are actually scepter wands that act as symbols of authority for the Father and the Son. This explanation bears no resemblance to the description offered in the writings of Johannes de Sacrobosco's *De Sphaera Mundi*, which was a medieval textbook on the introduction to the astronomy science.[43,44]

This work depicts an entirely out-of-place imagery that could fit within the time it was created. There was no basis within the knowledge possessed by the natives which could explain why the painting appears as it does. Two possible courses of analysis could explain that either Salimbeni used an extensive imagination to create it or he was either told or experienced an encounter himself in which this object was an integral part. Even today, we would have a difficult time relating the painting to some other symbolism except for attributing it to a depiction of an early space program orbital satellite. An example of this would be the Russian Sputnik satellite.

The Crucifixion of Christ (1600s) – This representation, residing in Svetishoveli Cathedral in Mtskheta, Georgia, is similar in scope to the depiction at the Visoki Decani Monastery. Two flying disc-shaped ships, shown under thrust power and propelling exhaust out the rear of their craft, flank the crucifix. The interior of both craft show pilot entity activity.[45]

Vallee, a highly-respected expert and prolific author of many ifology research books and journal articles, acknowledged and understood the importance of testing as many case studies as possible. He fully recognizes that people offer many different interpretations and visualizations as part of their perspective life experiences.[46] People can often only baseline their observations against what visualization parallels they have experienced in today's world. This human characteristic has been noted elsewhere in this book.

The main objective and Vallee's criteria for advancing the study to obtain conclusions is to make as many such incremental contributions as evidence to the overall study of the phenomena. The fact-finding and discussions in this book parallel those aims.

References

1   Quote Ref 1: Mack, John E., MD. 1999. *Passport to the Cosmos*, pg. 23. Copyright 1999 by John Mack. Published by Three Rivers Press, New York.

2   Hongshan Mysterious Artifacts. March 2009. Copyright 2008–2015 by Unsolved Mysteries in the World. Retrieved July 29, 2016, from: http://unmyst3.blogspot.com/2009/03/hongshan-mysterious-artifacts.html.

3   Baghdad Ref 3: Froelich, Paula. 2014. Proof of Ancient Aliens in the National Museum of Iraq. January 28, 2014. Published by The Huffington Post.

Retrieved August 27, 2016, from: http://www.huffingtonpost.com/paula-fro-elich/proof-of-ancient-aliens-i_b_4682168.html.

4   Lanzhou Ref 4: Stone from Outer Space Found in Lanzhou. June 29, 2002. Published by Lanzhou Morning News. Copyright Minghui.org 1999–2015. http://en.minghui.org/html/articles/2002/6/29/23603.html.

5   BBC News. 2/27/2003. "Riddle of Baghdad's Batteries." Archived. http://www.webcitation.org/66jb0KoWa.

6   Spiegel, L. Three-hundred-million-year-old tooth wheel found in Russian coal. http://Huffingtonpost.com/2013/01/23/.

7   Ostrovsky, N. Traces of crash "flying saucer" preserved in coal. http://tv.kp.ru/daily/26013/29368371.

8   Daems, Aurelie. A Snake in the Grass: Reassessing the Ever-Intriguing Ophidian Figurines. Copyright 2016 by Aurelie Daems and Academia.edu. Retrieved August 28, 2016, from: https://www.academia.edu/5150690/A_Snake_in_the_Grass._Reassessing_the_ever-intriguing_ophidian_figurines.

9   Head Shaping Ref 9: Trinkaus, Erik (April 1982). "Artificial Cranial Deformation in the Shanidar 1 and 5 Neandertals." Current Anthropology. **23** (2): 198–199. doi:10.1086/202808. JSTOR 2742361.

10 Head Shaping Ref 10: Agelarakis, A. (1993). "The Shanidar Cave Proto-Neolithic Human Population: Aspects of Demography and Paleopathology." Human Evolution. **8** (4): 235–253. doi:10.1007/bf02438114.

11 The Epoch Times. Kozhim, Russia Micro Machine Artifacts. Ancient Artifacts Found in Russia Point to ET Connection. October 16, 2009. www.theepoch-times.com/n3/1520473-russias-ancient-nanostructures/.

12 Ecognosis. Russia's Ancient Nanostructures. 2016. Epoch Times. www.ecogno-sis.org/show_news.php?n=6680/. Orignal reproduced from www.epochtimes.com/n2/content/view/23926.

13 The Epoch Times. Kozhim, Russia Micro Machine Artifacts. Ancient Artifacts Found in Russia Point to ET Connection. October 16, 2009. www.theepoch-times.com/n3/1520473-russias-ancient-nanostructures/.

14 Ecognosis. Russia's Ancient Nanostructures. 2016. Epoch Times. www.ecogno-sis.org/show_news.php?n=6680/. Orignal reproduced from www.epochtimes.com/n2/content/view/23926.

15 Rense.com. Ice Age Nanotechnology Discovered. 2.20.2001. www.rense.com/general8/nano.htm/. From Arthur Neumann. www.Arthur.Neumann@btinter-net.com.

16 Art Ref 16: Starr, Alex. 2010. Aliens in the Ancient World. December 10, 2010. Copyright 2010 by Tim Stouse. Retrieved August 28, 2016, from: http://www.timstouse.com/UFOs/ancientaliens.htm.

17 Art Ref 17: Lhote, Henri. 1958. *The Search for the Tassili Frescoes*. New York: E.P. Dutton.

[18] Art Ref 18: Astonishing Secrets of the Legendary Kachinas: Watchers of the Hopi. November 29, 2014. Retrieved August 28, 2016, from: http://www.messagetoeagle.com/kachinashopiancest.php#.V-HjqYWcG9o.

[19] Art Ref 19: Waters, Frank (1963). *Book of the Hopi*, pp. 333–334. Published 1963 by Penguin Books.

[20] Art Ref 20: Eight Cave Paintings Depicting Aliens. Ancient UFO, May 11, 2016. Retrieved August 28, 2016, from: http://ancientufo.org/2016/05/cave-paintings-depicting-aliens/.

[21] Art Ref 21: Explore Painting Ancient, Ancient Paintings and More! Retrieved August 28, 2016, from: https://www.pinterest.com/pin/355010383098127434/.

[22] Art Ref 22: Eight Cave Paintings Depicting Aliens. Ancient UFO, May 11, 2016. Retrieved August 28, 2016, from: http://ancientufo.org/2016/05/cave-paintings-depicting-aliens/.

[23] Art Ref 23: Eight Cave Paintings Depicting Aliens. Ancient UFO, May 11, 2016. Retrieved August 28, 2016, from: http://ancientufo.org/2016/05/cave-paintings-depicting-aliens/.

[24] Art Ref 24: Drawing of a...at Niaux Cave (Reseau Clastres) circa 13000 BCE. Retrieved August 28, 2016, from: https://www.pinterest.com/pin/384213411938772192/.

[25] Art Ref 25: Eight Cave Paintings Depicting Aliens. Ancient UFO, May 11, 2016. Retrieved August 28, 2016, from: http://ancientufo.org/2016/05/cave-paintings-depicting-aliens/.

[26] Art Ref 26: Eight Cave Paintings Depicting Aliens. Ancient UFO, May 11, 2016. Retrieved August 28, 2016, from: http://ancientufo.org/2016/05/cave-paintings-depicting-aliens/.

[27] Art Ref 27: Trayner, David. 2015. NASA to investigate if cave paintings prove aliens visited Earth 10,000 years ago. The Daily Star, UK, October 29, 2015. Retrieved August 28, 2016 from: http://www.dailystar.co.uk/news/weird-news/472750/nasa-investigate-Charama-cave-paintings-india-aliens-ufo-visited-earth.

[28] Art Ref 28: Eight Cave Paintings Depicting Aliens. Ancient UFO, May 11, 2016. Retrieved August 28, 2016, from: http://ancientufo.org/2016/05/cave-paintings-depicting-aliens/.

[29] Art Ref 29: Holloway, April. 2014. The haunting rock art of Sego Canyon: extra-terrestrials or spiritual visions. Ancient Origins.net, April 23, 2014. Retrieved August 28, 2016, from: http://www.ancient-origins.net/ancient-places-americas/haunting-rock-art-sego-canyon-extra-terrestrials-or-spiritual-visions-001584.

[30] Art Ref 30: Eight Cave Paintings Depicting Aliens. Ancient UFO, May 11, 2016. Retrieved August 28, 2016, from: http://ancientufo.org/2016/05/cave-paintings-depicting-aliens/.

[31] Art Ref 31: The Unsettling Tassili Cave Art. 2015. Ancient UFO, October 26, 2015. Retrieved August 28, 2016, from: http://ancientufo.org/2015/10/the-unsettling-tassili-cave-art/

32 Art Ref 32: Eight Cave Paintings Depicting Aliens. Ancient UFO, May 11, 2016. Retrieved August 28, 2016, from: http://ancientufo.org/2016/05/cave-paintings-depicting-aliens/.

33 Art Ref 33: Vallee, Jacques; Aubeck, Chris (2010). Wonders in the Sky: Unexplained Aerial Objects from Antiquity to Modern Times. Tarcher. ISBN 1585428205.

34 Art Ref 34: "Himmelserscheinung über Nürnberg vom 14. April 1561." NEBIS. Retrieved August 28, 2016.

35 Art Ref 35: UFO Fleet over Switzerland, 1566, Aerial Battle in the Skies. Educating Humanity, February 3, 2012. Retrieved August 28, 2016, from: http://www.educatinghumanity.com/2012/02/ufo-fleet-over-switzerland-1566-ufo.html.

36 Artwork Ref 36: Speigel, Lee. 2015. "Look! Is That a UFO Over Jesus's Head?" Huffington Post, December 23, 2015. Retrieved August 28, 2016, from: http://www.huffingtonpost.com/entry/ufos-in-renaissance-art_us_5679991de4b014efe0d7044b.

37 Listverse Ref 37: Poisuo, Pauli. 2013. Ten Mysterious Artifacts That Are Allegedly Alien. August 15, 2013. Copyright 2016 Listverse. Retrieved August 28, 2016, from: http://listverse.com/2013/08/15/10-mysterious-artifacts-that-are-allegedly-alien/.

38 Listverse Ref 38: Poisuo, Pauli. 2013. Ten Mysterious Artifacts That Are Allegedly Alien. August 15, 2013. Copyright 2016 Listverse. Retrieved August 28, 2016, from: http://listverse.com/2013/08/15/10-mysterious-artifacts-that-are-allegedly-alien/.

39 Listverse Ref 39: Poisuo, Pauli. 2013. Ten Mysterious Artifacts That Are Allegedly Alien. August 15, 2013. Copyright 2016 Listverse. Retrieved August 28, 2016, from: http://listverse.com/2013/08/15/10-mysterious-artifacts-that-are-allegedly-alien/.

40 Artwork Ref 40: Speigel, Lee. 2015. "Look! Is That a UFO Over Jesus's Head?" Huffington Post, December 23, 2015. Retrieved August 28, 2016, from: http://www.huffingtonpost.com/entry/ufos-in-renaissance-art_us_5679991de4b014efe0d7044b.

41 Listverse Ref 41: Poisuo, Pauli. 2013. Ten Mysterious Artifacts That Are Allegedly Alien. August 15, 2013. Copyright 2016 Listverse.
Retrieved August 28, 2016, from: http://listverse.com/2013/08/15/10-mysterious-artifacts-that-are-allegedly-alien/.

42 Listverse Ref 42: Poisuo, Pauli. 2013. Ten Mysterious Artifacts That Are Allegedly Alien. August 15, 2013. Copyright 2016 Listverse.
Retrieved August 28, 2016, from: http://listverse.com/2013/08/15/10-mysterious-artifacts-that-are-allegedly-alien/.

43 Listverse Ref 43: Poisuo, Pauli. 2013. Ten Mysterious Artifacts That Are Allegedly Alien. August 15, 2013. Copyright 2016 Listverse. Retrieved

August 28, 2016, from: http://listverse.com/2013/08/15/10-mysterious-artifacts-that-are-allegedly-alien/.

[44] Sphaera Mundi Ref 44: Thorndike, Lynn. 1949. *The Sphere of Sacrobosco and Its Commentators*, p. 118–142. Text in Latin, English translation, and commentary. Copyright 1949 by University of Chicago Press.

[45] Listverse Ref 45: Lowth, Marcus. 2016. Ten Historic Paintings That Clearly Show UFOs. April 24, 2016. Listverse. Copyright 2016 by Listverse. Retrieved August 28, 2016, from: http://listverse.com/2016/04/24/10-historic-divine-paintings-that-clearly-show-ufos/

[46] Artwork Ref 46: Speigel, Lee. 2015. "Look! Is That a UFO Over Jesus's Head?" Huffington Post, December 23, 2015. Retrieved August 28, 2016, from: http://www.huffingtonpost.com/entry/ufos-in-renaissance-art_us_5679991de4b014efe0d7044b.

CHAPTER 14

Physical Evidence of UFOs and Aliens Abound

We have on record many tens of thousands of UFO reports. Much of the UFO data are "hard," not necessarily as that term would be used by the physicist, but certainly "harder" than much of the data used in the social sciences and in the practice of law.

—Astronomer J. Allen Hynek, from a speech given at the United Nations General Assembly, November 27, 1978

Key word: phenotype

Ask most scientists what they think of the UFO enigma and you will almost certainly get a scoff and a brush-off such as this comment by astronomer Bernard Haisch:

> There's not one shred of evidence. That answer is simply not true. The problem is that this evidence does not follow our expected scientific logic, and so scientists dismiss what is, in fact, a huge number of accounts. Many sighting reports, as absurd as they sometimes appear, are probably real. Most professional scientists never bother to look at the evidence. Instead, the dogmatic dismissals by professional debunkers, which are often patently ridiculous, are simply taken at face value.[1]

This quote from the article *When UFOs Land*, written by Jim Wilson, is one of many quotes from interviews with scientists in reference to recent projects funded and led by Laurence Rockefeller and Peter Sturrock, PhD. Laurence Rockefeller is in the pedigree of the Rockefeller family, and Sturrock was the former director for space science and astrophysics at Stanford University.

The 1997 project involved a research investigation by twelve of the top U.S. scientists from science and academia with origins from Princeton, MIT, Stanford, and the Center for Space Research, located in France. The study examined physical evidence from hundreds of UFO sighting and encounter cases. The secret meetings were conducted at the Pocantico Conference Center on the Rockefeller Estate grounds north of New York City.

The *Popular Mechanics* article discussed four of the evidence cases involved in that Rockefeller study. These cases named for events that happened in Brazil in 1957, Iowa in 1977, the French Trans-En-Provence in 1981, and Florida in 1992. Here are details from that article and detailed descriptions of those events in chronological order.

The Ubatuba Incident:

In 1957, an unidentified flying craft exploded near the Brazilian town of Ubatuba. The craft was observed to explode upon hitting the coastal waters near Ubatuba. Metal debris from the explosion was gathered by witnesses, including a physician. Scientific tests determined the chemical composition of the debris to be an almost pure grade of magnesium. This evidence was of an isotope of magnesium that does not occur in nature. Among the samples acquired was a quantity found by UFO investigators. They sent a portion of it to the U.S. Air Force for separate analysis. The only follow-up received was that the samples were "destroyed by accident" before tests were completed or disclosed.

Many decades later, the Air Force declassified some of its documents. Some of them, though not related to the Ubatuba incident, described experiments they conducted in the 1950s and 1960s. The tests involved the theories of lift and propulsion on airframes. More specifically, electrostatic drives were used to create electric charges on the airframes that would repel surrounding atmosphere and affect the craft's lift and propulsion. These drives required immense electrical power to operate. Nuclear reactors small enough to fit inside an aircraft were supposedly in existence at the time, and, therefore, used to generate this required power. A separate set of testing experiments involved the operation of a device called a magnetohydrodynamic generator or MHD. A mass of fluid molten metal was used as a catalyst for the MHD to produce the power required for the in-flight operation of the craft. The potential of one of these units to become unstable in operation was not impossible, according to experts involved in the science. If it did, some of the fluid metal would have to be ejected from the craft.

Whether either of these situations ever involved any of the events and circumstances experienced by the main subjects of the Ubatuba incident was never concluded. What was concluded was sightings by multiple witnesses of a craft crashing into the coastal waters and that the craft exploded. A large quantity of this metal debris was recovered, and chemical testing concluded it was of a non-natural and almost pure isotope of magnesium that would be prohibitive for man to produce in such quantities. The test results exist, but the existence of more samples is unknown today. The U.S. Air Force was known to have destroyed additional samples investigators had asked them to study.

The Betty Hill Star Map:

Late on the night of September 19, 1961, an encounter involving Barney and Betty Hill and an alleged party of beings not from this Earth took place in rural New Hampshire. A well-documented and accredited star map was reported by Betty Hill as an artifact of the experience. This was recalled from the meeting with the leader who showed her this map as part of the conversation on board the craft. The map showed a sector of the local neighborhood from the perspective of the ship's home planet. It involved trade routes the craft was assigned to as its project activity.

A few years later, in 1968, Marjorie E. Fish, an elementary schoolteacher and amateur astronomer, first read about the existence of this three-dimensional map Betty Hill reported. It is said that she wondered whether she could create an accurate reproduction of the Hill hologram. Little did either Betty Hill or Marjorie Fish know until Fish delved further into the project that Betty would be the potential discoverer of two new members of the binary star system now known as Zeta Reticuli 1 and 2.

The project design began in 1968. Data collection did not begin until August 4, 1969, when Fish met with Betty Hill for the first time (Betty's husband, Barney, had passed away by that time). She obtained the hologram drawings and began her work. It would not be until January 1974 that Fish published her results in an issue of *Pursuit* magazine.

The details of the hologram map are the following: Betty Hill inspected a three-dimensional image, about three feet by two feet with stars that were "tinted and glowed." The map on the wall inside the alleged spaceship was flat. The hologram depicted "stars of different sizes, two of them much larger than the others." Those two had many solid lines connecting them. One of the two appeared to Betty Hill to be closer in a three-dimensional format than the other prominent one. This image that gave Betty the impression that successive layers of stars were further away continued. Solid lines also connected the next group of closest stars to the second of the largest pair. A third layer of stars, depicted to be further away still, had only dashed lines connecting to the second group of solid-line connected stars. There was also a "small distinctive triangle off to the left." Some other stars in the image had no lines connecting anywhere. In total, there were twelve connected stars. There were no grid lines or distance legends. The meaning of the lines was conjectured in the investigation conducted with Betty Hill by Dr. Benjamin Simon of Harvard University. The solid lines represented trade routes of the Zetas; the broken lines were expeditionary routes only.[2]

The logic was apparent to Marjorie Fish. As a society branches out and makes contact with others, they would need to explore for a time before economic trade or another type of more sophisticated relationship is established. The stars closer to the home star of the ship would have such a higher established relationship. Therefore, the connecting lines would be solid. The stars with only a younger relationship would typically be characterized by broken lines.

Fish ran many dozens of simulations and needed one piece of critical knowledge to lend credible and concluding proof to one of them. In early 1969, she had constructed a model which correctly depicted the schematic except for one of the two largest close-to-gether stars which were unknown at that time. Its existence was con-firmed later that year when the 1969 *Gliese Catalog of Nearby Stars* was published. The Gliese Catalog is named in honor of German astronomer Wilhelm Gliese, the creator of the first catalog.

With the existence of Zeta Reticuli (2) now confirmed, the logic and design attributions of the map were brought under a sharp-ly-focused lens. The Zeta Reticuli star system is a binary system in which members (1) and (2) are about 350 billion miles apart. When she applied other concepts of astronomical knowledge to the prob-lem, Fish determined that the hologram represented the star "neigh-borhood" as seen from the perspective of the Zetas. The two largest stars are the Zeta (1) and (2). The next farthest level of stars is all connected by solid lines. These are established trade partners with the Zetas. The next farther level out, which includes our sun as a member, are the exploratory routes denoted by dashed lines. The twelve primary contact stars are confirmed by earthly scientists to be the correct classification of stars that could support life as we know it.

Later studies and experiments, which simulated the existing parameters except the perspective was changed to that of looking at the neighborhood from Earth, confirmed the accuracy of Fish's model. Separate modeling simulations conducted within the astron-omy community confirmed the feasibility of the stated hypotheses in the Hill Star Map Study as "not being able to be disproved." When she published her results in January 1974, Fish received endorsements from such scientists as Walter Mitchell of Ohio State University, David R. Saunders of the University of Chicago, Mark Staggert of the University of Pittsburgh, Jeffrey L. Kretsch of Northwestern University, Kyle Cudworth of Yerkes Observatory, Frank B. Salisbury of the University of Utah, J. Allen Hynek, Stanton Friedman, and Walter Webb, all ufologists. One refuting critique came from cosmologist Carl Sagan, who together with associate Steven Soter, dismissed Fish's findings as a "product of a chance occurrence." This refutation came when many other scientists, including David R. Saunders, had already calculated the statistical probability of a "chance occurrence" of this particular pattern being anywhere from 1,000 to 10,000 to 1.[3] There is no additional input from Sagan as to the veracity of this determination.

Based on the corroboration of the analysis of all the collected data, application of the appropriate astronomical and astrophysical concepts and principles, and science peer review, it seems compelling that somebody within the general public domain (Betty Hill) could discover something to exist thirteen years before it can be proven by mainstream science because man's science knowledge was not yet sufficient to prove it otherwise.

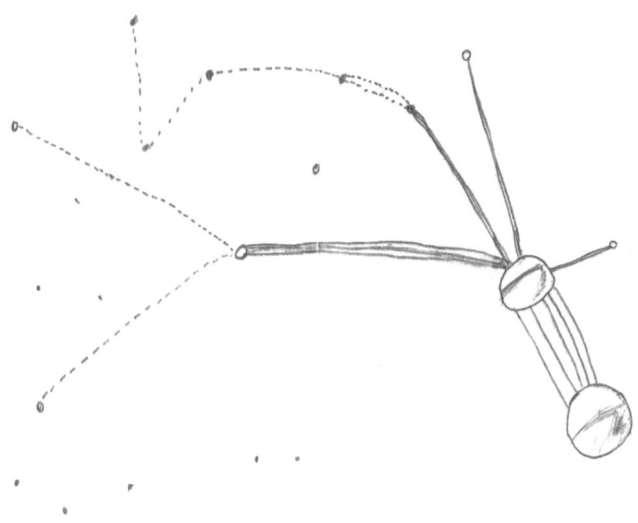

BETTY HILL
STAR MAP

Citation:
Courtesy of Kathleen Marden

The Tully Saucer Incident:

At about 9:00 a.m. on Wednesday, January 19, 1966, George Pedley, a banana farmer near Tully, New Queensland, Australia, was driving his tractor across the property of a neighboring cane farmer by the name of Albert Pennisi. The path Pedley took that day was alongside a pond called Horseshoe Lagoon. The weather was sunny and warm for early summer in Tully. The lagoon was about six feet deep and one hundred feet across. A thick growth of water reeds protruded about two feet out of the water. The stems, about half-an-inch thick, made up about half of this vertical protuberance.

Pedley heard a hissing noise from somewhere nearby. Thinking it was a tire (tyre in the literature), he stopped the tractor and leaned forward. Pedley was then shocked to see a large, saucer-shaped object rise from the lagoon, ascend vertically and slowly to tree-top level, tilted a bit to one side then, in a fantastic burst of speed vanish to the southwest. The object was shaped like two saucers with the tops placed together. It was about twenty-five feet in diameter, about nine feet thick, silver grey in color and made no sound when it rose out of the lagoon. The observation lasted about five seconds.[4]

Pedley ran to the water's edge. He observed a circular area, about the same size as the ship, of slowly-rotating water which was devoid of the reeds that were just there. The water in the circle was crystal clear. Around the perimeter of the circle was a patch of giant water couch grass that had been clipped short and the clippings removed. When he returned later that afternoon with Mr. Pennisi, Pedley and he noticed the impression the UFO left on the surface of the nest (a pond-like body of water). The hole was thrust downward and a perfectly circular protuberance of about eight feet in diameter. The circle of clear water which was there previously had been replaced by a circular floating mass of reeds wound in a radial clockwise pattern around the diameter. Curiously, the reeds had turned brown. This was strange because the species of reeds in question, like sword grass, normally stay green and do not change color for days after being uprooted. These uprooted samples were brown within a few hours.

One of Pedley's assistants, Christine Rounland, found an unexplainable set of tracks leading away from the lagoon from the shore and into the ploughed field adjacent. The tracks were shaped like teardrops, about four inches long and two inches across at their widest point. They were spaced at twelve-inch intervals in a straight line directly forward. An inspection of the disturbed water circle also found three large holes in the mud floor of the lagoon. Additionally, there was a strong sulphur odor around the liftoff site that persisted until the next day.[5]

Biochemical testing of samples of the reeds was performed at the Queensland University botany and physics departments. One of the samples taken from the center of the circle deposit revealed a small increase in beta activity, suggesting the reeds may have been placed under some intense but brief heat exposure. This suggestion takes on more support because in the weeks that followed, more lagoons were found from aerial and ground searches in the area that contained the same types of circular impressions and similar test results of the displaced reeds than those at the first site. The samples from the cores of those nests also showed evidence of fire-scorched reeds. Royal Australian Air Force information and explanations ruled out a helicopter landing because of the three deep impressions in the Horseshoe Lagoon and the atypical spiral patterns present. Helicopter blades would have made a counterclockwise impression, the opposite that were exhibited at Horseshoe and the other nests.[6]

Delphos, Kansas, Incident:

Chemical trace evidence was a basis for the inability to find suitable conclusions to this incident. On November 2, 1971, sixteen-year-old Ron Johnson was outdoors near his farmhouse when he first sighted a mushroom-shaped object about twenty-five yards away. This object was mushroom-shaped, about nine feet in diameter, with multicolored lights around its perimeter and was hovering motionless about two feet off the ground.

Johnson watched this craft for a minute or more. It made a sound similar to an "old washing machine which vibrates." His dog,

Snowflake, also observed this and fell silent. Johnson noted that "The sheep across the yard began to bleat nervously." After a time, it became very bright on its underside and began to rise vertically. The brightness of the craft's underside blinded Johnson for a few seconds. He ran into his farmhouse and retrieved his parents. They all returned outside to the scene. The ship was still visible but now only about the size of a full moon.

They inspected the landing site. There was a glowing ring on the ground more than eight feet in diameter and luminescence in the trees surrounding the site. Some had broken branches. Ron Johnson's mother, Erma, touched the ring. It felt crusty to her. She began to experience her fingers go numb. She then wiped her hands on her leg. Her leg began to go numb also.[7]

The next morning, the Johnsons returned to the landing site to find the ring still glowing luminescent in the daylight. They retrieved their Polaroid instant camera and took a few photographs of the site. The next day after they were visited by the *Delphos Republican*, newspaper reporter Thaddia Smith confirmed the status of the landing site by stating,

> The circle was still very distinct. The soil was dried and crusted. The circle was about eight feet across, the center of the ring and outside area was still muddy. The area that was dried was about a foot across and very light in color, much different in color than that of the surrounding ground.

The Johnson's also reported more details, such as,

> The object had crushed a dead tree to the ground and from appearance had broken a limb of a live tree. The broken limb was most unusual. It would snap and break as though it had been dead for quite some time, yet it was green under the bark, and the upper area still had green leaves

clinging to its branches. However, the lower area looked as though it had been blistered and had a whitish cast.

In addition to the other statements, Delphos Sheriff Enlow and two patrolmen offered this observation:

> We observed a ring shaped somewhat like a doughnut with a hole in the middle. The ring was completely dry with a hole in the middle. There was a slight, whitish discoloration on the trees.

Enlow's report also stated,

> A Mr. Lester Ensbarger reported that on November 2, he had observed a bright light descending in the sky in the vicinity of Delphos.[8,9]

The ring was just as visible over thirty days later in December when UFO investigator Ted Phillips visited the Johnson farm and started the case study. There was no degradation in the integrity of the deposits and the ground consistency during this time. The ring was entirely dry down to a depth of over twelve inches even though there was snowpack surrounding it. Multiple samples were tested chemically and biochemically. The results showed the samples were much higher in calcium and soluble salts, entirely resistant to water, and contained an unidentified hydrocarbon and an organic material composed of crystal-like fibers. Additional tests done in France and enabled by Ufologist Jacques Vallee identified the fibers as that of two organisms: a fungus of the order Actinomycetales and another of the order Basidiomysetes, which supports the luminescence observed around the ring.[10]

More samples were preserved in an attic until additional testing could be done. Ted Phillips and analytical scientist Phyllis Budinger were the scientists who performed further testing on the samples twenty-seven years later. This was done at Frontier Analysis Labs on August 9, 1999. The samples were conducted using more developed technology and advanced analytical techniques. The results showed an eighty-five percent mixture of a "humic substance" later concluded to be fulvic acid, anomalously high concentrations of oxalates and no nitrogen or potassium. The lack of nitrogen and potassium eliminates one of the popular debunking explanations that "there were sheep droppings in the exposed area." The oxalates caused the skin numbness and irritations Erma Johnson suffered. The humic substances have characteristics that make them effective for the bonding and removal of toxic substances and organic pollutants. This would have been evidence for the debunking theory, but "the lack of nitrogen, potassium, and sodium indicates a divergence from any purely organic processes."[11]

It was also noted that the samples tested in 1999 still retained the hydrophobic and unique chemical signatures and in the similar concentrations as those in the 1971–1972 samples. Even more unexplainable is when still more samples were tested more recently in 2014 by Dr. Erol A. Faruk, the chemical scientist who performed the first set of testing interventions back in 1971–1972, the same resultant properties of luminescence, hydrophobia, and chemical concentrations are comparable to the 1999 and to the 1971–1972 results.[12]

The Council Bluffs, Iowa, Incident:

On December 17, 1977, Mike and Criss Moore were approaching Council Bluffs in their automobile. They witnessed, in front of them, a large glowing red ball falling out of the sky toward an area around Big Lake Park. When they and other witnesses arrived at the crash site, they found a four-inch thick mass of molten but now red-orange metal on the frozen ground near the road.

Mike Moore's father, Jack, an assistant fire chief in Council Bluffs at the time, arrived on the scene fifteen minutes after the crash. He observed the molten mass was still glowing fifteen minutes later. A local astronomer, Robert Allen, among others, collected samples of the mass after it finally cooled some time later. He sent part of his sample to the U.S. Air Force Foreign Technology Division at Wright-Patterson Air Force Base and another part to Iowa State University Ames Laboratory.

The Air Force never made complete testing results known to the public but communicated with city officials much later that it was not a reentering spacecraft because "such craft does not impact and conclude its flight trajectory in a molten state." The results from the Ames Laboratory concluded the mass was not of any type of meteor ever known to have been observed before in the circumstances of its occurrence.

These first two events, while leaving unanswered questions that impact conclusions, were presented to the Pocantico Conference as just two of many other events with the same characteristics and circumstances throughout the years. Craft was observed flying through the atmosphere. Some of the events involved crashes of these craft, but all them experienced the ejection of various masses of molten substances that was observed by multiple witnesses. Physical evidence was tested and of the conclusions that were made public, the identity of these metallic substances was of isotopes that were not natural or man-made.

Two other incidents described in the article "When UFOs Land" did not involve exotic nonnatural isotopes of metals; instead, it focused on other physical evidence.

The Val Johnson Incident:

On August 27, 1979, Marshall County, Minnesota, deputy Sheriff Val Johnson was patrolling his district at about 1:40 a.m. when he saw a moving bright beam of light through his side window on County Highway 5. He knew the light was not situated on or near any road, so he turned onto State Highway 220 to try and get closer. He thought it may have been an airplane. The night was clear and starlit. The light appeared to have been about two to three miles in the distance.

When he made the turn, the light made a turn in midair and moved toward him, "Travelling so fast that it almost instantaneously was upon his car (covering the distance of about three miles)." The light came directly at him and passed through his patrol car. According to the police report,

> The light went at him, that his eyes hurt. Even when he closed his eyes, the light still seemed to be brilliant and hurting his eyes. All that he remembered was the light, the sound of glass breaking, and the vehicle braking, although Mr. Johnson does not remember putting on the brakes.[13]

When Johnson returned to consciousness at approximately 2:19 a.m., his car was stalled and had skidded across the highway. The patrol car had unusual damage. The driver's side inside headlight was smashed but not the outside headlight. A flat-bottomed circular dent appeared on the left side of the front hood, close to the windshield. The windshield right in front of driver Johnson was cracked from top to bottom with four holes. Both the car's clock and his watch had stopped for about fourteen minutes. The roof antenna was bent into a sixty-degree angle. The trunk antenna, directly behind the position of Johnson's driver seat, was bent at a ninety-degree angle.

A Ford Motor Company windshield designer, Meridan French, stated after examining the vehicle,

> Even after several days of reflection on the crack patterns and apparent sequence of fractures, I still have no explanation for what seem to be inward and outward forces acting almost simultaneously. I can only conclude that all cracks were from mechanical forces of unknown origin.

No explanation could be made for the clock and watch stopping, the antenna damage, or the other physical traces.[14]

Marshall County Sheriff Brekke and Officer Everett Doolittle noted in their report that Johnson suffered "welder's burns" to both eyes and had a protruding lump on his forehead. Johnson stated this about the emergency room treatment and diagnosis the doctor gave him: "He examined my eyes and said I had some irritation to the inner portions of the eye which could have been caused by seeing a bright light after dark."

The patrol car, a 1977 Ford LTD, has been preserved untouched after the incident at the Marshall County Museum with a plaque that says "UFO car."[15]

Trans-En-Provence Incident:

The Pocantico Conference was enlightened by Jean-Jacques Velasco, an investigator for GEPAN, the French Government agency that is the NASA equivalent in the United States. GEPAN is an acronym for Unidentified Aerospace Phenomena Study Group.

On January 8, 1981, witness Renato Nicolai, a retired contractor, observed a lead-colored object a few meters tall and more wide descend from the sky and come to a complete stop about six feet above the ground. The craft looked like two saucers with the bottoms touching. After about one minute, the craft rose into the sky again, leaving a trail of dust of an unknown kind. The dust trail left large circular traces that contaminated and discolored the ground. Accordingly, circular imprints outlined the exact spots with which the craft made contact. The next day, as is customary in the French protocol for investigation projects of this nature, the local police were contacted. They investigated and then contacted GEPAN to do a more comprehensive study.

Jean-Jacques Velasco represented GEPAN in the project. In the steps of the investigation, he first ruled out any military and commercial aircraft involvement. Many soil samples were taken and tested by different laboratories. The project's final report indicated the soil was contaminated with residue from highly pure metallic compounds. Surrounding vegetation was damaged, and new vegetation growth was prohibited in those areas.

Most of the puzzling biochemical mutations were discovered by Michel Bounias of the Institut National de la Recherche Agronomique (INRA), a French government and military agriculture research organization. Describing the young leaves to a journalist from France-Soir magazine, Bounias stated in 1983 that, "From an anatomical and physiological point, they [leaves] had all the characteristics of their age, but they presented the biochemical characteristics of leaves of an advanced age: old leaves! And that doesn't resemble anything that we know on our planet."

In a technical report published in the *Journal of Scientific Exploration*, Bounias concluded that,

> It was not the aim of the author to identify the exact nature of the phenomenon observed on the eighth of January 1981 at Trans-en-Provence. But it can reasonably be concluded that something unusual did occur that might be consistent, for instance, with an electromagnetic source of stress. The most striking coincidence is that at the same time, French physicist J.P. Petit was plotting the equations that led, a few years later (Petit, 1986), to the evidence that flying objects could be propelled at very high speeds without turbulence nor shock waves using the magneto-hydrodynamic effects of Laplace force action!

Out of a total of 2,500 reports collected officially in France since 1977 and investigated by GEPAN, this case and three other ground trace incidents (where strange ground traces were left after alleged UFO landings) continue to puzzle the original investigator, Jean-Jacques Velasco. At a meeting of the Society for Scientific Exploration (SSE) in Glasgow in 1994, Velasco summarized the four noteworthy cases with effects observed on vegetation.

These cases have all been the subject of enquiries by the police then GEPAN, also known as SEPRA. In each of these situations, a

UAP (Unidentified Aerial Phenomena) was observed in direct relation in a zone perturbed by the phenomenon:

1. Christelle case of 27/11/1979: Persistence of flattened grass several days after the observation. The samples taken and analyzed by a plant biology laboratory at Toulouse University did not give unequivocal evidence of chemical or biological disturbance of the samples taken from the marked area relative to controls. A study of the mechanical properties of grass tissue subjected to strong mechanical pressure showed that the duration is a more important factor than the mass.

2. Trans en Provence case of 8/01/81: Apparition of a circular print in a crown shape after observation of a metallic object resting on the ground. The vegetation, a kind of wild alfalfa, showed withering of the dried leaves in the central part of the print. The analyses revealed damage of a specific kind affecting the functional relationships of the photosynthetic system.

3. Amarante case of 21/08/82: Severe drying of the stems and leaves on a bush (amaranth) punctuated by the appearance of raised blades of grass before the phenomenon disappeared. Biochemical analyses revealed that no reported outside agent could be the cause of such effects. Only a corona effect due to powerful electromagnetic fields could partially explain the observations.

4. Joe Le Taxi case of 7/09/87: Leaf damage on a tree (birch) and functional disturbance of the photosynthetic system after an intense light and sound phenomenon had been observed. This case demonstrated the importance of good sample collection and preservation for biochemical analysis.

Of these cases, Trans-en-Provence still remains the best documented of the four. Velasco concluded that, after years of investigations, "The laboratory conclusion that seems to best cover the effects observed and analyzed is that of a powerful emission of electromagnetic fields, pulsed or not, in the microwave frequency range."

Sepra's latest thrust in the investigation has centered on "experimentally reproducing in the laboratory, continuous and pulsed emissions of microwave fields at various powers and frequencies so as to verify biochemical effects on plants." While the studies are still preliminary, Velasco concluded his SSE presentation with the following statement:

> However, these initial studies carried out to validate the hypothesis of microwave action on the biological activity of plants in relation with UAPs need to be extended if we are to understand the mechanisms involved at molecular scale. Similarly, an investigation of the frequency range, the power, and the exposure time would be useful to confirm the hypothesis of microwaves combined with other fields of electromagnetic forces coming into play in the propulsion of UAPs.

No terrestrial or mundane cause for the event could be determined. When the final report titled "Technical Note 16" was released, it reached the following conclusions:

1. Evidence indicates a strong mechanical pressure on the ground surface, probably due to a heavy weight, of about four to five tons.
2. At the same time or immediately after this pressure, the soil was heated up to between 300° and 600° C.
3. Trace quantities were found of phosphate and zinc.
4. The chlorophyll content of the wild alfalfa leaves in the immediate vicinity of the ground traces was reduced thirty percent to fifty percent, inversely proportional to distance.

5. Young alfalfa leaves experienced the highest loss of chlorophyll and, moreover, exhibited "signs of premature senescence."

6. Biochemical analysis showed numerous differences between vegetation samples obtained close to the site and those more distant.

This account has been taken from the book *The UFO Evidence: Volume II: A Thirty-Year Report* by Richard H. Hall (2000), Scarecrow Press. Further details can be found in the article by Jean-Jacques Velasco "Report on the Analysis of Anomalous Physical Traces: The 1981 Trans-en-Provence UFO Case," Journal of Scientific Exploration 4 (1) (1990): 27–48, and that by Michel Bounias, "Research Note: Further Quantification of Distance-Related Effects in the Trans-en-Provence Case," Journal of UFO Studies 5 [new series] (1995): 109–121.

Haines City, Florida, Incident:

When a member of a police department has an extended and more complex encounter, the event seems to gain more accreditation than is typical on many levels. So it was on March 19, 1992. Luis Delgado was patrolling the overnight shift in Haines City. While driving, he noticed a green light approaching behind him from the air. He continued driving for a while and then slowed down. The craft (because it was moving in the atmosphere in an organized trajectory) caught up to and flew over his vehicle. Patrolman Delgado observed the shape of the craft, which was dome-shaped. The craft was entirely silent when it approached and flew overhead. The craft also emitted a green glow, the same illumination as he noticed when it approached him.

He slowed the car to a stop. Upon doing this, the car's systems went entirely dead and would not work. Delgado exited his car and intently observed the fifteen feet in diameter craft for the next several minutes. It floated above him about ten feet in the air. The longer the craft stayed stationary, the more the surrounding atmosphere cooled.

Eventually, the air around the patrolman formed a cool fog. The craft emitted a glowing green luminance the entire time, and his car was completely dead.

After those several minutes and just as spontaneously as it approached and hovered overhead, the craft sped away with an acceleration that was never before observed or could not be matched by human technology of that day or today and especially when accelerating from a stationary physical position. Delgado eventually returned to his car and started it up. The car was now operating entirely normally.

Here is an official report of the incident:

> March 19, 1992, Haines City, Florida, 3:51 a.m. Glowing green dome-shaped object maneuvered around patrol car, E-M effects, officer badly frightened. Patrolman Delgado had just checked the doors on a local business (Alpine Industries) and had turned north onto 30th Street from State Road 544. After turning onto 30th Street, he stated he saw a green light in his rear view mirror. He thought the light was coming from a small plane that was about to crash. Just seconds after sighting the green light, the interior of his patrol unit (26) was illuminated with a green glow turning the color of his dark blue uniform to a purple color. The object began to pace his unit which was traveling at approximately 40 mph. The object moved from the right side to the front of the vehicle several times. When the object had moved to the front of the vehicle for the third time, Patrolman Delgado related that he slowed his unit and pulled off the roadway fearing he might collide with the object. The object was described as a green light (a color of green he had never seen before). The color on the object seemed to flow over the surface. He

described the shape as approximately fifteen feet long with a three-foot thick center. During the sighting, the object hovered approximately ten feet off the ground. Patrolman Delgado also stated that after he exited the roadway, the engine, lights, and radio on his patrol unit ceased to function. The object hovered in front of his unit and then shined a bright white light into the interior of his vehicle. At this time, he exited his unit and began to walk backward away from the object. He also tried to radio Haines City dispatch on his walkie-talkie which would not function. He reported he noticed the air around him had chilled to the point that he could see his breath fog when he breathed. The object was hovering approximately twenty feet northeast of his unit. He stated that after exiting his vehicle, the object sped away at a fantastic speed after approximately two or three seconds. The object departed the area in a northeasterly direction at approximately ten feet altitude. He stated he lost sight of the object in only seconds. It did not rise in altitude but hugged the tree tops as it departed. (Section V, Volume II, The UFO Evidence).[16]

This incident caught the particular attention of Michael Swords, PhD of Western Michigan University and a member of the Pocantico dozen study panel. Swords pointed out that encounters of this type have occurred many hundreds of times throughout his research during the last fifty years. This type of encounter is known as a CE-2 or close encounter of the second kind, according to J. Allen Hynek's Classification System for unidentified encounters. This level is only immediately below the first type of a direct encounter (CE-3) with an extraterrestrial biological entity or EBE.

Peter Khoury DNA Experience:

In *The Humaniverse Guide to Better Reasoning & Decision-Making* chapter, "What People Are Seeing and a Whole Lot More," there is a discussion on a case that originated from Australia. The witness, Peter Khoury, has lived in Sydney since emigrating there in 1973. On July 23, 1992, he allegedly experienced what is best described as a CE-4, a category which is an extension of the original Hynek Encounter Classification System. A CE-4 is a more extensive version of a CE-3 encounter. There was physical contact in the form of a sexual engagement. What resulted from the engagement is the subject of this discussion. Khoury recovered two body hairs from the female entity he had the experience with. They were saved and preserved by Khoury until some future time for possible testing. These hairs became the sample material that was tested using the then latest DNA genetic testing techniques.

Polymerase Chain Reaction (PCR) is a revolutionary genetic examination method developed by Kary Mullis in the 1980s. PCR is based on using the ability of DNA polymerase to synthesize new strands of DNA complementary to the offered template strand. The PCR reaction starts to generate copies of the target sequence exponentially. Only during the exponential phase of the PCR reaction is it possible to extrapolate back to determine the starting quantity of the target sequence contained in the sample.[17]

It was from the breakthrough in the perfection of this technique which made it possible for such an unprecedented testing analysis to be undertaken in the field of ufology. Chemist and science researcher Bill Chalker initiated a case testing study of the Khoury hairs in 1998. Step one of the testing was performed using the DNA-PCR techniques on mitochondrial DNA. This was done on the following: the hairs of a "tall, blond, genetically phenotypic Nordic entity as the female partner of the experience, Peter Khoury, and from his wife, Vivian."

Humans have two types of DNA: nucleic and mitochondrial. Mitochondrial is an extranuclear double-stranded DNA found exclusively in mitochondria that in most eukaryotes is a circular molecule and is maternally inherited—abbreviation *mtDNA*.[18] Therefore, mitochondria are present in and reside outside of a cell's nucleus. This is the most accurate and effective DNA material for testing and analyzing pedigree in an organism.

The tests for step 1 showed that five gene substitutions were present on the hair samples from the blond entity. The hair came from someone whose genetics were "close to normal human genetics" but of an unusual racial type A rare Chinese mongoloid type 1 of the rarest human lineages known. This type is as rare as African pygmies and aboriginals.

Step 2 of the testing was performed on a comparison of the cloned mitochondrial DNA with all the known worldwide DNA databases. Only four other humans were shown to contain a close but not exact match.

In summary, the blond hair "has a strange and unusual DNA sequence, showing five consistent substitutions from a human consensus (present in all cloned sequences), which could not easily have come from anyone else in the Sydney area except by the rarest of chances; is not apparently due to any sort of laboratory contamination; and is found only in a few other people throughout the whole world."[19] From this data, it is known that statistical probabilities of this occurrence are in the hundreds of fractions of millions or less.

It was then concluded that the DNA sample from the blond partner could not have come from anywhere else except what Peter Khoury described as a "tall, blond, fair-skinned female who does not have, use, or need any hair or skin coloring."

Since this testing regimen in 1998, the field of genetics testing has seen an exponential increase in man's knowledge and the development of new technologies and techniques. In 2005, Bill Chalker published his book *Hair of the Alien*, which provided a discussion of phase 2 of the then-updated study of the Khoury hair samples. Even as *Hair of the Alien* was being published, a phase 3 study was then being initiated. These studies continued to gather more knowledge as to the specific characteristics of the DNA sequences and translation of the genetic sequence of the DNA samples to the *phenotype* traits and behaviors. These phases also began to explore notions of potential purposes for hybrid DNA sequencing and the purposes alleged extraterrestrial entities may have for their utilization.[20,21]

The Writings of the Apkallu:

I.S. Shklovskii and Carl Sagan wrote about these scripts, glyphs, and artifacts in their book *Intelligent Life in the Universe*. The Apkallu were demigods reportedly to have been direct descendants of the Sumerian god Enki. The cuneiforms of ancient Mesopotamia and prior to even their emergence recorded this history were dated back to over 4000 BCE.

The Apkallu were sages, priests, and all-knowing super beings who imparted all the early wisdom and civilization of Enki to the humans. Shklovskii and Sagan attribute an ancient Babylonian priest, known as Berosus, to have decrypted the pre-Mesopotamian cuneiform texts and glyphs. Many versions of Berosus's scripts were translated but all refer to, according to one of these translators, Alexander Polyhistor, a demigod whom he called Oannes. Oannes maintained a teaching relationship with the Sumerian natives for a long time. According to the author,

> He gave them an insight into letters, and sciences, and every kind of art. He taught them how to construct houses, to find temples, to compile laws, and explained to them the principles of geometrical knowledge. He made them distinguish the seeds of the earth and showed them how to collect fruits."[22,23]

This account is evidential to parallel an evolutionary, if at a shockingly and far speedier rate, track of ancient man's technological explosion.

Summarily, Oannes was responsible for civilizing humankind. Sagan and Shklovskii analyzed all of the ancient scripts, glyphs, and artifacts of both Berosus and the later translators who included Alexander Polyhistor, Abydenus, and Apollodorus, among others.

They concluded that all the known accounts on record documented a description of a "sudden transition from chaos to civilization after the appearance of Oannes." Among the support of mainstream archaeological thinking in this area that Sagan and Shklovskii quoted was Harvard University Sumerologist Thorkild Jacobsen. He said, "Overnight, as it were, Mesopotamian civilization crystallizes. The fundamental pattern, controlling framework within which Mesopotamia is to live its life, formulate its deepest questions, evaluate itself and evaluate the universe, for ages to come, flashes into being, complete in all its main features."[24,25]

The writings of the Apkallu appear to have been reasoned documentation of a rational and plausible evolution of an ancient post-Enki society near the settlement at Eridu. The Eridu society was noted earlier in this text. Sagan and Shklovskii summarized the evidence for the requirement of further and deeper study of the ancient Sumerians as such:

> The gods are characterized by a variety of forms, not all human. They are celestial in origin. In general, each is associated with a different star. The cosmos is conceived as a state governed by a representative and democratic assembly of the gods, which made the great decisions on the fates of all beings, including a group of prominent deities called "The Seven Gods Who Determine Destinies."

Sagan and Shklovskii acknowledge the depiction and organization of the writings of Apkallu as describing an entirely plausible design of such a network of galactic confederacy that even today we could visualize and experience if given the opportunity to explore. The civilization of ancient Sumeria around Eridu was aided and developed by a broad and comprehensive intervention across only a small number of generations. The fantastic speed of technological and societal evolution of the ancient Sumerians was entirely anomalous and an out-of-place oopart when compared with any other set of histories recorded by any other ancient civilization.

In sum, we are left with this quote: "I feel that if Sumerian civilization is depicted by the descendants of the Sumerians themselves to be of nonhuman origin, the relevant legends should be examined carefully. I do not claim that is necessarily an example of extraterrestrial contact, but it is the type of legend that deserves more careful study."

References

[1] Haisch, Bernard, PhD. Popular Mechanics. Volume 178, Number 5. May, 2001, Pg. 64.

[2] Hill Star Map Ref 2: Clark, Jerome. 1998. *The UFO Encyclopedia: The Phenomenon from the Beginning*, Volume 1, p. 498–499. Copyright 1998 by Jerome Clark. Published 1998 by Omnigraphics Books, ISBN 0-7808-0097-4.

[3] Hill Star Map Ref 3: Dickinson, Terence. 1974. The Zeta Reticuli Incident. Astronomy Magazine, Volume 2, Number 12. December 1974, p. 2–17.

[4] Tully Ref 4: UFO Case Report. The Tully Saucer Nest. Retrieved May 5, 2016, from: http://www.ufoevidence.org/cases/case65.htm.

[5] Tully Ref 5: Chalker, Bill. 1997. Project 1947 Forum: The 1966 Tully Saucer "Nest" A Classic UFO Physical Trace Case. Retrieved October 20, 2016, from: http://www.project1947.com/forum/bctully.htm.

[6] Tully Ref 6: Chalker, Bill. 1997. Project 1947 Forum: The 1966 Tully Saucer "Nest" A Classic UFO Physical Trace Case. Retrieved October 20, 2016, from: http://www.project1947.com/forum/bctully.htm.

[7] Delphos Ref 7: 1971, The Delphos Kansas UFO Landing Ring. Retrieved May 5, 2016, from: http://www.ufocasebook. Com/Kansas.html.

[8] Delphos Ref 8: Clark, Jerome. 1998. *The UFO Encyclopedia: The Phenomenon from the Beginning*, Volume 1, p. 324–326. Copyright 1998 by Jerome Clark. Published 1998 by Omnigraphics Books, ISBN 0-7808-0097-4.

[9] Delphos Ref 9: Phillips, Ted. 1972. Landing Report from Delphos. FSR Case Histories 9, p. 4–10, February 1972. Copyright 1972 by Ted Phillips.

[10] Delphos Ref 10: Vallee, Jacques. 1988. *Dimensions: A Casebook of Alien Contact*, p. 164–166. Copyright 1988 by Jacques Vallee. Published 1989 by Ballantine Books.

[11] Delphos Ref 11: Budinger, Phyllis A. 1999. Technical Service Response No. UT001. Frontier Analysis: Analysis of Soil Samples Related to the Delphos, Kansas. November 2, 1971, CE2 Event.

12 Delphos Ref 12: Faruk, Erol A. 2014. The Compelling Scientific Evidence for UFOs. Copyright 2014 by Dr. Erol A. Faruk. Published November 25, 2014, by CreateSpace Independent Publishing Platform. ISBN-13: 978-1502715524.

13 Johnson Ref 13: Marshall County, Minnesota Police Supplementary Investigation Report, PIMV Accident, Marshall County Sheriff Deputy 407. Retrieved October 20, 2016, from: http://www.nicap.org/docs/790827mar-shallco_docs2.pdf.

14 Johnson Ref 14: UFO Casebook. Sheriff Blinded by Light from UFO; Minnesota, 1979. Retrieved October 20, 2016, from: http://www.ufocasebook.com/minnesotasheriff1979.html.

15 Johnson Ref 3: Enger, John. 2015. Whatever happened to the Marshall County cop who hit a UFO? MPR News, August 27, 2015. Retrieved October 20, 2016, from: htpps://www.mrnews.org/story/2015/08/26/Minnesota-deputy-squad-car-ufo-mystery.

16 Haines City Ref 16: Hall, Richard. January 3, 2001. UFO Evidence, Volume 2. Section V. Copyright 2001 by Richard Hall. Published by Scarecrow Press, Inc.

17 Khoury Ref 17: Polymerase Chain Reaction. National Center for Biotechnology Information. U.S. National Library of Medicine. Retrieved October 20, 2016, from: https://www.ncbi.nlm.nih.gov/probe/docs/techpcr/.

18 Khoury Ref 18: Definition of Mitochondrial DNA. 2016. *Merriam-Webster Dictionary*. Copyright 2016 by Merriam-Webster, Incorporated.

19 Khoury Ref 19: Chalker, Bill. 1999. International UFO Reporter, Spring 1999; Volume 24, Number 1, p. 16. Copyright 1999 by the J. Allen Hynek Center for UFO Studies.

20 Khoury Ref 20: Chalker, Bill. 2012. Peter Khoury and the "Hair of the Alien" – twenty years on. Retrieved October 20, 2016, from: http://theozfiles.blogspot.com.au/2012/07/peter-khoury-and-hair-of-alien-20-years.html.

21 Khoury Ref 21: Chalker, Bill. 2015. The Alien DNA Paradigm. Retrieved October 20, 2016, from: http://aliendnaparadigm.blogspot.com/.

22 Sagan Shklovskii Ref 22: Shlkovskii, I. . and Sagan, Carl. 1966. *Intelligent Life in the Universe*. Pgs. 457. Copyright 1966 by Holden-Day, Inc.

23 Sagan Shklovskii Ref 23: Bottéro, Jean (1995). Writing, reasoning, and the gods (Paperback ed.). Chicago [u.a.]: Univ. of Chicago Press. p. 247. ISBN 978-0-226-06727-8.

24 Sagan Shklovskii Ref 24: Shlkovskii, I.S. and Sagan, Carl. 1966. *Intelligent Life in the Universe*. Pgs. 460. Copyright 1966 by Holden-Day, Inc.

25 Sagan Shklovskii Ref 25: Jones, Lindsay, ed. in chief (2005). Encyclopedia of religion vol. 9 (second ed.). Detroit: Thomson Gale. p. 5964. ISBN 0-02-865742-X.

CHAPTER 15

What Scientists Are Saying

On the other hand, patterns that appear consistently in data derived
from several independent sources are far more significant than a
pattern that shows up in only one source. Strong facts of this type can
be obtained only be careful cataloging of data from as many reliable
sources as one can find. After a catalog has been compiled and patterns
supported by the weight of evidence in the catalog have been established,
one can then begin the comparison of evidence and hypothesis.
(An outstanding example of this process is the construction of the
Hertzsprung-Russell diagram in astrophysics, which provides the crucial
test for any theory of stellar evolution.) This procedure is complex, calling
for a careful organization of theoretical work and data reduction.

—Peter A. Sturrock, PhD, from *The UFO Enigma*[1]

Let's look at how other scientists and the science communities weigh in on science and ufology argument discourse.

Dr. Peter Sturrock, PhD, a British professor emeritus of applied physics at Stanford University, taught and researched astrophysics, plasma physics, and solar physics. He was also a practicing ufologist for much of that time. In a review of a 1995 book by Edward Ashpole, *The UFO Phenomena*, Sturrock listed the reasons scientists would not study such anomalies as UFOs and extraterrestrials. A recounting is provided below:

1.   The scientists' family, friends, and acquaintances might question his judgment on the subject.

2. He may become less respected by his peers and colleagues.
3. There would be no science funding available.
4. Publishing such works would be problematic.
5. Many people would think he would be wasting his time.
6. If in academia and without tenure, he would not attempt to even consider such a venture.

Sturrock also pointed out the massive grey area within in the world of science and the environment of science communities. Yet should the scientist proceed with his pursuit of ufology there would be many advantages, benefits, and potential breakthroughs.

1. The scientist would learn a great deal more about the scientific process as it exists.
2. The scientist would learn a lot about scientists and who his true friends are and who "has his back."
3. The scientist may start an entire scientific revolution, originating from Thomas Kuhn's philosophy.
4. It is possible that some financial gains, such as intellectual property copyrights and patents could be obtained.
5. Glory and honor could be achieved. As is often the case, though, in the school of honoring the achievements of a scientist, they may come after he has passed away.

In sum, this is a good starting point in surveying the landscape of memorable quotes from noted scientists. Some of these thoughts include the following:

- American Institute of Aeronautics and Astronautics (AIAA) Subcommittee, 1970, "Taking all evidence which has come to the subcommittee's attention into account, we find it difficult to ignore the small residue of well documented but unexplainable cases which form a hardcore of the UFO, a phenomenon which has such a high ratio of unexplained cases (about 30%) should arose scientific curiosity."

- Biot, Maurice, PhD, aeronautical engineer and former instructor at Harvard, Columbia, and Brown University. *Life Magazine*, April 7, 1952: "The least improbable explanation is that these things are artificial and controlled. My opinion for some time has been that they have an extraterrestrial origin."
- Von Braun, Werner. January 1959. *News Europa*: "We find ourselves faced by powers which are far stronger than we had hitherto assumed and whose base is at present unknown to us. More I cannot say at present. We are now engaged in entering into closer contact with those powers, and in six or nine months' time, it may be possible to speak with some precision on the matter." Stated with reference to observation and detection events during the re-entry phase of a Juno 2 test flight rocket.
- Breguet, Louis. Sometime before 1955: "The discs use a means of propulsion different from ours. There is no other possible explanation. Flying saucers come from another world."
- Chinese Academy of Social Sciences. One of the branches of the Chinese Academy of Social Sciences is the China UFO Research Organization (CURO). As of 1985, CURO had twenty thousand members and two publications: the *Journal of UFO Research* and *Space Exploration*. The journal's first issue in 1981 included an article by Comrade Bang Wen-Gwang of the Chinese Academy of Sciences' Beijing Astronomical Research Society. The article stated, in part, "In this field [ufology], prejudice will take you farther from the truth than ignorance. But with a topic such as UFOs, where does the scientific method begin? And where does it end? This grand endeavor would consist of the serious recording of the enormous available data and the use of all scientific procedures for the purpose of analysis. China is so vast, and UFOs are certainly being witnessed again and again all throughout China, and China most definitely will evolve her own indigenous school of UFO researchers.

This is our sincerest and deepest hope." Wen-Gwang, B., *The Aspirations & Hopes of the Chinese UFO Investigator*, *The Journal of UFO Research*, No. 1, People's Republic of China, 1981.

- Chop, Albert M., deputy public relations director, NASA and former spokesperson for U.S. Government Project Blue Book and the U.S. Air Force. *True Magazine*, January 1965: "I've been convinced for a long time that the flying saucers are real and interplanetary. In other words, we are being watched by beings from outer space." and, "The Air Force has never denied the possibility that interplanetary spacecraft exist. There are many people in the Air Force who believe in UFOs."

- Clerebaut, Lucien, 1989, secretary general for SOBEPS, the Belgian Society for the Study of Space Phenomena: "Scientifically, we eliminate the simple hypotheses: It's not a plane. It's not a helicopter. It's not a natural phenomenon because the descriptions don't match. Therefore, this global phenomenon resists any other explanation. The only remaining hypothesis is the hypothesis of extraterrestrial origin.""

- Czysz, Paul, PhD, formerly a professor of aeronautical engineering at Parks College in St. Louis. He spent eight years in the Air Force at Wright-Patterson Air Force Base and another thirty years working for McDonnell-Douglas in the field of exotic technologies. While at Wright-Patterson Air Force Base, he was involved in tracking UFOs over Missouri. November 2000: "When I was at Wright-Patterson Air Force Base, we had flying saucers that covered the zero-point energy represents about forty to fifty megawatts of power per cubic inch of space. That's a lot of power. If you could tap it at will, then no one would have to sell gasoline or oil anymore. Depending on the secrecy level, you have to go through a significant background check. When you do that, if you're in a very tight compartment, you sign a statement that you will not divulge the existence of the project or even answer a question that could acknowledge the existence of the project. I know people today that worked on one of the things I worked on, and if you asked them about it, they would say, 'No, I have no idea what you're talking about.' They're in their seventies now, but they still absolutely would never admit that they even know what you're talking about. If there were non-earthbound sources of information, the people who were doing the design or analysis work would never have any idea of where it came from."

- Dick, Steven J.; astronomer, astrobiologist, and former NASA chief historian, as quoted in the *Huffington Post* on September 25, 2014: "We're looking at all scenarios about finding life. If you find microbes, that's one thing. If you find intelligence, it's another. And if they communicate, it's something else, and depending on what they say, it's something else! The idea is not to wait until we make a discovery but to try and prepare the public for what the implications might be when such a discovery is made. I think the reason that NASA is backing this is because of all the recent activity in the discovery of exoplanets and the advances in astrobiology in general. People just consider it much more likely now that we're going to find something—probably microbes first and maybe intelligence later. The driving force behind this is from a scientific point of view that it seems much more likely now that we are going to find life at some point in the future."

- Federal Bureau of Investigation (U.S.)/Army Intelligence Report; July 1947, as declassified by the United States Freedom of Information Act (FOIA) in 1976. "The flying saucer situation is not all imaginary. Something is flying around."

- Haines, Richard F., PhD. NASA research scientist in Human Factors. 1998. CE-5: *Close Encounters of the Fifth Kind*. Published by Sourcebooks, March 1, 1999: "What I found (in doing research for the book Project Delta) was compelling evidence to claim that most of these aerial objects far exceeded the terrestrial technology of the era in which they were seen. I was forced to conclude that there is a great likelihood that Earth is being visited by highly advanced aerospace vehicles under highly intelligent control indeed."

- Haisch, Bernard, PhD, served at Max Planck Institute for Extraterrestrial Physics, Lockheed Martin, University California-Berkeley where he was deputy director of the Center for Extreme Ultraviolet Physics, editor-in-chief of *Scientific Exploration Magazine*, among others. *Popular Mechanics*, May 2001: "I propose that true skepticism is called for today: neither the gullible acceptance of true belief nor the closed-minded rejection of the scoffer masquerading as the skeptic. One should be skeptical of both the believers and the scoffers. The negative claims of pseudo-skeptics who offer facile explanations must themselves be subject to criticism. Moreover, just being a scientist confers neither necessary expertise nor sufficient knowledge. Any scientist who has not read a few serious books and articles presenting actual UFO evidence should out of intellectual honesty refrain from making scientific pronouncements. To look at the evidence and go away unconvinced is one thing. To not look at the evidence and be convinced against it nonetheless is another. That is not science."

In 1997, Laurance Rockefeller who has long been interested in UFOs and other scientific enigma asked Peter Sturrock, the former director of the Center for Space Science and Astrophysics at Stanford University, to address a meeting of a dozen top scientists to discuss UFO evidence. Researchers from places such as Princeton, MIT, Stanford, and the Center for Space Research in France focused on cases where physical traces were left behind, "While their findings were not conclusive," Rockefeller said, "I hope they will raise the level of the debate."

Former Lockheed scientist Haisch, who was on the Rockefeller panel, said, "We need to be skeptical of both the believers and the scoffers." He has created the website to encourage mainstream scientists to reconsider UFO evidence. "UFO sightings are not limited to farmers in backward rural areas." and, "There are astronomers and pilots and NASA engineers who have witnessed events for which there is no plausible conventional explanation." and, "Ask most scientists what they think of the UFO enigma and you will almost certainly get a scoff and a brush-off like, 'There's not one shred of evidence.' That answer is simply not true. The problem is that this evidence does not follow our expected scientific logic, and so scientists dismiss what is, in fact, a huge number of accounts. Many sighting reports, as absurd as they sometimes appear, are probably real. Most professional scientists never bother to look at the evidence. Instead, the dogmatic dismissals by professional debunkers, which are often patently ridiculous, are simply taken at face value."

In the article, *Be Skeptical of the Skeptics*, Dr. Haisch is also quoted as saying, "Cut through the ridicule and search for factual information in most of the skeptical commentary and one is usually left with nothing. This is not surprising. After all, how can one rationally object to a call for scientific examination of evidence? Be skeptical of the 'skeptics.'"

"Most scientists never look at the UFO evidence, which leads to their conclusion that there is no evidence."

- Halstead, Dr. Frank, astronomer and curator of the Darling Observatory, Minnesota, 1957: "Many professional astronomers are convinced that (flying) saucers are interplanetary machines."

- Hawking, Stephen, PhD, cosmologist, theoretical physicist, and research director, Cambridge University. Dr. Hawking headed the same research department that Sir Issac Newton headed over two hundred years ago, March

6, 1998, as quoted in *Imagination and Change: Science in the Next Millenium, C-Span Television*: "Of course, it is possible that UFOs really do contain aliens as many people believe, and the government is hushing it up."

- Heyerdahl, Thor, 1950, from his book titled *Kon Tiki*, published by Simon & Schuster, Inc.: "We saw the shine of phosphorescent eyes drifting on the surface on dark nights, and on one single occasion, we saw the sea boil and bubble while something like a big wheel came up and rotated in the air while some of our dolphins tried to escape by hurling themselves desperately through space" (p. 118).

Hynek, J. Allen, PhD, 1984, The UFO Cover-Up, Faucett and Greenwood. Fireside Books, Simon and Schuster: "For the government to continue to maintain that UFOs are non-existent in the face of the documents already released and of other cogent evidence presented in this book is puerile and, in a sense, an insult to the American people."

November 21, 1977, Newsweek Magazine: "It reminds me of the days of Galileo when he was trying to get people to look at the sunspots. They would say that the sun is a symbol of God. God is perfect. Therefore, the sun is perfect; therefore, spots cannot exist; therefore, there is no point in looking" (p. 97). And this from his 1972 book The UFO Experience: A Scientific Enquiry: "When I first got involved in the field, I was particularly skeptical of people who said they had seen UFOs on several occasions and totally incredulous about those who claimed to have been taken aboard one. But I've had to change my mind."

"I was there at Project Blue Book, and I knew the job they had. They were told not to excite the public, not to rock the boat. Whenever a case happened that they could explain, which was quite a few, they made a point of that and let that out to the media. Cases that were very difficult to explain, they would jump handsprings to keep the media away from them. They had a job to do, rightfully or wrongfully, to keep the public from getting excited."

- Jacobs, David M., PhD, Temple University history professor emeritus. From UFO Congress paper, February 1980: "Because few scientists have carefully studied the literature and conducted field investigations, most know practically nothing about UFOs. Their ignorance of the subject has much to do with their attitudes toward it."
- Jung, Carl Gustav, MD, PhD, physician and psychiatrist. *A Fresh Look at Flying Saucers Time Magazine*, August 4, 1967: "Unfortunately, however, there are good reasons why the UFOs cannot be disposed of in this simple manner. It remains an established fact, supported by numerous observations, that UFOs have not only been seen visually but have also been picked up on the radar screen and have left traces on the photographic plate. It boils down to nothing less than this: that either psychic projections throw back a radar echo or else the appearance of real objects affords an opportunity for mythological projections."
- Kaku, Michio, PhD, professor of theoretical physics at New York University and accomplished author. *ABC News*: "In my mind, there is no question that they are out there. My career is well established. My textbooks are required reading in all the major capitals on planet Earth. If you want to become a physicist to learn about the Unified Field Theory, you read my books. Therefore, I'm in a position to say: Yes, most likely, they're out there, perhaps even visited, perhaps on our moon."

  Dr. Kaku is often quoted with reference to his work, thoughts, and passions. In fact, an additional 192 can be accessed through this website: www.goodreads.com/author/quotes/18800.Michio_Kaku?page=1/.
- Katchen, Lee, NASA atmospheric physicist, June 7, 1968: "UFO sightings are now so common the military doesn't have time to worry about them. When a UFO appears, they simply ignore it. Unconventional targets are ignored

because apparently we are only interested in Russian targets, possibly enemy targets. Something that hovers in the air, then shoots off at five thousand miles per hour, doesn't interest us because (we know) it can't be the enemy. UFOs are picked up by ground and air radar, and they have been photographed by gun radar all along. There are so many UFOs in the sky that the Air Force has had to employ special radar networks to screen them out."

- Low, Robert J., project coordinator at University of Colorado for the Condon Committee. August 9, 1966, U.S. Air Force memorandum of instruction: "The trick would be, I think, to describe the project so that, to the public, it would appear a totally objective study but, to the scientific community, would present the image of a group of nonbelievers trying their best to be objective, but having an almost zero expectation of finding a saucer." As a footnote, the quote is reported to have been agreed as the prime goal of the Condon Project. That is, to either dismiss it without hurting any of the scientists' credibility or to comply with a rumored Air Force directive to produce a report showing UFOs to be unworthy of scientific study.

- Mack, John E., PhD, professor of psychiatry, Harvard University and Pulitzer Prize winning author, 1994, *Abduction: Human Encounters with Aliens.* New York: Scribners: "I will stress once again that we do not know the source from which the UFOs or the alien beings come (whether or not, for example, they originate in the physical universe as modern astrophysics has described it). But they manifest in the physical world and bring about definable consequences in that domain."

- MacKenzie, Dr. J.C., chairman of the Canadian Atomic Energy Control Board and former president of the National Research Council, January 1952: "It seems fantastic that there could be any such thing (UFOs and extraterrestrials). At first, the temptation was to say it was all nonsense, a series of optical illusions. But there have been so many reports from respectable observers that they cannot be ignored. It seems hardly possible that all these reports could be due to optical illusions."

- McDonald, James E., PhD, senior physicist at the Institute of Atmospheric Physics, University of Arizona. United States House of Representatives Hearings before the Committee on Science and Astronautics, July 29, 1968: "The type of UFO reports that are most intriguing and point most directly to an extraterrestrial hypothesis are close-range sightings of machinelike objects of unconventional nature and unconventional performance characteristics, seen at low altitudes, and sometimes even on the ground. The general public is entirely unaware of the large number of such reports that are coming from credible witnesses because ridicule and scoffing have made most witnesses reluctant to report openly such unusual incidents. When one starts searching for such cases, their numbers are quite astonishing. Also such sightings appear to be occurring all over the globe."

- McClelland, Clark, NASA aerospace engineer and technical assistant to the Apollo Lunar Landing Program manager, the Mercury and Gemini programs, NASA assistant in about six hundred launches at Cape Canaveral and the Space Shuttle program, 2013, *The Stargate Chronicles*: "As the Gemini capsule entered orbit, the RCA world tracking team began to realize that 'our' capsule was not alone as viewed through their incoming telemetry, visual theodolite and other high-powered optical data. Our capsule had four 'visitors.' The RCA team was ordered to run a recheck of the situation to be certain ghost images were not the cause. The Titan II stages were also excluded as causing the images. After much huddling and discussion, the intelligent determination was that we had other physical objects up there with our Gemini capsule. The official NASA determination was that the objects were the torn particles or remains of the Titan upper stage that apparently entered orbit with the Gemini capsule. I was at the news conference, and I nearly began to laugh. How could a broken stage overtake the capsule and stop slightly ahead of the capsule to accompany it an entire orbit around the Earth? But I held my laugh to save my job." This dialogue is from the April 9, 1964, unmanned launch of the Gemini-Titan II rocket.

- Mead, Margaret, PhD, world-renowned anthropologist and author of numerous books, such as *UFOs: Visitors from Outer Space? Redbook*, Volume 143, September 1974: "There are unidentified flying objects. That is, there are a hard core of cases—perhaps twenty to thirty percent in different studies—for which there is no explanation. We can only imagine what purpose lies behind the activities of these quiet, harmlessly cruising objects that time and again approach the earth. The most likely explanation, it seems to me, is that they are simply watching what we are up to."

- Meessen, Dr. Auguste, physics professor at the Catholic University in Louvain, Belgium. Interview with M.T. de Brosses, *F-16 Radar Tracks UFO*, Paris Match, July 5, 1990: "There are too many independent eyewitness reports to ignore. Too many of the reports describe coherent physical effects, and there is an agreement among the accounts concerning what was observed. But of course, there are also physical effects. The Air Force report [of the F-16 jet scramble incident on the night of March 30–31, 1990] allows us to approach the problem in a rational and scientific way. The simplest hypothesis is that the reports are caused by extraterrestrial visitors, but that hypothesis carries with it other problems. We are not in a rush to form a conclusion but continue to study the mystery."

In addition, as quoted by Dr. Meessen in, *SOBEPS, Vague d' OVNI sur la Belgique - Un Dossier Exceptionnel Brussels, 1991*, SOBEPS is the acronym for Belgian Society for Study of Space Phenomena: "On November 29, 1989, a large craft with triangular shape flew over the town of

Eupen. The gendarmes von Montigny and Nicol found it near the road linking Aix-la-Chapelle and Eupen. It was stationary in the air, above a field which it illuminated with three powerful beams. The beams emanated from large circular surfaces near the triangle's corners. In the center of the dark and flat understructure, there was some kind of 'red gyrating beacon.' The object did not make any noise. When it began to move, the gendarmes headed toward a small road in the area over which they expected the object to fly. Instead, it made a half-turn and continued slowly in the direction of Eupen, following the road at low altitude. It was seen by different witnesses as it flew above houses and near City Hall."

- Messel, Harry, PhD, professor of physics at Sydney University, Australia, 1965: "The facts about saucers were long tracked down, and results have long been known in top secret defense circles of more countries than one."

- Mitchell, Dr. Ed, aeronautical engineer and Apollo astronaut. *MSN* interview, October 1998: "I have been over the years very skeptical like many others. But in the last ten years or so, I have known the late Dr. Alan Hynek who I highly admire. I know and currently work with Dr. Jacques Vallee. I've come to realize that the evidence is building up to make this a valid and researchable question. Further, because my personal motivation has always been to understand our universe better and my own theoretical work has convinced me that life is everywhere in the universe that has been permitted to evolve, I consider this a very timely question."

- Oberth, Herman, a father of modern rocketry and NASA space scientist and engineer, 1972: "Today, we cannot produce machines that fly the same way UFOs do. They are flying by means of artificial gravity. This would explain the sudden changes of direction. This would also explain the piling up of these disks into a cylindrical or cigar-shaped Mothership upon leaving Earth. Because it is in this fashion that only one field of gravity would be required for all the flying saucers. We cannot take credit for our record advancement in certain scientific fields alone. *We have been helped.* And we have been helped by the people of other worlds."

A second quote from Dr. Oberth's can be found in *The American Weekly*, October 24, 1954: "UFOs are conceived and directed by intelligent beings of a very high order, and they are propelled by distorting the gravitational field, converting gravity into useable energy. There is no doubt in my mind that these objects are interplanetary craft of some sort. I and my colleagues are confident that they do not originate in our Solar System, but we feel that they may use Mars or some other body as a sort of way station. They probably do not originate in our Solar System, perhaps not even in our galaxy."

Finally, a third Oberth quote also appeared in 1954 in *The American Weekly*: "It is my thesis that flying saucers are real and that they are spaceships from another solar system. I think that they possibly are manned by intelligent observers who are members of a race that may have been investigating our earth for centuries. I think that they have been sent out to conduct systematic; long-range investigations first of men, animals, and vegetation; and more recently of atomic centers, armaments, and centers of armament production. They obviously have not come as invaders, but I believe their present mission may be one of scientific investigation."

- O'Leary, Dr. Brian, former NASA astronaut and Princeton University physics professor: "There is abundant evidence that we are being contacted, that civilizations have been monitoring us for a very long time. That their appearance is bizarre from any type of traditional materialistic western point of view. That these visitors use the technologies of consciousness, they use toroids, they use corotating magnetic disks for their propulsion systems, that seems to be a common denominator of the UFO phenomenon."

- Poher, Claude, PhD, astronomy and astrophysics and electronics and space research engineer at the French Space Agency (CNES), a statistical study prepared for CNES 1971: "The phenomenon seems to be real. The general coherence of sighting reports worldwide should not leave researchers indifferent. One does not conceive objective arguments to justify an attitude that would avoid at all cost these observations. The risk is, at worst, to confirm the existence of unknown vehicles appearing erratically into our atmosphere—a hypothesis that seems to explain nearly all reported aspects of the phenomenon and could be linked to the current (1970) exobiology branch of space research."

- Puthoff, Harold, PhD, director for the Institute for Advanced Studies at Austin and author of *Fundamentals of Quantum Electronics*, as quoted in, *Physics Essays*, Volume 9, Number 1, 1996: "The possibility of reduced-time interstellar travel either by extraterrestrial civilizations at present or ourselves in the future is not fundamentally constrained by physical principles."

- Rees, Lord Martin, cosmologist, astronomer and astrophysicist, and former President of the United Kingdom Royal Society. His July 2015 quote: "The chance of finding life has risen a billion fold when we realized that earthlike planets are not rare but that there are literally billions of them just within our own galaxy." From the article "(Very) Local SETI: The Launch of a New UFO Science" reported in the Huffington Post by Leslie Kean and updated December 6, 2017.

- Rich, Ben, director for Lockheed Skunk Works and regarded as the Father of Stealth. Led development of first mass production of the Stealth aircraft, 1993, in a speech addressed to fellow Alumni of UCLA: "Inside the Skunk Works (Lockheed's secret research and development entity), we were a small, intensely cohesive group consisting of about fifty veteran engineers and designers and a hundred or so expert machinists and shop workers. Our forte was building technologically advanced airplanes of small number and of high class for highly secret missions."

Rich continued in the presentation: "We already have the means to travel among the stars, but these technologies are locked up in black projects and it would take an act of God to ever get them out to benefit humanity. Anything you can imagine, we already know how to do. We now have the technology to take ET home. No, it won't take someone's lifetime to do it. There is an error in the equations. We know what it is. We now have the capability to travel to the stars. First, you have to understand that we will not get to the stars using chemical propulsion. Second, we have to devise a new propulsion technology. What we have to do is find out where Einstein went wrong."

- Sagan, Carl, PhD, astronomer, cosmologist and astrophysicist, director of Space Sciences at Cornell University, *Unidentified Flying Objects*, 1963, *The Encyclopedia Americana*: "It now seems quite clear that Earth is not the only inhabited planet. There is evidence that the bulk of the stars in the sky have planetary systems. Recent research concerning the origin of life on Earth suggests that the physical and chemical processes leading to the origin of life occur rapidly in the early history of the majority of planets. The selective value of intelligence and technical civilization is obvious, and it seems likely that a large number of planets within our Milky Way galaxy—perhaps as many as a million—are inhabited by technical civilizations in advance of our own. Interstellar spaceflight is far beyond our present technical capabilities, but there seems to be no fundamental physical objections to preclude, from our own vantage point, the possibility of its development by other civilizations." Another of many Sagan quotes appeared in *The Demon Haunted World in 1996*: "After I give lectures on almost any subject, I am often asked, 'Do you believe in UFOs?' I'm always struck by how the question is phrased, the suggestion that this is a matter of belief and not evidence. I'm almost never asked, 'How good is the evidence that UFOs are alien spaceships?'"

- Salisbury, Frank B. PhD, professor, plant physiology, Utah State University: "I must admit that any favorable mention of the flying saucers by a scientist amounts to extreme heresy and places the one making the statement in danger of excommunication by the scientific theocracy. Nevertheless, in recent years, I have investigated the story of the unidentified flying object (UFO), and I am no longer able to dismiss the idea lightly." (Paper on exobiology presented at the first annual Rocky Mountain Bioengineering Symposium, held at the United States Air Force Academy in May 1964.)
- Santorini, Dr. Paul, physicist and inventor of proximity fuse for first man-made atomic bombs and patents for guidance systems used in U.S. rockets and missiles, *UFOs: Interplanetary Visitors*, Fowler, R. New York: Bantam Books, 1974: "We soon established that they were not missiles. But before we could do any more, the Army, after conferring with foreign officials, ordered the investigation stopped. Foreign scientists flew to Greece for secret talks with me. A world blanket of secrecy surrounded the UFO question because the authorities were unwilling to admit the existence of a force against which we had no possibility of defense."
- Sathco, Dr. John, astronomer at the University of Southern California, 1973: "There are in excess of two hundred reports of the type that we had from down in Louisiana from people claiming that they have had direct contact with a spacecraft full of aliens. I mean two hundred reports from witnesses who are as reliable or more so than these people. I'm not counting the reports from the obvious crackpots that have an axe to grind. If you accept them at face value, then you're forced to accept that we have been visited."
- Slayton, Donald, U.S. astronaut, Mercury Program, 1951: "I was testing a P-51 fighter in Minneapolis when I spot-

ted this object. I was at about ten thousand feet on a nice, bright, sunny afternoon. I thought the object was a kite, then I realized that no kite is gonna fly that high. As I got closer, it looked like a weather balloon, gray and about three feet in diameter. But as soon as I got behind the darn thing, it didn't look like a balloon anymore. It looked like a saucer, a disk. About the same time, I realized that it was suddenly going away from me—and there I was, running at about three hundred miles per hour. I tracked it for a little way, and then all of a sudden, the damn thing just took off. It pulled about a forty-five-degree climbing turn and accelerated and just flat disappeared."

- Smith, Wilbert, electrical engineer and chief engineer for the Canadian government. From a Canada declassified government top secret memo dated November 21, 1950: "The matter (UFOs and extraterrestrials) is the most highly classified subjecting the United States government, rating even higher even than the H-bomb. Flying saucers exist. Their modus operandi is unknown but concentrated effort is being made by a small group headed by Dr. Vannevar Bush." And this from a speech dated March 31, 1958: "It soon became apparent that there was a very real and quite large gap between the alien science and the science in which I had been trained. Certain crucial experiments were suggested and carried out, and in each case, the results confirmed the validity of the alien science. Beyond this point the alien science just seemed to be incomprehensible."

- Sturrock, Peter A., PhD, professor of space science and astrophysics and deputy director of the Center for Space Sciences and Astrophysics at Stanford University, 1977 Report on a Survey of the American Astronomical Society concerning the UFO Phenomenon. Stanford University Report SUIPR 68IR, 1977: "The definitive resolution of the UFO enigma will not come about unless and until the problem is subjected to open and extensive scientific study by the normal procedures of established science."

An additional thought-provoking quote from a research article in the premier issue of *Journal of Scientific Exploration*, Volume 1, Number 1, 1987, in an article titled "An Analysis of *The Condon Report* on the Colorado UFO Project (i.e. Condon Report)": "In their public statements (but not necessarily in their private statements), scientists express a generally negative attitude toward the UFO problem, and it is interesting to try to understand this attitude. Most scientists have never had the occasion to confront evidence concerning the UFO phenomenon. Most scientists have never had the occasion to confront evidence concerning the UFO phenomenon. To a scientist, the main source of hard information (other than his own experiment's observations) is provided by the scientific journals. With rare exceptions, scientific journals do not publish reports of UFO observations. The decision not to publish is made by the editor acting on the advice of reviewers. This process is self-reinforcing: the apparent lack of data confirms the view that there is nothing to the UFO phenomenon, and this view works against the presentation of relevant data."

- Tombaugh, Clyde, astronomer who discovered Pluto in 1930. September 10, 1957, letter to Richard Hall: "The illuminated rectangles I saw did maintain an exact fixed position with respect to each other, which would tend to support the impression of solidity. I doubt that the phenomenon was any terrestrial reflection. I do a great deal of observing (both telescopic and unaided eye) in the backyard, and nothing of this kind has ever appeared before or since."
- USSR Scientific Commissions, Faminskaya, T. & Petukhov, A.: "At 4.10 Hours and After," Almanac Phenomenon 1989, Moscow Mir, 1989. The Soviet press was informed in the mid-eighties that the All-Union Council of Scientific and Technical Societies (now the Council of Scientific and Engineering Societies) had set up a nongovernmental Commission on Paranormal Events, headed by V.S.

Troitsky, a corresponding member of the USSR Academy of Sciences: "Of special value are the archives set up by the commission. They contain over thirteen thousand reports connected with PEs [Paranormal Events] and with UFOs in particular. UFOs have been seen to hover over ground objects, to chase or fly side by side with airplanes and cars, to follow geometrically regular trajectories, and to send out ordered flashes of light. In other words, such paranormals behave, from the viewpoint of human beings, quite often showing capabilities yet beyond the reach of the machines built on the Earth."

- Vallee, Jacques, PhD, astrophysicist, author, business owner, and UFO researcher. 1990, *Confrontations*, New York: Ballantine Books: "Skeptics, who flatly deny the existence of any unexplained phenomenon in the name of 'rationalism,' are among the primary contributors to the rejection of science by the public. People are not stupid, and they know very well when they have seen something out of the ordinary. When a so-called expert tells them the object must have been the moon or a mirage, he is really teaching the public that science is impotent or unwilling to pursue the study of the unknown."

   Another quote that appeared in *Challenge to Science: The UFO Enigma* written by Vallee with his wife, Janine, and published by Regenry: "The fact that since 1946 numerous persons in all countries have made detailed reports of events they regard as strange, mysterious, sometimes even terrifying, deserves attention. While many of the reports can be traced to natural events, we intend to demonstrate that, after the inevitable errors and the obvious hoaxes are eliminated, the reports reveal common characteristics, possess a high degree of internal coherence, and appear to be the result of the witnesses' exposure to a set of unusual circumstances."

- Webre, Dr. Alfred, Stanford Research Institute and Senior U.S. Government Policy Analyst: "I worked on the 1977

Carter White House Extraterrestrial Communication Project. It called for creation of central and regional databases under independent control on UFOs and EBEs—that is Extraterrestrial Biological Entities. The full management staff and the research institute had signed off knowingly on the proposal. I flew back from my meeting with the White House, at which this final approval had been given. And when I arrived back at my offices at SRI (Stanford Research Institute), I was called back into the office of the senior SRI official. The project was to be terminated. They had received direct communication from the Pentagon that if the study went forward, SRI's contracts would be terminated. These contracts were a substantial part of SRI's business at the time. The senior Pentagon liaison stated that the project was terminated because "there are no UFOs." Here we have a president of the United States who came to office under a pledge to open up the UFO issue and an open study in the White House, and that was squelched."

- *Yale Scientific Magazine* (Yale University), Volume XXXVII, Number 7, April 1963: "Based upon unreliable and unscientific surmises as data, the Air Force develops elaborate statistical findings which seem impressive to the uninitiated public unschooled in the fallacies of the statistical method. One must conclude that the highly publicized Air Force pronouncements based upon unsound statistics serve merely to misrepresent the true character of the UFO phenomenon."

- Zhousheng, Zhang, astronomer at the Yunan Observatory in Chengdu City, China. Zhousheng and others nearby watched a strange, glowing spiral object moving steadily across the sky for about five minutes on the evening of July 26, 1977: "What was especially important was that, at a distance of 180 kilometers apart, the records about the direction of movement of the strange aerial body in space, made independently by at least two different observers was basically the same. To the present time, this strange phenomenon has not been satisfactorily explained, yet there were thousands of good observers who had seen it."

- Zigel, Dr. Felix Y., professor of mathematics, cosmology, and astronomy at Moscow University and Moscow Aviation Institute. *Moscow Central Television*, November 10, 1967: "Unidentified flying objects are a very serious subject which we must study fully. We appeal to all viewers to send us details of strange flying craft seen over the territories of the Soviet Union. This is a serious challenge to science and we need the help of all Soviet citizens." Another quote appeared in *Soviet Life Magazine*, Volume 137, Number 2, February 1968: "Observations show that UFOs behave sensibly. In a group formation flight, they maintain a pattern. They are most often spotted over airfields, atomic stations, and other very new engineering installations. On encountering aircraft, they always maneuver so as to avoid direct contact. A considerable list of these seemingly intelligent actions gives the impression that UFOs are investigating, perhaps even reconnoitering. The important thing now is for us to discard any preconceived notions about UFOs and to organize on a global scale a calm, sensation-free, and strictly scientific study of this strange phenomenon. The subject and aims of the investigation are so serious that they justify all efforts. It goes without saying that international cooperation is vital."

In June 1947, a letter designated top secret was written by Albert Einstein and J. Robert Oppenheimer titled "Relationships with Inhabitants of Celestial Bodies." In summary, the letter states the reality of craft observed having flown over United States airspace under the piloting of intelligent means are generally accepted by the military and U.S. government. The briefing postulates about the origin of such ships and what purposes they may have had in visiting Earth. Additionally, if this scenario is generally accepted by official sources, then what could/should we do in response? Einstein and Oppenheimer also disclose the existence of nuclear warfare technology as built and tested by our military. Excerpts of this letter are in the "Disclosure: An Overview" chapter.

References

[1] Chapter Opening Quote: Sturrock, Peter A., PhD. 1999. The UFO Enigma, p. 39. Copyright 1999 by Peter A. Sturrock. Published 1999 by Warner Books Inc.

CHAPTER 16

What People (Many at One Time) Are Observing The Public

The methods of science—hypothesis, testing, rigor, experimentation, control—are valuable and essential for studying phenomena that reside primarily in the material world. But they may be inadequate for exploring matters that straddle the visible and unseen realms. They surely are insufficient for learning about realities beyond the manifest. Here, we must rely more upon experience, intuition, non-ordinary states of consciousness, and holistic or heart knowing, thoughtfully and rigorously applied.

—Dr. John E. Mack, Harvard University

Key words: flaps, UAPs, mass hysteria, experimental bias, hypnagogic, broadsheets, discreet and continuous thinking

As this and the next two chapter titles suggest, here arrives a presentation of case studies regarding UFOs or to reintroduce a contemporary corollary term, *unidentified aerial phenomena* or *UAPs*.

The acronym UAP originated from the British Ministry of Defence or MoD in its early history in at the least the 1970s. Archived files that were declassified and released to the public starting in 2006 regularly used this acronym and phrase. The viewpoint at the time was that the subject of these unidentified objects or craft was being taken seriously and pragmatically, and the term was better suited to that framework rather than UFO. UAP was not being used in any other slang or jargon like media or entertainment so as to negatively affect its purpose. Get used to it; whether its cause is faddish or generational, we will see both terms, as well as others, populate the literature as the semantics of our language takes over from time to time.

Precisely, this catalogue is an inventory of some sighting events in a historical context made famous and evidently noteworthy because they additionally contain the characteristic of simultaneous observation by many people. These next few chapters contextualize this common thread.

Contextualization is the frame of reference of an observation event. These chapters' frame of reference is defined as an event witnessed by a large number of observers. The reason for the discussion over multiple chapters is the analysis of the different characters of people and the catalogue of knowledge, unique and contextualized, each of them maintain and grow. Akin to this notion is that each individual could embody his or her own separate and different viewpoint or perspective on the observation. Most often, not everyone has such a uniquely different viewpoint. Around the occurrence of a singular event, at least the simple majority of the sample group will hold the same viewpoint and perspective. This is why there is an allure to an analysis of a mass observation sighting event. The more people observe the same thing at the same time, the more corroboration is bundled into the analysis. Credibility is incrementally added. A logical construct eventually forms that takes on a life of its own. The more this happens, the more toward reality and existence the paradigm moves. Communities begin to merge in their conclusions.

The chapters are organized according to certain criteria. The dimension of structuring the themes of these chapters ranges from this one, which embodies the heterogeneity of a sample group that includes the general public. The next chapter catalogues events that include only a highly defined sample group of individual observers. This group was/is specially trained and knowledgeable in conventions of extensive logic, technology, and/or in careers or education that embody both structures of learning. The later chapter acknowledges and catalogues a series of encounters that was observed by sample groups with which much was written and catalogued about after the specific events. A thematic characteristic at least as important as criteria is religious context derived from the event.

Primary selection inclusion criteria for this discussion include:

First, that it was a situational event observed by a large number of people.

Secondly, that the event was a singular one or at least one event that did not last longer than a few days or nights. This is a good time to introduce the concept of a *flap*. In ufology studies, a flap is a series of events, usually sightings but can be of other situations, which occur in one geographical area or region over an extended but condensed period of time. This is contextually different from what we are discussing. Here, a mass sighting definition is a one-time event or a situational series of events that develop, from beginning to end, in only a few days or nights. What the evidence here is trying to accomplish is to capture events that can include only one group of people at the same time. A flap could entail a series of events lasting months or more, and each event would be observed by a different subset of the sample group. Such flap sample groups would be tainted experimentally because there is a lack of immediacy and instantaneousnessthe mass sighting captures. Undue and contaminated influence among later observers could happen after the first sightings are communicated. The mass sighting eliminates this *experimental bias* from the analysis.

The third criteria is that enough details about the event have been recorded and made available and reported by enough observers to provide a warranty analysis of the occurrence. This warranty further establishes validity to the situation.

Fourth, that there is no single tipping point number of participants involved in selection to these lists except to say that it is more than a few dozen.

The fifth item is that any editorializing of all the events that you may come to see, such as what may have been done by representatives from the mass media or authored critiques and debunking of any kind through history, is factually and entirely eliminated from particulars of any discussed event. This catalogue is to be treated as reportable evidence free from any bias that was written about or alluded to in any literature. Additionally, these were real observation events and not perpetrated hoaxes. This catalogue criterion does not include any post-hoc opinions. If any of these situations became more than an observation event for any of the participants, it had been designated in the later chapter titled "What People Are Seeing Plus a Lot More."

Sixth, that there are numerous mass encounters which could be represented in more than one chapter. For example, the Belgium Sightings of 1989–1991, the Shag Harbour, Canada Event of October 4, 1967, and the Kecksburg, Pennsylvania Encounter of December 9, 1965, were all witnessed by a sizeable sample group of the general public. Additionally, a similarly sizeable sample group of military, civil police representatives, medical personnel, and many other individuals trained in some scientific, technologic, engineering, or mathematics discipline—otherwise known as the STEM disciplines—were all witness to their events.

Additionally, this catalogue includes participation of people from virtually all walks of life. These include prominent members of all the natural and social sciences such as the medical, political, legal, and financial fields, as well as artistic and athletic fields. Adults, as well as children, are included.

Here now is the short list and primer of eleven of the most popularly observed UFO/UAP events in history. Pursuing follow-up nuanced knowledge about any of these events can provide many reference portals.

Kecksburg, PA UFO Crash of December 9, 1965

The only fact that is agreed upon today by the two opposing sides of the debate in many cases of this nature is that some object landed or crashed in a then-forested thicket outside the town of Kecksburg, Pennsylvania. This conclusion only came about after initial postured reports from the factions representing the U.S. Government, military, police, and the public disagreed on what they observed.

Government and police search teams spent the night of December 9 investigating the subject area. Initial U.S. Army and Pennsylvania State Police statements were published in the Greensburg Tribune-Review newspaper on December 10, 1965, concluding nothing was found.[1] A story in the *Beaver County Times* newspaper on the same day took reports from numerous public witnesses to a red-orange fireball which crossed the local skies and crashed near Kecksburg. Reports from Wright-Patterson Air Force Base would not commit to anything. NORAD spokespeople said they "did not track the flash of light, but if it was a meteorite, it would have shown up on their scopes."[2]

The venue of this incident encompassed an extensive area from the Detroit, Michigan-Windsor, Ontario, Canada, corridor south and east to include northern-to-middle Ohio (Columbus) and east into Western Pennsylvania to "ground zero" near Kecksburg. Other initial reports from that night came from many airline pilots, astronomers, police, and the general public. Pilots also reported shockwaves buffeting their aircraft as they flew near the trajectory.

The general consensus from the mainstream public was unanimous and unwavering from the start. Some large object, twice as large as an automobile, traversed the skies from Michigan south into Ohio then changed course eastward and then southeast into western Pennsylvania. Air traffic controllers in Oberlin, Ohio, captured "only one object" on its radar. Debris was also found in the flight path further up in Michigan "near Lapeer, Michigan, two small stacks of shredded foil were found. After the sheriff's department reports of a fireball, the foil was different from the usual kitchen foil, made of lead and shredded in strips 1/16 inch wide."[3]

Sometime later (forty years), the government community changed its story. In 2005, NASA released a statement broadcasting that they had indeed examined parts of the Kecksburg object and now concluded it came from a Russian satellite.[4]

A local fire department was the first official representative on-site, along with public witnesses who observed then approached the fallen artifact. A blue glow could be seen along with smoke surrounding the aircraft. This was the search beacon available for approach. The fire department spent about fifteen minutes around the object before the U.S. Army and government officials arrived.

The "downed craft was acorn-shaped and about twice the size of a Volkswagen Beetle automobile and having hieroglyphic-like writing around its bottom ring."[5]

In the hours following the landing, many residents were able to gather in the neighborhood near the grove and at various points along the roads leading through Kecksburg to witness an acorn-shaped object covered in Army tarpaulin and filling an entire flatbed being transported out of the grove. One of the observers was John Murphy, a reporter for the area's local radio station WHJB. Many photos and aired public reports were done by Murphy.

A short time after that night, Murphy prepared a documentary for public broadcast called *Object in the Woods.* Before its scheduled to-air date, Murphy was visited by government officials. The conclusion of this meeting was that all the photographic and taped artifacts were taken and Murphy later aired an altered version of the broadcast. In it, he explained pressures from other witnesses forced him to modify the production and to retract many of his statements from that December 9 night.[6]

The other U.S. Government agency heavily involved in the proliferation of the case study was NASA. Beyond contradicting their own statements throughout the 50+ years Kecksburg is framed around, there have been multiple lawsuits litigated against them. A 2005 lawsuit filed by the courts emerged from the remnants of NASA's then latest multiyear effort to release their Kecksburg files. They smoke-screened the case by claiming the files could not be found and instead released approximately forty pages of unrelated material. That latest attempt for disclosure also stalled as NASA used another set of explanations for the doubtful nature of NASA even having had any files to begin with. They said that the personnel on site that night who identified themselves as NASA representatives were, in fact, Air Force personnel authorized to pose as NASA personnel as was common practice of the military in those days.

Additional attempts to explain the Kecksburg events as non-UFO related have used multiple scenarios, such as the Russian Kosmos 96 Venus space probe falling, a fireball, Air Force program activity of recovering space debris, a Corona satellite as part of a launch from California hours before the time of the incident, and a meteor—all which were described by adversarial opinion to have happened simultaneously on that day/night. Additionally, reports of the recovery of an acorn-shaped craft as big as a full-sized automobile have been contradicted to not have happened though many recordings of this event were intricately described. There is no shortage of inconsistency by these opinions in the description of facts and elements of the situation theater in Kecksburg.[7] If there was nothing to have been seen or nothing found, why are there a litany of different, incongruous, and incompatible explanations in the literature full of these inconsistent and illogical event constructs?

Ruwa, Zimbabwe, Africa Schoolyard
Event of September 16, 1994

Mass-sighting encounters have proven to be a worldwide phenomenon that is not biased against age, history, gender, or any other classification of human participants. This case involves a body of sixty-two primary school students, fifty administrators, and their alleged meeting with entities and craft in the rural location of the Ariel School locatednear the town of Ruwa, Zimbabwe.

This was another one-time event with no contamination of prior sightings, rumors, *flaps*, or such to taint an analysis. There were no accessible televisions or mass media for the children around that time or area to learn of any extraterrestrial or UFO/UAP information. In later interviews with noted psychiatrist Dr. John Mack—then actively researching, instructing, and practicing at Harvard University—most of the children had explained they had never even heard of a UFO.

The children, ages six to twelve, were at recess on a sunny Friday winter morning. Three silver-colored spheres appeared overhead then flew over the school. They each had red lights which flashed intermittently from underneath their craft. The children observed them moving around in the sky for a few minutes. They each flashed a very bright light at different times. When one of them did this, it would disappear from sight for a few seconds and then reappear in another part of the sky.

Eventually, one of the craft landed in a field of scrubs adjacent to school grounds. Within a few seconds, a "small man," approximately one meter in height, exited the ship and started walking across the ground. The children were interspersed at various distances near and around the part of the school grounds closest to the landed craft.

When the being first saw the children, it retreated to its ship. Seconds later, it reappeared and walked closer the children. The being was dressed in a shiny black suit which covered its body. Its head was not covered, and the children described facial features that included a long fluted neck and big round eyes.[8]

The little man held a conversation with the children, though not in the conventional use of mandibular speech; i.e. speaking in the same manner humans interact with each other. He was described as having only a slit opening for a mouth. The conversation, thus, was telepathic in delivery.

Later, many of the students told of the conversations they had with the little man. The central theme for all them was that of a warning about what humans are doing to hurt Earth's environment. Pollution and technology advancements that are not conducive to not harming the atmosphere, the land, and the oceans were told to the children as causes of the harm and that if they continue along this path, they will cause Earth irreparable harm. Some of them discussed at length how they were not talking with the little man with their mouths but rather in their minds.

Coincidentally, at the very time of the Ruwa mass sighting and encounter, Dr. Mack was traveling to South Africa to conduct research for a second book on the subject of alien abductions. He had just published his first book on the subject, a best-seller titled *Abduction: Human Encounters with Aliens*.[9] This was the then culmination of his first few years of research into alien encounters.

His trip was diverted to Zimbabwe upon contact from a long-time friend, Cynthia Hind, a local UFO researcher, literally as the September 16 day's events were unfolding. Within a couple of days, Mack arrived at Ruwa and began a lengthy intervention and research interview schedule. He first reached out to the community of parents to educate and advise them how to understand and handle ongoing and future interactions with their children about their encounter. Then he began investigating and interviewing all the sixty-two children and over fifty known adults who were at Ariel the day of the encounter.

What Dr. Mack, who was initially like most transformed scientific professionals, either ambivalent or not interested in the UFO and alien subjects, discovered and documented the exact same exceptional degree of consensus in their experiences as the other one hundred alien encounter participants he had studied to that time and were a major part of *Abduction: Human Encounters with Aliens*.

Dr. Mack had an extremely enlightening discussion with the Public Broadcasting System (PBS) television program *NOVA*[10] in 1996 about the experience, referencing his library of case history which included the Ruwa Ariel School Encounter.

Mack maintained an early-on neutral mind-set until he investigated a group of patients in the early 1990s who had had alien encounters in their past. With each case, Mack's investigation deciphered an emerging anomalous pattern; all these patients were describing the same scenario the same way! They had sustained genuine bodily injuries resulting from their experiences. This was an actual physical aspect that was supported by their discussions with Mack. A lot of patients' discourse evolved without the assistance of hypnosis or other medical practiced calmness exercises.

Mack categorically dismissed any denial of the phenomenon as due to neurological or *hypnagogic* hallucination causes, citing them as endogenous, arrogant, and wildly anthropocentric. He said, "We just can't accept the notion there could be another intelligence at work here."

He continued with a list of five proven factors that give credence to the alien experience. These include the consistency and congruency of the patient's discourse and that the subjects have no life experience to lay a baseline for their episodes. The third proof embodies the abundant real and demonstrated physical evidence—cuts, abrasions, ulcers, etc. —that the patients incurred. Fourth is the plentiful and nuanced association to UAPs/UFOs in most all the case studies, and fifth, the phenomenon has occurred in people of all ages, including school-aged children, as children as young as two years old.[11]

Which brings us back to the Ariel Encounter.

The *Mail & Guardian*, a Zimbabwean newspaper, published an interview by Sean Christie in 2014 with one of the then children who witnessed the encounter. The witness, now in her early thirties, recounted the messages the "little man" was trying to convey to her and all the other children. The main message circulated to the children was framed around humankind's destruction of Earth's atmosphere and environment by our runaway technology: "They (the aliens) weren't wrong, though, about the environmental stuff, were they? If you go out there now, you'll see the Miombo forests have disappeared for firewood."[12]

The interview wound down with a discussion about how an encounter can change a person, how they are perceived by others, and how those who choose to investigate for truth and the correct answers (Dr. Mack) with the following: "You want to know the real message is that this stuff can brand you for life. It undermined Mack's credibility, became this huge unending thing for others, they'll think you're a kook."[13,14]

Fiorentina, Italy Soccer Stadium Event of October 27, 1954

A football match between local rivals Fiorentina and Pistoiese was playing underway beneath a deep blue autumn sky in the Italian region of Tuscany at the Stadio Artemio Franchi, which is still being used today. A quote taken from the BBC News World Service eloquently described what occurred to alter the course of the enjoyment of both fans and players and soccer officials.

"Ten thousand fans were watching, but just after halftime, the stadium fell eerily silent. Then a roar went up. The spectators were no longer watching the match but were looking up at the sky, fingers pointing. The players stopped playing, the ball rolled to a standstill."[15]

In his game report, the head referee later indicated the match was suspended due to the observations of spectators of objects in the sky. What all them reported seeing was a group of objects moving slowly toward then over the stadium. The objects, which were described to be shaped like elongated eggs or stubby cigars, excreted a steady stream of glittery, silvery glass crystals as they traversed toward, and then over the stadium. The event in the immediate vicinity of the Stadio Artemio Franchi lasted a few minutes. From all reports, the objects followed a trajectory indicative of controlled flight and not affected by any wind or other weather factors.

Other reports from surrounding towns in the region described the observation as "rays of white light" that first day in the hours leading up to the stadium mass sighting. Shortly after seeing these rays, the local public then observed a strange, sticky substance falling from the sky after the solid objects flew over them. The unidentified substance landed on the ground and on surrounding buildings and other structures in its path. The environment was appeared as though it had snowed for a few minutes. The precipitation stayed intact for over an hour then evaporated. This occurred all over the Florence region.

The total number of participants in this encounter numbers in the tens of thousands by this time. All them were perplexed by the sighting of the craft moving through the Tuscany province and many even more so by the precipitation that had occurred in their locations. Numerous independent samples were gathered by individuals and submitted to authorities. One of these was the Institute of Chemical Analysis at the University of Florence. Among the tests administered on the samples were a series of spectrographic analyses. Conclusions and consensus drawn indicated the chemical composition of the samples contained elements of boron, silicon, calcium, and magnesium and was not organic in nature. The substance was deemed to not be found in its natural state on Earth and was not radioactive.

The facts about the probable nonnatural fabrication of this substance and its inorganic nature are significant because there were subsequent attempts to claim the source of the mass sightings and the substance precipitation was that of migrating spiders. The process that explains this hypothesis is known as *ballooning*. Certain species of spiders use this tactic as a legitimate witnessed process of movement from one place to another. A web of spider silk is made and properly aligned wind currents can take the web airborne.

Almost all the time, this movement was only a few meters from its origin and only achieved altitudes as high as the tops of trees. On exceedingly rare occasions, one or a few congruent webs can migrate higher and further than this.[16] Since 1954, numerous video analyses have captured this phenomenon. All the video recordings show spiders traveling with the balloons. This is logical as it is an instinctual mechanism of various spider species. Unfortunately for the spiders, ballooning is tantamount to a semi-suicidal activity as most do not survive the migration.

The literature reports archived from the hundreds or thousands who witnessed the encounter gave descriptions of these objects moving in straight paths for lengthy time periods very high in the sky. Unlike standard or typical spiderwebs, their appearance was similar to solid objects. Additionally, there were no reports of spider sightings, dead or alive, within any of these areas all around the Florence metropolitan area.

In reference to the chemical composition of the captured samples, it was shown by numerous experiments that the main composition contained inorganic compounds such as silicon and boron. While magnesium and calcium, other elements found in the samples, are elements that have an importance in organic chemistry, silicon and especially boron are sparse. Both of them, along with magnesium, have primary utilizations by humans in fields as technology and as far advanced as nuclear reactors and were the three most common elements found in all the tested samples.[17,18]

Citizens of Florence, whose municipal population was over one million at that time, reported the precipitation was widespread covering the district. When the total of the number of reports from the citizens who witnessed the encounter and the deposition of the snow are summed, the amounts of these deposits would exponentially dwarf any known, even to this day, events of ballooning by spiderwebs.

The Phoenix Lights; March 13, 1997

This encounter rivals the Fiorentina, Italy, soccer match in Florence in terms of the number of witnesses involved. Phoenix is the central hub of Maricopa County. Maricopa had a population of 3.2 million back then while the city of Phoenix had a population of 1.2 million. The best estimate when accumulating all the raw data of just the documented witnesses who reported their observation to some other party or authority indicates that more than fifteen thousand did so enough to report the event.

The objects arrived from the northwest from the Las Vegas, Nevada, area and further west and north before that from parts of California. The path of observations traveled into Prescott from the northwest, then south to Phoenix, and finally southwest to Tucson. The events of the northwest segment, known to many as the "first event," started after sunset at around 8:30 p.m. The segment after the objects traveled south then southeast started arriving in the Phoenix metropolitan area about an hour or so after. This segment has come to be known as the "second event." There were no reports of the objects landing anywhere. Reports from thousands of people that included engineers, astronomers, science researchers, police and military, and a widely heterogeneous cross section of the public concluded their velocities to be around a consistent 30 mph during the event. There were many occurrences during the two events where the objects stopped and hovered motionless overhead for a few minutes at a time.

The observations continued through most of the rest of the night as the objects continued their controlled flight trajectory out of Maricopa County and further southeast toward Tucson, about 120 miles away. The total number of witnesses from the venue upstream and downstream of the Phoenix flight path has never been accurately determined as it is difficult to quantify when increasingly more are involved and the tracking distance involved. As with the Florence, Italy, case, such numbers are most meaningful when they are left to a description of many thousands.

The encounter provided many visual perspective reports as the crafts held a steady flight path that indicated a controlled and maneuvered course, especially so at the times they stopped and hovered motionless. Unanimity among those witnesses who made reports of some kind for the first event indicated only one craft. Those who observed the second event indicated there were four such craft in the airspace over Phoenix. Temperature that night was in the low sixties with winds prevailing from the northwest or west, depending on where the observer was, at around five to eight miles per hour. Because the venue was so large, the conclusion is that the wind speeds were not high enough to significantly alter the flight characteristics of heavier-than-air airborne craft.

The Phoenix venue included among its observers, requisite law enforcement, military, science professionals, pilots, and engineers as part of the general public. Additional witnesses included those from the ranks of doctors, lawyers, scientists, baseball teams, and coaches who were engaged in sports game events at that time and celebrities. This case could have been included in the next chapter titled "The Especially Trained Public," as there were so many witnesses to the Phoenix Lights that night from all walks of life. Air traffic controllers at Sky Harbor International Airport, Deer Valley, and Mesa-Gateway confirmed the sightings. One anomalous fact, though, was no data was acknowledged to being captured on their radar equipment.[19]

The newspaper USA Today interviewed witnesses in a June 18, 1997, article. At the venue of the first event near Prescott one observer, a truck driver saw "two brilliantly lit orbs, shaped like spinning tops" while on the highway north of Phoenix. Another witness near Paulden, about sixty miles north of Phoenix, reported to police a cluster of five red lights underneath a craft headed toward Phoenix. One hundred and twenty seconds later another police report, directly from police officers themselves in Paulden and Prescott, indicated a craft with four white lights and one red light heading in the same direction. The truck driver then saw the formation of orbs enter into the airspace around Luke Air Force Base. Seconds later, as he witnessed a view right toward the runway, he saw three U.S. F-16 fighter jets take off and veer sharply right at the orbs. One of the orbs, in response to this maneuver by the F-16s, immediately shot vertically up into the air and left the F-16s in its wake. Sixty seconds after that, police telephone lines, as well as tele-connections to nearby Luke Air Force Base and media outlets, became overloaded with phone calls. This scenario continued unabated for about two hours. The flight of the F-16s was later confirmed by Lt. Col. Mike Hauser of Luke Air Force Base.[20]

The venue for the second event centered around the Phoenix metropolitan area and to the southeast. V-shaped objects with seven lights in a "trailing light" formation that never varied in distance from each other formed the general consensus. The observations indicated the craft did not make any noise as it flew overhead and past. It was extremely large. Some words used to report this characteristic were "huge, colossal, enormous, gigantic, mammoth." Computer analysis estimates taken from many photographic and videotape recordings place the craft at least the length of three football stadiums and upward to a mile wide. Many other videotapes and photos were examined to verify witness descriptions of characteristics when the craft was nearby. For example, many people described the entire night sky as clear and starlit, which suddenly became dark when the craft flew past. At times, it stopped completely and hovered. For people with a visual vantage point below the craft, no stars were visible.

Many observers described the outline of the main craft, as having seven main light protuberances in a "V formation" as a "grey distortion of the night sky, wavy, all you could see was the outline, as though something was blotting out the stars. The lights weren't bulbs. They looked like gas. The light didn't spill out or shine." Subsequent laboratory studies of videotaped samples concluded a unique quality of the light as "perfectly uniform with no variation." Further descriptions of the underside of the craft produced an appearance described "as if one was looking through water."[21]

Among the diverse community of observers that late winter night were two seminal participants, then Arizona governor Fife Symington, once active Air Force officer and pilot, and also a practicing medical doctor at the time, Lynne Kitei. Governor Symington explained his initial reluctance in admitting to his observation as not "wanting people to panic"[22] as he later discussed in an NBC News interview in 2007.

Later that year, in a separate interview with the CNN Network discussing an upcoming summit at the National Press Club in Washington, DC he continued,

> My office was besieged with phone calls from very concerned Arizonians. I decided to lighten the load at a press conference where my chief of staff arrived in an alien costume to lessen the sense of panic but upset many of my constituents. I would now like to set the record straight, never meant to ridicule anyone. My office made inquiries as to the origin of the craft, but to this day, they remain unanswered. This is indicative of the attitude from official channels, explanations that fly in the face of the facts—weather balloons, swamp gas, and military flares.[23]

The now retired Symington was moderator of the National Press Club summit on the Phoenix Lights topic that following November. Fourteen other high-ranking military officers and government officials from seven countries provided the discourse about the topic of unidentified flying objects or unidentified aerial phenomena. A key observation that has led to the frustration over many aspects of the Phoenix Lights is there never were any serious investigations utilizing a large manpower research community by experts from either science or military sources.

The U.S. Air Force, while taking their long-time official position that "they do not investigate anomalies of unidentified nature" (their position since the days following Project Blue Book ended in 1969), nevertheless, were quick to publically conclude that what all those people saw that night were simply flares dropped out of U.S. A-10 Warthog aircraft during a training exercise flown by members of the Maryland Air National Guard. The flares displayed the flight characteristics that permitted them to hover over a sighting point at

Luke Air Force Base and "wink out" of sight one by one as they then fell below the peaks of the Sierra Estrella mountain range southwest of Phoenix. This explanation, as described by the Air Force, who do not get involved in such explanations, was projected out to fit a description and scene depicted in one of the famous public broadcast sequences of the Phoenix Lights' case; the one where the lights are seen to "wink out" just like the Air Force said.

This video was taken by Lynne Kitei, MD, a then-medical physician at the Arizona Heart Institute in Phoenix. She captured the now famous video in question. The events of March 13, 1997, turned out to be a life changer for her, as she described in her documentary in print, *The Phoenix Lights*.[24]

Scientific laboratory tests were administered on Kitei's video and compared with the Air Force's explanation which indicates there were too many inconsistencies with the accurate fit of all the elements of the situation including situational, geographical, weather, and confirmed events of that night to give much credibility to what Air Force official reports stated.

Westall, Australia Schoolyard Encounter of April 6, 1966

The suburban Melbourne city of Clayton South in early autumn was ground zero for one of Australia's most disconcerting visits from something that flew in and over two schoolyards, stopped suddenly and hovered, then landed in a paddock grove adjacent to the campuses. When the first wave of human observers finally arrived on foot at the grove to observe and interpret their sightings, they saw the craft situated motionless on the ground for a while. After a few more minutes, the craft lifted off the ground silently, hovered for about ten additional minutes, and rose through the sky and out of sight within seconds toward the urban core of the city of Clayton.

The paddock grove, which is still there today, is called The Grange Reserve. Students from one of the schools, Westall Primary School, were outside at recess and were one of two groups attributed to the very first observation. Another group of students were in a science class led by Andrew Greenwood when all them spotted the ship start to rise slowly from the ground in the paddock.

Students from the other school, Westall High School (now called Westall Secondary College), were a second group who—together with teachers, administrators, staff, and professional and tradespeople, called "tradies"—summed the total number of witnesses to well over two hundred. Additional witnesses included residents and business people not connected with either school in any capacity and, who subsequently discussed their encounter with journalists for archival purposes, augmented the total by more than a couple hundred.

One of these witnesses was interviewed for *The Age* newspaper in Melbourne by journalist Stephen Cauchi.[25] Shaun Matthews was on the land where the paddock was situated that day because his family leased grazing land around the area. He was among the first to witness the events unfold. He recounted,

> I saw the thing come across the horizon and drop down behind pine trees. It was silvery with a purplish haze to it, the size of two automobiles, bright but not too bright that one couldn't look at it. It did not fly like any aircraft I knew of. When it finally lifted off, it went up and off very, very rapidly.

Instructor Andrew Greenwood reported the object was as like a classic silvery-grey saucer, which this was corroborated by many other witnesses. Greenwood's class witnessed the airborne ship wave then hover in a repeated mechanism for about twenty minutes. The *Herald Sun* newspaper, also of Melbourne, archived articles and interviewed other witnesses that day.[26] Students Terry Peck and Jacqueline Argent were both among the first to arrive at the paddock to see what teacher Greenwood and other students and administrators described. Some of the other students who were first on the scene were described by others as becoming sick and fainting, including one student who was transported to a hospital by ambulance.

They were also witness to the impression left on the grassy ground by the craft after it flew away. Along with others, Shaun Matthews noted the ground had a circle the size of the diameter of the landed craft "that had been cooked or boiled but not burnt."[27] The entire event lasted twenty minutes. After the craft departed, all observers saw at least five other aircraft that they could identify fly in the direction of the flying object. The Grange Reserve and the two school campuses are up range from Moorabbin Airport by six kilometers. All observer reports emphasized the difference in the appearance and flight characteristics between the unknown craft and the subsequent chase planes. The first unidentified craft was silent and maneuvered in ways they knew no human-made craft could accomplish. The others were exactly aircraft they had seen before.

After the event concluded, there was a requisite investigation and debriefing of the student bodies of both schools with threats of retaliation by both school administration and visiting "sharply-dressed men in dark suits," if word of this was discussed with anyone. Families of some of the students and other residents also received visits from these suits and were told not to discuss the day's events with anyone. Some students reported later seeing Royal Australian Air Force military scouring the Grange Reserve landing theater and eventually turning up the entire ground site. To add more high strangeness to the case, local television stations that had crews conduct stories and video of the event discovered their films were missing or lost from their archives sometime after their broadcasts. All film footage was thus lost.

The craft did not leave any physical artifacts behind on this occasion. Though its status remains "unknown," the craft is called a "craft" due to its corroborated physical flight and landing/ascending integrity characteristics. This analysis embodies the observations of hundreds of citizens and represents a very diverse cross section of demographics. Additionally, this encounter was not preceded by other encounters over a long period of time causing it to be known as a flap. This prevents the event from being biased and contaminated, as such flaps tend to be, by dissemination of prior events in the course of an evolution of mass hysteria, mass hallucination, or other irrationalities that are conjured up evidence by social scientists.

Tinley Park Lights of August 21, 2004

For a mass-sighting encounter that was witnessed by many thousands of people in suburban Chicago, Illinois, there has not been much information given publically. In ufology, Tinley Park is recognized as a top encounter of any type. There were four widely reported mass sightings that span over two years. This August 21 event was followed by one on October 31, another on October 1, 2005, and a fourth on October 31, 2006. The event theater was widespread. Tinley Park and neighboring city Oak Park, Illinois, tallied the most significant amount of raw sighting reports. The south and southwest Chicago suburbs are densely populated and Interstate 80 highway traverses through the area. On any of these particular evenings a theater estimate could exceed half a million. All four events occurred in the early evening prime time hours.

Only the event of August 21, 2004, will be addressed here to eliminate some of what can be referred to as "flap bias." Flap bias is defined as unadvisable guidance on a case when the public or any other potential eyewitnesses have prior knowledge of such a flap situation in prior days that could influence their judgment. Critical thinking could become adversely affected by this course of events. A complementary bias called "wave bias" is addressed elsewhere in *The Humaniverse Guide to Better Reasoning & Decision-Making*. The two terms are similar except that a flap can last for a couple of years; whereas, a wave typically lasts only a few weeks. It should also be noted there is a lot of ambiguity and interchanging of the time length of flaps and waves. Various literature contends that the author's particular piece will reverse the time length of the two terms.

A further factor that increased the sighting population was the duration of each event. Total event time of each was over thirty minutes. Additional augmenting logistical characteristics included a low to intermediate altitude platform which made observation details easier and more widespread. Later, mathematical research concluded visibility exceeded twelve miles in any direction. On all four nights, the local weather conditions were of clear skies. Stars were visible on each occasion.

Further geographical factors must be considered here. The cities of the south and southwest Chicago suburbs are situated within the general vicinity of both Midway and, further north, O'Hare airports. Both lie within forty-six miles to the north of Oak Park and Tinley Park. Many eyewitnesses indicated during their observations they could also see other identifiable aircraft, both commercial passenger jets and smaller propeller airplanes at the same time. These characteristics were reported by many to be of assistance in that the unidentifiable craft maintained anomalous flight paths and platforms from the identifiable aircraft also located in the area.

On August 21, 2004, three silent self-luminescent objects, red or red-orange in color and spherical in form, hung in the sky and moved slowly in formation along a controlled flight path. The duration of this occurrence was over thirty minutes.[28]

Other information of note refers to an abundance of still photographic and video films. This is due to the wealth of eyewitnesses at all four events. Unfortunately, as in the Phoenix Lights Incident, the mass sighting most akin to Tinley Park, there were no landings of any of the craft.

Even though thousands of still photographs and thousands of feet of video exist, this type of data frequently lacks some in substantiality. A popular type of camera used to obtain a significant portion of this visual recording was of an "off the shelf" variety used then in cell phones and in mass produced and more inexpensive cam and video recorders. The dynamic qualities inherent in technology of this level of quality correspond with less than excellent and fully reliable mechanics and output, in other words, of a lower quality. Also when dealing with photographical evidence, the ufology universe has had to historically cope with the availability of increased technology to produce hoax scenarios. Cameras and video recorders are readily available to just about anyone, and when the imagination of a hoaxer is inserted, it becomes a less than desirable outcome for a serious researcher to tolerate.

The shortcomings just described are countered by the sheer number of recordings that were made. Additionally, in the August 21, 2004, event and not found in the others are facts that it was a summer mass sighting; more people were outside, and an additional 30,000+ vehicles were stuck in gridlock in front of the Tweeter Center amphitheater in full view of that night's display.

The ships were reported by many to have the classic triangular shape of the Phoenix Lights craft. They were also very large, on a scale upward of 1,500 feet across. There was also a "persistent relative order between the objects,"[29] a triangulation of white lights,[30] and a total lack of conventional aircraft identification lighting patterns yet consistent flight characteristics, though anomalous from such conventional ships because of the turning patterns and the low speed displayed by these vessels. They were easily differentiated from the conventional aircraft seen by many of the eyewitnesses in the same sky.

Additional facts in reference to the proximity of both Midway and O'Hare airports support the anomalous nature of Tinley Park and Oak Park. The Federal Aviation Administration has designated the airspace surrounding this region as Class Bravo. This is a very controlled and tightly managed airspace which, by definition, would eliminate the military from any preoccupation into this area. The population density precludes this as being a valid element in the analysis. Also the parachuted flare exercises often used as an excuse for the Phoenix Lights were eliminated because of this feature of densely populated landscape.

Both the FAA and local military installations have not publically reported any investigations to date. Officials at both Midway and O'Hare airports summarized an accumulation of all the incoming reporting led to their conclusion of there being no impending risk to the safety of the general public to warrant expending resources to study the events.

Because of the favorable weather conditions that existed on everything the evenings within this flap, it is more than circumstantial that an advantageous opportunity existed to have a backdrop of stars with which to compare against the moving vessels.

Nuremburg, Germany, Mass Sighting of April 14, 1561

This entry is contextualized together with the next entry, "The Basel, Switzerland, Mass Sighting of 1565." Here is where the term of the concept *broadsheets* originates. Briefly, a broadsheet is the Middle-Ages equivalent of the newspaper of today. A broadsheet measured about fifteen by eleven inches. The craftsmen produced one of these every few weeks. They communicated the "news of the day" or enough of it that made the publication deadline to be included in the next broadsheet. They were posted in public places as the main public dissemination.

On April 14, 1561, residents of Nuremberg, Germany, one of the most prosperous cities in Europe at that time, arose around dawn to witness a fierce aerial battle that lasted over an hour. Hundreds of airborne objects of many designs engaged with each other in front of a backdrop of the rising sun. Spheres, cylinders, discs, globes, crescents, cross-shaped objects, and spear-shaped craft flew over the sky above. Many colored ships were seen. Particular emphasis was placed on blood-colored semicircular arcs, dull metallic globes and discs, red, black, orange, and blue-white globes and spheres.

The residents observed the cylinders dispensing many of the globes and spheres and interacting with the other aerial craft. In the broadsheet, mention was made of a delineation of adversaries; in particular, that they could recognize that the craft associated with the cylinders represented one side and the other arcs, discs, crescents, and some spheres the other. Many craft were glowing bright red before falling in a fiery trajectory to the ground some distance away, explode, and then fade from sight.

According to details from the broadsheet, "The sky was apparently filled with machines, clashing in battle. Comets and such were well identified. It is highly unlikely that what was witnessed was a meteor shower." So many different design shapes were reported to rule out a conclusion of any showers of celestial debris. "Afterward, a black spear-like object appeared"[31] and, "Accounts claimed the black orbs would sometimes turn red and fiery before fading to nothing. Science suggests these events were the result of meteor showers and comets as the most logical explanation. The fact that people believed a battle was taking place in the sky above, involving a number of differently shaped objects, makes it difficult to understand this mysterious phenomena as simply comets or a meteor shower."[32]

This broadsheet, which was authored by Hans Glaser, resides today in the Zurich, Switzerland, Central Library in the Wickiana Collection. In the aftermath of the battle, there were no reports of any artifacts being obtained in the historical records.

The contextualization of the witness' perspective on descriptions of the design of the battle theater must be addressed here. Specifically, a point should be made about how humans describe what they observe. It will occur in all situations of daily life. Whether in the battlefield, the science laboratory, a building, or a theater of any kind, people will describe what they observe relying on the mental record of shapes and contexts they are familiar with at the time. This notion cuts across and through time. If we observe something today that looks like an airplane, we will describe it according to the vocabulary we have retained. Five hundred years from now, our descendants will read our archives and wonder the exact same things about us as we wondered about our ancestors' observations at Nuremberg and Basel, Switzerland, four hundred fifty years earlier. These people were not lying or hallucinating. They were just trying to describe what they observed according to their vocabulary in use at that time.

Additionally, how will our descendants of five hundred years be able to access our archives? Will we use printed media to aid in this documentation? Will we document them in some current technology (for today at least) that our descendants will be able to successfully access? What would they think if they saw a flash drive or, heaven forbid, a DVD? Would they interpret the communication the same way as we meant it to be when we recorded it?

Human beings have not changed the way they contextualize observations or process their thoughts in general. People describe what they see, hear, feel, smell, and taste today literally using the same manner with which they did centuries ago. The thought processes are the same. Only the technology used to record the message has changed. It is a debunker's job to take this function of man's mentality and distort its contexts and manipulate it into a status of meaningless oblivion.

Author Colman S. Von Kevicsky wrote an article in *Official Ufo* magazine that gave a nuanced and profound description of the Nuremberg Battle of 1561.[33] Remember that an interpretation made a long time after the actual event tends to lose sight of the real meaning of what the original witnesses tried to communicate. It is this real meaning that is most important in its context and not what a modern interpretation is spun to produce. As the saying goes, "Real meaning is lost in the translation and also often in interpretation."

Basel, Switzerland, Mass Sighting of August 7, 1566

This was a mass-sighting event similar in scope and context to the Nuremberg encounter four years earlier. Because this event occurred after the Nuremberg sighting, the participants could be accused of copying Nuremberg for whatever purpose that could be described as rational and logical. Modern interpretations aim this in many directions.

What is significant is the descriptions portray similar details of the Basel battle theater.

A further statement regarding an earlier point made about the nature of the human characteristic that defines our mental thought process as it pertains to our communication with others was written in an article "UFO over Switzerland 1566, Aerial Battle in the Skies": "It should be pointed out that the descriptions and woodcuts are attempts by artists nearly five hundred years ago that depict an event that they potentially could not comprehend, the event sounds very much like a modern day 'dogfight' between planes of opposing forces."[34]

The Basel broadsheet reports one further observation, in a different design, than the Nuremberg visualization depicted. Where the effects of fired weapons hitting their targets in the Basel woodcut are generally more circular in design, the Nuremberg representation shows a more laser-like explosion smoke trail.

Later in the "UFO Fleet Over Switzerland" piece, this comparison was made about these black circles: "The black circles are very similar to explosions in the sky as photographed by World War II reporters. The cross-shaped vehicles look very similar to the profile of a World War II fighter ascending on a steep climb." The literature takes a detour with this passage: "Some have proposed, witnessed were actually World War II battles. Some sort of slip in space/time allowed the sixteenth century town to witness an event that would not occur for another four hundred years."[35]

Here is another perspective on the notion of train of thought processes. There is no attempt to compare the broadsheets of Nuremberg and Basel encounters or anything of what they represent. This is a very alive and enduring characteristic of humankind that spans the history of our time on Earth. Man's brain function and thought processes recurrently process events with objectives of categorization and classification. When two or more things are compared or measured against each other, our brains tend to categorize and/or classify them. Everyone who is human does this. We have done this since recorded historical literature depicts. What the citizens of Basel reported that August morning was not compared to Nuremberg or any other event. They were just recounting what they saw. Thousands of people saw essentially the same presentation in the sky. This improves the credibility of the observation. Any categorizations, classifications, or comparisons were and are being done after the event has already taken place.

Next is the application of two concepts that are themselves the very foundation of the calculus mathematics, those of discreetness and continuity. Maybe we could frame this notion of train of thought around differently from the very discreet and measured stepwise dimension that causes us to obsessively categorize, classify, and compare; in other words, the notion of "the three c's." Maybe more of a continuous train of thought, one that eliminates the discreetness of the three c's above, may assist us in effective observation and analysis and lead us to the conclusion of situations we face each day. It takes some discipline and continuous practice for each of us to become and stay proficient. Scientists, skeptics, professionals trained in logic, laypeople, and debunkers are all guilty and could be helped by recognition and practice.

Dalnegorsk, Russia, January 29, 1986

January 29, 1986, was a date the thousand or so residents of the small far eastern Russian mining town of Dalnegorsk, near Vladivostok, have not forgotten, as we continue our global survey of this category of mass encounters. In the early evening at approximately 7:00 p.m., a spherical object descended into the low airspace about 2,800 feet above the town. Its trajectory brought it down at about a sixty-degree angle and then straight into view. The craft continued silently on a straight and level flight path through the town with a velocity calculated later by science and engineer investigators of about fifty-five kilometers per hour.

As the craft traversed the airspace overhead, it was described by all witnesses as having a stainless steel-colored sheen and somewhat shiny but not extremely so. At the outset but only for a portion of the sighting, it carried a blue aureole around it. As the craft proceeded overhead through and then away from the town, the blue aura was slowly replaced by a reddish glow aureole, which is an important point that will be explained next.

A ridge of mountains is situated at the northern end of Dalnegorsk. One particular mountain, known as Izvestkovaya Mountain, was the ultimate landing site during the encounter. Known also as Height 611 or Hill 611, Izvestkovaya Mountain has an elevation of about 611 meters.

As the craft entered the mountain's airspace, it began to rise in altitude or at least tried to. It attempted this six different times. Each time it rose, the craft began to pulsate in color of a far deeper reddish hue than when the witnesses first observed the color as it left Dalnegorsk airspace and headed directly toward Hill 611. The craft rose steadily for a few seconds and then began to descend again. When it descended, the reddish glow mostly dissipated. As it cycled through the sixth time, upon its apex, the craft had a sudden jerking motion to change its direction. It descended in a straight line and, within a couple of seconds, hit the top of Izvestkovaya Mountain, which is commonly known as Height or Hill 611 with a loud thud sound.

Observers watched the events for about sixty minutes as the craft was continuously visible on the cliff because of the reddish glow that radiated from the ship. The area around the landing site burned intensely in this reddish glow the entire time. The burning intensified each time it attempted to take off. For that period of time, the craft attempted to lift off only to fall back to the site. Eventually, at the end of the sixty minutes and a few tries, the craft did lift off successfully and fly away.

The times after the event were overflowing with activity as a series of geophysical investigations started arriving at Hill 611. On February 3, the first of these teams led by Valeri Dvuzhilni, then head of the Far Eastern Committee for Anomalous Phenomena started their investigations. [36]Here is what was found at that first venue investigation.

An irregularity of significance occurred that winter in the region. Daily temperatures remained consistently below zero. The mountains were covered in a meter of snow or more. However, at the crash site, there was no snow in the ring around the landing. Many splintered rocks made of silica were found and collected along with sprayed silvery appearing debris in two forms—sprayed—and in solid balls that gave an appearance of ball bearings. Many small mesh-like pieces, more balls made of lead and other elements, and glass substances were also collected. The investigating team collected many tens of kilograms of material. Overall, the debris cache was discovered to have gone through a process that took on the appearance of being vitrified.

Soviet scientists from a variety of facilities examined and experimented with assortments within the collection. Included in this group of research teams were scientists from the Soviet Academy of Sciences, Omsk Branch. In their book, *UFO Case Files of Russia*, authors Philip Mantle and Paul Stonehill give the most comprehensive and pragmatic single accounts of the evolution of this case study.[37,38]

The unusual evidence discovered in this first cache of many to follow was plentiful. Some of the silica rocks were pure silica, not unusual in and of themselves but also found to have magnetic properties. They were, in essence, magnets. Silica is not magnetic when found in nature. Other silica rocks were alloys and composed of many other substrate elements, including iron, aluminum, manganese, nickel, chromium, tungsten, cobalt, and copper, among others. The sprayed silvery compounds and solid spherical balls had their own unique chemical compositions that exemplify the same diverse variety of elements as the silica-based debris. Other elements found among the cache included cerium, vanadium, sodium, and others. The mesh pieces contained multilayered entanglements of wirelike filaments that measured seventeen microns wide, less than one-third the width of a human hair. Within these layers were thinner layers of differing metallicity. Also within these layers were additional layers of gold filament threads. The main point is that these substances have never been found in nature on Earth; they were somehow manufactured and constructed. They came from the object craft that crash-landed that January 29. They were retrieved only days later.

Other experiments within this first study included spectroscopy, roentgenological (a form of radiology analysis), and heating of samples inside a vacuum-closed system environment. The heterogeneity characteristics of the cache were evident upon the analysis of these tests. Some of the samples melted and spread out in layers and others coalesced into balls. Some of the balls then formed glass-like crystal layers while others did not. Additional Soviet research teams, including the USSR Academy of Sciences, Far Eastern Department, could not isolate a precise and consistent formation mechanism that could synthesize only one distinct chemical composition. Stranger still was the observation that came from the experiments which showed that, upon melting the gold, silver, and nickel elements of the debris substrates completely disappeared and the appearance of other elements such as molybdenum and beryllium sulphide occurred.

Dr. Dvuzhilni published an article shortly after the initial investigation in *NLO Magazine*, a Uzbekistan, Soviet Union periodical, and also in reprint from a March 1990 article in *FENOMEN Magazine*.[39]

The anomalies continued to accrue at Dalnegorsk as time passed after the encounter. Among these were many reports from all research teams of electronic equipment failing to operate after being tested and verified functional before using at the site. Also noted was the complete absence of any biological life at or near the venue including any insects for years afterward. Additionally, many of the scientists became ill with various afflictions including blood disorders, high blood pressures, and sensory abnormalities. Other findings that conclude irregularities in nature from Dalnegorsk included tree branches where the wood was welded together yet not charred or burned.

Three additional Soviet academic and eleven research institutes provided analyses leading to enigmatic conclusions. Commonalities included the infeasibility of any of these substances. These substances were not there before the events of January 29, 1986.

One of the theories that circulated for a time around the case was that the craft was American-made and exemplary of the advanced technology it enjoyed over the USSR. Dr. Dvuzhilin helped coordinate further studies that included American, British, and Japanese science research teams, among others in the community. All researchers agreed in their conclusion that there was no American or any other man-made technologies that could replicate any mechanism that produced the debris.

The January 29, 1986, event was, as it turned out, the very first of a defined flap of activity over the next few years. This indicates the initial mass sighting was not tainted by social media, rumors, or communications at risk for inflammatory ramifications.

Shag Harbour, Canada of October 4, 1967

This experience began at about 11:00 p.m. along the shore of a small fishing town near Halifax, Nova Scotia. A very large round craft, lighted on its bottom by four linear bright orange lights and flashing in sequence, appeared over the shoreline and was witnessed by the first two groups of witnesses, including fishermen in boats and residents on shore along the highway that runs parallel with the shoreline. A Royal Canadian Mounted Police officer who was among the first observers on Highway 3 near the scene, estimated its dimensions to be about sixty-five feet long, ten feet high and dome-shaped. The craft was observed to be in controlled flight initially at an altitude of only a few hundred feet. While over the shallow waters close to shore at about six hundred meters out, the low-flying lighted ship banked at a forty-five-degree angle and descended slightly into the water. Many other local residents, in addition to the eyewitnesses, reported loud bangs and whistling noises. It was observed to stay just below the water's surface for a few minutes and move horizontally in the water. The orange lights slowly turned yellowish. All along its path in the water a yellowish thick foam floated to the sea's surface. The ship then sank slowly deeper into the water until the rays of light slowly disappeared.[40]

The onshore residents, as well as the fishermen as eyewitnesses, placed calls to local police as they had feared an airplane may have crashed. A large group of Royal Canadian Mounted Police (RCMP's) arrived at the scene within minutes and were able to witness the craft just as it was beginning its descent into deeper water and its lights were slowly starting to fade.

Minutes after the arrival of the initial group of RCMP officers, a fleet of additional fishing boats sailed to the crash site searching for survivors. They were joined by the Canadian Coast Guard in a possible rescue situation. Up to this time, only about an hour had elapsed since the very first sighting.

By the next day a, Canadian government spokesmen from the rescue coordination center in nearby Halifax concluded no aircraft was missing that could have been in the region. For days after, the Royal Canadian Navy and Royal Canadian Air Force investigated every cubic inch of the area and divers scoured the sea bottom. The search area extended as far as the Gulf of Maine waters. Technology that assisted in the investigation included divers into the three-hundred-foot depths and the most advanced SONAR and other electronic equipment the Canadian NAVY, Air Force, and Coast Guard possessed. The trace evidence found throughout the three-day first session included a very thick and glittery yellow foamy residue which reportedly could not be identified from later laboratory analysis. That night of the crash, the Canadian Coast Guard ships and the local fishermen reported, "Sailing through a thick yellow foam that indicated that something submerged."[41] No other physical evidence was reported by the Canadian Armed Forces according to the Canadian government.[42,43,44]

This last sentence is important. Some years later, two MUFON case investigators, Chris Styles and Doug Ledger, dedicated many years of research to uncover a cover-up perpetrated by the Canadian and U.S. governments. It was proven from hidden documents that later surfaced and then revealed the Canadian military did, in fact, track the ship for over twenty-five miles to a site called Government Point, a U.S. advanced military base.[45,46,47]

The U.S. Navy had detected and positioned a fleet of its own ships directly over the submerged craft for over three days. The Navy fleet observed and waited. The submerged craft did not move during that time. Without any warning, on the fourth day, a second unidentified craft submarined in and joined the first craft. These two craft were "SONAR-ed" together for an entire week by the U.S. Navy. On the eighth day, the two unidentified craft started to move toward the Gulf of Maine. The Navy pursued but fell behind their targets. Uncounted numbers of military officers and enlisted men witnessed the ships eventually break the water's surface and accelerate skyward ultimately vanishing in seconds.[48]

The archives for the Shag Harbour incident hold another surprising element to the discourse. When talking about the UFO/ET phenomenon, society has had reason to voraciously criticize the government (United States and foreign governments) for doing anything they can to downplay, discredit, or deprive the citizens of the truth and of denying reality. The Shag Harbour incident was intensively investigated by thousands of officials of not one but two governments. Both initially reported and continued to maintain for some time that it was an airplane which crashed into the sea, despite having left boxes of reams of defined archived reports, analysis, and documents, available free within the public domain that say otherwise. Shag Harbour is another of these scenarios where the public, which has been rebuked many times through history for not accepting the government's conclusions about the encyclopedia of UFO/alien events, is left without closure.

References

[1] Kecksburg Ref 1: Author unknown. 1965. Unidentified Flying Object Falls near Kecksburg-Army Ropes off Area. Greensburg Tribune-Review. December 10, 1965. Retrieved from: Jump up to: [ab]Greensburg Tribune-Review headline story, December 10, 1965.

[2] Kecksburg Ref 2: Author unknown. 1965. Countians See Mystery Fireball. Beaver County Times. December 10, 1965. Pgs. A-1, A-4. Retrieved from: https://news.google.com/newspapers?id=82AyAAAAIBAJ&sjid=4rIFAAAAI-BAJ&pg=3668,2274894&hl=en.

[3] Kecksburg Ref 3: Author unknown. 1965. Countians See Mystery Fireball. Beaver County Times. December 10, 1965. Pgs. A-1, A-4. Retrieved from: https://news.google.com/newspapers?id=82AyAAAAIBAJ&sjid=4rIFAAAAI-BAJ&pg=3668,2274894&hl=en.

[4] Kecksburg Ref 4: Author unknown. 2008. NASA under Pressure Over UFO. CBC News. May 12, 2008. Retrieved from: http://www.cbc.ca/cp/world/051208/w120872.html.

[5] Kecksburg Ref 5: Kiger, Patrick J. 2012. Top 10 Mass Sightings of UFOs. National Geographic Magazine. June 11, 2012. Retrieved from: channel.nationalgeographic.com/.../top-10-mass-sightings-of-ufos.

[6] Kecksburg Ref 6: Wilson, Patti A. 2011. UFOs in Pennsylvania: Encounters with Extraterrestrials in the Keystone State. Pgs. 53–55. Copyright 2011 by Stackpole Books.

[7] Kecksburg Ref 7: Oberg, Jim. 2008. NASA Lawsuit Over Kecksburg UFO Documents. Published January 7, 2008, by Jim Oberg. Retrieved from: http://www.jamesoberg.com/statement_nasa_kecksburg.pdf.

[8] Ruwa Ref 8: Cynthia Hind. 1994. The Children of Ariel School – Case No. #96. Ruwa, Zimbabwe. UFO AFRINEWS 1994. Retrieved from: African Schoolchildren See Landed UFO and Occupant (Ariel School Sighting in Ruwa, Zimbabwe). September 16, 1994. www.ufoevidence.org/cases/case127.htm.

[9] Ruwa Ref 9: Mack, John E., MD. 1994. *Abduction: Human Encounters with Aliens.* Copyright 1994 by John E. Mack, MD. First published on January 1, 1994, paperback by Pocket Books – NY.

[10] Ruwa Ref 10: Author unknown. 1996. Interview with John Mack Psychiatrist, Harvard University. NOVA Television episode. Public Broadcasting System. February 27, 1996. http://www.pbs.org/wgbh/nova/aliens/johnmack.html.

[11] Ruwa Ref 11: Author unknown. 1996. Interview with John Mack Psychiatrist, Harvard University. NOVA Television episode. Public Broadcasting System. February 27, 1996. http://www.pbs.org/wgbh/nova/aliens/johnmack.html.

[12] Ruwa Ref 12: Christie, Sean. 2014. Remembering Zimbabwe's Great Alien Invasion. *The Guardian* newspaper. September 5, 2014. Retrieved from: mg.co.za/.../2014-09-04-remembering-zimbabwes-great-alien-invasi. Ruwa Ref 4.

[13] Ruwa Ref 13: Christie, Sean. 2014. Remembering Zimbabwe's Great Alien Invasion. *The Guardian* newspaper. September 5, 2014. Retrieved from: mg.co.za/.../2014-09-04-remembering-zimbabwes-great-alien-invasi. Ruwa Ref 4.

[14] Ruwa Ref 14: Mack, John E. 1999. Passport to the Cosmos. Crown Publishing. Published November, 1999. Copyright 1999 by John E. Mack, MD.

[15] Italy Ref 15: Padula, Richard. 2014. The Day UFOs Stopped Play. BBC World Service Sport Magazine. October 24, 2014. Retrieved from: http://www.bbc.com/news/magazine-29342407.

[16] Wikipedia.com. Ballooning (spider). May 12, 2016. Retrieved from: www.wikipedia.org/wiki/Ballooning_(spider).

[17] Italy Ref 17: Pirro, Deirdre. 2010. Unexpected Fans: The day extraterrestrials attended a Viola game. The Florentine. June 17, 2010. Retrieved from http://www.theflorentine.net/lifestyle/2010/06/unexpected-fans/.

[18] Italy Ref 18: Padula, Richard. 2013. The Day UFOs Hovered over Fiorentina's Stadio Artemio Franchi. British Broadcasting Company. January 4, 2013. Copyright 2013 by the BBC. Retrieved from: http://www.bbc.com/sport/football/20917594.

[19] Phoenix Ref 19: Price, Richard. 1997. Arizonans Say the Truth about UFOs Is out There. *USA Today* newspaper. June 18, 1997. Retrieved from: http://www.ufosnw.com/history_of_ufo/phoenixlights1997/usatodayarticle06181997old.pdf.

[20] Phoenix Ref 20: Price, Richard. 1997. Arizonans Say the Truth about UFOs Is out There. *USA Today* newspaper. June 18, 1997. Retrieved from: http://www.ufosnw.com/history_of_ufo/phoenixlights1997/usatodayarticle06181997old.pdf.

[21] Phoenix Ref 22: Unknown author. 1997. Phoenix Lights, 1997. Mutual UFO Network (MUFON). Copyright 2014 MUFON International. Retrieved from: http://www.mufon.com/phoenix-lights---1997.html.

[22] Phoenix Ref 22: Unknown author. 2007. Former Ariz. Governor Boosts UFO Claims. Associated Press. March 23, 2007. Retrieved from: http://www.nbcnews.com/id/17761943/.

[23] Phoenix Ref 23: Symington, Fife. 2007. Symington: I saw a UFO in the Arizona Sky. CNN News Network. November 9, 2007. Retrieved from: http://www.cnn.com/2007/TECH/science/11/09/simington.ufocommentary/index.html.

[24] Phoenix Ref 24: Kitei, Lynne D., MD. 2000. The Phoenix Lights. Published by Hampton Roads Publishing. Copyright 2000, 2004, 2010 by Lynne D. Kitei, MD.

[25] Australia Ref 25: Cauchi, Stephen. 2005. Academic Throws Light on Forty-Year-Old UFO Mystery. *The Sunday Age Newspaper*, Melbourne, Australia. October 2, 2005. Retrieved from: http://www.theage.com.au/articles/2005/10/01/1127804696941.html.

[26] Australia Ref 26: Brown, Terry. 2010. Clayton South Residents Remember the 1966 Day They Saw Flying Objects. *Herald Sun* newspaper, Melbourne, Australia. June 3, 2010. Retrieved from: http://www.heraldsun.com.au/news/clayton-south-residents-remember-the-1966-day-they-saw-flying-objects/story-e6frf7jo-1225874696667.

[27] Australia Ref 27: Cauchi, Stephen. 2005. Academic Throws Light on Forty-Year-Old UFO Mystery. *The Sunday Age Newspaper*, Melbourne, Australia. October 2, 2005. Retrieved from: http://www.theage.com.au/articles/2005/10/01/1127804696941.html.

[28] Tinley Ref 28: Unknown author(s). 2016. Wikipedia. Tinley Park Lights: Wikis. Referenced 6/14/2016. Retrieved from: www.thefullwiki.org/Tinley_Park_Lights.

[29] Tinley Ref 29: Unknown author(s). 2016. Wikipedia. Tinley Park Lights: Wikis. Referenced 6/14/2016. Retrieved from: www.thefullwiki.org/Tinley_Park_Lights.

[30] Tinley Ref 30: Adams, Josh. 2008. UFO Investigator Discusses 2004 Tinley Park Event. UFO Updates, 12/9/2008. Retrieved from: www.ufoupdatelist.com/2008/dec/m15-004.shtml.

[31] Nuremberg Ref 31: Unknown Author. 2014. April 1561. A War in Heaven? Originated in Rense.com. April 4, 2014. Retrieved from: http://www.ufoevidence.org/cases/case486.htm.

[32] Nuremberg Ref 32: Author unknown. 2014. UFO Battle over Nuremberg Germany in 1561: Medieval Woodcut Shows. Cosmos Television. June 2014. Retrieved from: http://www.cosmostv.org/2014/06/ufo-battle-over-nuremberg-germany-in.html.

[33] Nuremberg Ref 33: Von Kevicsky, Colman S. 1976. The Ufo Sighting over Nuremberg in 1561. *Official* UFO. January 1976, pp. 36–38, 68. The translation is by Ilse Von Jacobi.

[34] Basel Ref 34: Unknown author. 2012. UFO Fleet over Switzerland 1566, Aerial Battle in the Skies. Educating Humanity. February 3, 2012. Retrieved from: http://www.educatinghumanity.com/2012/02/ufo-fleet-over-switzerland-1566-ufo.html.

[35] Basel Ref 35: Unknown author. 2012. UFO Fleet over Switzerland 1566, Aerial Battle in the Skies. Educating Humanity. February 3, 2012. Retrieved from: http://www.educatinghumanity.com/2012/02/ufo-fleet-over-switzerland-1566-ufo.html.

[36] Russia Ref 36: Huneeus, J. Antonio. 1990. Great Soviet UFO Flap of 1989 Centers on Dalnegorsk Crash. Reprinted from the Foreign News Tribune, New York, NY. June 14, 1990. Retrieved from: http://www.cseti.org/crashes/066.htm.

[37] Russia Ref 37: Mantle, Philip and Stonehill, Paul. 2010. UFO Case Files of Russia. Published by Eleventh Dimension Publishing, February 20, 2010. Referenced and excerpts reprinted and retrieved from: http://blog.seniorennet.be/peter2011/archief.php?ID=1724456.

[38] Russia Ref 38: Unauthored. 2010. The Dalnegorsk UFO Crash: Roswell Incident of the Soviet Union. *Pravda* newspaper. May 2, 2010. Retrieved from: http://www.pravdareport.com/society/anomal/05-02-2010/112049-dalnegorsk_ufo_crash-0/.

[39] Russia Ref 39: Mantle, Philip and Stonehill, Paul. 2010. UFO Case Files of Russia. Published by Eleventh Dimension Publishing, February 20, 2010. Referenced and excerpts reprinted and retrieved from: http://blog.seniorennet.be/peter2011/archief.php?ID=1724456.

[40] Shag Harbour Ref 40: Author unknown. 1967. The Shag Harbour Incident. Original Source: MUFON Canada. Case ID: 166. Edit: 166. Retrieved from: http://ufoevidence.org/cases/case166.htm.

[41] Shag Harbour Ref 41: Author unknown. 1967. The Shag Harbour Incident. Original Source: MUFON Canada. Case ID: 166. Edit: 166. Retrieved from: http://ufoevidence.org/cases/case166.htm

[42] Shag Harbour Ref 42: Department of National Defence. 1967. Shag Harbour memo. Library and Archives Canada. Unknown publication date. Retrieved from: http://www.collectionscanada.gc.ca/ufo/002029-1500.01-e.html.

[43] Shag Harbour Ref 43: MacLeod, Ray. 1967. Could Be Something Concrete in Shag Harbor UFO – RCAF. *Chronicle Herald Newspaper*, Halifax, Nova Scotia. October 7, 1967. Retrieved from: http://www.roswellproof.com/Shag_Harbour/Shag_Harbour_articles.html#anchor_4.

[44] Shag Harbour Ref 44: Unknown author. 1967. UFO Search Called Off. Halifax *Chronicle Herald Newspaper*. October 9, 1967. Referenced from: http://www.roswellproof.com/Shag_Harbour/Shag_Harbour_articles.html#anchor_4.

[45] Shag Harbour Ref 45: MacLeod, Ray. 1967. Could Be Something Concrete in Shag Harbor UFO – RCAF. *Chronicle Herald Newspaper*, Halifax, Nova Scotia. October 7, 1967. Retrieved from: http://www.roswellproof.com/Shag_Harbour/Shag_Harbour_articles.html#anchor_4.

[46] Shag Harbour Ref 46: Unknown author. 1967. UFO Search Called Off. Halifax *Chronicle Herald Newspaper*. October 9, 1967. Referenced from: http://www.roswellproof.com/Shag_Harbour/Shag_Harbour_articles.html#anchor_4.

[47] Shag Harbour Ref 47: Kimball, Paul. 2007. The Other Side of Truth. Redstart Films, written and directed by Paul Kimball. March 31, 2007. Retrieved from: http://redstarfilms.blogspot.com/2007/03/shag-harbour-ufo-incident-sui-generis.html.

[48] Shag Harbour Ref 48: Author unknown. 1967. The Shag Harbour Incident. Original Source: MUFON Canada. Case ID: 166. Edit: 166. Retrieved from: http://ufoevidence.org/cases/case166.htm.

What People (a Lot of Them at One Time) Are Seeing The Especially Trained Public

It was the purpose of this essay to treat the UFOs primarily as a psychological phenomenon. Unfortunately, however, there are good reasons why the UFOs cannot be disposed of in this simple manner. UFOs have not only been seen visually but also have been picked up on the radar screen and have left traces on the photographic plate.

—Dr. Carl G. Jung, from the book, *Flying Saucers: A Modern Myth of Things Seen in the Skies*[1]

Key words: wave bias, psychosocial hypothesis, gendarmes

This chapter focuses on the credibility and validity of observations by hundreds, if not thousands of witnesses, of a single UFO/UAP encounter. Instead of a sample of the heterogeneity of a large amount of participants in the general public, this body of evidence segregates the entire population into groups who work in occupations with specialized training that recognizes and acknowledges their credibility and validity to the UFO/UAP extraterrestrial hypotheses.

The details of some of these cases could have been included elsewhere in this book. For example, the 1952 Washington, DC Encounters and the 2006 Chicago O'Hare Airport Incident could have been presented in the chapter titled "What People...The Public" and vice versa.

The main intention was to differentiate among those cases in which the public reports were numerous and notable and also in which reports from any official government, military, or trained civilian police, airline personnel, or scientists and engineers was primarily lacking.

The cases in this chapter differ in that the primary sample group of human participants consists of witnesses from these professionally trained occupations. Additionally, their observation and participation has been documented and is part of the public domain. It is also noted that in these situations there were an abundant number, many thousands or more, of general public witnesses. Contributions to their respective cases were not at all insignificant. They brought their own special competencies to the investigations. This community of professionally trained observers and researchers share a credible and competent perspective from many other technology and learning disciplines and perspectives.

Critics of anything pertaining to the UFO/UAP extraterrestrial hypotheses are prone to assailing all reports from the general public. These same critics, whether scientists or not, seem to have something to say in any case they find. In their manifestation, they attempt to insert any connotation or generalization that could possibly fit, no matter how irrelevant and/or illogical. In the case of the general public, there are recurring thematic strategies that try to excuse perceptions away. For example, psychological explanations for observations and encounters of these themes are often used. Author Frederick Bailes made this statement: "Man's power of choice enables him to think like an angel or a devil, a king, or a slave. Whatever he chooses, the mind will create and manifest."[2]

It is of this premise that many critics dismiss any sighting, encounter, and/or abduction due to these psychological presentations. That seems sufficient to them to debunk any contribution by a member of the general public. But what about "The Especially Trained Public?" As previously mentioned, everyone is human and has the same brain wiring. Is anyone susceptible to this cognitive process?

What about all these trained individuals including scientists, the military, government and civilian engineers, pilots, and police personnel? Is it possible they were trained away from this seemingly hardwired function so they would not have these sensory episodes? If this was the case, it would seem desirable to have everyone, no matter what the occupation or walk of life, receive training to dissuade them from having illusory experiences such as these.

Let's say this is possible. An individual could be discouraged from creating illusions about witnessing UFOs/UAPs, alien encounters, and such. If this is the case, then in the context of these two hypotheses, what did all these trained professionals witness in their encounters as recounted below?

We can also say, however, that avoidance training of this type is not possible. Then it is concluded that there is no unimpeachable member of the human race left with which to credibly and validly include in any analysis of these hypotheses, at least not one who in the constructs of today's science could then investigate. Yet the mass sightings, the encounters, the electronic and visual evidence, and the preceding identification of many thousands of top secret and public domain of reports filed by humans from all walks of life and occupations throughout history are still there. All of this evidence cannot be eliminated from reality.

Psychiatrist and creator of the field of analytical psychology, Carl G. Jung, wrote about such encounters: "The possibility of a purely psychological explanation is illusory, despite its contradictory statements, the American Air Force, consider the sightings to be real and have set up special bureau to collect the reports. The view that the disks are real is so widespread."[3]

Michigan Events of March 14–20, 1966

As with most mass-sighting encounters, multiple eyewitnesses from the general public and the especially trained public could influence placement into either of the relevant chapters. Residents and police officials both became intimately involved in the event of March 14, 1966, and subsequent days thereafter. The scene in Southeastern Michigan that late winter month became a well-documented wave which quickly involved the Air Force and a future U.S. president, and it changed the literal life course of one government scientist, just to name a few of the favored credible advocates involved.

Only the event of March 14, 1966, will be addressed here to eliminate some of *wave bias*. Wave bias is defined as unadvisable guidance on a case when the public or any other potential eyewitnesses have prior knowledge of such a situation on days prior that could influence their thought process. Critical thinking could become adversely affected by this course of events. Not coincidentally, wave bias sounds familiar as compared to a similar "flap bias," a term described in the last chapter. They are complementary terms. A wave normally only lasts a few weeks; whereas, a flap can last a couple of years. It should also be noted there is a lot of ambiguity and interchanging of the time length definitions of flaps and waves. Notably, each author's literary piece will reverse the time length of the two terms.

This wave lasted about two weeks. It also involved sightings of similar craft displaying comparable flight and landing characteristics at a variety of sites around the world both a couple of days before and weeks after the morning of March 14. The flap rose to be known as "The Michigan Sightings/Swamp Gas Case." It was later included in the U.S. Air Force study, Project Blue Book. Two of the famous principal participants, Dr. J. Allen Hynek and future President Gerald R. Ford, gained a significant measure of popularity from this experience.

Around 3:45 a.m. on March 14, sheriff's offices and multiple police officers, both on patrol and near their barracks in Southeastern Michigan Washtenaw County and Lima Township, Michigan, and Sylvania, Ohio, saw airborne craft that maintained active flight control and performed controlled airborne maneuvers. Witnesses also included personnel at Selfridge Air Force Base, located approximately fifty miles to the east of Washtenaw and Lima. Sylvania, Ohio, is approximately forty-three miles south of Washtenaw ground zero, though a separate ground zero could be established over Selfridge also.

Four UFOs/UAPs were seen by patrolling Washtenaw officers and the sheriff's headquarters. They were disc-shaped moving at "fantastic speeds and making sharp turns, diving and climbing and hovering." At one point, four UFOs in straight-line formation were observed. Selfridge AFB confirmed radar tracking of the UFOs over Lake Erie one hour later. The sightings continued for another ninety minutes after the Selfridge first contact. Police patrol logs of that event were made part of the public domain. Police sheriffs and deputies from nearby Livingston, Ypsilanti, and Monroe counties reported their own events and described their theaters comparably with those of Washtenaw police and Selfridge. Hundreds of phone calls from the public were also reported.[4,5]

Just from this event alone, there were many credible participant perspectives. The sheriff of Washtenaw County, Douglas J. Harvey, never wavered in his reporting of the encounter and from the start never offered any input that could be considered having any explanation due to natural causes. He was among the first officials who called for more top-level investigation of the matter immediately.[6]

Another credible participant was Gerald Ford. He was U.S. representative from Michigan in 1966. Ford tried very intensely to elicit help from Congress for initiating a full-scale investigation of the Michigan Sightings. He made six news releases, statements, radio broadcasts, and official congressional memos in late March and early April, incidentally when more encounters were transpiring all over Southern Michigan and elsewhere.

He reiterated a general recurring theme and encouraged congressional, U.S. government, and science agencies to get involved: "I believe Congress should thoroughly investigate the rash of reported sightings of unidentified flying objects in Southern Michigan."[7]

"Because maybe substance to reports and because I believe the American people are entitled, more thorough explanation than, has been given by the Air Force. I am proposing that either the science and astronautics committee or the armed services committee schedule hearings."[8]

Another statement supplied by Congressman Ford later in April was, "Apart from the pranks and natural phenomena, some of these products of experimentation by our own military. If this is so, why doesn't the Air Force concede it and in this way reassure the American people?"[9]

Ford also communicated via official Congressional Memorandum about the Michigan Sightings. In an April 1, 1966, memo, he stated, "It is proper for the Federal government to look into, causing alarm to the people of our nation. It is for this reason that I have called for the investigation."[10]

Along with Ford, another Michigan congressman, Weston Vivian, was campaigning hard to motivate an investigation at that time. He obtained a positive response from the Air Force. Around March 24, an investigator from their Project Blue Book, Dr. Hynek, was detached to Washtenaw County. On that visit, it should be mentioned that Hynek was recovering from a broken jaw and had his mouth wired shut. The reason for this is feedback made public from interviews Hynek conducted indicated a critical review of his visit, more specifically, "Hynek's interviews were perfunctory, and he didn't seem interested in what they had to say." He was preoccupied, as just noted.[11]

On March 25, Hynek was pressured into making a speedy report of his findings. After his fact-finding and interview tour, Hynek met with Sheriff Douglas Harvey. According to Harvey, "He (Hynek) has no idea of what it is. Then he makes one phone call to Washington and comes out and gives a statement that..."[12]

That statement was Hynek's press conference presser. The broadcast informed a shocked public that the flap was nothing more than a series of swamp gas episodes. Or they could have been "the moon and stars" and "kids pulling pranks with flares." In reality, the flares prank did happen as, on March 24, the day of his official visit, Hynek was escorted to the site near the northern end of the University of Michigan campus where fraternity students conducted a prank that elicited a rash of phone calls to police from anxious nearby residents.

Other explanations of the Michigan sightings, from various people of alleged credibility, included the essential secret military operation, which was immediately denied by Selfridge AFB. The Northern Lights were also offered as an excuse.

For Hynek, this week was a seminal moment of his professional career. As long-time astronomy professor at Northwestern University and long-time U.S. government science consultant, Hynek risked a lot of controversy and loss of professional prestige at that presser. Privately, he knew there was more to this subject than what was being told to American citizens. He was as involved as any insider could be and trying to reconcile his personal critical reasoning with that of his employers.

To a lot of observers, the Michigan sightings swamp gas determination was also the favored conclusion for the Project Blue Book investigation. This explanation synergized well with the overall philosophy of the investigation. Later, in 1969, Project Blue Book was terminated. Also in 1969, this preferred conclusion was published in *The Condon Report*. Forever embedded in the lore of the U.S. government became the use of swamp gas as an official debunking mechanism for anything UFO/UAP and/or extraterrestrial related.

The Battle of Los Angeles; February 25, 1942

Sometime on February 24, U.S. Naval Intelligence commented that an attack could be expected within the next ten hours. Later on that evening, reports of many flares and blinking lights were seen in the vicinity of the manned defense manufacturing plants along the coast. Alerts were called after 7:00 p.m. and lifted three hours later.

Early on the twenty-fifth, military radars detected an unidentified target about 120 miles west of Los Angeles, out at sea. Antiaircraft batteries were alerted at 2:15 a.m. Six minutes later, a blackout was ordered in the Southern California coastal region by the military regional controller. From that moment on, reports flooded in about sightings of enemy planes. The unidentified object originally picked up on radar, though was not among any of the reports, that were acknowledged and part of the public domain.[13]

Over twelve thousand air-raid wardens mustered out to their posts in the dark at about 2:25 a.m., only minutes after the long network of air raid sirens began piercing Southern California overnight. The responding civil air servants were from the California Civilian Air Warning System and the Thirty-Seventh Coast Artillery Brigade.

The number of citizens awakened from this separate flurry of activity approached two million.

At approximately 2:43 a.m., planes were reported to be flying near Long Beach. Minutes later, a coast artillery colonel spotted over two dozen planes flying directly over the Los Angeles metropolis. Shortly after 3:00 a.m., a balloon carrying a red flare was spotted over Santa Monica. This seemed to have been the seminal moment, according to the montage of archived information, when the first battery of the reported 1,430 rounds of ammunition were fired amidst a matrix of one million candlepower floodlights illuminated the sky. The shells were 12.8 lbs. Army Air Force issue antiaircraft type.

Gunfire was continuous until 4:15 a.m. then intermittently until after sunrise. During the barrage, the craft was observed to oscillate in altitude within a range of from nine thousand feet to eighteen thousand feet. The shells were exploding upon contact with the vessel, but none penetrated the ship's exterior. Shrapnel rained down for well over two hours. This is significant as some death casualties were reported from as being caused by falling shrapnel and debris.[14,15,16]

Some of the data was unanimously agreed upon, such that exactly 1,430 rounds of ammunition were fired. Exact times of many of the separate events occurring that night were agreed upon by the abundance of reports from credible sources throughout Southern California. The U.S. Navy, Army, Civil Defense, and Coast Artillery were among the many departments of U.S. military that reported data, interviewed eyewitnesses, and/or intimately participated. No bombs were dropped from the air, no military troops were killed, no planes were shot down, and no U.S. Army or Navy planes were deployed into the air. Neither branch gave orders for any of their planes to sortie into action.

But amazingly, much could not be agreed upon. Initially, as previously noted, the theater of operations in this encounter transgressed over forty miles of coastline around Los Angeles. The stationed posts of the thousands of military officials scattered over this area suggested that there may have been multiple things going on in the skies and that were observed from many viewpoints.

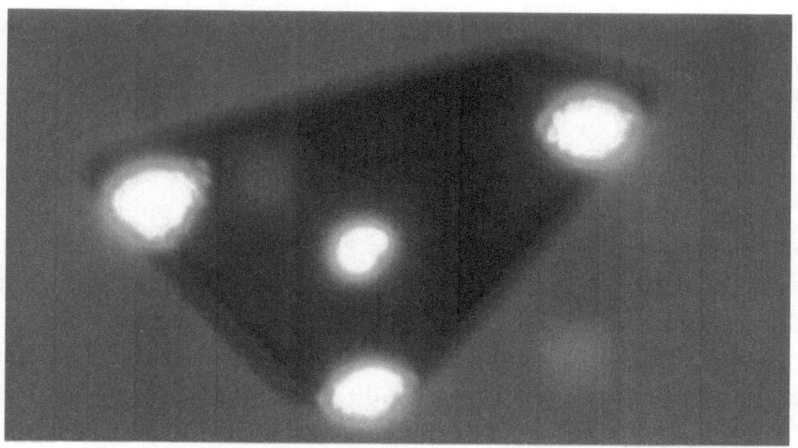

BATTLE OF LOS ANGELES

Citation:
LA Times / Page B of the February 26, 1942, Los Angeles Times
/Public Domain

Quote from J. Edgar Hoover about the LA Case:
"We must insist upon full access to disks recovered. For instance, in the LA case, the Army grabbed it and would not let us have it for cursory examination." (J. Edgar Hoover)

Operation Mainbrace Encounters over Yorkshire,
England; September 13–25, 1952

In September 1952, the united forces of the North Atlantic Treaty Organization (NATO) and other nonaligned countries such as New Zealand conducted one of the largest multilateral military operations since World War II. The best estimated number of active personnel present during the thirteen day exercise was upward of eighty-five thousand. Over one thousand planes and two hundred ships, including aircraft carriers and destroyers, took part, as well as a large number of military bases and personnel from England to Ireland, Germany, Denmark, and Norway.

Operation Mainbrace was designed and developed by Major General Dwight Eisenhower who was also reported to be present in the naval theater of operations during the exercise. Commanding operations were Admiral Lynde McCormick (U.S. Navy) and Commander Matthew B. Ridgeway (U.S. Army).

Seven detailed incidents occurred within the public domain. No one knows an exact number of eyewitnesses to all the events, but one of the documented incidents involved forest workers in Kirknes, Norway. One of the incidents was observed by the members aboard the U.S. aircraft carrier, *USS Franklin D. Roosevelt*, and others happened around various military air bases scattered around Northern Europe. Countless numbers of radar and other electronic equipment recorded the incidents, and a documented series of photographs and videos was taken by both newspaper journalists and military reporters. Only a couple of the journalist's photographs ever made it into the public press. It was confirmed most of this data was buried within military files presumably with "top secret" or "above top secret" classifications.

As has been the recurring theme with this presentation, one effort to reduce biases of what some call "mass hysteria" and synonymous terminology has been to concentrate just on the investigations that involve one event when, in fact, more than one episode occurred in some of these cases. Only the first chronological event in the flap or wave was investigated to eliminate those notions of bias. Some of these multiple incidents have involved many especially trained members of our society from civilian to military to professional. The discipline in this area has been maintained.

The investigation of Operation Mainbrace, while overwhelmingly documented from the lens of involvement of the military, provides an immense argument in refutation of the mass hysteria family of biases. The combined number of participants from diverse geographical areas reporting a simultaneous wave of observations necessitates being mentioned here.

A general overview of the magnitude of the incident(s) was provided by William Maguire of the then Royal Air Force stationed at RAF Sandwich, Kent, England as the following: "A huge, unidentified aerial object was being tracked on the radar scopes high over the English Channel. Every single instrument on the base was showing this enormous object sitting up. It was the size of a warship and it just stood there."[17]

This analysis concludes it may have been a mother ship or central ship of some sort that exists similarly to the NATO aircraft carriers taking part in the military exercises in the theater of operations during that time.

While civilian participants filed reports for documentation, the vast majority of chronicled and openly accessible reports came from military sources. The encounter chronology occurred in this fashion:

September 13 was the first date of maneuvers. That afternoon reports of unidentified flying objects came from naval personnel in the theater wing inclusive of Ireland and Iceland. All reports confirmed observation of three flying triangular-shaped craft emitting a blue-green hue and a white light emission at the rear. Velocity was calculated to be roughly 1,500 mph. In addition to this report another one came from the Danish destroyer ship *Willemoes*, documenting the same style of ship moving to the southeast. The trajectory would place this telemetry straight down range of the earlier sighting over the coast of Ireland. Lieutenant Commander Schmidt Jensen and personnel under his command calculated the velocity of this ship to be "over 900 mph."[18]

On September 18, a Harstad, Norway, newspaper published an article about a sighting from civilian forest workers in which a translated copy was discovered in of all places, Central Intelligence Agency files. The article reported that, "At 1400 hours, three forestry workers, outside Kirknes, noticed a flat, round object hovering motionless at about five hundred meters altitude with a diameter of fifteen to twenty meters. After observed it for a while, it suddenly flew away at great speed in a northwesterly direction."[19]

September 19 was the date referenced when a second body of observation reports flooded into military officials. Aerial traffic became much heavier from this date forward. An interesting analysis would be to decipher if there were any precise activities undertaken by the fleet maneuvers that could pinpoint a strong positive correlation between the type of maneuver on the specific date in question and the presence of the unidentified aerial ships. The answer probably resides within the military complex files.

Late that morning, British Meteor jet aircraft, normally based at RAF Ballykelly, Ireland, were returning to the Topcliffe RAF field in Yorkshire, England. Flight officers and ground-based officers, including Flight Lieutenant John W. Kilburn, who prepared the Topcliffe report, indicated sighting and recording of the episode, "The object was silver in colour and circular in shape. It descended from upward of twenty thousand feet altitude and five miles behind, swinging in a pendular motion during descent. The object then stopped behind the meteor and began to rotate about its own axis. Suddenly, it accelerated at an incredible speed before disappearing to the southeast."[20]

This trajectory again follows a vector toward the English Channel and points further downrange into mainland Northern Europe. This military document was written and signed by Flight Lieutenant Dolphin and was sent to Topcliffe Group Captain J.A.C. Stratton. Stratton also received multiple reports from civilian witnesses regarding this episode. He forwarded the bundle to the British Air Ministry at Whitehall, London.

At 4:00 p.m. on the afternoon of September 20, another episode was documented. This theater section was located around the area in which the *USS Franklin D. Roosevelt* was stationed at the time. Ship personnel witnessed a "silvery, spherical" object off the stern moving laterally across the sky. An American press photographer named Wallace Litwin spotted flight deck personnel paying attention to that portion of the sky and immediately began snapping pictures of episode.

Litwin acquired three color photographs of the object as it traversed by the *Franklin D. Roosevelt*. Captain Edward J. Ruppelt later stated, "[The pictures] turned out to be excellent. Judging by the size of the object, one could see that it was moving rapidly." Litwin was able to capture the deck of the *Franklin D. Roosevelt* in all the photos which allowed naval personnel to identify the size and velocity of the craft to about 1,000 mph and over 150 feet in diameter. Investigations by all ships in the fleet were taken repeatedly to determine who if anybody launched any balloons that day. The unanimous response from TBS radio was, "Nobody."[21] Intelligence officers aboard the *Franklin D. Roosevelt* studied all of Litwin's photographs. Though Ruppelt referenced them heavily later in Project Blue Book reports, none of Litwin's photos ever reached the public. Only one poor print appeared in the Project Blue Book files with no analysis undertaken.

A few hours later, at 7:30 p.m., three Danish Air Force officers observed a UFO move around Karup Field in Denmark. The object was described as a "shiny disk with a metallic appearance" that flew overhead of the airfield and continued in an easterly direction.[22]

The next day, September 21, a formation of six British jet fighter planes encountered a shiny spherical craft approach from the west where a big part of the Mainbrace fleet was situated. The formation then became a sortie and gave chase to the craft. The sortie was aborted a few minutes later when the unidentified craft eluded them. However, when the formation was returning to their base, one of the pilots spotted and watched the UFO chasing him from behind. The pilot broke formation and turned toward the craft, but it also turned to avoid him and eventually maneuvered away.

The RAF Topcliffe airfield again experienced an incident on September 22. Another British Meteor jet was sortied in close pursuit of a UFO, which was "round, silvery, and white and seemed to rotate around its vertical axis and sort of wobble."[23]

In the days following through Sunday, September 28, many observations—mostly from the civilian public from Western Germany, Denmark, and Southern Sweden—were reported. Descriptions of bright luminous white craft were alternatively speeding across the sky then stopping and hovering and then speeding away again. Other witnesses near Hamburg and Kiel, Germany, documented three of these objects rotating around a much larger similarly-shaped craft. Craft with a cylindrical cigar shape moving silently eastward was also reported.

No less than seven separate military agencies took part in the Mainbrace incident in some official and substantial way. The U.S. Navy and U.S. Air Force Intelligence agencies, the R.A.F., NATO, the British Air Ministry, the U.S. Navy, and U.S. Air Force were some of the main participants.

A post-note statement that came from Edward Ruppelt following the incident provides the most compelling reference to acknowledgement and admission of the anomalous nature of the UFO and extraterrestrial hypotheses. In his book, *Report on Unidentified Flying Objects*, Ruppelt revealed that conversations with a British Royal Air Force intelligence officer enlightened him to the conclusion that it was the Mainbrace incident that led the British government to officially admit and recognize the existence of the UFO.[24]

The New York-to-Nellis Air Force Base Incident of April 18, 1962

It appears the Air Force is profoundly the most visible and engaged branch of the U.S. military services with regard to the ufology subject. Not to be discounted in quantity and significance is there have been many additional studies that engage other branches, like Operation Mainbrace. Logically, it is conclusive that, with an abundance of mass sighting encounters occurring over land, it is the Air Force and not the Navy or Army (post 1949) that becomes the engaging party.

This incident follows that protocol. Again, this is an example of a case that involves thousands of people from the general public but also many Air Force, municipal police, airline, and other professional officers and personnel. A multitude of radar and other electronic equipment detection and tracking of this flight exist.

The object in question was identified as "lengthy long, maybe cylindrical" in nature. Around 9:45 p.m. EST, residents, municipal officials, and the Air Force began observing and tracking the object over Oneida, New York. It was described as a "glowing red ball" maintaining a "high" altitude and moving very fast; individual observers saw it disappear within seconds.

Heading then in a westerly direction, the object continued to be tracked by the Air Force and NORAD radar, as well as commercial airport equipment, across the continental United States spanning eleven states. This cross-country trajectory was accomplished in less than thirty minutes. The Air Force scrambled fighter jets from multiple sites in a chase operation.

About the time the object appeared to eyewitnesses in Utah, around Nephi and then Eureka, the craft was alternately speeding up and slowing down in the sky while gyrating up and down in altitude. Witnesses also reported a "red-glowing ball" that emitted rumbling noises from its rear structure. As these noises grew louder, the craft started emitting an extremely strong, blinding, piercing white light. An Air Force cargo flight crew was navigating in the airspace near the craft's flight path at about 8,500 feet altitude. It was reported the craft remained motionless above the airplane at that time.

Captain Herman Gordon Shields, chief pilot of the C-119 plane, witnessed the intense white light illuminate the sky from behind his aircraft: "The light intensity increased until we could see objects on the ground as bright as day for a radius of five to ten miles from the aircraft. After the light decreased in intensity, this object which I saw had a long, slender appearance, comparable to a cigarette. Fore part was very bright, intense white, the aft section, yellowish color."[25]

The ship flew past Nephi, turned northwest, and then began losing altitude more suddenly as it neared Eureka. In Robinson and Silver City, witnesses observed the object spurting forward and hesitating numerous times as it flew past. It began to descend again toward the ground and turned a bluish color while descending. The rumbling noises became louder. The electrical power in Eureka suddenly shorted out extinguishing all the town's lights. Some witnesses reported their motor vehicles stopped operating. Other witnesses observed the craft landing on the ground outside of Eureka near the power plant that services the Eureka area. All power from the plant shorted out.

The sheriff of Nephi, Raymond Jackson, witnessed the lights go out sequentially in Nephi right after he heard "a succession of twenty booms" coming from the rear of the craft. Other residents also heard the booms. To them, they sounded like "a series of explosions, twenty or thirty, like rocket engines or artillery shells flying over."[26]

While driving on a local highway, two other witnesses, Bob Robinson and Floyd Evans, observed the object descending over

Eureka. They observed the craft at an altitude of "no higher than five hundred feet, and he thought he could see a series of square windows on the craft almost hidden in the glow of it."

As the two witnessed the craft fly overhead, it slowed then sped up again and again, then their truck sputtered and died. Their truck headlights went out. As the craft exited their airspace, their truck restarted and the headlights turned back on.[27]

Other residents, including Utah state sheriffs, local police chiefs, and officers witnessed the unfolding events. All saw the ships' trajectory take it down to ground level somewhere outside of town. It was down for only a brief time then took off again. The craft changed course and flew west into Nevada toward Reno.

The ship moved over and past Reno and into California airspace for a brief time. In a similar fashion, the human and electronic tracking and observation continued just as it had for all the witnesses in Utah. Additional witnesses and equipment included radar and airport radio control tower apparatus at Reno Airport, and later in Elko and Las Vegas, and in-flight commercial planes who were flying at an altitude of about eleven thousand feet. These crews reported their observations "took place below them."

Then the ship changed its course and started heading southeast toward Las Vegas, Nevada. By that time, additional fighter jets that were scrambled from Nellis and Luke Air Force Bases in Nevada and Phoenix, Arizona, respectively were on the chase. Once in the general vicinity of Las Vegas, the ship again changed course and flew northeast to where numerous reports of a fiery crash occurred in the Mesquite Mountains near Nellis.

The New York-to-Nellis incident received abundant press coverage in newspapers such as *The Las Vegas Sun, Los Angeles Tribune, Desert News and Telegram, Salt Lake Tribune, Eureka Reporter and Nephi News-Leader,* among others. Project Blue Book archived many files and wrote extensively about the case. NICAP and the Air Force also have many archived files documenting that night's events.

Additionally, a short investigation was headed by J. Allen Hynek. At that time, Hynek was still employed by the Air Force and Project Blue Book. His public conclusion was the object was of a rare type of "bolide meteor." These objects are very bright and sometimes explode in the atmosphere.

The short length of the investigation, which only lasted two days, led the rush to judgment of the bolide meteor concept. Nothing was ever investigated or discerned as to any explanation of how the meteor was able to display the following flight characteristics:

First, that it was able to maintain a level flight path across the continental United States. Second, that numerous altitude and course compass changes in direction were observed and documented not only by civilians but by commercial aviation radio and radar and numerous military and defense radar and electronic recording apparatus. This documentation is extended to in-flight airline flight crews and military fighter jet sorties. Third, given that the duration of this incident, as evidenced by the Air Force, NORAD, and triangulation of thousands of reports from eyewitnesses from around the flight path is equated to about thirty-two minutes, an average velocity of the meteor calculates to about 4,600 miles per hour. This is the average velocity over the entire thirty-two minutes with no accounting for the speeding up and slowing down reported by many witnesses.

All extraterrestrial meteors, including bolides, have to achieve an entry velocity in order to be able to successfully traverse through the earth's atmosphere. When the velocity of the earth itself is accounted for, which is about thirty kilometers per second, the meteor must have a functional velocity of around seventy kilometers per second in order to successfully penetrate the atmosphere and to traverse a portion of it.

The atmosphere itself can supply a drag force on the meteor, as it does on the manmade orbital space craft such as the International Space Station shuttle craft, the U.S. Space Shuttle, or Apollo craft from yesteryear. When applying this science to the New York-to-Nellis meteor and given the overall average velocity of about 4,600 miles per hour for a 32-minute duration traversing about 2,500 miles, this meteor was traveling at about 2 kilometers per second. This is far slower than a meteor travels in nature or even unnatural occurrences.

Additional dysfunction present was the recurring attitude and protocol of government and military operations to offer a study and proceed to conclusion as fast as and for as little a fiscal expenditure as possible. This surfaced again and was even more evident with the whirlwind effort and duration of the Hynek study. In the philosophical and scientific world of logic, the fallacies of cherry picking and rush to judgment are evident of this recurring thematic within government and military agency operations and design.

Here are some descriptive statements from Air Force files regarding activities in the Utah/Nevada flight theater the last few minutes that the ship was still airborne. At 0319Z on April 19, 1962, a brilliant midair illumination appeared over central Utah and witnessed by many. After that, as many as twenty to thirty explosions were heard and an object was seen falling toward the ground. Observers at Jericho, Utah, stated the object was emitting a gasping sound, retarding forward movement, and surging ahead three or four times. The object was thought to have landed in the Rush Valley in the vicinity of the McIntire Ranch (39* 29' N, 112* 23' W).[28]

The Air Force Defense Command was puzzled by an aerial object that had exploded and seemed to be a meteor but had the unique distinction of being tracked by radar seventy miles northwest of Las Vegas, Nevada, as well as being chased by jets before it exploded in Nevada in a blinding flash. An Air Force Defense Command alert reported the object was tracked and traced over New York, Kansas, Utah, Idaho, Montana, New Mexico, Wyoming, Arizona, and California.

NORAD had tracked the object which had covered ten states in thirty-two minutes but changed course over Utah. Now detected on GCA radar at Nellis, the object was moving NE.

Project 10073 Record Card and report of radar sighting Blip, "Speed of object varied. Initial observation at 060 no elevation. Disappearance at 105 az 10,000' altitude. Heading tentatively NE, however, disappeared instantly to S."[29]

Belgium Sightings of 1989–1991

The country of Belgium is situated in the northwest of the mainland European continent. Its land mass measures about 31,500 square kilometers, which is a little smaller than the U.S. state of South Carolina. The population of Belgium, in the late 1980s, was about 9.605 million.[30] The population density, therefore, is among the densest in all Europe.

The Belgium wave, as it is popularly called, lasted officially from November 29, 1989, well into the middle of 1991. Some literature even extends the time as more as fewer sporadic encounters were reported. As in previously discussed cases, the semantics of flaps and waves continues. It should be noted there is a lot of ambiguity and interchanging of the time length definitions of flaps and waves. You will note in the various literatures that the author's particular piece will reverse the time length of the two terms.

Noted is the case classification component where the Belgium Wave could have sufficiently fit into the "What People Are Seeing... The Public" chapter. Nonetheless, the facts and evidential observations were documented by a heterogeneous cross section of the general public, as well as especially-trained public witnesses. For the Belgium Wave, only the events of the first day of any defined wave or flap are being detailed. This date is November 29, 1989. The number of recorded witness reports totals 143.[31]

Most of this selection of cases involved multiple events on multiple days or nights, either a flap or a wave. The very first occurrence of a mass sighting is wonderfully displayed in *The Belgium Wave and the Photos of Ramillies* by Auguste Meessen.[32] A graphical illustration, as well as explained written conclusion, details raw data of the Belgium wave as contextualized with the time length of the wave and what psychologists and such researchers call the "psychosocial hypothesis."

In ufology, PSH, *psychosocial hypothesis*, argues that at least some UFO reports are best explained by psychological or social means. The psychosocial hypothesis builds on the finding that most UFO reports have mundane explanations like celestial objects, airplane lights, balloons, and a host of other misperceived objects seen in the sky which suggest the presence of an unusual emotional climate. This emotional climate is one that distorts perceptions and the perceived significance and peculiarity of merely terrestrial stimuli.[33]

This example demonstrates an absolute elimination of what can be referred to as "the psychosocial hypothesis bias." It explains that on the first day/night of the Belgium wave, our designated event theater of November 29, 1989, the psychosocial hypothesis bias was nonexistent. Only after the flap or wave has been well underway does the analysis show that the measure of raw data for the psychosocial hypothesis, when graphed, indicate some reported observations being classified as erroneous. It does not happen on the first occurrence of the flap or wave! Also the analysis proves that in the realm of mass observations, there are still many (in the case of the Belgium Wave, almost a thousand) valid and credible reports in which this bias does not exist. So while some reports documented later on in the time length of the flap or waves indicates erroneous observations, there are still hundreds of data points of good observations on a particular day.[34]

Meessen further explains that the contagion process crucial to the psychosocial hypothesis argument "did not result from media reports."[35]

Here are the documented facts of Wednesday, November 29, 1989:

The event theater spanned a five-kilometer square area. The local weather demonstrated clear skies as the event transpired over a three-hour time frame. Some of the reports took note that, among the observations, various other objects in the sky were visible throughout the theater time. For example, the bright stars Venus and Fomalhaut occupied different locations near the horizon. These were accounted for in observations from many people from varying vantage points. While it is possible one or a few witnesses had an appropriate vantage point that could put either of these celestial objects in their direct line of sight with or as to impede or hide the unidentified objects, the palette was being viewed from many different angles. The collection of these numerous vantage points dispelled the doubts that Venus and Fomalhaut were what was actually being observed. In this instance, reality cannot be negated.

Reports came from over thirty groups of witnesses within the general public and from three separate groups of police officers known as *gendarmes*. The craft were of a flat, triangular shape, with lights underneath and flying at low altitude. The ship was silent and maintained controlled flight characteristics throughout, including lengthy times where it was motionless.[36]

The size of the triangular-shaped craft, documented in reports from one group of gendarmes, was about thirty-five meters in length and twenty-five meters in height. When the observation window by one group of two gendarmes began, the ship was motionless and silent in the sky at an altitude of about 150 meters. At each point of the triangle, a bright, sharply defined cylinder of white light, a few meters wide, beamed down to the ground. In the midsection of the undercarriage, a red light also beamed down to the ground, alternating a sequence of pulses with a blinking characteristic. The local weather conditions were of below-freezing temperatures and dry with clear skies. The intensity of the lights was another atypical feature to many eyewitnesses with regard to the weather conditions. The boundaries of the light beams were extraordinarily sharp; there was no diffusion or gradual degradation of the light beam either in the direction of the beam or peripherally.

Reported and catalogued sighting times during the event, all in the p.m. hours, from a widespread geographical landscape of eyewitnesses are listed thusly: 4:00, 4:40, 5:20, 6:00, 6:10, 6:30, 6:40, 6:45, 6:50, 7:20, and 7:23. A more detailed list appears below.

The 4:00 p.m. sighting was made by public citizen eyewitnesses. The trajectory of the cigar-shaped ship was downward at a consistent angle and was silent. About fifty meters above the ground, it stopped descending and continued on a level path. After a distance, it just stopped completely and remained stationary for about thirty seconds then began rising in altitude with about the same angle as the downward trajectory. It should be noted this report was clocked during sunlight hours, though the sun was only about three degrees above the horizon meaning dusk was to occur within the next hour.

The 5:20 p.m. sighting was encountered by one of the groups of gendarmes stationed in the town of Eupen. Hubert von Montigny and Heinrich Nicoll encountered a stationary triangle less than the 150 meters in the distance and with the dimensions noted earlier.

This encounter lasted two hours! Within this time frame, among the observations, there were two narrow intense red light beams that lengthened and shortened as the object passed over the terrain below. At the ends of these red beams were red-orange-colored balls. In the course of the event, the beams suddenly and immediately turned off leaving the balls stationary at their positions at what would have been the ends of the beams had they still been there. This process continued every five minutes for the event time.

During this time, the gendarmes contacted the Eupen barracks where it was confirmed there were no maneuvers being undertaken from nearby military stations or bases. Even though the installations surrounding the countryside in this region were heavily restricted, prohibited, and posted as dangerous, there is still peripheral access because the various sites are not a continuous risky landscape.

The 5:20 p.m. observation was reported by other citizens from different positions within the theater of operation. From these viewpoints, additional reports of rectangular windows were seen. According to the reports, they had a luminescent characteristic to them. The 4:40 p.m. sightings were made by several groups of public witnesses. These sightings occurred again at 5:20, 5:30, and 5:50 p.m. by other groups of public witnesses. They corroborated the observations made by gendarmes von Montigny and Nicoll. Another group of gendarmes, posted in the border town of Lichtenbusch on the Germany border, observed the same craft as the other groups, though from the opposite (east) side. The ship silently approached, flew overhead, and continued west at about the same altitude at which the others had observed.

The 6:00 p.m. and 6:30 p.m. sightings were made from yet other observation angles and contained descriptions of the red light beams and red-orange balls in the same general synchronization as reported from the gendarmes. This series of reports were from additional groups of gendarmes near the barracks in nearby Rotenberg. The 6:45 p.m. sightings documented another airborne triangular-shaped craft with a dome and a series of windows backlit from the inside.

The 6:10 p.m. report also detailed four objects with very large white lights approaching the town of Liège from the east and German airspace. The 6:50 p.m. outline again reported intense, non-diffuse, and large beams of white light contacting the ground along with the red beams. Additionally, an electric faint humming sound was audible from this site. A color observation was made here also and was disclosed to be of a very dull grey aluminum. Additional public witnesses near the area saw the same ship at approximately 7:00 p.m.

The 7:20 p.m. surveillance was documented by yet another pair of gendarmes, Peter Nicholl and Dieter Plumans. It was reported by Plumans that the red-orange balls descended from the ship on the end of the red light beam toward the ground. They then became unattached from the beams and vectored horizontally at a right angle then followed a parallel course with the ground.

No radar from either ground-based or AWACS aircraft, none of which was confirmed or disclosed by any officials in any of the regions. Because of a low altitude flight path, the UFOs were too low to be able to be detected by radar. The noise characteristics, the hum of electric motors, were reported by many, though not all the groups.

Investigative activities in the aftermath of the Belgium Wave included abundant inquiries of various military officials for case files. All attempts were met with claims of "no such files exist." Even then, thirteen years later, further attempts for file disclosure were answered with the following new response: "The files concerning flights older than five years are not preserved anymore."[37]

Contributors to official report research papers came from sources such as General De Brouwer then head of operations of the Belgian Air Staff.

The Chicago Airport Event of November 7, 2006

The twelve eyewitnesses of C-17 is a suitable nickname to give to one of the more recent mass sightings of credible substance. A substantive breakdown of the elements of the nickname go like this: twelve employees of United Airlines, including pilots, officials, and ground personnel were performing their work assignments at gate C-17 at Chicago's O'Hare airport on the afternoon of Tuesday, November 7, 2006. At 4:15 p.m. Central Standard Time, all these personnel observed a dark grey metallic saucer disk-shaped craft hovering motionless and silently directly above the stretch of landscape over this section of O'Hare.

The local weather conditions were of a solid overcast cloud cover at an altitude of about 1,900 feet. Visibility beneath that cloud cover was for three to five miles, a typical standard under only those descriptions of "ten miles" or "unlimited." The air mass below the cloud floor was moist and drier above 1,900 feet. Precipitation was not actively falling in that hour but clouds, of course, consist of water droplets so it could be said the air mass in the clouds was also moist.

Additional local weather conditions included an air temperature of 53°, wind coming from the west at about 7 knots, relative humidity of 83%, and a dew point of 48°. A comprehensive description was determined to be of "calm" conditions.[38,39]

The encounter was viewed from varying altitude and geographic locations from the tarmac around Concourse C where ground personnel were preparing a United Airlines flight for later takeoff to the cockpit level of the Boeing 777 jet Flight 446 when the two pilots on duty were performing a preflight checklist. An airplane mechanic in a separate Boeing 777 jet cockpit who was taxiing his plane into a maintenance hangar further down the tarmac also witnessed the event. Officials of United inside airport headquarters at both Concourse C and Concourse B had yet additional viewpoints of the ship throughout the episode. In summary, the witnesses observed this event from all angles below the event theater, including from directly underneath.

All witnesses, from compilation of information in later interviews, saw a dirty-aluminum-colored ship about twenty-two feet in diameter hover silently and motionlessly for ten minutes. A curious situation within the study parameter was that the object was not visible to traffic controllers in the tower nor was it detected by radar. Because of this, there were studies conducted to help determine why there was no confirmation from these aspects due to the object being later classified as "a definite potential threat to flight operations."[40]

This data was confirmed as described by the two pilots doing their preflight check, one of which had more than 13000 hours of flight time in service. This first officer described the ship as "a dirty-aluminum color, very stable, and without any optical distortions near it. It was perfectly round and silent."[41]

There were communications between the United Airline's offices of Concourse C and B which alerted various airline officials to observe the events taking place outside their windows. There were also alert calls to the control tower that initiated activities by air traffic personnel. The control tower personnel did not witness what was occurring above Concourse C or B.

After about ten minutes, the craft instantaneously and without warning rose almost vertically up and through the cloud bank above. The cloud bank consisted of only water droplets up to an altitude of ten thousand feet because the freeze level was determined to be above that altitude at that time. The sudden acceleration of the craft left a cylindrical-shaped hole through the entire cloud bank just a little larger in diameter than the craft itself. It was observed like a "cookie cutter" chunking a precisely chiseled cylinder out of the clouds. Blue sky appeared subsequently observed through this hole by eyewitnesses. The hole remained intact for an additional ten minutes.

An interesting anomaly determined from a subsequent study by NARCAP was that a particular antenna, an ASR-9 ORD #1 type, should have recorded the ship when it left the airspace but did not. Separate bodies of information were obtained from a variety of airport recorded taped conversations, meteorological data around the theater event, triangulation of the ship from the dozens of vantage points of observation and measurements of the various artifacts involved. A process of elimination was used to rule out natural or man-made explanation as they were served up for investigation.

Some interesting conclusions were gathered from the NARCAP investigation. First, given the recorded conditions within the theater the craft was determined to be about twenty-two feet in diameter and somewhat spherical in shape more rounded on top and flatter on the bottom. No lights were observed emanating from the vehicle.

Second, regarding the action of creating the cylindrical "hole," the ship was observed to have risen through a vertical altitude vector of over 8,500 feet in a second or so. The craft was hovering at an altitude of approximately 1,500 feet, about 400 feet below the cloud floor. Cloud ceilings around O'Hare were from two thousand feet and above. Visible through the cylindrical hole was clear blue afternoon sky.

An investigation of the process that creates such a hole in the cloud layer gave some factual data. The ship created a hole that completely evaporated all water droplets to create the blue sky imagery by a precisely delineated mechanism. For a vessel of that type and size to create the type of hole, the power it needed would have been more than one hundred megawatts. With the loud layer estimated to be over three hundred meters tall (over one thousand feet) and a time elapse from motionless to the first sight of blue sky being about one second seems unexplainably unusual. By comparison, a Boeing 747 Jumbo Jet accelerating with an airspeed of Mach 0.4 (about 306 mph) manufactures a power output of about 60 megawatts. In addition, it cannot create the hole this vessel had generated.

So the inferential takeaway conclusion from this point is that: 1) the object and phenomenon was real; and 2) conclusions of natural and/or man-made causes cannot explain what happened.

A fourth aspect of the theater of events investigated concluded that, coincidentally or not, the UAP was in the only precise position and vector of a blind spot that airport control tower radar would not be able to spot or identify. This was discovered and determined from radar archived files and schematics and recorded as follows: any objects "above an altitude of about 1,438 feet were invisible to any radar and tower occupants without leaning forward over the equipment and looking up into the sky."[42]

As for the inferences pertaining to the evaporation of water vapor that created the cylinder hole, there are three known methods science can use to explain how to make a cloud disappear. One is to freeze the cloud into ice particles. This was impossible because the temperature gradient for that measurement was over ten thousand feet.

The second method is to create a mechanism to greatly expand the size of the drops into larger raindrops. These drops would then fall immediately, unless the wind was extreme. The winds measured that afternoon never rose above 7 knots (8 mph+). It was also not raining that afternoon. The humidity never reached one hundred percent. Finally, the third method is to evaporate the droplets. This is what seemed to have been accomplished in an immediate reaction and as witnessed.

The logistics of all known geophysical and chemical processes given the data excludes explanations such as weather balloons; in fact, balloons of all kinds. Also no human built aircraft could synthesize the cylinder hole in the way it was accomplished. Man-made vessels cannot elicit the power necessary to successfully create the phenomenon this ship generated. There were no lenticular clouds nor could any gaseous reactions have provided an explanation. If any military operation sortie could have been held responsible, the middle of O'Hare International Airport was not a safe or productive theater at all. If they were, then they were keeping secrets. In an opinion that is reported by some skeptics, this is something the United States or any government should not be capable of orchestrating successfully.

A post-hoc debate ensued in the time following the O'Hare incident still resonates today. An incongruity exists within the Federal Aviation Administration between the body of rules and regulations and making a more effective investigative study of the events of Tuesday, November 7, 2006. A complicating issue was the FAA at first denied there were any reports of a witnessed sighting which existed. This post-hoc created some highly toxic interactions, whereby a Freedom of Information Act request program was successful in demonstrating there were indeed reported sightings by a mass of pilots, airline mechanics, company officials, and members of the general public.

The FAA finally had to conduct an investigation, or at least they said they did. FAA spokeswoman Elizabeth Isham Cory was tasked with the conferring of summary information to the public. While no specific data was released by the FAA, she did say the FAA's investigation was concluded and "their theory was that the sighting was caused by a weather phenomenon."[43]

In the FAA mandate regulations, it is their responsibility to dedicate complete and thorough investigation activities which involve a possible security and/or safety compromise within or over any airspace that is zoned for commercial and public activities and thoroughfare. They did not do this, at least to the satisfaction (or knowledge) of the public they are entrusted to serve and protect.

As a closing comment, United Airlines did not perform any post-hoc investigation of the O'Hare Incident either.

The White House Washington, DC
Events of July 29–August 1952

This is one of the more complex mass-sighting events for analysis. It is because some explanations for the activities were attributed to a public mass hysteria. Other excuses attributed much of the observations to the wave bias noted earlier. Wave bias is where, after the first reported encounter, the public becomes aware of the situation. Their later observations are deemed biased because of prior knowledge of the first event. This is argued repeatedly by social scientists with no overarching validity across all similar situations. The events which make up any particular situation contain so many unique situational elements to defeat any universal attribution and validity or credibility to their argument. The time line of the Washington, DC flap, as it is commonly known as indicates that the first mass sighting was on Saturday, July 19, more than a week after a group of isolated sightings took place.

Despite this flurry of events dated Thursday, July 10, Sunday the thirteenth, Monday the fourteenth, and Wednesday the sixteenth, what became evident is none of these sightings were reported to any media. The July 10 sighting was by the flight crew of a National Airlines jet airborne around Quantico, Virginia. The July 13 and 14 incidents were observed by other airline crews flying over Northern Virginia and Newport News, Virginia, which is home to the United States's largest naval port.[44] The July 16 sighting was made by scientists from the Aeronautics Laboratory at Langley Air Force Base near Hampton Roads.

All these observations were made by airline industry professionals and aeronautical engineers and physicists. None of this knowledge became available to the public until Captain Edward J. Ruppelt documented them in his book *The Report on Unidentified Flying Objects*.[45] Ruppelt was the head of the U.S. Air Force Project Blue Book in 1952. These sightings were made by some of the most widely compelling of the especially trained public. The general public did not have any knowledge of any of these sightings until the night of Saturday, July 19, public. It is for these reasons that the eyewitnesses around Washington, DC and Northern Virginia were not subject to mass hysteria. This includes members of the general public, as well as members of the especially trained.

At about 9:30 p.m., a U.S. Army artillery officer named Joseph Gigandet encountered the first sighting that was reported and investigated. From his home in Alexandria, Virginia, Gigandet detected a "red cigar-shaped object" traverse over his house, then turn around, and fly over his house a second time before moving closer to and into the Washington, DC city itself. He described the craft's size as comparable to a DC-7 airliner at about a ten-thousand-foot altitude. The ship had two series of illuminated lights closely set together on its sides. The craft had enough thickness for the lights to be on the sides. It moved slowly over his house and circled around back again after a few minutes. The total observation trajectory time was about ten minutes. On the second pass around and beyond, the ship turned a deeper colored. Gigandet was joined in the observation by his neighbor who was an FBI agent.[46]

Less than two hours later, at about 11:40, separate radar equipment installation stations at Washington National Airport (which is now Ronald Reagan Washington National Airport) and at Andrews Air Force Base started tracking eight unidentified objects south of Washington, DC in the area of Andrews. These craft were tracked by the first of two separate work shifts of tower personnel at all radar locations to vector at roughly 130 mph and then immediately changed direction and accelerate swiftly across the radarscopes. As stated on record, "One of these targets was timed at seven thousand miles per hour."[47] These anomalies gained the attention of personnel from both stations. While radar operators observed the on-screen maneuvers, others checked functionality of all the equipment and other officials had visual contacts and tracking from control tower windows.

The lights radiated a very bright orange color. Tracking observations continued throughout the night. The ships would crisscross the radar grids as they traversed the Washington, DC metropolitan area. Flight altitudes varied a bit from 1,700 feet to 3,000 feet. Air transportation was disrupted throughout the events. One pilot, flight crew, and passengers of a Capital Airlines flight awaiting takeoff observed different formations of six and then eight unknown objects close in and fly over their position on the runway. The ships all had bright white lights and moved silently over the airport at varying but fast speeds. The ships then made radical ninety-degree turns while traveling over the airport. The pilot of the Capital plane was in continual radio contact with the control tower at National who verified the positioning, velocity, and maneuverability of the craft through the event theater. This series of flyovers lasted in excess of twenty minutes.[48]

A lot of these were dual observations. The objects were observed to maneuver across radar screens while the air traffic controllers simultaneously observed them flying the exact same vectors. The controllers' experiences were duplicated by officers at all the installations.

Tracking operations like this continued for a few hours by the two work shifts. At some point, the still unidentified ships made vectors over the U.S. government buildings, including the White House, the Capitol Building, the Lincoln Memorial, and Washington Monuments, among others. This motivated the U.S. Air Force to send in two fighter planes from Delaware before 3:00 a.m. Their appearance caused all ships and their radar signatures to immediately disappear.

The Air Force fighter jets ran low on fuel soon after and returned to Newcastle AFB in Delaware. When the jets vectored off, the radar screens at all the tracking installations the fleet of the unidentified ships returned on-screen. The craft continued to be sighted and tracked by Air Force officers and personnel, civil aviation officers and personnel, airport personnel, airline pilots and crew, all the general public surrounding the airport theater, and eyewitnesses from the DC metropolitan area until after 5:30 a.m. around sunrise.

Sunday's national news media captured the story and spread it across North America. A well-known headline came from the *Cedar Rapids Gazette* (Iowa): "Saucers Swarm over Capital."[49] One surprising expert, Captain Ruppelt, was actually in Washington, DC at the time but did not hear of the July 19 mass sighting until the following Monday. Ruppelt read the headlines in *The Washington Post* that day.

Ruppelt was the active head of the U.S. Air Force Project Blue Book at the time. His office was responsible for accumulating reported information from UFO sightings from across the United States in 1952. In addition, the project was investigating sightings from the recent past. Little did Ruppelt know, he would witness and be chief investigator in another series of mass sightings by United States and worldwide, military officers, naval fleets, and service people who performed their duties during Operation Mainbrace later that October.

Ruppelt wanted to investigate, firsthand, the Washington events of July 19. He was informed by U.S. Pentagon officers that he could not use any official transportation or other equipment during his time in DC that week. This was because he rank was not high enough to be allowed to do so. In anger, Ruppelt participated in a few interviews himself without any assistance and then returned to his office at Wright-Patterson AFB in Ohio. This became part of the material for his book *Report on Unidentified Flying Objects*.

Ruppelt also could not anticipate what would happen that next weekend after he left Washington, DC. On Saturday–Sunday, July 26–27, the sightings and radar trackings and recordings returned. Shoot-down orders had been given upon encounters with any UFO around Washington, DC and nationwide. There is controversy as to whether the Air Force acted on their own to give the orders or whether the orders came from the White House (President Truman). It is known that Truman interviewed Captain Ruppelt on the matter.

The orders were published in newspapers all around North America on July 28 and thereafter.

The July 26–27 encounters were the smoking gun that catapulted U.S. Air Force Major Generals John Samford and USAF Director of Operations Roger Ramey once again into the UFO controversy and the public domain. Roger Ramey was involved in the Roswell, New Mexico, encounter in 1947. Samford was the attending head of U.S. Intelligence and involved in the start-up of Project Blue Book earlier in 1947.

General Samford led the longest and largest press conference any U.S. military branch had conducted since World War II. On July 29, the Pentagon was packed with press representatives from all over North America as Samford and Ramey were regarded by the press to be among the most respected experts on the subject of UFOs.

Both Samford and Ramey led a burgeoning, though unofficial, strategy of utilizing whatever tools necessary to downplay and deter the public from pursuing any aspect of UFOs and aliens. A formal strategy policy such as this would not be documented until the following year with the recommendations of the Robertson Panel.

So Samford and Ramey used a variety of explanations to deflect the UFO flap of Washington, DC and recent waves away from the public. This was their intent from the beginning. Among the explanations from the Air Force's study was that UFOs posed no national threat, so there was no need to study these further. Samford reasoned that all the radar readings were not caused by solid metal targets. Then they explained they were misidentified as meteors and other cosmic projections. After that, they were explained away by different weather phenomena such as temperature inversions and double temperature inversions. Lastly, they were explained as mirages. Samford admitted there were "hundreds of radar exposures" that the Air Force could not justify by any other means. So they used all these excuses and more of the not impossible ones that could take place in a real situation.

The Pentagon press conference was exemplary of how many such attempts to discourage society from holding to a dogma of UFO existence were undertaken. This developed in future press conferences when such events occurred, in more lengthy summaries of matters of unidentified anything, like the 1969 *Condon Report*, the various Roswell, New Mexico, incident books, and so on. This tactic is still being used by the U.S. government today.

The National Aviation Reporting Center on Anomalous Phenomena (NARCAP) is a nongovernment-funded organization dedicated for a twofold purpose: first, to be an access portal enabling members of the aviation industry to report their encounters with unidentified aerial phenomena (also known as UAP); the second purpose is to catalogue and provide serious investigation parameters for this collection of encounters. The NARCAP allows anonymity for the witnesses and offers public benefit to help make the airways safer for to perform these investigations.

When asked why they do not study these themselves, to fill a void in responsibility left by the government and military to such UAPs, their responses tend to be "to not affect the national security or public safety." These events happen, aviation industry personnel witness them, and reports of these events having been mentioned here date back to at least the 1940s.

A NARCAP study titled "Aviation Safety in America: Underreporting Bias of Unidentified Aerial Phenomena and Recommended Solutions" was published in July 2004. The three-part study uncovered an environment of underreporting bias, fear of ridicule, employment risk, the complementary phenomenon of lack of open discussion opportunities, lack of effective reporting functionalities, lack of investigation, and lack of public dissemination of reporting among the variety of aviation industries present in the United States, including public commercial, military, and NASA. All these entities have acquired and maintained information, data, and knowledge throughout their histories.

This landscape in the United States is representative elsewhere on Earth but not universal. While real events have happened all over the globe since at least the 1940s, the learning opportunities vary within a complex organizational structure and proliferation process of such events. The NARCAP takes individual reports, catalogues lists of natural phenomena that are observed to be used as possible explanatory summaries of some if the science fits the encounter, surveys eyewitness attitudes and post-experience interest and epistemology, conducts widespread surveys to indicate ongoing interest levels, seeks ways to increase participation in reporting events, and looks for ways to improve reporting mechanisms. This last point has uncovered a woefully deficient documentation discourages witnesses from coming forward.

Part of this reporting and underreporting bias originates from programs announced by the U.S. military and airline industries back in 1954. The joint Army, Navy, and Air Force publication, JANAP 146, summarizes another functional program, the CIRVIS or Communications Instructions for Reporting Visual Intelligence Sightings. JANAP and CIRVIS were mandatory reporting systems for the commercial airline industry.

Reporting requirements to all reporting personnel included being "sworn to secrecy" under threat of fines and imprisonment.[50] While it is true JANAP and CIRVIS were created in 1954 against a backdrop of the Cold War in the United States and could have served as one meta level of motivation for its creation, this also served as a protected and covert information funnel into the sphere of government secrecy of other non-Cold-War-related sighting events, including UFOs/UAPs. Thus, the first elements of the under-reporting environment were born.

This provided the government with a valuable tool to proliferate its efforts of eradicating the UFO phenomenon from the public. JANAP and CIRVIS operated until 1977 when the program's restrictive policies were relaxed. The airline industries were then left with no effective reporting mechanism and no guidance in how to conduct one.

The NARCAP study uncovered an industry-wide incapability and unwillingness of witnesses to report their encounters for reasons due to lack of knowledge about UAP characteristics and no mechanism for learning or being taught about any of them. Also the continuing proliferation of a skeptical environment and the nature of the anomalous experience itself are factors indulgent of the underreporting landscape.[51]

The NARCAP study calls for all the participating aviation governing bodies, including the FAA, National Transportation Safety Board (NTSB), and NASA to implement a rigorous educational program among member and affected personnel which includes more streamlined, secure, safe, and open-reporting program. Additionally, a central data collection system should be formed which would integrate currently scattered information and knowledge about the subject.[52]

References

[1] Ref 1: Jung, C.G. 1978. *Flying Saucers: A Modern Myth of Things Seen in the Skies*. Copyright 1978 by Princeton University Press. Published by MJF Books.

[2] Ref 2: Bailes, Frederick. 1941. *Your Mind Can Heal You: A New Thought Classic*. Retrieved from 2007 version. Originally published in 1941. Copyright 2007 by The Book Tree.

[3] Ref 3: Jung, C.G. 1978. *Flying Saucers: A Modern Myth of Things Seen in the Skies*. Copyright 1978 by Princeton University Press. Published by MJF Books.

[4] Michigan Ref 4: Unknown author. 1966. Sheriffs Watch High-Performance Discs (The Michigan Sightings/Swamp Gas Case). UFO Case Report 778 from UFO Evidence.org. March 14, 1966. Retrieved from: http://www.ufoevidence.org/cases/case778.htm.

[5] Michigan Ref 5: Unknown author. 1966. Close Encounters in Washtenaw County. Old News, Ann Arbor District Library. Retrieved from: www.oldnews.aadl.org/features/1966_UFO_Sightings.

[6] Michigan Ref 6: Unknown author. 1966. Close Encounters in Washtenaw County. Old News, Ann Arbor District Library. Retrieved from: www.oldnews.aadl.org/features/1966_UFO_Sightings.

[7] Michigan Ref 7: Ford, Gerald R. House Minority Leader, R-Mich. 1966. Press Secretary and Speech File, Ford Congressional Papers 1947–1973, Box D37. March 25, 1966. Radio broadcast. Retrieved from: Gerald Ford UFO Talk, written by Grant Cameron. http://www.presidentialufo.com/gerald-ford/89-gerald-ford-ufo-talk.

[8] Michigan Ref 8: Ford, Gerald R. House Minority Leader, R-Mich. 1966. Press Secretary and Speech File, 1947–1973, Box D9. March 28, 1966, News Release. Retrieved from: Gerald Ford UFO Talk, written by Grant Cameron. http://www.presidentialufo.com/gerald-ford/89-gerald-ford-ufo-talk.

[9] Michigan Ref 9: Ford, Gerald R. House Minority Leader, R-Mich. 1966. Jerry Ford papers, 1947–1973, folder UFOs 1966 Press Secretary and

Speech File Box D9. April 21, 1966, Statement. Retrieved from: Gerald Ford UFO Talk, written by Grant Cameron. http://www.presidentialufo.com/gerald-ford/89-gerald-ford-ufo-talk.

10 Michigan Ref 10: Ford, Gerald R. House Minority Leader, R-Mich. 1966. Congressional memo dated April 1, 1966. Retrieved from: www.files.above-topsecret.com/forum/thread7493371/pg1.

11 Michigan Ref 11: Glenn, Allen. 2014. Ann Arbor Versus the Flying Saucers. April 13, 2014. Retrieved from: http://michigantoday.umich.edu/ann-arbor-vs-the-flying-saucers/.

12 Michigan Ref 12: Glenn, Allen. 2014. Ann Arbor Versus the Flying Saucers. April 13, 2014. Retrieved from: http://michigantoday.umich.edu/ann-arbor-vs-the-flying-saucers/.

13 Los Angeles Ref 13: Battle of LA. U.S. Military Museum. Retrieved from: http://www.militarymuseum.org/BattleofLA.html.

14 Los Angeles Ref 14: Blum, Ralph and Blum, Judy. 1974. *Beyond Earth: Man's Contact with UFOs*, Pg. 68. Published 1974 by New York, New York: Bantam Books, 1974.

15 Los Angeles Ref 15: Army Antiaircraft Journal. Retrieved from: http://sill-www.army.mil/ada-online/antiaircraft-journal/_docs/1949/5-6/May-June%201949%20Screen.pdf.

16 Los Angeles Ref 16: San Francisco Museum. Retrieved from: http://www.sfmuseum.net/hist9/aaf2.html.

17 Mainbrace Ref 17: Redfern, Nick. 2014. UFOs & NATO: The Mainbrace Affair. April 22, 2014. Retrieved June 4, 2016 from: www.mysteriousuniverse.org/2014/04/ufos-nato-the-mainbrace-affair.

18 Mainbrace Ref 18: Hall, Richard. 1952. Operation Mainbrace Sightings. Published by National Investigations Committee on Aerial Phenomena (NICAP). Retrieved June 4, 2016, from: www.nicap.org/5209XXmainbrace_report.htm.

19 Mainbrace Ref 19: Redfern, Nick. 2014. UFOs & NATO: The Mainbrace Affair. April 22, 2014. Retrieved June 4, 2016 from: www.mysteriousuniverse.org/2014/04/ufos-nato-the-mainbrace-affair.

20 Mainbrace Ref 20: Redfern, Nick. 2014. UFOs & NATO: The Mainbrace Affair. April 22, 2014. Retrieved June 4, 2016 from: www.mysteriousuniverse.org/2014/04/ufos-nato-the-mainbrace-affair.

21 Mainbrace Ref 21: Ruppelt, Edward J. 1956. Report on Unidentified Flying Objects. London: Victor Gollancz, 1956. 2nd, expanded edition New York: Ballantine, 1960.

22 Mainbrace Ref 22: Hall, Richard. 1952. Operation Mainbrace Sightings. Published by National Investigations Committee on Aerial Phenomena (NICAP). Retrieved June 4, 2016, from: www.nicap.org/5209XXmainbrace_report.htm.

[23] Mainbrace Ref 23: Ruppelt, Edward J. 1952. The Mainbrace Sightings. Published by National Investigations Committee on Aerial Phenomena (NICAP). Retrieved June 4, 2016, from: www.nicap.org/5209XXmainbrace_ruppelt.htm.

[24] Mainbrace Ref 24: Ruppelt, Edward J. 1956. Report on Unidentified Flying Objects. London: Victor Gollancz, 1956. 2nd, expanded edition New York: Ballantine, 1960.

[25] Nellis Ref 25: Ridge, Fran. 1962. National Air Defense Alert of April 18, 1962: From New York to Las Vegas; a Ten State Incident. Retrieved from: http://www.nicap.org/620418adc_alert_dir.htm main article reference.

[26] Nellis Ref 26: Ridge, Fran. 1962. National Air Defense Alert of April 18, 1962: From New York to Las Vegas; a Ten State Incident. Retrieved from: http://www.nicap.org/620418adc_alert_dir.htm main article reference.

[27] Nellis Ref 27: Ridge, Fran. 1962. National Air Defense Alert of April 18, 1962: From New York to Las Vegas; a Ten State Incident. Retrieved from: http://www.nicap.org/620418adc_alert_dir.htm main article reference.

[28] Nellis Ref 28: Wilson, Daniel; Ridge, Fran. 2006. Letter Briefing. National Investigations Committee on Aerial Phenomena. October 5, 2006. Retrieved from: http://www.nicap.org/reports/620418nellis_rep3.htm.

[29] Nellis Ref 29: Ridge, Francis. 2006. Radar Sighting at Nellis Confirms Object Changed Course. National Investigations Committee on Aerial Phenomena. October 8, 2006. Retrieved from: http://www.nicap.org/reports/620418nellis_rep8.htm.

[30] Belgium Ref 30: Population Pyramid.net. Population Pyramids of the World from 1950–2000. Retrieved on 6/15/16 from: https://populationpyramid.net/belgium/1989/.

[31] Belgium Ref 31: Meessen, A. 2011. The Belgium Wave and the photos of Ramillies. Pg. 4. Retrieved from: http://www.meessen.net/AMeessen/Ramillies.pdf.

[32] Belgium Ref 32: Meessen, A. 2011. The Belgium Wave and the photos of Ramillies. Pg. 4–5. Retrieved from: http://www.meessen.net/AMeessen/Ramillies.pdf.

[33] Belgium Ref 33: Wikipedia. Psychosocial Hypothesis. Retrieved 6/15/16 from: https://en.wikipedia.org/wiki/Psychosocial_hypothesis.

[34] Belgium Ref 34: Meessen, A. 2011. The Belgium Wave and the photos of Ramillies. Pg. 4–5. Retrieved from: http://www.meessen.net/AMeessen/Ramillies.pdf.

[35] Belgium Ref 35: Meessen, A. 2011. The Belgium Wave and the photos of Ramillies. Pg. 5. Retrieved from: http://www.meessen.net/AMeessen/Ramillies.pdf.

[36] Belgium Ref 36: Unknown author(s). The Belgium UFO Wave. Pg. 1. Retrieved 6/15/16 from: http://www.ufoevidence.org/documents/doc404.htm.

37 Belgium Ref 37: Leclet, Renaud. 2008. The Belgian UFO Wave of 1989–1992: A Neglected Hypothesis, Pg. 16. The CNEGU Files. October, 2008. Retrieved from: http://gmh.chez-alice.fr/RLT/BUW-RLT-10-2008.pdf.

38 Chicago Ref 38: Haines, Richard F., PhD, Efishoff, K., Ledger, D., et al. 2007. Report of an Unidentified Aerial Phenomenon and Its Safety Implications at O'Hare International Airport on November 7, 2006: Case 18. Published March 9, 2007, and revised July 24, 2007. Retrieved from: http://www.narcap.org/reports/TR10_Case_18a.pdf.

39 Chicago Ref 39: Cohen UFO. 2006. Results of NARCAP Investigation concerning a UFO incident at O'Hare International Airport (USA) November 7, 2006. Retrieved 5/27/2016 from: http://www.cohenufo.org/ohare2006.htm.

40 Chicago Ref 40: Cohen UFO. 2006. Results of NARCAP Investigation concerning a UFO incident at O'Hare International Airport (USA) November 7, 2006, Pg. 4. Retrieved 5/27/2016 from: http://www.cohenufo.org/ohare2006.htm.

41 Chicago Ref 41: Cohen UFO. 2006. Results of NARCAP Investigation concerning a UFO incident at O'Hare International Airport (USA) November 7, 2006, Pg. 5. Retrieved 5/27/2016 from: http://www.cohenufo.org/ohare2006.htm.

42 Chicago Ref 42: Haines, Richard F., PhD, Efishoff, K., Ledger, D., et al. 2007. Report of an Unidentified Aerial Phenomenon and Its Safety Implications at O'Hare International Airport on November 7, 2006: Case 18. Published March 9, 2007, and revised July 24, 2007. Retrieved from: http://www.narcap.org/reports/TR10_Case_18a.pdf.

43 Chicago Ref 43: Hilkevitch, Jon. 2007. In the Sky! A bird? A plane? A UFO? Chicago Tribune, Getting Around. January 1, 2007. Retrieved from: http://articles.chicagotribune.com/2007-01-01/travel/chi-0701010141jan01_1_craig-burzych-controllers-in-o-hare-tower-united-plane.

44 Wash Ref 44: Ruppelt, Edward J. 1956. Report on Unidentified Flying Objects. London: Victor Gollancz, 1956. 2nd, expanded edition New York: Ballantine, 1960.

45 Wash Ref 45: Ruppelt, Edward J. 1956. Report on Unidentified Flying Objects. London: Victor Gollancz, 1956. 2nd, expanded edition New York: Ballantine, 1960.

46 Wash Ref 46: Clark, Jerome. 1998. *The UFO Book: Encyclopedia of the Extraterrestrial.* Visible Ink, Pg. 657. ISBN 1-57859-029-9.

47 Wash Ref 47: Ruppelt, Edward J. 1956. Report on Unidentified Flying Objects. London: Victor Gollancz, 1956. 2nd, expanded edition New York: Ballantine, 1960.

48 Wash Ref 48: Unidentified author. 1952. UFOs Fly Directly over Washington, DC, 1952. *The Washington Post* newspaper. July 28, 1952. Retrieved from: http://www.ufoevidence.org/documents/doc892.htm.

[49] Wash Ref 49: Michaels, Susan. 1997. *Sightings: UFOs*, Pg. 22. Published 1997 by Simon and Schuster. ISBN 0-684-83630-0.

[50] NARCAP Ref 50: Roe, Ted. 2004. Aviation Safety in America: Underreporting Bias of Unidentified Aerial Phenomena and Recommended Solutions. Pg. 9. Revised Edition, July 20, 2004. Retrieved from: http://www.narcap.org/files/narcap_TR-8_2002.pdf.

[51] NARCAP Ref 51: Roe, Ted. 2004. Aviation Safety in America: Underreporting Bias of Unidentified Aerial Phenomena and Recommended Solutions. Pg. 10. Revised Edition, July 20, 2004. Retrieved from: http://www.narcap.org/files/narcap_TR-8_2002.pdf.

[52] NARCAP Ref 52: Roe, Ted. 2004. Aviation Safety in America: Underreporting Bias of Unidentified Aerial Phenomena and Recommended Solutions. Pg. 14. Revised Edition, July 20, 2004. Retrieved from: http://www.narcap.org/files/narcap_TR-8_2002.pdf.

CHAPTER 18

What People (a Lot of Them at One Time) Are Seeing The Military

The means by which we live have outdistanced the ends for which we live. Our scientific power has outrun our spiritual power. We have guided missiles and misguided men.

—Martin Luther King, Jr.

Key words: green fireballs, flap, atmospheric ducting

There are untold numbers of incident reports that tell of unexplained and unidentified craft of some kind that are encroaching on global restricted airspace specifically designated as nuclear weapons facilities. According to one such tally compiled back in 1998, author Larry Hatch of the Nuclear Connection Project lists 193 sightings in or around nuclear sites and related areas.[1] The context includes a series of sightings within a close time frame of about a month as one sighting incident or wave.

The accumulation of evidence appears to be indicative of some of a relationship between or among UFO sightings or encounters and in the geographic proximity to where some nuclear presence is situated. A question that comes to mind is, "Is there an intensified interest from somewhere that precludes these events to occur at nuclear sites in such quantities?"

A few years ago, Donald A. Johnson, PhD of Sun River Research conducted a study for NICAP which explored the question, "Is there any statistical significance that indicates a heightened attention to nuclear sites?" He studied UFO sighting and encounter data within the public domain reported since World War II. Johnson studied two sample groups; the populations of 164 counties that contained nuclear sites of some kind and 164 counties that did not. He matched comparison groups by population size and geography being very aware to carefully matching data from a comparison of individual regions that had equal numbers of both population and that were situated geographically in the same area of the country; i.e., northeast, southwest, and so on.

The results did provide statistical significance that more sightings or encounters occur in those counties with a nuclear presence than those that did not. The data concludes that the ratio is approximately from 1.5 times to over 2.6 times increase in counties with a nuclear presence.[2]

Included here is a quantity of sighting and encounter cases that are of a higher profile and in which case reports and history are readily available for inspection. As is the recurring characteristic for all the chapter subject areas covered in this book, the events of the military sector of society included here is only a small proportion of the total. For data on other occurrences, there are some resources that log reports and compile databases for study. Among these are the National Investigations Committee on Aerial Phenomenon (NICAP), the National UFO Reporting Center (NUFORC), and the Mutual UFO Network (MUFON).

With regard to the arguments of credibility and validity in any research study from education to legal, liberal arts, scientific, or the social sciences, several factors need to be considered. These are the individual's characteristics, such as his occupation, integrity, trustworthiness, believability, honesty, and other human traits that define his worth as a witness to something. The military has historically been among the most accredited with these honorable attributes. The total of this equation is that resident members are to be credited at their word to be realistic in their rationale, in part, because of their job training. For example,when comparing the accuracy of citing an event and then analyzing the facts between a member of the military and the general public, a person in the military is given more credibility. In this chapter, in addition to the case history of UFO experiences reported by members of the classification known as the general public, a separate account is made for this special class of witnesses.

The following script is designed sequentially from the context of the date of the first sighting at the site theater. Here are a few cases of unexplained experiences encountered by members of the highly trained military forces from the United States and around the world.

Green Fireballs, New Mexico

The United States entered the atomic age in the mid-1940s. The creation and development of the design platforms needed as architecture for the construction of the final physical apparatus were started years before. The U.S. government instructed their military to set up research bases in central New Mexico. Many U.S. and allied scientists, including some imported ex-German researchers, were relocated to the regions around Albuquerque, Los Alamos, Sandia, Alamogordo, White Sands, and Las Vegas, New Mexico. Subsequently, they became historically renowned as the creators of the atomic age.

Green fireballs are a type of unidentified flying object which have been sighted in the sky since the late 1940s. Early sightings primarily occurred in the southwestern United States, particularly in New Mexico. They were once of notable concern to the U.S. government because they were often clustered around sensitive research and military installations, such as Los Alamos and the Sandia National Laboratory, then named the Sandia base.[3]

Historically, this phenomenon has been characterized by spheres of an intensely-glowing green signature, a flight trajectory commonly used to explain natural phenomena such as meteors, bombs, weather patterns, etc. Green fireballs have appeared too small to be considered as piloted craft in the sense we consider planes being flown by pilots. However, others of the unusual features, including the organized flight paths which have been observed as a flat, level path, rising in straight vectors, arcing gradual turns and vectors that allow them to fly to match the velocity of the observing aircraft in those types of sightings. It is with these attributes that they are known as unidentified flying objects.

The earliest green fireball sighting occurred in March 1944 near Carlsbad, New Mexico. An Air Force B-17 Super fortress crew observed a "fast-moving, glowing green object which lit up the cockpit, and moved out of sight over the horizon." Carlsbad is southeast of the main venue in New Mexico where the primary history of sightings is documented. A reported and investigated sighting took place on January 19, 1946, near Brest, France.

The first *flap* began in late November 1948 when citizens around Albuquerque saw these objects streak through the sky, low to the horizon but did not seem to land. Almost every night since, more reports came in describing the same events. It is noted and not noticed by others and unexplained that most all the reports in the archives are of nighttime events. Very few of all the green fireball sightings occurred during daylight hours.

On December 5, 1948, an Air Force C-47 cargo crew and a separate civilian DC-3 commercial jetliner sighted a green fireball at about twelve thousand feet just west of Las Vegas, New Mexico, at 9:05 p.m. Both parties reported the green light rise from the ground to about five hundred feet altitude on the east side of the Sandia Mountains. Twenty minutes later, another C-47 crew sighted a green fireball ten miles east of Albuquerque about one hundred miles southwest of Las Vegas. The pilot, USAF Captain Goede, reported an intense green color and a flight path that arched up from a position near the ground. It then leveled out. At 9:35 p.m., the flight captain of a Pioneer Airlines commercial Flight 63 radioed Kirtland tower to report a green fireball just east of Las Vegas. He was approaching Las Vegas from the east. The fireball was approaching Las Vegas from the west. It almost hit the Pioneer flight as it sidled up alongside. The captain said, "The light was definitely larger and more brilliant than a shooting star, meteor, or flare. As the green ball of fire was abreast of them, it began to fall toward the ground, getting dimmer and dimmer and leaving a trail of fragments redish-orange until it disappeared."[4]

Daily sightings commenced and have continued from December 5. On December 6, one was sighted "almost directly overhead above

Sandia Base nuclear weapons assembly site." On December 8, an Air Force flight flying at 11,500 feet saw a fireball about 2,000 feet above approach the plane, level out for a few seconds, then "dropped off rapidly."

On December 12, a sighting from the ground was encountered by the ultimate chief investigator and historian of the green fireball phenomenon, Dr. Lincoln La Paz, USAF Captain Charles L. Phillips, and USAF Lieutenant Allen B. Clark. Their sighting was of a magnitude -4 intensely-lit green fireball on a level flight about ten miles in altitude for one minute, thereupon breaking into three or four pieces.

Dr. La Paz was an astronomer and mathematician from the University of New Mexico who became the chief expert on this subject. He conducted numerous separate conferences and investigations with audiences such as the scientific community of Los Alamos, White Sands, and military communities of all these areas, in addition to larger global communities.

Before February 16, 1949, when Dr. La Paz would conduct the first conference, almost daily sightings were encountered by military personnel, officers, and the New Mexico nuclear scientific community, including such names as Dr. Edward Teller and geophysicist Dr. Joseph Kaplan. Additional attributes to this wave series included objects that "moved with very low velocity," "had noticeable but controlled acceleration," "demonstrated turning characteristics," "moved either very fast or very slow," "exhibited some lifting off the ground and rising in altitude," and "presented no sound heard or audio reverberations heard." Including events where even with a precise known landing site where a sighting observed a fireball wreckage descend and land or crash, investigation results not being able to find any debris of any kind either physically, or via electronic equipment measurement.

On February 16, the conference on aerial phenomena was conducted. Dr. La Paz was trying to determine what conclusions the attendees could offer through analysis. All the attendees had their own sighting(s), and there was a split consensus as to the explanation of natural or man-made. The consensus points produced were that they were not naturally-occurring phenomena, and they were not man-made to their knowledge.[5]

The green fireball phenomenon persisted almost entirely over New Mexico skies for the next decade in a regular regimen. Ultimately, there were sightings reported from Texas, Nevada, and California and eventually from other locations both domestically and internationally. Characteristically, most all them were experienced near nuclear facilities of some type. Other scientists such as Clyde Tombaugh (Pluto), Dr. Donald Menzel (noted overt UFO skeptic but covert acknowledger), J. Allen Hynek, Wernher von Braun, Edward Ruppelt, and many military officers and nuclear scientists confirmed their own encounters and acknowledged the theoretical conclusion of the nonnatural attributes.

Dr. La Paz would head an Air Force Project *Twinkle* that was approved in December 1949. The study lasted about two years but never received much fiscal or human resource support. For two years, no stronger or more affirmed consensus could be obtained from the study. Project Twinkle was terminated in December 1951.

Dr. Edward J. Ruppelt interviewed many of the Los Alamos scientists in 1952 and determined that there was no natural explanation for them: "The scientists speculated that they were extraterrestrial probes projected into our atmosphere from a 'spaceship' hovering several hundred miles above the earth." Ruppelt responded with, "Two years ago (1950), I would have been amazed to hear a group of reputable scientists make such a startling statement. Now, however, I took it as a matter, of course."[6]

More contemporary theories on the nature of the green fireballs include a summary in *The Condon Report*. This theorized that they might be explained by lunar materials ejected during meteor impacts on the moon's surface.[7] Another theory was introduced in Robert Hastings's book *UFOs and Nukes*. The correctly-labeled hypothesis postulated that radioactive fallout from clouds formed at the site of nuclear explosion testing in Nevada drifted through the atmosphere in the 1950s. Most of the prevailing winds took the fallout over New Mexico. One in particular mentioned the example was the Buster series of tests in November 1951 which produced such fallout and corresponded chronologically with a series of fireball sightings in New Mexico a couple of days later and in other states downrange a few days after that.[8]

Oak Ridge, Tennessee Sightings

During World War II, special new complexes were needed to accommodate the emergence of nuclear technologies being created and developed in New Mexico. These were strictly wartime pursuits as more commercial projects, such as nuclear power plants, would not be considered until after the War, though it was feasible at that time to accomplish this. The projects undertaken were to mine and manufacture or both radioactive metals needed for the construction of atomic weapons.

Oak Ridge was the site of the production of radioactive uranium. The complex, situated about ten miles west of Knoxville, was built only a few years before the first sighting but had attained a population of over 75,000 by 1945. The uranium manufacturing and processing facility site was over forty-four acres, the largest single enclosed building in the world at the time.[9]

The first sighting was reported in September 1944. An object that looked like a metal tube was seen to have been elevated motionless over the diffusion plant. In the summer of 1947, an Air Force Colonel Gasser reported to the FBI that photos of a UFO had been taken near the Oak Ridge complex. On May 25, 1949, multiple witnesses saw a "strange, flat metallic object" passing over the complex. It made a "cracking noise" as it vectored past. A month later, on June 20, several people witnessed three objects, two rectangular and one circular in shape, fly over Oak Ridge.

The first extended wave around Oak Ridge began in 1950. On October 15, AESS Troopers Captain Zarzecki and Rymer J. Moneymaker saw two shiny silver bullet-shaped objects dive from an altitude of twelve thousand feet to about six feet off the ground then stop and hovere about fifty feet away from them. The objects flew away briefly then returned and repeated this sequences several more times. Knoxville radar control tracked two objects along this flight path. The next day, other observers from the Oak Ridge Security Patrol Force saw a "silver-white spherical object" approach them in a similar manner then fly away only to repeat this sequence.

Noting that a pattern was emerging, the Army's Air Technical Intelligence Center (ATIC) and the Counter-Intelligence Corps (CIC) met to map a strategy for dealing with this situation. They considered natural phenomena and prank activities but could not find plausible explanations. Characteristically, the witnesses were mostly credible and gave detailed and similar descriptions of their encounters. The CIC report concluded that "the objects are a physical phenomenon which have a scientific explanation (undetermined source), electrical or an operation of harassment. The fantastic is generally rejected."[10] The last sentence gave a lot of reliability and validity to the credibility of the many witnesses to the Oak Ridge series.

The sightings continued on a daily basis. Visual sightings were made, radar trackings were recorded, and fighter jets were sortied. On October 24, among the witnesses were William Fry, the assistant chief for the National Environmental Policy Act (Project NEPA). The craft "changed color rapidly from a reddish hue to a bright orange and again to a brilliant light blue." Also witnessing the craft was Air Force Major Lawrence Ballweg, who described similar events. Radar also recorded the events.

Sightings continued through January 1951. Among the characteristics reported were triangular-shaped craft, objects that changed shape in midair, post-sighting investigations that recorded abnormal increases in gamma and alpha radiation (from nuclear decay activity) not attributable to any process at Oak Ridge, and atmosphere ionization anomalous to Oak Ridge. Some of the fighters intercept sorties that reportedly obtained photographic data but was never released to the public. Many of the sightings indicated alternating sequences of craft slowing down then stopping in midair and just as suddenly accelerating to a vector with exceptional velocity not replicable by human technology.[11]

Hanford, Washington Nuclear Plant

The Hanford Engineer Works, the first plutonium production facility in the United States, lies in southern Washington State, close to the Oregon border. It was commissioned in 1943 to become a wartime facility for production of such nuclear materials. The then municipality of Hanford was effectively eliminated by the federal government for construction of this site. When operational only months later, Hanford manufactured the combustible plutonium that was used in the Trinity atomic tests and the Hiroshima and Nagasaki deployments.

In February 1945, the first of many sightings took place. Navy Lieutenant Clarence Clem was called to the flight line and a waiting F6F hellcat fighter at the end of officers mess meal to intercept a bogey hovering in a stationary airborne position directly over the Hanford plant. Pasco Naval Air Station in Washington is about sixty miles from the Hanford plant.

Lieutenant Clem was joined by two other pilots, including base Lt. Commander Richard Brown, in the sortie. Radar at the base tower had been tracking the UFO since it had arrived in local airspace a half hour before the three pilots were airborne. It was now hovering stationary over the reactor building. Lt. Commander Brown was the first to obtain a visual sighting of a "bright ball of fire." The fighters took chase in the F6F fighters up to their flight ceiling of thirty-seven thousand feet. The UFO continued to hover over the plant for a time not allowing the planes to get close to it. Radar and the pilots then observed the craft accelerate at a fantastic speed to the northwest toward Seattle. According to Commander Brown, "It was so bright that you could hardly look directly at it. It took off at a high rate of speed. No maneuvers really, just a straight-line course." The signature on the radar scopes disappeared within a couple of seconds. The pilots lost visual sighting within a couple of seconds.

For this night and into the next day, there was discussion among the base personnel of a possible explanation of a type of weapon the Japanese had been using since the beginning of the war. They flew incendiary balloons across the Pacific with the hope that they would land and explode at U.S. territory. These balloons had no mechanical means of flight control characteristics. They were inexpensive and could feasibly traverse the ocean via upper atmospheric wind patterns, but they were entirely at the mercy of such wind patterns for their flight path. In summary, the technology was very crude, and besides not having any known success in their stated mission of incendiary destruction during World War II, the Pasco Naval Station personnel had too much contrary evidence of the radar tracking and the pilots' encounters to seriously consider this possibility.[12]

Later that year, on July 15, 1945, another Navy pilot, Rolan Powell, was stationed at Pasco for carrier operations training when at noon scramble alerts were sounded. Radar tracking detected a fast-moving object that came into radar observation at thousands of miles per hour. It stopped suddenly and hovered directly over the Hanford plant.

Powell and his group of six attempted to intercept in their Grumman F-6F Hellcats. Their visual sighting included the description of a craft with a "saucerlike appearance, is bright, extremely fast, and very high." Powell later described the craft to be "as large as three aircraft carriers side by side, oval-shaped, very streamlined like a stretched-out egg and pinkish in color. It emitted a kind of vapor around the outside edges from portholes or vents."

One pilot yelled over the radio, "What the hell is that?"

Another answered, "Nothing I've ever seen before."

Orders were sent to the crew to go higher. "Blow the engines if you have to, but use full military power, full throttle injection, maximum, continuous. Go for it!" They did this. The pilots estimated the UFO to be stopped and hovering in position at an altitude of sixty-five thousand feet. Powell said, "It doesn't make any overt moves, gives no signals, just hovers there as if observing."

The six planes did manage to get as high as forty-two thousand feet before their engines gave out and they had to glide back to base. All this time, the UFO stayed in a hovering position, for well over an additional thirty minutes after the sorties achieved first sighting before it sped off straight up in a vertical trajectory and out of radar and visual range.[13,14]

According to Air Force declassified intelligence reports, there were many later Hanford sightings. For example, on May 21, 1949, a silvery, disc-shaped object was simultaneously sighted by F-82 pilots and tracked by radar from nearby Moses Lake AFB. Before the fighter pilots could intercept it, the unidentified craft accelerated out at a rate of speed far faster than any aircraft. The Air Force report then diverged in its explanation, offering two different scenarios. The official version included pilot statements that "the sighting involved flying saucers." The public version said the sighting was that of a conventional aircraft.[15]

Another example includes another Air Force intelligence memo dated August 8, 1950, documenting Hanford as a "continued place of interest." "Since July 30, 1950, objects, round in form, have been sighted over the Hanford AEC Plant. Air Force jets attempted interception with negative results. All units including the antiaircraft battalion, radar units, Air Force fighter squadrons, and the Federal Bureau of Investigation (FBI) have been alerted for further observation. The Atomic Energy Commission states that the investigation is continuing."[16]

Still another example was introduced to the public by USAF Project Blue Book Chief Captain Edward J. Ruppelt. In his 1956 book *The Report on Unidentified Flying Objects*, he wrote: "On December 10, 1952, near another atomic installation, the Hanford plant in Washington, the pilot and radar observer, spotted a light while flying at twenty-six thousand feet. Ground Control station told of no planes were known to be in the area. They closed on the object and saw a large, round, white 'thing' with a dim reddish light coming from two 'windows.' When they attempted to close and lock radar, it would reverse direction and dive away. Several times, the plane altered course because collision seemed imminent."[17]

Ellsworth AFB UFO Sighting

As observed through reading these case study summaries, the sighting phenomena categorically does not discriminate as to certain times of the year. These summaries also do not predominately discriminate as to prevailing weather conditions at the times of the incidents. Encounters are made in clear, cloudless skies. Also they occur under cloud cover. No general reports of general precipitation episodes are mentioned in the literature. Although a significant minority of these events occurs in the daytime, most occur at night.

The night of August 5–6, 1953, was one of those clear, starlit, moonless nights in western South Dakota. According to a summary in *The Condon Report*, "The event theater from Rapid City, South Dakota, to nearby Blackhawk and up north toward Bismarck, North Dakota, had stable conditions, excellent visibility, temperature inversions and radio surface ducts present."[18]

The Condon Report used these latter two characteristics often in their case summaries, but there was never an account of who or where these descriptions were analyzed and concluded from nor were any accuracy checks made. The concepts of temperature inversion and what is known as *atmospheric ducting* (radio surface ducts) were often used in both Project Blue Book and Air Force explanations of sightings. In addition, often, these meteorological characteristic mechanisms were not accurate factors when inserted into the specific case study they were analyzing. Atmospheric ducting is a mode of propagation of electromagnetic radiation, usually in the lower layers of the Earth's atmosphere, where the waves are bent by atmospheric refraction.[19]

That night, at approximately 8:00 p.m., a civilian observer from Blackhawk (about ten miles west of Ellsworth AFB), Miss Kellian, spotted a "red-glowing light making long, sweeping movements." Radar controllers at Ellsworth confirmed a tracking and visual sighting of a bluish-white light at elevation of sixteen thousand feet making a "wide sweep around Rapid City and returning to its original position in the sky."

After moving circularly around Rapid City, a Master Sergeant, Harry, and other observers at Air Defense Command watched the moving light as it "danced between telephone lines. It would remain stationary then hop up several degrees very quickly, almost simultaneously. It would stop, move to the left, and then swerve in a sort of slanting motion, repeating these maneuvers several times." This target continued for about three hours.

One of the warrant officers called to scramble an F-84 jet to intercept the object. The Air Defense Filter Center in Bismarck, ND was also notified and began monitoring the events. The first sortie of a few that night began the intercept. Although the F-84 Thunderjet was the first USAF fighter capable of carrying an atomic weapon on board, there is no verification in the record that this sortie had such a weapon.

A cat-and-mouse game ensued. When the fighter closed to within about three miles, the UFO would begin to move ahead to increase the distance between the two. Again, the fighter closed to within three miles, and the UFO would repeat its maneuver. Gun camera and radarscope photos were also taken of the chase and are filed somewhere in the Air Force documents.[20]

The chase continued until near Bismarck; the pilot had to return to base due to low fuel. Bismarck Air Defense Filter Center secured radar tracking of the entire chase. As the pilot was returning to Ellsworth, all radar installations and on board the F-84 recorded the UFO beginning to chase the plane. Back at Ellsworth, the conversations were overheard, and two more F-84's scrambled up at about 11:00. The UFO broke off from the one plane and began playing cat-and-mouse with the new fighter. The single pilot, Lieutenant Needham, was now at fifteen thousand feet. The UFO colors changed from white to green as it "jumped in elevation instantaneously by a few thousand feet." The pilot also noted new objects below him and from behind. A pair of red and white lights were about seven thousand feet below the fighter's position.

The other F-84 pilot established visual contact for about thirty seconds. He described the bright light emitted from the craft as "brighter than the brightest star I've ever seen." When that plane gave chase, the light disappeared from view in front of him.

Meanwhile, at this point, Needham was unsure of what he was seeing, so he turned off all the plane's lights, so the cockpit went dark. The lights in front of and behind him were still on and visible. They were not reflections or inversions. Then Needham engaged his radarscope cameras to capture photos. Those photos were unusable. Meanwhile, radar at Ellsworth, Bismarck, and Fargo, ND were tracking or were about to track the four airborne craft.[21]

One other observation was made by an Air Force Globemaster II plane as it flew into Bismarck airspace. At 1:09 a.m., on August 6, the flight crew observed four airborne objects nearby. The craft closest to the Globemaster II then began a sequence flashing of red and green lights. Within a second or two, the flashing pattern was repeated by the other three craft in succession. One of the eye witnesses described as "it was as if a 'wave' passed from one to the other."[22]

Investigations by Project Blue Book and others did eliminate weather balloons as a possibility. The Rapid City Municipal Airport sent up a balloon around 10:00 p.m. on the fifth, but the prevailing winds took it south, away from the event theater. But later in the Blue Book report, it mentioned two upper-air research balloons were tracking around the town of Lowry at some unknown time (no day or time was given). These balloons were lost and "could have been in the area at the time of the sighting." Upper-air research usually means in the stratosphere or higher which is well over four thousand feet altitude. Also there were only two that were mentioned as lost. None of them were known to have lighting apparatus on board that blinked in patterns.

Additional explanations of the incidents were made as observations of meteors, the star Capella, and the temperature inversions summary. The report also notes the "data on inversions was not available as it had been forwarded to Ashville, North Carolina."[23]

Other incidents of note at Ellsworth occurred first in 1966 and then in 1992. In 1966, when at the Juliet-5 silo complex, an Air Force Security Alert team came within the proximity of a few yards away from a "mysterious object [that] was round, apparently metallic, and resting on a tripod inside the security chain-link fence that secured the missile silo. That missile silo had gone off-line prompting the investigation."

A second Security Alert Team, including USAF (Ret) Staff Sgt. Albert Spodnik, was observing the field where this object occupied a Juliet-3 missile silo some distance away and monitoring the radio communications between the Juliet-5 site and Ellsworth Missile Command Post. The Juliet-3 missile had also gone off-line, and the Security Alert Team was repairing Juliet-3.

The Flight Security Controller at Juliet-5 called the Missile Command Post who ordered a helicopter to the landing site. When the helicopter was spotted near the landing site, both Security Alert Teams watched the craft produce a "brilliant white light" directly over the Juliet-5 missile, then launch off the ground vertically, and vector vertically up and out of sight within a couple of seconds. Juliet-3 observers described the light as having "the appearance of an inverted flashlight beam."

Spodnik and his team were unexpectedly met by the missile maintenance commander who promptly inquired about what the two men saw. Spodnik, who later justified his response by referencing a department of defense regulation at the time, known as the "Personnel Reliability Program" (PRP), reported the team saw nothing unusual out in the field. The Personnel Reliability Program governs the behavior of personnel who work around nuclear weapons. Severe job and legal consequences are bestowed on a person who is judged by his superiors to be psychologically unstable and therefore unfit for work clearance.

At the venue of the 1992 encounter, USAF (Ret) Tech. Sgt. Jeff Goodrich compiled reports from missile maintenance personnel of "a group of white lights moving rapidly in rigid formation. The fact that the lights did not vary in their positions relative to one another concluded that they were arranged across the surface of a very large but unseen craft. The light formation moved directly toward the Minuteman missile maintenance hangar, hovered over it momentarily, and then moved away." The airmen provided drawings that resembled lights appearing like a string of pearls around the perimeter of a kidney-bean or boomerang-shaped craft that hovered motionlessly three hundred to five hundred feet directly atop the missile hangar.[24]

Kirtland AFB UFO Sighting

As noted, there has been a long history of UFO sightings in New Mexico since the modern era of ufology took shape. The proximity of installations with a nuclear presence such as Los Alamos, White Sands, Alamogordo, Roswell, and now Kirtland AFB in Albuquerque, as evidenced in the encounter report documentation.

November 4, 1957, was the date of another detailed sighting by two air traffic controllers. At approximately 10:45 p.m., two civil aeronautics officers in the Kirtland tower, R.M. Kaser and E.G.Brink, noted a white light traveling east across the airfield. Radar contact was established as the two continued their observations. They saw a dark egg-shaped object descend almost vertically and stop at an elevation level about three thousand feet away from the tower. It stayed stationary for a few minutes. The officers described the craft as being from fifteen to twenty feet tall and elongated vertically with a single white light at its base and oval in appearance, hence, initially described as egg-shaped.

The UFO stopped motionless near the nuclear weapons storage area. After hovering there for a few minutes, the craft started to move eastward until it reached within the boundary of the base. It stopped briefly then accelerated vertically at a high velocity through an overcast sky. Kaser and Brink contacted the Albuquerque Radar Control Tower who then confirmed these movements. The radar tracked the object moving south to the Albuquerque Low-Frequency Range Station for a few moments. It then moved back north toward Kirtland and hovered over the main north-south runway outer marker for a few minutes. The craft then allowed a C-46 transport plane to lift off before following it out of Albuquerque airspace after about fourteen miles.[25,26]

Project Blue Book took up an investigation of the over twenty-minute encounter and published their results in the *Condon Committee on UFOs* report. The explanation the study presented was that a "small, powerful private aircraft was confused and attempted to land at the wrong airport."[27]

In his 1969 paper, *Science in Default*, noted physicist and ufologist Dr. James E. McDonald refuted the Project Blue Book conclusions citing no analysis was provided, and a false interpretation of the only witness statement investigated by the Blue Book committee was included. The witnesses later corroborated the accurate version to McDonald with that of their original statement to Captain Shere from Ent Air Force Base in Colorado. Their version of the flight characteristics and dynamics of the craft refuted a description of any manmade aircraft and also contested the weather dynamics of the night and time of the event in question with those that were written in Blue Book reports.[28]

Walker Air Force Base, New Mexico

The Walker Air Force Base was known as the Roswell Army/Air Force Base, home of the 509th Strategic Air Command Composite Group and the first atomic weapon launch capability in history. The facility launched the Superfortress flights that delivered the Hiroshima and Nagasaki atomic devices that ultimately ended World War II were launched from the facility.

Walker AFB was renamed on January 13, 1948, and since that time, additional UFO sightings have occurred. On November 7, 1957, three days after the Kirtland incident, according to an Air Force declassified Project 10073 Record Card, a sighting of "a black object with a light blue tail of two hundred to three hundred yards in length" was the first of two that day. The second sighting was a "long and round large stove pipe, bluish-white. Long trail 200–300 yards long. The object made a 360-degree turn. The pattern of turn was horseshoe shaped." These two sightings of several objects continued for about two minutes. The Record Card contains descriptions with many similarities to the Kirtland sighting of November 4, 1957.[29,30]

Toward the end of 1963, on several occasions, security and missile technicians and officers, including First Lt. Jerry C. Nelson, reported a "silent, very bright UFO hovering over the site. These objects would hover over the silo and shine lights down on them without making any noise, probably more than three but fewer than ten such incidents."

In the autumn of 1964, one of the silo sites, Site 7, received communications from sister sites. First Lt. Philip Moore reported that he was informed of "an extremely bright light that was repeatedly hovering directly over the site, racing away, returning, and hovering again." Silo technicians who were summoned to the surface to witness the sightings also reported, "The UFO zooming from the direction of Site 6 to the direction of Site 8 and hovering for a while. A silent light that moved extremely rapidly-instant go and instant stop, no getting up to speed or slowing down. It was specifically interested in those sites."

In December 1964, numerous additional encounters occurred. One event enlisted the use of jets to intercept the two craft observed that day. According to Airman First Class Thomas Kaminski, "[He] observed the jets attempting to approach the unidentified lights, which then put on a burst of speed and outran the interceptors." As noted elsewhere, the Air Force had a habit of keeping some things secret. Kaminski reported that the next morning his missile team was in a debriefing when the sighting subject was raised. His captain asked about the fighters that were sent to intercept the UFOs. The response from the missile team was, "What fighters? We didn't send up any fighters." The people in the room then knew covert activity was in progress.[31]

Cape Canaveral, Florida

An unidentified flying object altered the vector trajectory of a U.S. Navy Polaris missile test flight on January 10, 1961. Video of the launch shows a change in the flight trajectory of the Polaris missile as it lifted off from Cape Canaveral. The disc was described as about twenty-five feet in diameterwith a height of six to eight feet at its core. It is shown to alternately hover then accelerate swiftly while in a leadership pattern in front of the missile. Air Force radar tracked this UFO from its first penetration into airspace as it was also tracking the launch of the Polaris missile.

The Navy also filmed the flight of the missile. The UFO was tracked both with the naked eye and from electronic observation. The craft altered the missile tracking for about ten minutes. The flight patterns of both were not in a straight line but rather of alternating curved trajectories, and the missile appeared to be following the UFO. After the UFO exited the area, the missile's vector altered back to its originally intended course. Later testing of the radar equipment and the camera equipment showed no malfunctions; however, no other explanations were offered.[32,33]

Moscow, Russia

A few months later in the summer of 1961, the city of Rybinsk was the venue for a sighting event. Rybinsk bases were situated about 150 kilometers from Moscow and served the Soviet Union's ICBM purposes in the same manneras U.S. missile bases, both concepts stemming from nuclear technology.

A massive disc-shaped object allegedly appeared at an estimated altitude of twenty thousand meters, surrounded by many smaller objects. A nervous battery commander panicked and gave, unauthorized, the order to fire and "the missiles were fired. All exploded, creating a fantastic spectacle in the sky." The third salvo was never fired, for at this point, the smaller saucers went into action and stalled the electrical apparatus of the whole missile base. When the smaller discoidal UFOs had withdrawn and joined the larger craft, the electrical apparatus was again found to be in working order.[34]

This incident was initially disclosed and published in an Italian scientific research journal. In Timothy Good's book titled *Above Top Secret*, he states the Italian science journal is a respected source, and it would be unlikely that they would write such a story if there was not any substance behind it.[35]

F.E. Warren Air Force Base

Situated on 5,866 acres in a region that grids the regions of southeast Wyoming, northeast Colorado, and southwest Nebraska, the main complex of F.E. Warren Air Force Base is responsible for 50 Peacekeeper missiles among its weapons inventory. The human resources include about four thousand military and over six hundred civilian personnel who maintain presence on the main base and peripheral missile facilities that spread over twelve thousand square miles in the same region.

The base was originally commissioned by the U.S. Army in 1867 and known as Fort David Allen Russell, named after a U.S. Civil War general killed at the Battle of Opequon in Winchester, Virginia, in 1864.[36] The Air Force officially commissioned and renamed Fort David Allen Russell as F.E. Warren AFB on June 1 and September 18, 1947, respectively. Its operational and inventory history is known to consist of the Air Training Command from 1947–1958, the Strategic Air Command from 1958–2005, and currently the Twentieth Air Force. Primarily, a strategic Intercontinental Ballistic Missile base (ICBM) and one of three in the United States, Warren does not possess busy landing runway facilities; therefore, the same routine aircraft traffic is not the same as most other Air Force installations. For the most part, historical inventory consisted of various Atlas series and Minuteman nuclear missiles.[37]

Warren had a prolific history of UFO incidents. The first of the most unusual is actually a series of events. On August 1, 1965, numerous active military personnel, including the base commander Colonel Johnson, called Project Blue Book headquarters at Wright-Patterson AFB in Dayton, Ohio, to report "a large circular object emitting several colors but no sound, sighted over the city (Cheyenne, WY)." Colonel Johnson also indicated that other Army military officers at Sioux Army depot in Nebraska reported spotting five objects minutes after the Cheyenne sighting. An hour later, at 2:50 a.m., nine more UFOs were sighted at Sioux. Following that, at 3:25 a.m., six UFOs were sighted stacked in a vertical formation. Then at approximately 4:05 a.m., Colonel Johnson called Wright-Patterson again to report nine UFOs in groups surrounding the depot. The Strategic Air Command entered the episode by declaring a "white oval UFO directly overhead over Warren, descending at a high rate of speed and a flashing red light in its center moving east, reported to have landed ten miles east of the site."[38] The UFO maneuvers lasted over three hours and were witnessed to have alternately joined together in formation both while in motion and while hovering in place while airborne, risen, and descended in altitude and vectored very swiftly at times.

Over the next month, numerous sightings and tracking events took place in this area of the United States. Tinker AFB in Oklahoma City reported a similar sighting the next evening. Back at Warren, a few nights later, a security crew reported a UFO hovering over a missile silo and above their security vehicle. They indicated that "the vehicle shook while the aerial vehicle was near it." USAF Airman Second Class Robert R. Thompson, upon witnessing the event, stated, "I wasn't sure what we were seeing until I reported back to the launch commander. When he told me of the report of UFOs from Cheyenne Mountain, he wasn't joking."[39]

According to Thompson, two members of his unit reportedbeing stationed at a Security Alert Team camper unit located a few miles from the missile silos of the site known as the Quebec Flight. One of his unit members reported the camper was shaking violently. He quickly leaned his head out of the window and saw a large, very bright light hovering without a sound directly above the camper. After a few seconds, the shaking ceased, and the light rapidly departed. The silo launch facility confirmed that they observed and tracked the light hovering over the location and acting uncharacteristically as the security team members observed and reported.[40]

Another unexplained set of occurrences at Warren that summer was learned from reports by local ranchers and the public. They reported many similar sightings, including the disappearance of many of their cattle.

These episodes were also reported to the then still-operating Project Blue Book. In 1965, many Air Force personnel present at the base, including the base commander Colonel Johnson informed the Air Force's Project Blue Book that their base was inundated by sightings of unknown aerial objects, including a large, white oval-shaped craft.[41,42]

A second series of Warren incidents began on April 15, 1966, when the Air Force Strategic Air Command was housing upwards of 150 Minuteman II missiles at the time when a sighting was made at about 11:00 p.m. According to Air Force personnel observers, "Above us was a formation of eight objects. When I reported back to the Missile Combat Crew Commander, he told me that Cheyenne Mountain had just told him of eight UFOs over our location. Cheyenne Mountain had radar confirmation of these objects."[43]

A third wave began in the autumn of 1973 and lasted into the following year. On that occasion, all twenty of the Launch Control Centers (LCCs) were notified when one of the LCCs, the India crew, reported alarms going off at the silo. This security team watched a brightly-lit UFO hovering directly over the silo. It remained stationary over the missile compartment for over a minute then sped up vertically to an altitude of a few thousand feet, hovered a few seconds, then sped out of sight. The next day investigations were made and disclosed that the target tapes which directed the missile to its respective target were erased and had to be reprogrammed.

At the end of 1973, another incident occurred in which a UFO stopped in midair and watched a missile maintenance crew reassemble a job they had disassembled earlier that day. Each party watched the other for over five minutes before the UFO sped away.

A third event was reported but without much-released detail. In the spring of 1974, a security crew at Charlie LCC witnessed a UFO land near the center. There were no reports released pertaining to craft occupants. The three military observers watched the ship for ten minutes before it lifted off and sped away.[44]

A more current incident at Warren that received widespread attention occurred October 23 and 24, 2010. Active Air Force personnel scattered around the Warren complex after observing a set of huge cigar-shaped objects. Upon the receipt of initialsighting reports, missile maintenance technicians began to notify their superiors that their Minuteman-III nuclear missiles went completely off-line. All communications and functional control of fifty such missiles was completely lost. According to *The Atlantic* magazine, an unidentified military officer who had been briefed on the event said, "We've never had something like this happen. We can deal with maybe five, six, or seven at a time, but we've never lost complete command and control functionality of fifty ICBMs."

The Air Force reported, via their public affairs director for Air Force Global Strike Command, Lt. Col. John Thomas at Barksdale Air Force Base in Louisiana, the failure lasted slightly under an hour. A controversy exists, though, in that some of the missile technicians on-site the night of October 23/24 report that the knockout of the missiles lasted over twenty-four hours. Ufology researcher, author, and expert Robert Hastings said, "I have detailed information about the events. It was intermittent and involved a very specific sequence of these five missile alert facilities going on and off-line. It was not a commercial blimp (the cigar-shaped object). It had no passenger gondola and no advertising on its hull. Its aspect ratio was long and thin."

Hastings said the witnesses said, "The object was seen in the sky above the field, throughout the weekend, both during the (missile) disruption and the following day."[45]

Hastings explained he often saw the Air Force threaten witnesses and anyone not within the Air Force circle of command if they claimed to have observed this and other such incidents at military and nuclear facilities. Warren witnesses who were on active duty during and after their episodes were often warned of the prosecutorial consequences of discussing any such matters with anyone.

Robert Hastings is an expert on the history of military involvement with UFOs. He has researched the field for over forty years and has given lectures at over five hundred universities. The much more factual material of the hundreds of cases he has investigated can be discovered in his recent books *UFOs and Nukes: Extraordinary Encounters as Nuclear Weapons Sites, UFOs,* and *Nukes: The Documentary* or on his website www.ufoevidence.org.

Minot Air Force Base

Minot was a main Strategic Air Command base responsible for the operational performance of a bomber-based and ballistic missile nuclear arsenal. These were the B-52 Fifth Bombardment Wing and the Ninety-First Strategic Missile Wing. The SAC entered into the current Air Force Global Strike Command in 2009.[46]

August 24, 1966, became the first of a flap of sightings over and near Minot AFB. This flap, a long-term sequence of unidentified or unexplained or both sightings and experiences that last longer than a wave, continued until 1970. Both military (U.S. Air Force and some Army) and civilians, workers in military employment and the general public reported encounters on a regular basis.

That night at about 10:00 p.m., an Airman observed a multicolored light high in the sky. A second light was sighted passing under intermittent cloudcover as it was watched continually. Three Minuteman ICBM missile sites confirmed visual and radar tracking of both objects. The objects hovered at times at ground level, and all encounter locations encountered radio interference (Minot Ref 2).

Later in March 1967, an unidentified object descended from high in the atmosphere to a position over Minuteman ICBM silos of the Ninety-First Strategic Missile Wing. As the events played out on the fifth of March, a metallic, disc-shaped craft ringed with bright laser-like flashing lights moved slowly toward, maneuvered, and stopped over each silo at an altitude of about five hundred feet. Then the ship circled over that particular launch control facility before it moved to the next one. Air Force fighter jets were scrambled to intercept the object. When the jets came within visual range, the ship climbed vertically into the atmosphere and disappeared almost instantly.[47]

On October 24, 1968, Minot AFB radar tracked an orange-glowing UFO heading west-northwest. According to this investigation, the incident started at around 2:30 a.m. on October 24, 1968. An investigation of the Minot incident was conducted afterward by scientists, and the findings were later published in *Physics Forums*. An examination revealed new photographic and radar evidence from the Air Force unknown previously to the public domain.

Missile maintenance team members Airman First Class Robert O'Connor and A1C Lloyd Isley spotted the first unexplained light to the east of the base. The airmen reported the UFO was "pacing their moving vehicle while growing brighter." It was also reported by security personnel at the three other LCCs that "the object(s) separated into two parts and went in opposite directions and returned and passedunder each other." An independent team near the object reported seeing an object land, and then the bright light extinguished. Upon this, the witnesses saw nothing. Other groups reported an illuminated aerial object that started emitting alternate colors from brilliant white to amber and green. The craft would slow, abruptly change direction, stop and hover, then accelerate at a fantastically high rate of speed.

The location of the first tracking trajectory was about thirty miles northwest of Minot. An Air Force B-52H bomber was in the vicinity at the time and began pursuit. Other radar tracking stations soon picked up the electronic signatures on their radar equipment.

The B-52H came to a vector about three miles away. During the time of approach to intercept, the crew of the B-52 also observed the orange-glowing craft. At about 3:58 p.m., the plane's radio transmitter failed when it was circling within that six-mile diameter. The UFO paced the aircraft off its left wing for about twenty miles. The B-52 video recorded the entire incident.

Four minutes later, the transmitter reestablished communications with the other tracking stations. In addition to the film, the B-52 retreived telemetry and velocity data. After one failed landing attempt by the B-52, its pilot was ordered to reestablish intercept activities and continue photographing and tracking the UFO. From an initial tracked velocity of over 3000 mph, the UFO landed on the ground at a location the Air Force labeled "AA-43" in less than sixteen miles of airspace. That location was where, when the B-52 was on final vector intercept approach the second time, they acquired another visual. The pilot, Major James Partin, compared the UFO at that point to "a miniature sun placed on the ground below the aircraft."

They reached the grounded USO (unidentified stationary object, not to be confused with USO; unidentified *submerged* object-one underwater) and flew around it. The copilot, Captain Bradford Runyan, described "a huge egg-shaped object with a surface that gave off a dull reddish color like molten steel—a smooth metallic tubular section extending [and] connecting to a curved crescent-shaped protuberance, not unlike a bumper [with] a greenish-yellow glow from its interior back." Also as before, all communications with any ground-based personnel failed.[48]

The reference for this citation includes links to over 145 pages of declassified documentation, located in the "documentation" section of the website: www.minotb52ufo.com/introduction.php.

As reported by on-scene witnesses, including Air Force Major Bradford Runyan, the copilot of the B-52 flight, Captain Robert Salas (whom you will read about later), and other officers, additional sightings of other UFOs at other Minuteman ICBM silo sites were made simultaneously as the aerial sortie. Alarms were set off uncontrollably, silo doors opened and closed, and combination locks of inner door enclosures unlocked and later locked again as unusual and unexplained features of the scene played out on the ground theater.[49]

Malmstrom Air Force Base

Situated in the northern region of Montana near the Canadian border, Malmstrom has had more activity than any other site with a nuclear presence. In the realm of disclosure, additionally, there have been more details released to the public domain than most others. Here is a chronological history of some of the more prominent encounters.

The morning of March 16, 1967, was seminal toward the involvement of some of the witnesses who have since documented their stories to investigative research and the public. One of these, Robert Salas, was on-site at one of the Minuteman missile complexes on this day. The Echo-Flight complex was the first to experience the shutdown of all ten missiles at 8:30 a.m. Multiple eyewitnesses reported a single UFO stopping while airborne, directly over two silos. This occurrence was not a power failure; it was a complete shutdown of all guidance and control systems. The missiles were rendered inoperable and would remain so for over twenty-four hours.

About twenty miles southeast of the Echo-Flight complex, Robert Salas was the deputy crew commander present in one of the silos below ground of Oscar-Flight complex that day. The flurry of activity commensed when Salas began fielding multiple calls from personnel above reporting the activities of a fleet of UFOs over the silos.

Security personnel topside phoned Salas: "Sir, there's one hovering outside the front gate!"

Salas: "One what?"

Security: "A UFO! It's just sitting there. We're all just looking at it. What do you want us to do?"

Salas: "What? What does it look like?"

Security: "I can't really describe it. It's glowing red. What are we supposed to do?"

Salas: "Make sure the site is secure."

Security: "Sir, I have to go now, one of the guys just got injured."

Airmen in the same complex saw a "star begin to zigzag across the sky," which was followed by another. Then they got larger and closer. Other personnel from the Launch Control Center for Oscar-Flight came out and watched them "stop, change directions at high speed, and return overhead."[50]

The last craft sighted was described to be saucer shaped with a red glow. Salas confirmed that up to eight of Oscar-Flight's missiles had been defeated to a "No-Go" inoperable flight status. Subsequent investigations by the Air Force, Strategic Air Command, and contractors for the Minuteman missile projects, including the Boeing Company, concluded their only known possible explanation was that an Electro-Magnetic Pulse or EMP could have reacted the effects seen that day at Malmstrom. No technology humankind had at that time could have caused this event. It remains unexplained despite the Air Force public claim that there was no effect on national security.

The sightings at Malmstrom have not ended. Many observations of unexplained aerial craft have been described as follows: "This light would move at incredible speeds, make right-angle moves, and continue for hours."; "It (UFO) simply outran the helicopters. We heard that it zipped out to Belt and back to the base in no time."; "The 'helicopter' was tracked by radar equipment adjacent to and within visual sight of the craft to travel at approximately one thousand knots, far faster than any known helicopter."[51]

The dynamics of sightings at Malmstrom continued in a series of waves for decades. In October 1975, a new wave started. Reports of ships, "having a white light with a red flame behind and a green light on top then changing color arrangements," were the norm. Shapes of the craft varied somewhat but reported as either egg-shaped or triangular. One report on October 20 by civilians near Malmstrom described such an egg-shaped ship landing near their yard. The yellowish-gold craft then extended two appendages from its sides. The arms moved in a circular motion for over five minutes. They retracted back into the craft, and it lifted off and accelerated out of sight within a few seconds.[52]

November 1975 was a particularly eventful month for sighting encounters at Malmstrom. From the seventh through the ninth, the skies over northern Montana were as busy as ever, despite the fact that airspace overhead is highly restricted or forbidden by the government. Many radar equipment stations on military sites and civilian installations in the region tracked unidentified craft both in the air and on the ground at various times throught that week. On November 7, hundreds of military personnel saw craft with orange, red, and/or yellow glowing lights perform aerial and stationary maneuvers. Multiple reports indicated "an object that was tremendous in size. They were reviewing a brightly glowing orange, football-field-sized disc that illuminated the missile site." North American Aerospace Defense Command (NORAD) radar tracked the object until it landed, where some of the eyewitnesses observed the huge craft again take off and rise vertically and very swiftly to over two hundred thousand feet when NORAD tracking lost signature on it.[53]

Other reports from that first night recorded observations of an "object to their northeast [that] seems to be issuing a black object from it, tubular in shape." These encounters carried over into the next day, November 8, and prompted additional comments, such as,

"As the sun rose, the UFOs disappeared." Later that day and night, the sightings resumed. As many as nine at a time were observed intermittently accelerating very fast then slowing to less than 10 mph or stopping and hovering in midair altogether.

It was later on the eighth that base targeting teams discovered, investigated, and confirmed one of the missile sites had been penetrated by some instrumentality by the UFO to the extent the numbers on the target tape, designed to direct the launched missile to a predetermined target, had been changed. This meant the missile in question was reprogrammed to fly in a different direction and to a different target.[54]

Activity continued that day and into the next, November 9, when the fleet of UFOs engaged Air Force fighters and ground-based personnel in a game of cat and mouse. While a sortie of F-106 fighter jets pursued the unidentified ships, the pilots obtained visuals, and then suddenly, the UFO lights would shut off. All radar tracking was also lost. When the planes retreated, the UFOs returned to engage again. The exchange continued throughout the evening.

On November 11, Cascade County Sheriff Department Captain Keith Wolverton gave a firsthand account of incidents during October and November. On that night, he was on patrol when he and the deputy riding with him witnessed a craft with orange lights descending from out of nowhere and passing directly over the cruiser at only about two hundred feet. The ship accelerated away and off into the horizon within seconds.

Captain Wolverton was part of the nonmilitary community which became involved with another peripheral phenomenon that beset that region during those years. The emerging incidence of cattle disappearances and cattle mutilations was beginning to appear on the landscape. Local ranchers had their own encounters with UFOs, unexplained cattle mutilations of such a technologically precise nature not reproducible by human efforts and even encounters where the animals were physically raised into or dropped from hovering airborne ships. During 1975–1976 alone, over 130 reports of UFOs or unidentified helicopters were logged.[55]

Malmstrom has had sighting reports into the 1990s and even after 2010. Air Force declassified documents acknowledge continuing unexplained aerial activity and even disclose events of September 19, 2012, when a civilian, Jennifer Styer, reported UFO sightings just southeast of the Oscar-Flight missile complex at Malmstrom.[56]

"For many years, the Air Force has maintained that no reported UFO incident has ever affected national security. It is a fact that a large number of Air Force personnel reported sighting UFOs at the time many of our strategic missiles became impossible to launch. The incidents described above clearly had national security implications."

References

[1]   Ref 1: Hatch, Larry. 1998. NCP Report NCP-03: UFO Sightings and Nuclear Sites. NICAP, September 13, 1998. Retrieved October 20, 2016, from: http://www.nicap.org/ncp/ncp-hatch1.htm.

[2]   Ref 2: Johnson, Donald A. Do Nuclear Facilities Attract UFOs? Nuclear Connection Project. NCP-11. NICAP. Retrieved October 20, 2016, from: http://www.nicap.org/ncp/ncp-johnson1.htm.

[3]   Fireballs Ref 3: Wikipedia, Green Fireballs definition. Retrieved October 20, 2016, from: htpps://www.en.wikipedia.org/wiki/Green_fireballs.

[4]   Fireballs Ref 4: Ridge, Francis. 2005. The New Mexico Sightings: The Nuclear Connection Project. Retrieved October 20, 2016, from: http://www.nicap.org/nmexico/newmexicosightings.htm.

[5]   Fireballs Ref 5: Transcript of Los Alamos Conference. Retrieved October 20, 1969, from: http://www.project1947.com/gfb/cap21649.html.

[6]   Fireballs Ref 6: Ruppelt, Edward J. 1956. The Report on Unidentified Flying Objects, Chapter 4. Copyright 1956 by Edward J. Ruppelt.

[7]   Fireballs Ref 7: Condon, Edward U. 1968. Scientific Study of Unidentified Flying Objects, Chapter 2. University of Colorado. Published in 1969 by E.P. Dutton. ASIN B000CBOMXI.

[8]   Fireballs Ref 8: Hastings, Robert L. 2008. UFOs and Nukes, p. 64–84. Copyright 2008 by Robert L. Hastings. Published by Author House.

[9]   Oak Ridge Ref 9: For Your Information: A Guide to Oak Ridge (United States Engineering Department – Community Relations Section, September 1946), p. 3.

[10] Oak Ridge Ref 10: Ridge, Francis. 2005. The Oak Ridge Sightings: The Nuclear Connection Project. Retrieved October 20, 2016, from: http://www.nicap.org/oakridge/oakridgesightings.htm.

[11] Oak Ridge Ref 11: Ridge, Francis. 2005. The Oak Ridge Sightings: The Nuclear Connection Project. Retrieved October 20, 2016, from: http://www.nicap.org/oakridge/oakridgesightings.htm.

[12] Hanford Ref 12: Hastings, Robert. 2013. UFO Activity at the Hanford Plutonium Plant. Reprinted from the May 2013 MUFON UFO Journal. Retrieved October 20, 2016, from: http://www.mufon.com/ufos-and-nukes/ufo-activity -at-the-hanford-plutonium-plant.

[13] Hanford Ref 13: Powell, Rolan D., Varner, Byron D., Andrus, Walter. 1996. UFO Sighting over Nuclear Reactor 1945. Retrieved October 20, 2016, from: http://www.nicap.org/reports/hanford.htm.

[14] Hanford Ref 14: Dolan, Richard. Richard Dolan Blogsite. April 30, 2013. Retrieved October 20, 2016, from: http://richarddolan.tumblr.com/post/50498484338/dolan-chd-statement-3-ufos-and-nuclear-.

[15] Hanford Ref 15: Ruppelt, Edward J. 1956. The Report on Unidentified Flying Objects. Copyright 1956 by Edward J. Ruppelt.

[16] Hanford Ref 16: Dolan, Richard. Richard Dolan Blogsite. April 30, 2013. Retrieved October 20, 2016, from: http://richarddolan.tumblr.com/post/50498484338/dolan-chd-statement-3-ufos-and-nuclear-.

[17] Hanford Ref 17: Ruppelt, Edward J. 1956. The Report on Unidentified Flying Objects. Copyright 1956 by Edward J. Ruppelt.

[18] Ellsworth Ref 18: Condon, Edward U. 1968. Scientific Study of Unidentified Flying Objects, p. 212–213. University of Colorado. Published in 1969 by E.P. Dutton. ASIN B000CBOMXI.

[19] Ellsworth Ref 19: Navy Supplement to the DOD Dictionary of Military and Associated Terms (PDF). Department of the Navy. August 2006, pp. 2–25. NTRP 1-02.

[20] Ellsworth Ref 20: Ridge, Francis. The Rapid City/Ellsworth AFB Incident (RV). NICAP/RADCAT Case Directory. Retrieved October 20, 2016 from: http://www.nicap.org/530805ellsworth_dir.htm.

[21] Ellsworth Ref 21: Ruppelt, Edward J. 1953. The Ellsworth AFB Case (RV). Reproduced for NICAP. Retrieved October 20, 2016, from: http://www.nicap.org/reports/530805ellsworth_ruppelt.htm.

[22] Ellsworth Ref 22: Emenegger, Robert. 1978. *UFOs Past, Present, and Future.* Published as Fifth Printing July 1978 by Ballantine Books.

[23] Ellsworth Ref 23: Ellsworth UFO Document 1: Secret. PBB85 901-904 docs. pdf. Retrieved October 20, 2016, from: http://nicap.org/reports/530805ellsworth_rep3.htm.

[24] Ellsworth Ref 24: Hastings, Robert L. UFO sightings at ICBM sites and nuclear Weapons Storage Areas. Copyright 2006 by Robert L. Hastings. Retrieved October 20, 2016, from: http://www.nicap.org/babylon/missile_incidents.htm.

[25] Kirtland Ref 25: McDonald, Dr. James E. 1969. Science in Default. The American Association for the Advancement of Science, December 1969.

[26] Kirtland Ref 26: Ridge, Francis. 2006. The Kirtland UFO Incident; Albuquerque, New Mexico; November 4, 1957. Nuclear Connection Project, February 5, 2006. Retrieved October 20, 2016, from: http://www.nicap.org/kirtland57dir.htm.

[27] Kirtland Ref 27: Condon, Edward U. 1968. Scientific Study of Unidentified Flying Objects, p. 212–213. University of Colorado. Published in 1969 by E.P. Dutton. ASIN B000CBOMXI.

[28] Kirtland Ref 28: McDonald, Dr. James E. 1969. Science in Default. The American Association for the Advancement of Science, December 1969.

[29] Walker Ref 29: Condon, Edward U. 1968. Scientific Study of Unidentified Flying Objects, p. 212–213. University of Colorado. Published in 1969 by E.P. Dutton. ASIN B000CBOMXI.

[30] Walker Ref 30: Hensley, Nicole. 2015. USAF "Project Blue Book" details UFO reports in new archive. *New York Daily News*, January 18, 2015.

[31] Walker Ref 31: Hastings, Robert L. 2006. UFO sightings at ICBM sites and nuclear Weapons Storage Areas. Copyright 2006 by Robert L. Hastings. Retrieved October 20, 2016, from: www.nicap.org/babylon/missile_incidents.htm.

[32] Canaveral Ref 32: McClelland, Clark. 2006. UFO "Alters" Tracking of U.S. Rocket, Cape Canaveral, Florida, January 10, 1961. National Investigations Committee on Aerial Phenomena. Retrieved October 20, 2016, from: http://www.nicap.org/reports/610110canaveral.htm.

[33] Canaveral Ref 33: NICAP. 1962. UFO Blocks Tracking of U.S. Rocket. Retrieved October 20, 2016, from: http://www.nicap.org/reports/blovmiss1.htm.

[34] Rybinsk Ref 34: Ridge, Francis. Nuclear Connection Project. 2006. Missile Battery Fires at Disc, Summer 1961; UFO Stalls Electrical Apparatus Near Moscow. Retrieved October 20, 2016, from: http://www.nicap.org/babylon/61sumr_moscow.htm.

[35] Rybinsk Ref 35: Good, Timothy. 1989. Above Top Secret, p. 227. Published September 1989 by Quill.

[36] Warren Ref 36: Courtesy of the National Park Service, U.S. department of the Interior. Retrieved October 20, 2016, from: https://www.nps.gov/history/hps/abpp/battles/va119.htm.

[37] Warren Ref 37: F.E. Warren History. U.S. Air Force, F.E. Warren Air Force Base. Retrieved October 20, 2016, from: http://www.warren.af.mil/ and www.warren.af.mil/AboutUs/FactSheets/Display/tabid/3813/Article/331281/fe-warren-history.aspx.

[38] Warren Ref 38: Hastings, Robert L. 2006. UFO sightings at ICBM sites and nuclear Weapons Storage Areas. Copyright 2006 by Robert L. Hastings. Retrieved October 20, 2016, from: www.nicap.org/babylon/missile_incidents.htm.

[39] Warren Ref 39: Hastings, Robert L. 2006. UFO sightings at ICBM sites and nuclear Weapons Storage Areas. Copyright 2006 by Robert L. Hastings. Retrieved October 20, 2016, from: www.nicap.org/babylon/missile_incidents.htm.

[40] Warren Ref 40: Hastings, Robert L. 2006. UFO sightings at ICBM sites and nuclear Weapons Storage Areas. Copyright 2006 by Robert L. Hastings. Retrieved October 20, 2016, from: www.nicap.org/babylon/missile_incidents.htm.

[41] Warren Ref 41: Hynek, J. Allen. 1972. The UFO Experience: A Scientific Inquiry. Published in 1972 by Marlowe. ISBN 156924782X, 9781569247822.

[42] Warren Ref 42: Speigel, Lee. 2011. UFO Researcher Claims Air Force Not Revealing Truth about Communication Outage at F.E. Warren Missile Site. Huffington Post, November 15, 2011. Retrieved October 20, 2016, from: http:/www.huffingtonpost.com/2011/07/07/ufos-at-nuclear-missile-site-site-eyewitnesses-afraid-to-talk_n_881802.html.

[43] Warren Ref 43: April 15, 1966, F.E. Warren AFB, WY, UFO over missile silo. NICAP Report October 3, 2006. Retrieved October 20, 2016, from: http://www.nicap.org/fewarren660415dir.htm.

[44] Warren Ref 44: Hastings, Robert L. 2006. UFO sightings at ICBM sites and nuclear Weapons Storage Areas. Copyright 2006 by Robert L. Hastings. Retrieved October 20, 2016, from: www.nicap.org/babylon/missile_incidents.htm.

[45] Warren Ref 45: April 15, 1966, F.E. Warren AFB, WY, UFO over missile silo. NICAP Report October 3, 2006. Retrieved October 20, 2016, from: http://www.nicap.org/fewarren660415dir.htm.

[46] Minot Ref 46: "Air Force Global Strike Command (USAF)." Air Force Historical Research Agency. July 17, 2009.

[47] Minot Ref 47: The Nuclear Connection Project. 1975. Missile Base Incidents. Product of NICAP. Retrieved October 20, 2016, from: http://www.nicap.org/babylon/missiles.htm.

[48] Minot Ref 48: The Minot AFB UFO Case. Physics Forums. Linked from: www.minotb52ufo.com/introduction.php. Retrieved October 20, 2016, from: www.physicsforums.com/threads/ufo-study-at-minot-afb-october-68.517042.

[49] Minot Ref 49: The Minot AFB UFO Case. Physics Forums. Linked from: www.minotb52ufo.com/introduction.php. Retrieved October 20, 2016, from: www.physicsforums.com/threads/ufo-study-at-minot-afb-october-68.517042.

[50] Malmstrom Ref 50: Salas, Robert and Klotz, Jim. 1999. The Malmstrom AFB UFO/Missile Incident. Copyright 1999 by Robert Salas and Jim Klotz. Retrieved October 20, 2016, from: http://www.cufon.org/cufon/malmstrom/malm1.htm.

[51] Malmstrom Ref 51: Hastings, Robert L. 2006. UFO Sightings at ICBM sites and nuclear Weapons Storage Areas. Copyright 2006 by Robert L. Hastings.

Retrieved October 20, 2016, from: http://www.nicap.org/babylon/missile_incidents.htm.

52 Malmstrom Ref 52: Fawcett, Larry and Greenwood, Barry. 2007. "Faded Giant?" – Intrusions at Malmstrom – 1975. NICAP 97 Initial Report. Retrieved October 20, 2016, from: http://www.nicap.org/articles/CI-Malmstrom.htm.

53 Malmstrom Ref 53: Fawcett, Larry and Greenwood, Barry. 2007. "Faded Giant?" – Intrusions at Malmstrom – 1975. NICAP 97 Initial Report. Retrieved October 20, 2016, from: http://www.nicap.org/articles/CI-Malmstrom.htm.

54 Malmstrom Ref 54: Moulton-Howe, Linda. 1980. A Strange Harvest: Thoughts beyond the Scenes. Denver Magazine, September 1980.

55 Malmstrom Ref 55: Fawcett, Larry and Greenwood, Barry. 2007. "Faded Giant?" – Intrusions at Malmstrom – 1975. NICAP 97 Initial Report. Retrieved October 20, 2016, from: http://www.nicap.org/articles/CI-Malmstrom.htm.

56 Malmstrom Ref 56: Hastings, Robert L. November 4, 2012. UFOs Reported Near Malmstrom AFB's Nuclear Sites in September 2012. Retrieved October 20, 2016, from: http://www.ufohastings.com/articles/ufos-reported-near-malmstrom-afbs-nuclear-missile-silo.

CHAPTER 19

What a Lot of People Are Seeing Plus a Lot More

The idea of benign or hostile space aliens from other planets visiting the earth [is clearly] an emotional idea. There are two sorts of self-deception here: either accepting the idea of extraterrestrial visitation by space aliens in the face of very meager evidence because we want it to be true or rejecting such an idea out of hand, in the absence of sufficient evidence, because we don't want it to be true. Each of these extremes is a serious impediment to the study of UFOs.

—Carl Sagan from a 1969 lecture[1]

Key words/Terms: hypnagogia, trace evidence, vitrified, amnesia/ missing time

Science is currently unable to address issues of nature or inquiry beyond those that occur in three dimensions for the most part. As we have seen elsewhere, answers to questions concerning matters involving the fourth and additional dimensions, the ones Einstein and other theoretical physicists and cosmologists have been hard at work researching for centuries, are very elusive and generate even more questions for each one question that is answered. A large part of this example and more generally any such matter of science inquiry requires inventing new measuring instruments and other technologies to be able to address the questions being asked. In the theoretical physics field, the CERN Hadron Collider in Switzerland needed to be invented and built before any such research could be undertaken.

J. Allen Hynek created the Scale of UFO Classification as noted in the chapter titled "Who Has Seen Them?" This chapter discusses the situations that, according to the Hynek Scale, are determined to be "Close Encounters of the Fourth Kind" or CE-4s: a person(s) claims to have been taken aboard a non-earthly craft in the presence of otherworldly occupants.

Kelly Cahill's (and Others) Experience

The Melbourne, Australia, suburb of Narre Warren North was the venue of only one episode but was experienced by four known separate groups of families. The conclusions offered from the investigation produced a heightened level of validity thanks to the Phenomena Research Australia organization. This is because interviewing and investigative interventions conducted on each of the four families was conducted independent of any of the other families. While, as we will see, the families did observe other people and their cars at the site that night and they had no contact with them that would bias any type of valid investigation. This is one advantage research teams have when forming an investigation.

On August 7, 1993, the first event started for one of the groups: twenty-seven-year-old Kelly Cahill and her husband and three children. About 7:00 p.m., they were driving to Narre Warren North, in the foothills of the Dandenong Mountain Range, Victoria to attend one of Kelly's friend's daughter's birthday celebration. They sighted a "ring of orange lights" in the air over a field of a flat fielded highway. There was a brief discussion about a UFO sighting, but it was dropped. The one-way trip to Kelly's friend's house took one and a half hours.

The visit involved a trip to play bingo which ended before 11:00 p.m. They returned to the girlfriend's house for only a short time due to the friend's conflict with an ex-boyfriend. So the Cahill's started their return trip home. Shortly after midnight on August 8, about halfway home, they both noticed a "round shape with some sort of glass around or what looked like windows and lights around the bottom." As they got closer, they noticed there was no sound coming from the craft. Then suddenly, "it just shot off to the left as fast as it could go." A kilometer ahead, they then noticed a "really bright light in front of us." That was the last they remembered about the experience until weeks later.

When they returned home, a few peculiarities unfurled. First, Kelly noticed a triangular mark on the skin of her navel. Second, they all noticed an intense odor of vomit. Thirdly, they noticed their arrival time was about two-thirty. That meant the trip home would have lasted over three hours or twice as long as normal.

Over the next few weeks, both Cahills (not the children) complained of intensely-growing stomach pains. Three and a half weeks later, Kelly was hospitalized. The doctor diagnosed Kelly as having been hemorrhaging for a few weeks (she had her menstrual cycle reset one week before the experience). He told her, "You must have been pregnant. The only way you can get an infection in the womb is if you've been pregnant or it's been caused by an operation." Kelly returned to the hospital six weeks after and had a laparoscopy performed. The incision was made next to the triangular mark which was still present.[2]

It was only weeks after when the Cahill's again visited the same friend and took the same route. When they approached the site of the August 8 encounter, Kelly began remembering, and the entire episode was pieced together gradually in sprints: "What we had actually done, we had driven into the light, but the road curved, and the light was actually to our right. It was in the field, and it was massive. There was orange lights and this blue stuff underneath." The diameter of the ship was about 150 feet. It had landed about one hundred yards in the field.

Kelly continued to recall without the assistance of any hypnosis or other intervention: "We crossed the road and walked up. There was another car—a light blue car—pulled up. I'm standing there, conscious, and we are looking at this thing. All of a sudden, there is a black figure on the field about seven feet (sic) tall." She continued, "Its eyes seemed to turn to a red fire. I started screaming to my husband, 'They've got no souls.'"[3]

Then there were "heaps of them" in the field. They broke off into groups and approached all the parties that had pulled up to the site, three automobiles in total. "The next thing I know, I felt this umph! In my stomach, I was thrown right back. I sat up with my head between my knees." She then felt nauseous. This may explain the vomit odor sensed later. Kelly then attacked one of the beings, described as a female that was interacting with her husband. The leader placed a hand on her shoulder and led her back to the ship. She later recalled feeling a sucking sensation and a twinge on her naval. When she looked over, she saw the same tall being dressed in a floor-length hooded cloak.

In the time after the experience, Phenomena Research Australia started an investigation. The project lasted a few years but conducted interviews, ran tests, and collected drawings and images from all four parties involved. This data included drawings from all experiencers.[4] This was done without having any knowledge of the others, unless one of them had observed another. This did occur when Kelly sighted one other car "a hundred yards up the road." Another driver, Jane, along with her husband Bill and a female companion, also observed the Cahill car at the same distance down the road. These versions correlated in their description. A third group of experiencers, Glenda and her husband, an employee of the Victorian government law department, corroborated the facts through their own independent descriptions. It was the car owned by Glenda and her husband that had its headlights on and shining on the car owned by Jane's group. The Cahill's had seen Glenda's car headlights on and not Jane's car because her car headlights were off. Jane explained they turned their car off and exited the vehicle like the Cahill's had. Jane and Bill and Glenda corroborated their own experiences. Bill was violently sick, and Jane and Glenda described medical examination procedures being conducted on them. The other females reported triangular marks on their abdomens similar to that of Kelly Cahill. In summary, it was the females that the visitors were interested in and not the males.[5,6]

All the women reported the same sequence of events, the same descriptions of the artifacts and the same descriptions of the UFO and their experiences with the entities. Those who experienced it also described what they observed of the sequence of their fellow recipient's interactions with the beings. Phenomena Research Australia also claimed that findings from two independent research laboratories confirmed many physical and chemical anomalies present at the site. These included alteration in the soil chemistry, such as above-average sulfur content, pyrene (a derivative of coal) and tannic acid, and a triangular patch of dead vegetation on the ground at the location where the witnesses said the craft had landed. Further, a fourth family reported, under protection of confidentiality, their own encounter at the site that night.[7]

The Cape Girardeau Experience

The Roswell Incident is the single case more people have heard and know at least some details about than any other case in history. There was one, with a Hynek Classification Scale rating of CE-3, which allegedly occurred in 1941, six years before Roswell.

Cape Girardeau County was the location of a retrieval effort that involved the Army, local sheriffs, fire department, press, and a Baptist minister. Near the city of Jackson in September 1941, Reverend William Huffman, pastor of the Red Star Baptist Church in Cape Girardeau County, received a telephone call from local police at about 9:00 p.m. to accompany them to a crash scene, "possibly that of an airplane." When Reverend Huffman arrived with the police officers, there were already other police, fire, the FBI, and photographers on the scene. Army personnel would arrive soon after Huffman did. Among the officials at the crash scene were Cape Girardeau County Sheriff Ruben Schade, Chief of Police Marshall F. Morton, Cape Girardeau Fire Chief Carl J. Lewis, Jackson city coroner Norval B. Short, unknown FBI agents, and Army personnel.[8,9]

According to granddaughter Charlette Mann, Reverend Huffman returned after the meeting and told his wife and children, including Charlette Mann's father, the details of what had happened. "It was not a conventional aircraft, as we know it. It was a disc-shaped object, broken open in one portion, the interior contained a small metal chair, gauges, dials, and hieroglyphic-like inscriptions and writings around the inside. It [the saucer] was metallic in color, no seams. What impressed him the most were the (hieroglyphics). Then he saw three entities or nonhuman people, lying on the ground. Two were just outside the craft, a third somewhat farther away."[10,11]

Charlette Mann continued, "He described the bodies as hairless with large heads, big eyes, small mouths, and very small ears about 4 feet tall, no hard bone structure but very long arms and fingers and wearing a suit that appeared like soft aluminum foil." There were many people at the scene by that time. Reverend Huffman was asked to administer last rights to them, which he did. Then two men lifted one of the entities up by each shoulder thus propping the entity up and took a series of photographs.

Just then, the Army detachment arrived on the scene and surrounded the area. Reverend Huffman was instructed by Army personnel, "This didn't happen, you didn't see this. This is national security and is to never be talked about again." He finished his duty then returned home.[12]

About two weeks later, a member of Reverend Huffman's church, Garland D. Fronabarger, gave him one of the photographs of the dead entity being propped up by two of the recovery team. The photograph was later given to Reverend Huffman's son, Guy Huffman, Charlette Mann's father. He later offered the photo as proof of the encounter to a Walter Wayne Fisk. Fisk is still said to have possession of the photograph but has declined all attempts to be contacted.[13,14]

Once the identities of some of the recovery team on-site that night were discovered, researchers then tracked down log records from the Cape Girardeau City Fire Department and Cape Girardeau Sheriff's Department. These records confirm the activities of that night involving a crash and recovery of wreckage. The brother of Sheriff Ruben Schade, Clarence, also confirmed the activities of that night.

There have been investigative attempts to link some of the known events of the Cape Girardeau experience to other recovered and released documents from the U.S. government regarding ufology history. For example, the *Majic Eyes Only-Mission Assessment of*

Recovered Lenticular Aerodyne Objects-White Hot Intelligence Estimate presidential briefing dated September 24, 1947, references a conjectured link to the Cape Girardeau experience. Other government declassified documents show redacted passages and conversations, but the individual document's references and attributions appear to link the government's knowledge to that of the Cape Girardeau case.[15,16]

The Linda Napolitano Experience

Research and investigation literature identifies Linda Napolitano as using the alias Linda Cortile. As the nature of this case has been lengthy and continues to add information, it is further exemplified by the expanded list of independent eyewitnesses in corroboration with each other and the alleged events of Thursday, November 30, 1989. The list includes a Roman Catholic Cardinal, a then-secretary general of the United Nations, and a famous newspaper reporter, among others.

At about 3:15 a.m., Linda Cortile participated in an encounter with one then four more "small figures." After finishing the family laundry, she retired to bed in her lower Manhattan apartment where her husband, Steve, was already asleep. A short time after, she began to "feel a disturbingly familiar sense of numbness moving from her feet up her body. She sensed a presence in the room." When she was unsuccessful in waking her husband, she turned and spotted a "large-headed but small creature with huge black eyes approach the bed." The paralyzing effect had not progressed further.

So in a panic, Linda Cortile grabbed a heavy decorative pillow and threw it at the figure and immediately after became entirely paralyzed. She then remembered that four different beings approached the bed and moved her first into the living room then outside through a closed window and into a bluish-white beam of light. As she recounted her story, Linda remembered standing vertically, although in midair outside her twelfth story apartment. The five then floated in an upward direction through a circular opening and into a large craft hovering stationery above the building.

Cortile also remembered bits of a "physical examination on a table, including the soft, methodical tapping of tiny alien hands along her vertebra." She also remembers the later return to her apartment. She felt herself "being dropped on her bed from what seemed to be a foot or two above it." She tried to wake her husband but could not. She then raced to her sons', Johnny and Steven's, room and with a mirror placed underneath, discovered the telltale "fogging" of the mirror from atop Johnny's nose. Then she noticed her other son, Steven, was now breathing audibly and could hear her husband snoring in the other room.

She did not know what to make of this encounter. Cortile was frightened and had thought she was dreaming the whole thing. But she felt far worse than what would be experienced following a nightmare. She did not, at the time, have knowledge of a type of sleep onset condition called *hypnagogia*, the experience of the transitional state from wakefulness to sleep; the hypnagogic state of consciousness during the onset of sleep. Mental phenomena that occur during this threshold consciousness" phase include lucid thought, lucid dreaming, hallucinations, and sleep paralysis.[17]

The next day, she contacted Budd Hopkins, noted psychology, and ufology researcher. Cortile knew of Hopkins's alleged past history of experiences with extraterrestrials. Linda was asked to and joined Hopkins's support group in later months. The process of collecting facts and data would be a slow, arduous one. Hopkins would conduct six regression sessions with Linda and conduct dozens of interviews, taped recordings, and collected written correspondence and signed statements from a witness pool that would increase to over twenty-two.

The first of these witness contacts came in a letter sent to Hopkins postmarked February 1, 1991. It was a detailed letter from Richard and Dan, inquiring about the status of the experiencer. Excerpts from the letter contained,

> I looked up through the windshield, a strange oval hovering over the top of an apartment building two to three blocks away. Its lights

burned from a bright reddish orange to a very bright whitish blue, coming from the bottom of it. She was floating in midair in a bright beam of whitish blue light. She was then brought up into the bottom of that very large oval. After she was escorted up and in, the oval turned reddish orange again and whisked away, above us. It then plunged into the river behind us, not far from Pier 17, behind the Brooklyn Bridge.

Richard and Dan, in addition to a third man in the car, not originally identified, also saw a domed-shaped structure on top of the main hull of the oval ship.[18,19,20]

Richard and Dan were the first two witnesses to become known. They were driving a limousine near the Brooklyn Bridge that morning when they observed the incident. The identity of the passenger would himself make contact with Hopkins a few months later and confirm the events of that morning. In a letter to Hopkins dated December 23, 1991, then active and sitting secretary-general of the United Nations, Javier Perez de Cuellar, was that person. He wished at the time to have the contacts with Hopkins remain confidential, to which Hopkins agreed. More significantly, de Cuellar was a participant in Linda's abduction, as well as a later abduction of Linda and her son, Johnny.[21]

Why were Richard and Dan associated with de Cuellar? Because they were later discovered to have been CIA agents on assignment as de Cuellar's driver. Their car had died at the Brooklyn Bridge that night, as did the car of another witness. Both of them were very sympathetic and pursued contact with Linda for a while. But it became known to Hopkins, Linda, and her family that the experience affected Dan much more severely than Richard. He became psychotic as a result and had episodes of stalking Linda in the years that followed.

Two witnesses from the *New York Post*, worker Yancy Spence, and investigative reporter Steve Dunleavy, and a third worker "Bobby" (a pseudonym) met with Hopkins much later and issued reports to him regarding "the procession of limousines stopped on South Street along with the Rolls Royce." Their sighting sequence and subsequent drawings corroborated the reports from all other twenty-one witnesses' to date.

Then incumbent Catholic Cardinal John O' Connor confirmed his witness status later in 1993. He maintained a friendship and became a confidant to Linda Cortile. Another woman, Janet Kimball, was the code name for a firsthand witness whose car also died (along with the limousine carrying UN Secretary-General de Cuellar, Richard and Dan) on the Brooklyn Bridge. Janet observed the ship over Linda Cortile's apartment building and the abduction. "Erica," code name for a firsthand witness who lived in Linda Cortile's apartment building, was also concluded to have been abducted on November 30, 1989, and witnessed the actual incident of Linda Cortile.[22]

Linda's husband, Steve, and her two sons, Johnny and Steven, were all alleged experiencers in a separate intervention on Sunday, May 24, 1992. In that encounter, the four Cortile's and Brian, a friend of Steven's, all awoke simultaneously with incidents of "bleeding from their right nostrils." Later regressions confirmed that Linda, Johnny, Richard (the limo driver-CIA agent), and another individual named Melody experienced an intervention. Dr. Lisa Bayer, podiatric surgeon, performed an examination and x-rays of Linda Cortile's head on November 12, 1991. The x-rays showed an object with curled metallic flanges protruding from it.[23,24]

A woman named Marilyn Kilmer reported she had a similar experience and communicated with Linda, son Johnny, and the "third man" (de Cuellar) during an intervention after the November 30, 1989, incident. Another by the name of Cathy Turner was a firsthand witness to the encounter while driving over the elevated FDR Drive near the Brooklyn Bridge. Yet another Francesca was a firsthand witness who lived in the same apartment building as Linda Cortile. It was reported that, "She was awoken in the middle of the night in her apartment, and then all of a sudden, everything lit up outside like it was daylight."[25]

The Betty and Barney Hill Experience

As is typical of all case studies in this chapter, many books have been written analyzing facts about the histories. In fact, for most but not all these cases, movies depicting similar events have been made to create entertainment for its viewers. I say create because movies about ufology do not translate accurately to facts and data. Their ultimate production and portrayal are under the control of the movie studios sponsoring the project. The studio's prime directive is to make a profit. Often, the real persons who are the principle subjects of the movie are directly told their movie is being significantly altered to fit the profit paradigm more than the canon of certainty. The storyline that attracts the most viewers and the most money is the only consideration of the studio. Therefore, it is most often correct to reflect on the written book when accuracy of information is the objective and to be entertained by the movie and not much else.

As mentioned earlier, very few movies such as this have been produced other than the film based on the case involving Betty and Barney Hill. The modern era of ufology is entering its seventieth year. This incident may rank second in popularity only to that of the Roswell incident.

Originally known as the *Hill Abduction*, this allegedly took place back on September 19, 1961. The couple was returning home from a late summer vacation via Niagara Falls and Montreal, Quebec, Canada, and back to their home in Portsmouth, New Hampshire. As with a majority of experience cases, this developed in the midnight hours.

About 10:30 p.m., while driving on U.S. Route 3 in middle New Hampshire, Betty first spotted a bright point of light moving upward through the sky through the backdrop of stars that included Jupiter and the moon. The light was moving upward and erratically and was growing bigger. They stopped and using binoculars Betty first noticed an odd-shaped object flashing multicolored lights fly across the moon then continue to get larger. The craft was now getting a lot closer. This may have been attracted by the headlights of the Hill's auto. Barney realized, "It was a very funny feeling seeing this large thing coming to land, and I thought of that when I was standing on the highway that this thing is coming right toward us and I said, 'Come on, Betty, let's go.'"[26]

A little further down the road, the craft came right down and hovered silently and stationery on the highway about eighty feet above and three hundred feet in front of the Hill automobile. The ship was sixty to eighty feet in diameter; it filled the entire windshield. Through the array of windows on the ship were from eight to twelve humanoid figures with glossy black uniforms and blackcaps looking down at Betty and Barney. After the ship tried to maneuver in front of the car, Barney and Betty turned around and sped away. The ship approached immediately, and they heard beeping and buzzing sounds and electrical vibrations through the car. The Hills later reported they entered an altered state of consciousness and that they had no control over independent movement. It was only after they recalled a second repetition of these beeping, buzzing sounds and electrical vibrations that they regained full consciousness and then discovered they were thirty-five miles south and driving on the highway once again.[27]

They were silent and returned home that morning only to find that they could not account for three hours of time. The next day, Betty placed her damaged clothing into her closet. Her dress was torn and had pink powder stains on it. She later hung it on an outdoor clothesline, and the pink powder blew off. Barney inspected the car. There were concentric circle rings emblazoned into the trunk of their car. Barney tested the auto with a compass and found that it rotated more and more rapidly when placed closer to a series of circles. This had meant there had been an electromagnetic disturbance causing atomic activity within the metallic structure of the trunk, which had not been present the day before.

Betty and Barney both had recurring dreams while sleeping and awake during the months that followed. Slowly, the events started to take on a dynamic and organized format. The Hills endured interviews from NICAP investigators, return trips to the event theater to try and evoke more details about that night, and serious contemplation about undergoing hypnosis. On November 23, 1962, they met

Air Force Captain Ben H. Swett and discussed their thoughts. Again, Swett suggested hypnosis.

The Hills first discussed their experience with the public at their church on March 3, 1963. They made contact with an amateur UFO study group later that November. On December 14, 1963, they first met with psychiatrist Dr. Benjamin Simon of Harvard University. Finally, on January 4, 1964, the hypnotism sessions began. What was revealed was to become the substance of one of the more seminal studies and contribution of information to the extraterrestrial phenomenon since the modern era of ufology began.

Barney Hill was first to undergo the testing intervention. The car had stalled, and three men approached the car leaving three others behind them. He was told not to be afraid that they would not be hurt. Here were some of his statements: "I felt like the eyes had pushed into my eyes. All I see are these eyes. I'm not even afraid that they're not connected to a body. They're just there."[28]

They were separated once on board the ship. Barney and Betty also reported that the communication was by telepathic method. He had semen extracted and his skin scraped. A tube was inserted into his nose and the examiners inspected his back, vertebrae, ears, and mouth. A bit later, the examiners burst into the room where the humanoid "leader" was tending to Betty. They had discovered the dentures Barney had and were entirely confused by their presence.

Barney did consciously remember being escorted from the ship to his car. Back together with Betty, they watched the ship leave and then experienced the second round of buzzing noted earlier. They then regained consciousness at that point, though they would not discover and assimilate until later the missing time they experienced.

Betty had more traumatic hypnotic sessions than Barney, though both sets were traumatic for each of them. One of Betty's sessions was ended early due to her emotional distress. Betty had more communication with the leader than Barney. When she was being examined, a needlelike tube was inserted into her naval. This was painful and she asked it be removed. The leader directed the examiners to do so and it was. She later asked for a book to serve as proof that she was visited by them. The leader gave it to her, though later. After a heated discussion among the workers, the leader took the book back from Betty.

Also Betty asked the leader where they were from. The leader caused a star map image to emerge from the wall. Betty provided the following description: "The oblong map with dots scattered all over it. Some were little, just as pinpoints, and others were as large as a nickel. There were curved lines going from one dot to another. Then there was one large circle with several heavy, solid lines that connected it to another slightly smaller circle."

She asked him where his home port was located on the map. The leader said, "Where are you on this map?"

She laughed and said, "I don't know."

The leader said, "If you don't know where you are, there wouldn't be any point in my telling you where I am." He then put the map away.

Betty also recalled the incident with Barney's dentures. Her examiner went into the hall, where Betty had overheard Barney's examiners conferring. She asked the Leader if there was a problem with Barney. One of his examiners came in holding Barney's dentures in his hand. Betty's examiner returned and inspected Betty's mouth, pulling at her teeth to see if they came out. She explained what dentures are, what they are used for, and that not all people had them.[29]

Dr. Benjamin Simon summarized the study in separate statement reports. The first dated March 8, 1965, indicated that Simon "seemed willing to accept the idea that the Hills's UFO sighting was real." However, some of Simon's case details leaked out to the public. Quotes about "Freudian tendencies" and initial inconsistencies pertaining to Betty's descriptions, such as her description of the landing walkway, weighed on his final determination of the abduction aspect of the experience. He maintained somewhat of an opinion that this determination was not entirely conclusive, though, "he told them that anything was possible." Simon stressed the more evidence of a physical nature that could be obtained, the more support for a positive. They eventually found the exact landing spot that also corroborated results from Dr. Simon's hypnosis interventions. This was more evidence that would sway Dr. Simon's opinion that the abduction aspect of the experience was also indeed real.[30]

From facts and data corroborated by the Mount Washington Observatory, weather conditions that night were reported as, "The evening and nighttime conditions of the nineteenth were quite tranquil. Visibility was 130 miles throughout the night." An intermittent presence of thin cirrus clouds did not affect visibility nor did the weather conditions propagate or cause any of the known manifestations often used as explanations such as weather inversions, gasses of any kind, or other geophysical phenomena.[31] The Air Force did corroborate radar tracking of an unidentified aerial object on that night at the time and in the area the alleged Hill experience took place. Two separate installations, Pease in New Hampshire and North Concord in Vermont, confirmed the vector tracking and flight characteristics as concurring with what the Hill's had reported.[32]

One aspect of evidence not discussed here but explored in the chapter "Physical Evidence of UFOs and Aliens Abound" pertains to the star map shown by the leader showed Betty. She remembered the details of the map's contents during her hypnotic sessions. The accuracy of these recollections was tested by later recall sessions. Teacher and amateur astronomer Marjorie Fish first learned about the star map in the year 1968. Intrigued, she contacted Betty Hill, and they began a friendship, and Fish began a five-year investigation of the star map. After a thorough scientific investigation and years of peer review, it was Marjorie Fish who determined the leader's home was in the Zeta-Reticuli star system and was credited with such a discovery. Zeta-Reticuli was unknown at the time of the abduction. This was Betty Hill's discovery as much as Marjorie Fish.

The University of New Hampshire library houses the Betty and Barney Hill Papers. These contain correspondence, personal journals, essays, manuscripts, photographs, slides, films, and audio tapes about both the Hill experience history and the ufology subject collected by them during their lifetimes.[33]

The Peter Khoury Experiences

Numerous types of physical evidence exist and have been studied in various alien encounter investigations. The existence of this evidence runs contrary to the comments of some skeptical inquirers. One type is a significant collection of *trace evidence*—tangible artifacts that can be tested under the parameters of inquiry via scientific method. Another type is that of various forms of electronic investigation from use of instrumentation technology to obtain chemical, geophysical, or spectral data with which to analyze and further test.

A third form of evidence that, from an inspection of its paradigm and meaning, would fall within a marginal acknowledgement from skeptics yet has immense potential as a main source of incontrovertible evidence of the extraterrestrial (and UFO) hypotheses. This evidence, more specifically, would fall under the heading of biological sciences.

This is the first case study in which the attributes of the right time and the right circumstance coincide with regard to trace evidence. A man named Peter Khoury migrated to Sydney, Australia, from Lebanon at the age of nineteen in 1973. Khoury was employed in the building trades. He eventually married his wife Vivian in 1990. A few years later, in 1993, he became founder of a UFO community support group called "UFO Experience Support Association" (UFOESA). He continues to be its coordinator today. Here is his reason for perseverance and longevity to this project as well as the connection to the topic of biological evidence.

Khoury is a multiple experiencer, although he did not become aware of this for some time. The first (eventually, Khoury came to know this was not his first) experience was brief and inconsequential. In February 1988, Vivian and he had a sighting of a group of "unusual moving lights." They confirmed their encounter and then moved on.

Later, on July 12, Khoury experienced a much more profound incident. Early that morning while in bed, he felt "something grab my ankles." He remained conscious but felt a "strange numbness, tingling and churning sensation" moving up his body to his head.

He was able to see four small hooded entities alongside him. Their faces were wrinkled and shiny black in color. On the other side of the bed were two thin tall figures with gold-yellow-colored skin. Khoury was told telepathically not to be afraid because "it would be like last time." He then saw them with a long needle and passed out.

The next morning, he and fiancée Vivian noted a puncture mark and traces of dried blood on the side of his head.[34]

The next and more profound experience came to Khoury on July 23, 1992. At seven o'clock that morning, he had just returned from dropping wife Vivian off at work and went back to bed. About thirty minutes later, he awoke and sat up. Two human-looking females were sitting in front of him, both completely naked. They looked mostly human except that their eyes were somewhat larger than typical. One with the appearance of an Asian heritage had black hair, dark skin and eyes, and appeared average height. The other appeared Scandinavian with very long blond hair, light skin, a long face, high cheekbones, narrow chin, and was very tall. Except for the shape of her head and an "exotic" hairstyle, she was human looking in other respects. They both appeared to be age thirty or so.

Khoury soon gained an impression they were not entirely human. The Asian-looking female was watching the other one throughout the encounter. The blond female then cupped both hands around his neck and drew his face into her breasts. He resisted and this repeated three times. Khoury noted the "stiff and blank, clinical" characteristics exhibited by both females that confirmed earlier thoughts about his doubts of their humanity. Khoury then bit the blond's nipple, and a piece of it was severed, and he swallowed it.

He noticed that she did not cry out in pain. Rather, she gave an impression when she looked at the Asian, then back at Khoury, that this "was not supposed to be the way it happens." Khoury got out of bed and started a coughing episode from the tissue being swallowed. He went to the bathroom and drank a glass of water. When he returned, they were gone.

A few minutes later, Khoury returned to the bathroom to satisfy the urge to relieve himself. He then noted that his penis began to feel very painful. He inspected and found two thin long strands of blond hair wrapped tightly around. As he removed them, the pain turned into an intense burning sensation. Khoury successfully extracted the two hairs and sealed them in a plastic bag. Later on in his recounting when discussing the sequence of events and the retrieval and saving of the samples, he explained, "There was no way that hairs that size (over four inches long) and wrapped around the way it was should have been there, and thinking of something bizarre had just happened."[35]

Khoury called Vivian at her place of employment thirty minutes after the encounter. He mentioned the event but said he wanted to discuss (the event) with her. He was still in a coughing fit. This lasted for three days despite all his efforts (he did not seek medical attention for this) to extinguish it. On the fourth day, the coughing spells disappeared. He did not tell Vivian until August 14. After their discussion and their own investigation, they both determined it was not a dream nor that Vivian was involved. The hairs were not identifiable as hers as they are different in appearance and color and that they did not engage sexually before the encounter.

Peter Khoury and wife Vivian had additional experiences in later years. In 1994 and 1995, two separate incidents occurred while both were in bed. On one occasion, bursts of light and explosions were seen and heard by both of them. On the other occasion, Peter felt a sensation of a pins and needles paralytic progression form not unlike when blood is returning to the arteries of your arm or leg after they have been deprived of it for some length of time. Only after some difficulty was he able to touch Vivian to wake her. As soon as he woke her, the feeling of paralysis disappeared.

Peter nor Vivian had any additional experiences in the years following the last episode. In 1996, Dr. John E. Mack, Harvard psychiatrist and ufologist, traveled to Sydney to meet with Khoury in order to clarify his 1988 experience. Under hypnosis, Khoury was able to describe being taken into a brightly-lit room after his last conscious recollection of the moments before the entities were ready to insert the needle into the side of his head.

He was on a table, and one of the entities was speaking to him. The entity sounded like "fifty birds chirping." During that time in the session, Peter was thinking, *How am I going to remember what you are telling me?* His thoughts became entangled in the regression, and Mack eventually terminated it. A couple of other sessions were attempted, with Mack attending, but no additional significant data was discovered.[36]

Peter and Vivian Khoury had not experienced any other events since 1995. Nevertheless, this case study does not end at that point. Thanks to the efforts of scientist and researcher Bill Chalker, a research group called the *Anomaly Physical Evidence Group* (APEG) was created shortly after 1996. The team of genetic and biological scientists was formed to study and experiment with the hair samples Chalker learned that Khoury had saved. The field of genetic testing was emerging in the mid-1990s, and they were then capable of producing analytic and meaningful results from then new types of testing techniques. A method first used back then included DNA Polymerase Chain Reaction (PCR) analysis.

A more thorough discussion of this history can be found in the chapter titled "Physical Evidence of UFOs and Aliens Aboun."

Pascagula, Mississippi, October 11, 1973

The evening of Thursday, October 11, 1973, was a typically benign evening for shipyard foreman Charles Hickson and his fishing companion, Calvin Parker. They were recreating near their place of employment, Walker Shipyards. The weather was similarly benign with clear skies above the town of Pascagoula, a coastal town about twenty miles east of Biloxi, Mississippi.

A little after 7:30 p.m., the men allegedly heard "a whirring/whizzing sound" then saw two flashing blue lights coming from an oval-shaped craft about ten feet round and eight feet or more in height. The ship stopped twenty-five yards from them and hovered motionlessly a couple of feet off the ground.

Hickson gave these details in all his interviews:

> A little buzzin' sound—it didn't hit the
> ground. It hovered. And all of a sudden, right in

the end of it, this opening was laid up there, and three of them just floated out of the thing. They wasn't on no ground.

And this:

> They didn't have toes. But they had feet shape, a round-like thing on a leg. They just glided up there to me. Just zzzzz. Two of them just floated around behind me and lifted me off the ground. By my arms with their pincher things. No force they didn't hurt me. I didn't feel nothin'."

Hickson then noted that his fishing partner, Parker, had already passed out by explaining that

> When I got in there (the ship), they just kind of had me there. I just floated. There was some kind of instrument, not an x-ray machine. It looked like an eye; a big eye with an attachment to it. And it went all over my body, up and down. And then they left me. They were about five feet tall. Best I remember they looked pale like to me. It (clothing) looked kind of like a skin fit (below the nose) like a slit, and I never saw that opening move. And they had something on each side of the head that resembled ears but didn't look like ears that we know. And the head—I didn't see any neck. It look like it just sit there on the body.[37]

Jackson County Chief Deputy Barney Mathis and Captain Glen Ryder, who both interviewed Hickson and Parker after the experience, dismissed alcohol as a motivation for their story. Mathis said, "Hickson appeared to be a reasonable man and not a heavy drinker." This was verified by Hickson's wife and his employer. "Both men said they were not drinking when the incident occurred but admitted they went to have a drink or two after it was over. They had to have something to settle their nerves." Mathis quoted Hickson as saying, "I was so damned scared I didn't know what it was."[38]

Other witnesses to the sighting included a crane operator at the shipyard and a person named Larry Booth, a service station operator in the nearby Pinecrest subdivision of Pascagoula. Booth was at home slightly after 9:00 p.m. while closing his front door for the evening he noticed a "huge object of some sort hovering five to eight feet above a nearby street."

Booth reported to investigators:

> This object was standin' still, it wasn't movin' at all when I seen it. But all the lights around the outside of it were turnin' clockwise motion. And they were red. It was larger than the props on a helicopter. The lights all the way around it, a lot of them, (were) close together, circling, slower than an ambulance light turns, about half that fast. I couldn't hear a sound. A helicopter would've jarred everybody in here out of the house."[39]

There were also three other people in a car nearby: R.H. Broaduw, a probation officer; E.P. Sigalas, the city councilman; and their companion, a musician. They were on the way to a concert at a religious rehabilitation center. Councilman Sigalas said, "[At] about twenty to eight, we saw it. First, I thought it was a helicopter, (but) it wasn't doing anything. It looked perfectly cylindrical. I'd say it had an oblong shape."

The next morning (Friday), a shipyard worker, foreman Jim Flynt from Walker, the place where Hickson was also a foreman, observed "a radar plane and a flight of F-111s (that) swept around over several times. We ain't seen nothing like that around here before."[40]

During the investigation and interviews at the Pascagoula Sheriff's offices, Hickson and Parker were left alone in an office between meetings and surveillance recording devices were activated to attempt to see if incriminating statements or confessions would be offered. This technique was conducted a handful times by Jackson County Sheriff Fred Diamond. None of the recordings offered any confessions, incriminating evidence, or statements contradictory to their story. According to Diamond, "Instead, they continued to talk the voices of the terribly distressed."[41]

The following is part of one of the transcribed recordings compiled by NICAP and reproduced by MUFON:[42]

Calvin (Parker): "I can't sleep yet like it is. I'm just damn near crazy."

Charlie: "Well, Calvin, when they brought you out, when they brought me out of that thing, goddamn it, I like to never in hell got you straightened out."

Calvin: (His voice rising) "My damn arms, my arms, I remember they just froze up, and I couldn't move. Just like I stepped on a damn rattlesnake."

Charlie: "They didn't do me that way."

Calvin: "I passed out. I expect I never passed out in my whole life."

Charlie: "I've never seen nothin' like that before in my life. You can't make people believe—"

Calvin: "I don't want to keep sittin' here. I want to see a doctor."

Charlie: "They better wake up and start believin'. They better start believin'.

Calvin: "You see how that damn door came right up?"

Charlie: "I don't know how it opened, son. I don't know."

Calvin: "It just laid up and just like that those son's bitches, just like that they come out."

Charlie: "I know. You can't believe it. You can't make people believe it."

Calvin: "I paralyzed right then. I couldn't move."

Charlie: "They won't believe it. They gonna believe it one of these days. Might be too late. I knew all along they was people from other worlds up there. I knew all along. I never thought it would happen to me."

Calvin: "You know yourself I don't think."

Charlie: "I know that, son. When I get to the house, I'm gonna get me another drink, make me sleep. Look, what we sittin' around for. I gotta go tell Blanche (Hickson). What we waitin' for?"

Calvin: (panicky) "I gotta go to the house. I'm getting' sick. I gotta get out of here."

Charlie got up and left the room, and Calvin was alone.

Calvin: "It's hard to believe. Oh, God, it's awful. I know there's a God up there."

Police were skeptical of their story initially. Hickson and Parker offered to take a lie detector test to prove their honesty.

J. Allen Hynek was present during most of the authority questionings, including those from Air Force representatives from Keesler AFB in nearby Biloxi, Pascagoula Sheriff's Department, and representatives from Walker shipyards, medical doctors from as far away as Berkeley, California. Hickson and Parker were subjected to intense medical and radiation examinations, including the hypnosis that unfortunately was not successful. Also chemical, electromagnetic, and other physical and radiological tests were conducted around the experience site.

Hynek had these statements to offer:

> There's simply no question in my mind that these men have had a very real, frightening experience, the physical nature of which I am not certain about, and I don't think we have any answers to that. Under no circumstances should these men be ridiculed. They are absolutely honest. They have had a fantastic experience. It should be taken in context with experiences that others have had elsewhere.[43]

The excitement and crowds of investigators and data seekers continued well after October 11. While Calvin Parker was having difficulty adjusting to the attention, Hickson, being older and a Korean War veteran, was more trained to accept his new role. Many more medical and diagnostic tests were administered to both in the weeks that followed. On October 30, Hickson took a polygraph administered by Scott Glasgow of the New Orleans Pendleton Detective Agency. Glasgow's statement concluded, "I am convinced that he believes he saw a spaceship and that he believes he was taken into the spaceship by three creatures."[44]

Hickson and Parker shunned the public spotlight for a long time. Parker spent some time in a hospital recovering from an emotional breakdown[45] and was later hypnotized by Ufologist Budd Hopkins. He spent a little time in a ufology business venture.

Immediately after their experience, both stayed away from any publicity. Hickson later appeared on some television talk shows and gave some lectures, interviews, and published a book based on their encounter titled *UFO Contact at Pascagoula*. He reported three more UFO encounters in 1974. Hickson passed away on September 9, 2011, at the age of eighty. Parker suffered a stroke in 2010 and has since recovered.[46]

Rendlesham Forest/RAF Woodbridge Air Force Base Encounters of December 25–29, 1980

This case is another in the series of multiple governments' involvement in extraordinary encounters of ufology investigation. This mass-sighting encounter is populated by a higher percentage of military witnesses-to-general public than many others. While both constituencies shared in the observations, as well as the local wildlife, the military participants contributed by far the most evidence.

The United States occupied and helped maintain operations at two Royal Air Force bases in Suffolk, Eastern England in 1980. These were RAF Bentwaters and RAF Woodbridge. Situated geographically between them was Rendlesham Forest, the landing sight of a craft on the first night of observations.

Late on the night of December 25, an out-of-place light was seen moving downward from the sky into Rendlesham. A reconnaissance team of U.S. servicemen were sent to investigate what they thought was a plane crash. No sound to accompany a crashing airplane was noted.

The first night's investigation traversed the early morning of December 26. The servicemen pursued toward and into Rendlesham for over two miles before making initial contact. Sergeant James Penniston and servicemen John Burroughs approached into a clearing in the forest. During this time, some local citizens had called the two bases asking for explanations for what they had just seen. The local wildlife and farm animals were "going into a frenzy."

Penniston and Burroughs moved right up to a triangular craft, about eight feet tall with a triangular landing system. Penniston observed peculiar writing on the side closest to his position. He then reportedly made physical contact with the craft by touching and moving his hand across the side panel of the object. Burroughs was next to Penniston during this series of events. The two soldiers next reported a period of time when they could not totally account for all their actions.[47]

What was additionally entered into evidence about these events was later contributed mostly by Sergeant Penniston, as the lead officer at the scene. The writing on the side panel was described something quite similar to Egyptian hieroglyphics. When Penniston stroked that panel and the adjacent ones, he felt a mental absorption of what he later documented as binary code: the "0's and 1's" we know of as computer digits. He recorded this whirlwind communication into a notebook and kept it for a period of years. When he regained interest in the case years later, Penniston had the recording deciphered by computer scientists. The code was translated to contain a transcript with references to a futuristic mission to which the ship that had landed was a part.

The first nights' events were documented to have ended after 4:00 a.m., and because it was still dark, not much could be observed or deciphered by the police.

A detachment of servicemen, including Penniston, returned and researched the landing area early that morning. Trees had broken branches and scorch marks around them. Three depressions into the ground forming a triangular pattern identified the landing spot of the recently witnessed ship. The local police were called to join the servicemen in the research. Together, they reported what information they gathered. Reports were filed by the servicemen, including the most detailed by Penniston and Burroughs, which included drawings and sketches.

Thirty-six hours later, the UFO returned to Rendlesham on December 28. This time, a much larger reconnaissance and intercept team and an abundance of equipment were rushed out into the forest. The mission was led by the bases' U.S. Deputy Commander-in-Chief Lieutenant Colonel Charles Halt. Testing equipment brought into the area included radiation and magnetic detection instruments, floodlight fixtures specially engineered for military field use, military mobile generators, many different styles of radio communication equipment, and audio recording devices, one of which Halt carried to record his voice.

When they arrived on the field, they encountered the UFO which this time was airborne at the tops of the trees about forty or so feet in the air. Colonel Halt began his tape recording diary of the events. While the recordings survived the mission intact and were made public record shortly thereafter, some of the rest of the equipment encountered significant operational interference. The floodlights, called "light alls," started to fail. Intermittent static and cracking noises interfered with radio communications throughout the time the ship was in the area. The radiation and magnetic instrument readings were registered and added to other data taken both in the days between December 26 and 28 and afterward. All of it was analyzed.

That craft hovered around the encounter theater for over one hour and fired light beams in patterns on the ground below the soldier's feet, including Halt's. Three other craft hovered in the sky around the Bentwaters base before, during, and after the main event in Rendlesham that early December 28 morning. All them were recorded to have performed similar light transmissions patterns down from their craft into various buildings in the Bentwaters complex. After the encounter, it was disclosed by the military that Bentwaters base operations were partly engaged in warehouse storage of a variety of nuclear weapons. Bentwaters base was situated close to the eastern coastline of England and was a most logical asset during the Cold War operations in the early 1980s. These warehouses were targeted by the craft.

Radar screens at RAF Watton in nearby Norfolk showed an unexplained, anomalous tracking signature near Bentwaters. When the radar target moved directly over Bentwaters, the track disappeared. A separate radar-tracking incident occurred at Bentwaters RAF which corroborated the sequence of events recorded at Watton. Additionally, radar track recordings at RAF Woodbridge also confirmed the sequence of events as those of Watton and Bentwaters.[48,49,50]

Colonel Halt's recording was telling in and of itself. It logged the whole encounter in Rendlesham Forest that morning. As noted, the recordings of Charles Halt have been released in the public domain. The evidence obtained in the literature from his documentary is both powerful and indisputable. Most important, it is an actual account of the events of that morning. Colonel Halt distributed a memo that has come to be known as *The Halt Memo*. An access link to the entire memo can be found in the "References" section. Here are excerpts of that memo:

> Early in the morning of 27 Dec 80, two USAF security police patrolmen (Penniston and Burroughs) saw unusual lights outside the back gate at RAF Woodbridge. The individuals reported seeing a strange glowing object in the forest; metallic in appearance and triangular in shape, approximately two to three meters across the base and approximately two meters high. It illuminated the entire forest with a white light and had a pulsing red light on top and a bank(s) of blue lights underneath. The object was hov-

ering or on legs. As the patrolmen approached the object, it maneuvered through the trees and disappeared. Animals in a nearby farm went into a frenzy.

Two nights later, from another excerpt:

A red sunlike light was seen through the trees. It moved about and pulsed and then broke off into five separate white objects and then disappeared. Immediately thereafter, three starlike objects were noticed in the sky; two objects to the north and one to the south.[51,52]

The leader of the incursion, Colonel Halt, was the most baffled of the party. Later, he commented, "Here I am, the senior official who routinely denies this sort of thing and diligently works to debunk them, and I'm involved in the middle of something I can't explain."[53,54] This is often what happens to a prior skeptic. From former U.S. presidents to high-ranking military leaders and representatives from all areas of society, these officials either never heard or knew anything about the UFO and alien phenomena. Then become involved in an event. All the elements and facts of the event sum to an encounter that leaves evidence and the participant wondering in a whole new framework.

Roswell, New Mexico, Experience

The story of Roswell was generally accepted as motivation for the genesis of the term "the modern era of UFOs." It has only been modified today to refer to "the modern era of ufology." When the topic of the origin of this term is discussed, it is vital to include the Kenneth Arnold sighting of June 24, 1947, as a motivating force. Kenneth Arnold's incident occurred only about two weeks before Roswell took place.

There has been more research, documentation, and review of the Roswell incident preserved than any other experience in history to date. These notions of research, documentation, and review can take on more than one dimension; there is the archive available in the public domain and the alleged archive that is stored somewhere within the confines of the U.S. government.

The Fourth of July was on a Friday in 1947. The early summer in the southwest United States is the rainy season. There is often heavy and sometimes violent precipitation and lightning activity. In south-central and southeast New Mexico, the first part of July experienced a few of these weather events, particularly on July 3. Here is a sequence of events that bring the alleged Roswell incident together.

In the twilight hours of Wednesday, July 2, Mr. and Mrs. Dan Wilmot of Roswell, New Mexico, reported to the *Roswell Daily Record* a "large, glowing object, about fifteen to twenty feet in diameter (that) glowed as though light were shining through from inside." On the evening of Thursday, July 3, a witness, James Ragsdale, and his female companion were camping about forty miles northwest of Roswell during a violent lightning storm. They noticed "a flash of bright light moving toward the southeast." They noted a noise sounding "like a crash of something."[55]

On Thursday morning, July 3, William (Mac) Brazel, a rancher working as foreman for the Foster Sheep Ranch on the land that was the alleged recovery site, and seven-year-old Dee Proctor, granddaughter of Brazel's neighbor Loretta Proctor, first discovered debris of an object scattered around the landscape in clumps and individual pieces. He inspected some of this material which was spread out over an area that stretched for a mile or more. Brazel collected some pieces and took them back to his cabin. He also took some to show to Floyd and Loretta Proctor when he took Dee Proctor back to her home.

In a statement Loretta Proctor made afterward, Brazel and Floyd tried to "burn and cut" pieces of the debris. She explained that:

> The piece he brought looked like a kind of
> tan, light-brown plastic. It was very lightweight,
> like balsa wood. It wasn't a large piece, maybe four

inches long. We cut on it with a knife and would hold a match to it. We knew it wasn't wood. It was smooth like plastic, it didn't have any real sharp corners, kind of like a dowel stick, kind of dark tan. It didn't have any grain, just smooth. I hadn't seen anything like it.[56]

On Sunday, July 6, Brazel travelled into Roswell and called Sheriff George Wilcox to report the incident. Wilcox arranged a meeting with Brazel, U.S. Army-Air Force Captain Jesse Marcel Sr., and Captain Sheridan Cavitt in Wilcox's office. Marcel and Cavitt left and started out to the location of Brazel's sighting. On this day, information was communicated to the Army Air Force through Wilcox's office as well as from Marcel who was to recover some of the debris later that day.

On Monday, July 7, Glenn Dennis, owner of the Roswell Ballard Funeral Home, received a call from Roswell AAF Headquarters with inquiries about the detailed characteristics of embalming fluid. They also inquired about the availability of a quantity of child-sized caskets. Dennis completed the interview. Later that day, he was required to go to the Roswell Air Field on an unrelated matter. When he arrived inside the medical building, he was confronted by military intelligence security officers who threatened him with the consequences of telling any of what he saw to the people back in Roswell. He was escorted off the base.

The next morning, Dennis received a call from Matilda McAvoy, a nurse acquaintance from the base. She was frantic and encouraged him to meet with her later that day. At the base restaurant club, Dennis interviewed with the nurse who described her activities in the autopsy of beings not from Earth and also included drawings of some of the beings with whom she assisted the doctors in the autopsy procedures.

Later, on Tuesday, July 8, Sheriff Wilcox had heard nothing back from Army Air Force officials. Mac Brazel, at that time, had not been visited by anyone from the Army-Air Force. Meanwhile, there were local reporters, including radio station *KGFL in Roswell* reporter John

McBoyle, investigating the story. Jud Roberts, the radio station part owner, contacted Brazel, and they met for an interview later that day.

At about 4:30 p.m., radio teletypist and secretary, Lydia Sleppy, received a story from McBoyle to transmit to the Associated Press. At 4:50 p.m., she began the transmission. About five minutes later, warning bells that were attached to every teletype transmitter back then began ringing. A message was received on Sleppy's machine that read, "LLLine Interrupt-Attention Albuquerque. Cease transmission immediately. Do not transmit this story. Repeat. Do not transmit this story. Authority FBI, Dallas, Texas, 070547.996277." Again, bells rang at the end of this transmission signifying the end of the communication. This confirmed the government was also monitoring society back in 1947.[57,58]

While this particular transmission had been intercepted and successfully blocked by the FBI, Sleppy was told to write up the story in shorthand. In any event, word that the Army had "recovered a flying disc" had already been disseminated. The story had already hit the newspapers nationwide in time for their July 8 edition. The Roswell Army Air Force Commanding Officer William Blanchard ordered public information officer First Lt. Walter Haut to release their own press release that would become the cause for the world to hear of the incident. The world would hear of the incident from their press in the days that followed.

Meanwhile, Marcel had been summoned to Fort Worth, Texas, on July 8 to participate in his own press conference with Army brass headed by General Roger M. Ramey. Army cargo planes shipped four boxes of debris to Fort Worth that morning. Reports of the shipping containers by the pilot and crew were made that confirmed they were "extremely light weight for the containers" suggesting that there may not have been anything inside of them.

The press conference was conducted that afternoon, and four photographs were taken by Army Major Charles A. Cashon, public intelligence officer, and transmitted in the newspapers and elsewhere. The photos Marcel posed for contained debris from assorted weather balloons. He had been ordered into being a participant in an activity that was problematic for his future Air Force career.

At 6:17 p.m., a teletype was made from the FBI: Dallas, Texas, bureau office titled cease message was sent to FBI Director J. Edgar Hoover informing him that "a weather balloon was responsible—Roswell. The debris was prepared to be moved to the Army Air Forces Technical Base at Wright Field." There are documented reports from Brig. General Arthur Exon, commanding officer at Wright-Patterson in the 1960s, but who was lt. colonel in 1947 and worked in the Foreign Technology Division, that cargo planes delivered "some type of debris" to Wright Field and were stored initially in Hangar P-3 (Building 84 today). Exon's research team studied the debris. His initial conclusion statements included that "the tested debris were from space."[59]

The Army Air Force, meanwhile, had heard of the Brazel interview given to Jud Roberts of radio station KGFL earlier on the eighth. They searched for and confronted Brazel later that day. According to the Army Air Force, Brazel was not arrested but was escorted back to the Roswell base and interrogated for a period of five days before being allowed to leave.

On Wednesday, July 9, at 8:00 a.m., some boxes of debris allegedly left Roswell AAF and traveled to Kirtland Field in Albuquerque then was later diverted to Wright Field. Later that morning, a tractor trailer with a tarpaulin draped over the top traveled through Roswell down Main Street and turned left onto Second Street on its way to Fort Worth, Texas. Other debris from Roswell traveled to Los Alamos Laboratories.

Newspapers and radio broadcasts later that day started the official retractions by the Army Air Force of the reports earlier made about the recovery of flying discs and bodies near Roswell. Washington, DC military headquarters had allowed the conclusions of the mistaken retrieval of weather balloons as the (first) official explanation of the debris recovered. This would be the first official explanation protocol that would attach later explanations of "anthropomorphic dummies," "secret radar and weather balloon projects," "surveillance balloons," "disks (of earth origin) suspended from balloons," and more.[60]

Meanwhile, as noted, some of the evidence allegedly found its way to Wright Field. Air Force Brig. General Arthur E. Exon (in

1947 a lt. colonel) was a decorated pilot and prisoner of war in Nazi Germany during World War II. He was assigned to Wright Field in early July 1947. The following are statements from Exon as to the landscape of event that would occupy his early time at Wright:

> "We heard the material (Roswell) was coming to Wright Field." Hangar P-3 was the disembarking platform for at least some of the paraphernalia. From that landing site, it was assigned to various labs within the Wright complex for testing. "Everything from chemical analysis, stress tests, compression tests, flexing. It was brought into our material evaluation labs. I don't know how it arrived, but the boys who tested it said it was very unusual. (Some of it) could be easily ripped or changed. There were other parts of it that were very thin but awfully strong and couldn't be dented with heavy hammers. It was flexible to a degree."

There has been some ongoing speculation that the Roswell incident may have involved two separate crashes at two different times. General Exon left open a possibility that this could be the case. He also offered that it was "probably part of the same accident, but (there were) two distinct sites, as I remember flying the area later, that the damage to the vehicle seemed to be coming from the southeast, but it could have been going in the opposite direction. The bodies were outside of but near the main craft."

General Exon continued, "Some of it was flimsy and was tougher than hell and other almost like foil but strong. It had them pretty puzzled. A couple of guys thought it might be Russian, but the overall consensus was that the pieces were from space." When asked about the bodies, he said, "There was another location where they did say there were bodies. They were all found, apparently, outside the craft itself but were in fairly good condition. In other words, they (the bodies) weren't broken up a lot."[61]

Exon confirmed there were no weather balloons at that time. He later explained General Ramey was the first person of high authority who became involved in the incident. The initial confusion and lack of direction was cured once the Pentagon and then Chief of Staff Dwight Eisenhower became involved. This all happened on July 8. By the time of Ramey's original entry into the activities, the Army Air Force back in Roswell had already retrieved substantial wreckage and cleaned up most all the debris field.

The John R. Salter Experience

"When I think of that night of March 20, 1988, I have only positive feelings (as does John III, his son) about the not-so-different-from-us people from afar whom we met and with whom we spent well over an hour" (From the personal papers of John R. Salter II).

John Salter II was a college professor on a lecture tour with his then twenty-two-year old son that took them through Wisconsin and eventually down to New Orleans for his lecture at the Popular Culture Association/American Culture Association. The senior Salter was a social justice activist of Native American heritage.

There was a period of about two hours in the late afternoon when they turned off their mapped route on Highway 61 at 4:30 p.m. and onto an alternate Route 14 in southern Wisconsin. They proceeded on this detour for over eighty-five miles. In that time, neither of the Salters could remember being on that stretch of highway. The next morning, they left Bettendorf, Iowa, from where they spent the night and continued past Peoria, Illinois. At 10:14 a.m., they sighted a "bright, expanding light" on a flying object over two-thirds the size of the four-lane highway on which they were traveling. The saucer-shaped craft with a dome atop it flew over the truck and continued on out of sight.

John II and John III continued on their trip without another incident, except that the events of the previous evening and that next morning were always on their minds. They initially explained the time erroneously spent on Highway 14 in Wisconsin as some sort of amnesia. Senior Salter did not have any recall of that situation until the following June; the junior Salter not until November of that year.

The recall was gradual over the next couple of years. John Sr. began taking notes. The amnesia was redefined as "missing time." During those times of recollection, he first remembered thinking, *They should not be taking this route, it is a mistake, but they could not help themselves away from it*. He was also compelled to revisit the site. The composite diary then developed into the following series of events.

They turned off Route 14 onto a narrow, rough road where the pickup stopped. Both Salters got out and stood near the passenger door. Three small humanoid figures appeared at and climbed on the back bumper of the truck. Salter described them as "four and one-half feet tall, thin bodies, limbs and conspicuously large, quasi-slanted eyes." They were joined by three others of the same type, as well as a taller figure about six feet tall and appearance of a human.

They communicated with the tall humanoid telepathically. They walked back to the UFO. Along the way, John Sr. stumbled and fell backward, but "an immediately cushioned force" prevented him from falling down the slope as they were walking uphill at the time. Once inside the ship, they described a "brightly-lighted room, a kind of white light, and a deep, blue glowing panel." An implant was positioned in his nose. Similar probes were placed in his neck and central upper chest. John Jr.'s face and body were scanned with a "flashlight-light" type instrument. The testing continued for some time, then the tall humanoid escorted them back to the ship carrying a flashlight with them to see through the heavily-wooded, hilly terrain.[62]

Throughout the "interception," as John Salter II later referred to it, his mind was experiencing unexplained flashbacks to a time in May 1957 when he was twenty-three and driving in the Arizona forest near the small town of Mayer, Arizona, near Prescott. The sequence of events for March 20, 1988, was eerily similar to that night back in Arizona.

As his memory recall increased, Salter was able to reconcile the changes and better explain the physiological changes that were happening to him as well as the unexplained events and observations they made subsequent to that actual night. For example, when their truck was serviced a few days after the encounter, the mechanic questioned the Salters about the damage done to the underneath suspension and wheel and the bent misshapen bumper. The truck had only sixteen thousand miles on it, and John Sr. claimed had only been driven on paved roads. The mechanic described the truck as having done so recently. It was also very muddy.

By the spring of 1988, Salter Sr. noticed some changes happening to his body. His hair, fingernails, and toenails were now growing three times the normal rate. His eyebrows and body hair also grew faster. When he suffered a cut, the wound would clot and heal almost immediately. Denture inserts that caused daily blood since first inserted four years prior stopped permanently. He stopped smoking without any of the withdrawal symptoms he had heard and read about. Age spots and wrinkles disappeared. Incidence of colds became almost nonexistent. He felt more energetic than at any point in his fifty-four years and now required far less sleep than before. A walking gait abnormality suddenly disappeared. Facial disfigurement from a 1963 auto crash in Mississippi disappeared. He also noticed an increased psychic capacity. Lastly, Mr. Salter became a little more sensitive to bright sunlight than before.

Son John III also noted future physiological effects that appeared beneficial. He experienced a growth spurt that made him one and a half inches taller. He had not grown since age eighteen.

An incident that occurred early in 1990 would offer validity to some of his physiological improvements. He stated that on Saturday, February 6, "At 10:55 a.m., I was working with a screwdriver. Things slipped and the blade cut significantly into the tip of the finger next to my thumb on my left hand. My blood, extremely dark and rich looking, welled to the surface and clotted immediately, in such a fashion that a nubbin rose a bit over the surface of the skin. I rinsed the nubbin off, and again, with nothing bleeding, a healthy dark scar formed immediately. Within the hour, the scar grew smaller and smaller, and things just faded away into normalcy."[63]

Salter eventually pieced together one aspect of a particular situation within the experience. He had asked himself many times, "Why were the implants made in the places they were?" The nose implant was made to assimilate and activate some homeostasis change within his pituitary gland. This explained the growth modifications to his hair, nails, and some of the epidermal (scars) features and disfigurements that disappeared. The implant in his neck was to produce respectively similar physiological changes to his thyroid gland, which changed his metabolism. This explained his generally better health, increased appetite, and healing of previous injuries. The implant in his chest was similarly enacted to stimulate his thymus gland which explained his improved immunity to colds and other foreign bacteria and viruses.

Also significant was the emergence of his recall of some of the situation in Arizona in 1957. Salter remembers the entities as "the humanoids involved are of the same basic racial group as those in March 1988." That recall also allowed him to remember a sighting event in 1941 near his Kansas farmhouse. He remembered having had a scar above his kneecap with which he could not link a cause. A similar recall of a 1952 event back in Arizona produced no physical CE-4 manifestation and no physical injuries or marked tangible evidence.[64]

Eventually, Salter recounted more details about the nature of the interception. The tall humanoid was the only communicator with both Salters, and this was done telepathically. They were from the Zeta-Reticuli double star system. This was the same region as the home of the visitors in the Betty and Barney Hill experience. It was explained to Salter that their history took them first from their home planet to establishing civilizations on some of the others in their solar system in a stepping-stone colonization. Their planets are closer situated than of our own solar system. Some were not habitable, but others were. The humanoids, which both Salters described as generally human in phenotype, were of one racial stock but evolved minor phenotypic variants due to the diverse living environments present on the different colonies. The humanoid then gave Salter a visualization of the Aleutian Island chain in Alaska. This, the humanoid said, was the appearance of the Zeta-Reticuli system when viewing it from a ship.[65]

There were purposes and ramifications from the interception besides the physiological ones described. "Their actions are quite positive. I believe the basic thrusts focus on helping some of us (directly) 'keep on keeping on' in the business of edging humanity and sensitizing humanity with respect to the relatively nearby presence of other forms of intelligent life."[66]

This explanation continued,

> I see the so-termed alien humanoids as friendly and with positive motivations and beneficial effects, essentially one race, quite similar to ourselves in many ways. I am convinced that social justice work is a critical piece of the alien agenda.[67]
>
> I believe these extraterrestrial persons (are) similar to ourselves and perhaps even related or at least the results of a parallel evolutionary course. They are solid and physical and "all-around" tangible entities, sharply intelligent, and their range of emotions is comparable to ours. I categorically do not see them as angels/devils/psychic manifestations.[68]
>
> But I am convinced that social justice work is a critical piece of the alien agenda—cosmic citizenship sensitization and good works—for us make up the only logical explanation for ET involvement. I see altruism on their part, and I also see enlightened self-interest. I have a strong feeling the ETs can offer us some valuable insights into avoidance of "super collective" hive mentalities or cut-throat individualism and can offer insights into the balancing of collective and individual well-being.[69]

Travis Walton Experience

The Walton Experience is one of the very few alleged cases of abduction that are evidenced by multiple eyewitnesses with reports that correspond to each other, multiple examinations and tests that corroborate other data results, and some events post experience that add a measure of controversy. Conclusions of the many people from all walks of professional life involved are also substantial but mixed as well.

On November 5, 1975, a logging crew of six was retreating from their worksite in the Apache-Sitgreaves National Forest near Snowflake, Arizona. It was late dusk on a clear, cool autumn evening. After 6:00 p.m., the pickup truck carrying the crew slowed on the unpaved dirt road to observe the gold-colored glow of an object forty feet in diameter hover motionlessly about fifteen feet off the ground in a clearing about one hundred feet from the truck.

Travis Walton, then twenty-two, called for the driver, friend Mike Rogers, to stop the truck. He hurriedly exited the front passenger door to gain a closer look. The other members of the crew slowly began to emerge from the truck, anxiously encouraging Walton to return back to them. Walton stopped about thirty feet from the craft. He described its shape as "two pie pans placed rim to rim." All members saw this and heard a beeping sound.[70]

Walton retreated a couple of steps. He had to look up at a sixty-degree angle to watch the ship hover and rotate. Walton reported, "I jumped for cover and then jumped up to run back to the truck, and that's when this blast of energy hit me, and I just felt this numbing shock go through my body." A blue-green laser-like beam of light and particles was emitted from the bottom of the craft and struck Walton's upper body. The beam lifted him off the ground and then threw him yards away from the craft into a wooded thicket, injuring him into unconsciousness.

The crew reported that Walton was below the craft when it began making noises that sounded like turbines rotating. The disc wobbled back and forth. When Walton saw this, he retreated. The beam then shot out, hit Walton, and threw him about ten feet. "He landed on his right shoulder, and his body sprawled limply over the ground."[71]

Rogers and the crew then reentered the truck and sped off for over a one quarter mile then stopped. When they looked back, the five remaining members observed a rounded craft with blue and white lights rise over the trees and swiftly streak to the northeast and out of view. They returned to the site to look for Walton but did not find him.

Walton's conscious account of the experience was reported later after his return. He awoke on a table in a room that was "warm and damp." As he gradually awoke, he noted three "creatures, less than five feet tall, very pale with large, domed heads, large eyes, small nose, mouth and ears, and their bodies, encased in tannish, orange, seamless jumpsuits, and were very thin."

Walton rose off the table, and the entities approached him. He grabbed a pipelike rod and attacked them. The entities fled the room. Walton left and turned down the opposite hallway and into a separate room. He sat in the only chair in the room that had consoles on each arm. He touched one and the ceiling began to move with a mural of a star-filled sky moving above him. He was then met by a human described as being "about as tall as himself, about six feet tall, with brown hair and golden-brown eyes."

The stranger called for Travis to come with him, and eventually, Walton did. He tried to ask questions, but the stranger would not answer him. Instead, they walked to another room where a wall without any apparent opening just opened. They walked down a ramp and into another much larger room somewhat resembling a hangar. Walton observed a number of oval-shaped metallic craft secured in place. Walton was then led by the stranger, who was dressed in a blue jumpsuit and a helmet, out of the hangar and into another room where he was met by three other humans—two men and a woman. These strangers looked like the first one except they were not wearing helmets.

Walton was coaxed onto a table where an apparatus that, to Walton resembled an oxygen mask with a black ball attached, which was placed over his face. He then lost consciousness until the fifth night afterward when he awoke on the highway near Heber, Arizona.[72]

Meanwhile, Rogers and the crew returned to Snowflake and called police from nearby Heber. Sheriff Marlin Gillespie and Deputy Sheriff Chuck Ellison met the loggers at a coffee shop where they gave initial statements. Police search crews granted Mike Rogers's one request for a search of the area. Rogers wanted search dogs to investigate also but were told by Gillespie none were available.

Police investigated the scene but found no evidence. Suspicions heightened. The police withdrew. Meanwhile, Walton's brother, Duane, and Mike Rogers returned to the scene the next morning to find no search being conducted. They returned to the sheriff's office and confronted Gillespie about the lack of search teams. Gillespie acknowledged Duane and Rogers's concerns. By that afternoon, more searches with helicopters were investigating the venue.

In the days during the investigation and before Travis Walton's return, reports from other investigators said that Duane had witnessed a UFO similar to the one Rogers described about twelve years prior in another place. When the sheriff heard this, Snowflake Town Marshal Sanford Flake declared that "the entire affair was a prank engineered by Duane and Travis. They lighted a balloon and released it at the appropriate time." The newspapers later reported statements from Flake's wife suggesting that her husband's story "was just as far-fetched as Duane Walton's."[73]

On Monday, November 10, the remaining five logging members were given polygraph examinations administered by Cy Gilson of the Arizona Department of Public Safety. Gilson concluded that four of the five men passed the test and that the results were conclusive. Only member Allen Dallis's test resulted in an inconclusive determination. From Gilson's official report, "These polygraph examinations prove that these five men did see some object they believed to be a UFO and that Travis Walton was not injured or murdered by any of these men on that Wednesday. There's no doubt they're telling the truth."[74]

Later on that night, Walton's sister in Snowflake received a call from a frightened Travis. He told her to go to a phone booth in Heber to find him. Together with Travis's brother Duane, they retrieved Travis and brought him home. When he was discovered, Walton had a five-day beard growth and had lost some weight but appeared okay except for the frightened mood he was experiencing.

Many medical, physiological, psychological, and polygraph examinations administered to Walton in the months post experience. One examination was a polygraph given by John J. McCarthy who was hired by gossip tabloid *The National Enquirer*. McCarthy's qualifications came into immediate question, and the test was ridiculed by many people afterward. The method, some of the questions and emotional intervention by McCarthy during the test, were all brought into question enough so that *The National Enquirer* agreed to suppress his concluding report.

The existence of the test, though, was learned by Philip J. Klass, an aviation journalist and most active UFO debunker of the time. He engaged in a decade's long smear campaign against Walton and all the people sympathetic to his case. Klass's main early efforts were to suggest that because of the five-thousand-dollar reward offered to Walton, *The National Enquirer's*, for the most fantastic UFO story of the year for its publication efforts, this was his motivation to concoct such an unbelievable hoax.

Eventually, life settled into an uneasy truce for the other members of Mike Rogers's logging crew. They all continue to deal with their plight. Travis Walton settled back in Snowflake and married best friend Mike Rogers's sister, Dana, and raised several children while working as a foreman at a lumber mill. He also continues spreading his knowledge and continuing to research the UFO/extraterrestrial hypotheses. In 1978, he published his first book *The Walton Experience*. That book was modified and made into the 1993 movie, *Fire in the Sky*, funded by Paramount Pictures. Paramount did not like the first version of the script and hired screenwriter Tracy Torme to rework the script. The documentary became a fictional account of a "flashier more provocative abduction story."[75]

In 1997, Walton released *Fire in the Sky: The Walton Experience*, a new edition of the 1978 book. In 2016, Walton released a documentary, not involving Paramount Studios, but by Onwinges Productions, LLC, titled, *Travis: The True Story of Travis Walton*.

Walton has, from time to time, returned to the site of the 1975 experience. This was both for personal reasons and for research activities. Some of the research has analyzed unexplained growth anomalies among the trees and vegetation in the area of the hovering platform. The trees nearest the UFO hovering position have been shown to be growing over thirty-five times the rate the trees had in eighty-five years prior to that. Also a more specific discovery was that the growth was most pronounced on the trees directly facing the hovering platform and closest to it.[76]

Lonnie Zamora Experience

This incident would be classified in J. Allen Hynek's UFO Classification System as a "CE-3" or "Close Encounter Level 3." This means the witness(es) observed living entities as well as an unidentified aerial (or submerged) craft. Level 3 also indicates that there was no direct contact with the entities.

Socorro Police Department patrol officer Lonnie Zamora, then thirty-one years old, was on duty on the late afternoon of April 24, 1964. The weather noted to have clear, sunny skies. At about 5:50 p.m. local time, he was chasing a speeder down the highway outside of town. In the middle of the pursuit, he noted a bluish-orange cylinder of flame attached to a cylinder-shaped object in the sky off to his right. It was about half a mile or a bit longer from where he stood. Thinking it might be a dynamite shack in the desert that had just exploded, he broke off chase with the speeder and started to pursue the flame and the object emitting it.

Zamora watched the craft settle down behind a ridge separating the two. As his patrol car climbed the steep hill to approach, he came upon a "shiny whitish-aluminum color craft that had just landed" in the only gully now separating the two. Zamora later reported it as an oval long axis vertical craft settled on the ground on four landing legs. Two small adultlike entities in white coveralls were outside beside the ship inspecting some external feature. One of the entities turned and saw Zamora. The two beings then began jumping up and down. Zamora pulled up to a distance of about one hundred feet and exited his car. He then heard two or three "loud thumping noises" like a door shutting hard.

Zamora then walked around his car toward the craft. Suddenly, the ship ignited its engines, and he heard a very loud low frequency roaring sound. This changed to a high frequency roar and not a blast then back to a low frequency roar. He saw a smokeless blue-orange flame emanating from beneath the ship.

> Hardly turned around from the car, when heard, very loud roar—at that close was real loud. Not like a jet. Started low frequency quickly, then rose in frequency. At same time as roar saw flame. Flame was under the object. Object was starting to go straight up, slowly up. Flame was light blue and at bottom was sort of orange color.[77]

As the ship rose slowly off the ground, he noticed a red insignia, about two and a half feet wide, adorned on the side of it. As the ship rose, Zamora thought it might explode. He then ran around the other side of his car and away from it. He fell once and then glanced back to see the ship rise to about twenty-five feet in elevation. He further explained that "After fell by car and glasses fell off, kept running north, with car between me and object. Glanced back couple of times. Noted object to rise to about level of car, about twenty to twenty-five feet guess." The noise had subsided, and the flame was subsiding as well. Another five or six seconds later, he looked back again to observe the craft head off to the southwest at a level height about the same as the last time Zamora looked at it. It was now moving "very fast" and just missed hitting the same dynamite shack in its path.[78]

He radioed this experience back to Ned Lopez at the sheriff's headquarters while still observing the ship retreat into the late afternoon sky. Investigation teams were on site almost immediately. The first of these was Socorro Sergeant M.S. Chavez. Chavez found a frightened Zamora and was the initial corroborator of some of the trace evidence that would be thoroughly investigated in subsequent weeks after.

Together, they observed and started to collect these physical artifacts. Zamora then drew a representation of the insignia he noted on the side of the ship. Other police and an initial round of investigators soon started arriving and obtained trace evidence that included actively-burning bushes and brush and indentations in the ground that were corroborated to be from the landing legs of the craft. Several other eyewitnesses not at the event theater also made reports of sightings within the first couple of hours after the experience before dusk.[79]

The next morning at daylight, additional police and sheriff's investigators continued piecing together details and data. A surprising piece of evidence noted was that the bushes and vegetation were still burning. Within two days, representatives from the U.S. Air Force, Hector Quantanilla and David Moody from Project Blue Book, Ray Stanford from NICAP, James and Coral Lorenzen of the civilian UFO research group APRO, and J. Allen Hynek, U.S. Army Intelligence Officer Captain Holder, FBI agent Arthur Byrnes, and Major Connors of Kirtland AFB, among others arrived for investigation activities.[80]

Major Hector Quantanilla told the CIA journal,

> There is no doubt that Lonnie Zamora saw an object. There is also no question about Zamora's reliability. He is a serious police officer, a pillar of his church, and a man well versed in recognizing airborne vehicles in his area. He is puzzled by what he saw and frankly, so are we. This is the best-documented case on record (at that time), and we still have been unable, in spite of thorough investigation. To find the vehicle...[81]

Ray Stanford documented witness reports and corroborated and cross-referenced their stories, which checked out in logistics, temporality, and representation. Among the reports were two tourists, Paul Kies and Larry Kratzer, who were entering Socorro from the southwest, about a mile from the landing site. They allegedly witnessed the takeoff and saw the flame and dust being kicked up. The Dubuque, Iowa, *Telegraph-Herald* reported their story a few days later. In all various reporting agencies recorded dozens of sighting encounters from different angles surrounding the event site.[82]

The information studied from the investigation was varied and substantial. The weather that day noted strong winds were blowing from the south-southwest within a low pressure system. This would eliminate a balloon sighting as the ship flew off into the wind.[83] An investigation of the velocity and flight pattern characteristics determined the craft had achieved a velocity of about 760 mph within about 20 seconds.[84] Also there were studies of sand samples that were concluded to be *vitrified* (converted into glass or a glasslike substance, typically by exposure to heat). The area of such vitrified sand deposits was roughly two and one half feet square, confirmed by Zamora as about the size of the wake of the flame exhaust. The research investigator, Mary G. Mayes of the University of New Mexico, also concluded the thickness to be about one-fourth inch. Mayes also concluded that the samples looked "as if a hot jet hit it."[85]

Among the curious events that ensued after the wake of the incident had subsided was correspondence NICAP member Robert Barrow had with the Department of the Air Force. On January 8, 1965, he received a reply to an inquiry asking for the Air Force clarification on the UFO classification status. The reply received was only two sentences and said the following, "The Socorro, New Mexico, sighting is still unsolved. However, no evidence was found which indicated that the vehicle was from outer space. Sincerely, Maston M. Jacks, Major USAF."

The curious element was the Air Force's use of the word *vehicle*. This has not been used often in government and/or military investigation or correspondence regarding UFOs and extraterrestrials. This episode was the reason the variety of explanations offered by refuters and debunkers in this case does not include those of natural phenomena or of "mirages, delusions, etc."[86]

References

1   Quote Ref: Sagan Carl, Page Thornton (1972), "UFOs: A Scientific Debate." Cornell University Press, ISBN 0-8014-0740-0.
2   Cahill Ref 2: Chalker, Bill. 2005. *Hair of the Alien: DNA and Other Forensic Evidence of Alien Abduction*, p. 50–53.
3   Cahill Ref 3: The alien abduction of Australian Kelly Cahill. Prufon, July 27, 2011. Retrieved October 20, 2016, from: http://www/prufon.net/2011/07/summary-following-story-is-about.html.
4   Cahill Ref 4: Booth, B.J. 2002. The Kelly Cahill Encounter. Published November 16, 2002. Retrieved October 20, 2016, from: http://www.ufocasebook.com/Cahill.html.
5   Cahill Ref 5: Miletic, Daniella. 2016. The Oz Files: Who Is Victorian "Abductee" Kelly Cahill and What Did She See in Narre Warren? Sydney Morning Herald. February 1, 2016. Retrieved October 20, 2016, from: www.smh.com.au/entertainment/the-oz-files-who-is-victorian-abductee-kelly-cahill-and-what-did-she-see-in-Narre-Warren.
6   Cahill Ref 6: Cahill, Kelly. 1997. Encounter. Copyright 1997 by Kelly Cahill. Published August 29, 1997, by HarperCollins Publishers (Australia) Pty Ltd.
7   Cahill Ref 7: Chalker, 2005, p. 64–66.
8   Cape Girardeau Ref 8: Wickersham, Bill. 2015. The 1941 Cape Girardeau UFO Crash. Columbia Tribune, February 3, 2015. Retrieved October 20, 2016, from: http://www.columbiatribune.com/opinion/oped/the-cape-girardeau-ufo-crash/article_687f4.
9   Cape Girardeau Ref 9: Wood, Ryan S. 2001. The First Roswell: Evidence for a Crash Retrieval in Cape Girardeau Missouri in 1941. MUFON 2001 Symposium Proceedings. Retrieved October 20, 2016, from: http://www.ufocasebook.com/pdf/crashmufon2001.pdf.
10 Cape Girardeau Ref 8: Wickersham, Bill. 2015. The 1941 Cape Girardeau UFO Crash. Columbia Tribune, February 3, 2015. Retrieved October 20, 2016, from: http://www.columbiatribune.com/opinion/oped/the-cape-girardeau-ufo-crash/article_687f4.
11 Cape Girardeau Ref 11: Booth, B.J. Charlette Mann's Testimony for TV documentary. Retrieved October 20, 2016, from: http://www.ufoevidence.org/Cases/CaseSubarticle.asp?ID=861.
12 Cape Girardeau Ref 12: UFO Casebook. 2010. UFO Crash and Retrieval-Missouri, 1941, October 17, 2010. Retrieved October 20, 2016, from: http://www.ufocasebook.com/missouricrash.html.
13 Cape Girardeau Ref 13: Wood, Ryan S. 2001. The First Roswell: Evidence for a Crash Retrieval in Cape Girardeau Missouri in 1941. MUFON 2001 Symposium Proceedings. Retrieved October 20, 2016, from: http://www.ufocasebook.com/pdf/crashmufon2001.pdf.

14 Cape Girardeau Ref 14: UFO Casebook. 2010. UFO Crash and Retrieval-Missouri, 1941, October 17, 2010. Retrieved October 20, 2016, from: http://www.ufocasebook.com/missouricrash.html.

15 Cape Girardeau Ref 15: Wood, Ryan S. 2001. The First Roswell: Evidence for a Crash Retrieval in Cape Girardeau Missouri in 1941. MUFON 2001 Symposium Proceedings. Retrieved October 20, 2016, from: http://www.ufo-casebook.com/pdf/crashmufon2001.pdf.

16 Cape Girardeau Ref 16: Wood, Ryan S. 2005. MAJIC Eyes Only: Earth's Encounters with Extraterrestrial Technology, pp. 38–39. Copyright 2005 by Ryan S. Wood. Published 2005 by Wood Enterprises.

17 Cortile Ref 17: Wikipedia. 2015. Hypnagogia, Definition. Last updated September 28, 2016. Retrieved October 20, 2016, from: www.en.wikipedia.org/wiki/Hypnagogia.

18 Cortile Ref 18: Hopkins, Budd. 1996. Witnessed: The True Story of the Brooklyn Bridge UFO Abductions, p. 4. Copyright 1996 by Budd Hopkins. Published by Pocket Books.

19 Cortile Ref 19: Hopkins, Budd. 1992. Linda Cortile Case. MUFON Journal, September 1992. Retrieved October 20, 2016, from: http://www.mufon.com/looking-back-columns/linda-cortile-abduction -case.

20 Cortile Ref UFO Casebook: Booth, B.J. 2013. The Manhattan Abduction (Linda Cortile Napolitano). Retrieved October 20, 2016, from: http://www/ufocasebook.com/Manhattan.html.

21 Cortile Ref 21: Meers, Sean F. 2012. The Witnesses: The Linda Cortile UFO Abduction Case. Published March 13, 2012. Copyright 2012 by Sean F. Meers.

22 Cortile Ref 22: Hopkins, 1996, photographs, x-ray photo, p. 144–145.

23 Cortile Ref 23: Meers, Sean F. 2012. The Witnesses: The Linda Cortile UFO Abduction Case. Published March 13, 2012. Copyright 2012 by Sean F. Meers.

24 Cortile Ref 24: Meers, Sean F. 2012. The Witnesses: Introduction.

25 Cortile Ref 25: Meers, Sean F. 2012. The Witnesses: Introduction.

26 Hill Ref 26: Friedman, Stanton and Marden, Kathy. 2007. Captured! The Betty and Barney Hill UFO Experience, p. 102. Copyright 2007 by Kathleen Marden and Stanton Friedman.

27 Hill Ref 27: MUFON Journal. Betty & Barney Hill Abduction 1961. 2014. Copyright 2014 by MUFON International. Retrieved October 20, 2016, from: http://www.mufon.com/betty--barney-hill---1961.html.

28 Hill Ref 28: Clark, Jerome. 1998. The UFO Book: Encyclopedia of the Extraterrestrial, p 284, 291. Published by Visible Ink, 1998.

29 Hill Ref 29: Marden and Friedman, p. 129–133.

30 Hill Ref 5: Marden and Friedman, p. 155–166.

31 Hill Ref 31: Marden, Kathleen. Betty and Barney Hill: Where the Debunkers Went Wrong, p. 7.

32 Hill Ref 32: Marden and Friedman, p. 47–48.

[33] Hill Ref 33: University of New Hampshire. Guide to the Betty and Barney Hill Papers, 1961–2006. Retrieved October 20, 2016, from: http://www.library.unh.edu/find/archives/collections/betty-and-barney-hill-papers-1961-2006.

[34] Khoury Ref 34: UFO Casebook. 1988, DNA Sample from Khoury Abduction Raises Big Questions. Retrieved October 20, 2016, from: http://www.ufocasebook.com/khouryabduction.html.

[35] Khoury Ref 35: Chalker, Bill. 1999. Strange Evidence. International UFO Reporter, Volume 24, Number 1, p. 3–16, 31.

[36] Khoury Ref 36: Chalker, Bill. 1999. Strange Evidence. International UFO Reporter, Volume 24, Number 1, p. 3–16, 31.

[37] Pascagoula Ref 37: Blum, Ralph and Judy. 1974. Beyond Earth: Man's Contact with UFOs, p. 31–34. Copyright 1974 by Ralph Blum. Published April 1974 by Bantam Books.

[38] Pascagoula Ref 38: Blum and Blum, p. 10.

[39] Pascagoula Ref 39: Clark, Jerome. 1998. *The UFO Encyclopedia: The Phenomenon from the Beginning*, Volume 2, p. 715–716. Copyright 1998 by Jerome Clark. Published 1998 by Omnigraphics Books, ISBN 0-7808-0097-4.

[40] Pascagoula Ref 40: Blum and Blum, p. 14–19.

[41] Pascagoula Ref 41: Clark, Jerome. 1998. *The UFO Encyclopedia: The Phenomenon from the Beginning*, Volume 2, p. 447. Copyright 1998 by Jerome Clark. Published 1998 by Omnigraphics Books, ISBN 0-7808-0097-4.

[42] Pascagoula Ref 42: MUFON. 2014. Pascagoula Mississippi Case 1973. Reprinted from NICAP transcriptions. Retrieved October 20, 2016, from: http://www.mufon.com/pascagoula-ms-case---1973.html.

[43] Pascagoula Ref 43: Blum and Blum, p. 24–25.

[44] Pascagoula Ref 44: Clark, p. 717.

[45] Pascagoula Ref 45: Clark, p. 449.

[46] Pascagoula Ref 46: Associated Press. 2013. Man says 1973 UFO incident turned life upside down. Friday, October 11, 2013. Retrieved October 20, 2016, from: http://www.nydailynews.com/news/national/man-1973-ufo-incident-turned-life-upside-article.

[47] Rendlesham Ref 47: MUFON. Rendlesham Forest Incident 1980. Copyright MUFON International, 2014. Retrieved October 20, 2016, from: http://www.mufon.com/rendlesham-forest---1980.html.

[48] Rendlesham Ref 48: UFO Landings-Rendlesham Forest, 1980. UFO Casebook. Retrieved October 20, 2016, from: http://www/ufocasebook.com/Rendlesham.html.

[49] Rendlesham Ref 49: Tamblyn, Thomas. 2015. Rendlesham Forest UFO Sighting: New Radar Evidence Could Finally Prove It Was Real. Huffington Post, July 14, 2015. Retrieved October 20, 2016, from: http://www.huffingtonpost.co.uk/2015/07/14/rendlesham-forest-ufo-sighting-new-radar-evidence.

[50] Rendlesham Ref 50: Warren, Larry and Robbins, Peter. 1997. Left at East Gate: A Firsthand Account of the Rendlesham Forest UFO Incident, Its Coverup, and

Investigation. Copyright 2005 by Larry Warren and Peter Robbins. Originally published in 1997 by Marlowe & Company.

51 Rendlesham Ref 51: The Colonel Charles Halt Memo of January 13, 1981. Retrieved October 20, 2016, from: http://siriusdisclosure.com/wp-content/uploads/2013/03/Rendlesham-Forest-Halt-Memo.pdf.

52 Rendlesham Ref 52: The Rendlesham Forest Incident. Copyright 2011 by ufo-evidence.org. Retrieved October 20, 2016, from: http://www.ufoevidence.org/documents/doc663.htm.

53 Rendlesham Ref 53: MUFON. Rendlesham Forest Incident 1980. Copyright MUFON International, 2014. Retrieved October 20, 2016, from: http://www.mufon.com/rendlesham-forest---1980.html.

54 Rendlesham Ref Nick Pope: Pope, Nick. Unknown Date. The Bentwaters UFO Incident. Retrieved from: www.nickpope.net/bentwaters-ufo-incident.htm.

55 Roswell Ref 55: Roswell Daily Record. July 8, 1947, Volume 56, Number 189. Copyright 1947 by The Roswell Daily Record, Inc.

56 Roswell Ref 56: Berliner, Don and Friedman, Stanton. 2004. Crash at Corona, p. 72. Copyright 2004 by Don Berliner and Stanton T. Friedman. Paraview Special Edition published from original by Marlowe & Company in May 2004.

57 Roswell Ref 57: Berliner, Don and Friedman, Stanton. 2004. Crash at Corona, p. 72. Copyright 2004 by Don Berliner and Stanton T. Friedman. Paraview Special Edition published from original by Marlowe & Company in May 2004.

58 Roswell Ref Museum: Courtesy of Roswell International UFO Museum, 2100 N. Main Street, Roswell, New Mexico, USA.

59 Roswell Ref Exon: Randle, Kevin. 2007. July 1947: The Wright Field Connection/General Arthur Exon. Retrieved October 20, 2016, from: http://www.nicap.org/roswell/roswell_exon.htm.

60 Roswell Ref Museum: Courtesy of Roswell International UFO Museum, 2100 N. Main Street, Roswell, New Mexico, USA.

61 Roswell Ref Exon: Randle, Kevin. 2007. July 1947: The Wright Field Connection/General Arthur Exon. Retrieved October 20, 2016, from: http://www.nicap.org/roswell/roswell_exon.htm.

62 Salter Ref 62: A Short Statement by John R. Salter, Jr. dated February 6, 1990. Retrieved from the John R. Salter Papers Archive.

63 Salter Ref 63: A Short Statement by John R. Salter, Jr. dated February 6, 1990. Retrieved from the John R. Salter Papers Archive.

64 Salter Ref 64: Salter, John R. Jr. 1989. An Account of the Salter UFO Encounters of March 1988, p. 1–16. Copyright 1989 by John R. Salter.

65 Salter Ref 65: Salter, John R. Jr. 1990. Recall from the 3/20/88 UFO encounter in Wisconsin-Re; Zeta Reticuli planetary systems. Published January 9, 1990. Copyright 1990 by John R. Salter Jr.

66 Salter Ref 66: An Account of the Salter UFO Encounters of March 1988, p. 20.

67 Salter Ref 67: Salter, John R. Jr. 1992. Extraterrestrial Contact and Human Social Justice Sensitivity and Action, p. 2–12. Copyright 1992 by John R. Salter Jr.

68 Salter Ref 68: An Account of the Salter UFO Encounters of March 1988, p. 20.

69 Salter Ref 69: Salter, John R. Jr. 1992. Extraterrestrial Contact and Human Social Justice Sensitivity and Action, p. 11–12. Copyright 1992 by John R. Salter Jr.

70 Walton Ref 70: Lorenzen, Coral, Aerial Phenomena Research Organization (APRO). The Walton Abduction. Retrieved October 20, 2016, from: http://www.nicap.org/reports/751105snowflake_report.htm.

71 Walton Ref 71: Clark, Jerome. 1998. *The UFO Encyclopedia: The Phenomenon from the Beginning*, Volume 2, p. 628–629. Copyright 1998 by Jerome Clark. Published 1998 by Omnigraphics Books, ISBN 0-7808-0097-4.

72 Walton Ref 72: Lorenzen, p. 3.

73 Walton Ref 73: Clark, p. 632.

74 Walton Ref 74: Clark, p. 633.

75 Walton Ref 75: Murphy, Ryan. 1993. Reworking Fire in the Sky. Entertainment Weekly. March 19, 1993. Copyright 2016 Entertainment Weekly, Inc. Retrieved October 20, 2016, from: http://www.ew.com/article/1993/03/19/reworking-fire-sky.

76 Walton Ref 76: Speigel, Lee. 2015. UFO-Alien Abduction Still Haunts Travis Walton. Huffington Post. April 23, 2015. Copyright 2015 by Huffington Post, Inc. Retrieved October 20, 2016, from: http://www.huffingtonpost.com/2015/04/23/travis-walton-still-haunted-by-ufo_n_7119910.

77 Zamora Ref 77: Project Blue Book case number 8766. Retrieved October 20, 2016, from: http://www.ufocasebook.com/Zamorareport.html.

78 Zamora Ref 78: Project Blue Book case number 8766. Retrieved October 20, 2016, from: http://www.ufocasebook.com/Zamorareport.html.

79 Zamora Ref 79: Sparks, Brad. Lonnie Zamora/Socorro Landing Case. National Investigations Committee on Aerial Phenomena. Retrieved October 20, 1964, from: http://www.nicap.org/640424socorro_dir.htm.

80 Zamora Ref 80: Barrow, Robert. An Incredible Admission: What Did the Air Force Mean? NICAP Pursuit Winter, 1979. Retrieved October 20, 2016, from: http://www.nicap.org/reports/640424zamora6.htm.

81 Zamora Ref 81: Sparks, Brad. Lonnie Zamora/Socorro Landing Case. National Investigations Committee on Aerial Phenomena. Retrieved October 20, 1964, from: http://www.nicap.org/640424socorro_dir.htm.

82 Zamora Ref 82: Socorro El Defensor Chieftain. April 28, 1964. Retrieved October 20, 2016, from: http://www.theufochronicles.com/search/label/Socorro%20Incident.

83 Zamora Ref 83: Map of winds and discussion. Retrieved October 20, 2016, from: http://www/roswellproof.com/Socorro/SocorroWinds_April_24_1964.

[84] Zamora Ref 84: Stanford, Ray. 1976. Socorro "Saucer" in a Pentagon Pantry. p. 34. Published in 1976 by Blueapple Books. ISBN 0-917092-00-7.

[85] Zamora Ref 85: Druffel, Ann. 2003. Firestorm: Dr. James E. McDonald's Fight for UFO Science, p. 219. Published 2003 by Wild Flower Press. ISBN 0-926524-58-5.

[86] Zamora Ref 86: Lorenzen, James and Coral. 1976. Encounters with UFO Occupants. Published 1976 by Berkeley Publishing Corp.

CHAPTER 20

Disclosure: An Overview of What We Don't Know

*I am a firm believer in the people. If given the truth,
they can be depended upon to meet any national crisis.
The great point is to bring them the real facts.*

—Abraham Lincoln[1]

Key words: pragmascience, proprietary, post hoc and proper hoc fallacy, principle of noncontradiction fallacy, men in black

Dr. Carl Jung, world-renowned psychiatrist, also tries to understand the reason for the accusations of the military and their reluctance for Disclosure when he says,

> What astonishes me most is that all the American Air Force, despite all the information it must possess and despite its alleged fear of creating a panic similar to the one which broke out in New Jersey, Welles's radio play[2] is systematically working toward that very thing by refusing to release an authentic and reliable account of the facts.[3]

There have been twelve serving U.S. presidents since the generally-acknowledged beginning of the Atomic Age. In terms of a timeline, not coincidentally when substituting Atomic Age for the modern UFO era, the answer is the same. The first one in service was Harry S. Truman who inherited the presidency from Franklin D. Roosevelt in 1945.

Truman was the first president to have his name attributed to the subject of the "modern UFO era" or the "modern age of ufology" in any way, more exactly, in a significant way. Every U.S. president since the modern era has been drilled by the press and other representatives of the general public to find what they know about UFOs/UAPs and extraterrestrials or what they will do about disclosure: "letting us know." So the accurate application of one of the phrases to any U.S. president would be the "in any way" phrase.

Harry Truman, thus, was generally known to be involved in a significant way on the subject today known as *ufology*. When the question is asked, "How many presidents could be linked in a significant way to ufology?" a scholar on the subject would reason to answer two or three. There, of course, is no universally absolutely correct answer to this question. Identification of the two being Truman and Dwight D. Eisenhower. A conjecture may be made for Ronald Reagan only because he had a little substantive attribution in that he reported multiple UFO/UAP sightings in his time in public office and hinted at the subject in some of his public speeches.

Dwight Eisenhower was the other of the two presidents with the most valuable ufology attributes. These came from both activity within the public's eye and his own eye as he was also positioned as an eyewitness to a UFO mass sighting at Operation Mainbrace in September 1952. Eisenhower was conjectured to have participated in multiple meetings with representatives not of this Earth.

The correlation between the timeline of chronology and presidential meaningful involvement with ufology fits really well. This is achieved when thinking critically about what governmental actions took place from the end of World War II, aka the beginning of the Atomic Age, through the decades and leading up to where we are now in society's quest for disclosure. The events within government activity which we know are real and existed because of our collection of documents are abundant in quantity and authenticity. We also know there have been a large number of disclosed government documents that have been "whited out" before public dissemination through the decades.

There has also been an assortment of hoaxed perpetrated documents publicized throughout the modern age of ufology. Their existence is extremely loathsome to those of us aligned with a pragmatic approach to ufology or what is known as *pragmascience*.

There is a lot more to the disclosure paradigm than the participation of two presidents, given that there is a dearth of chronological events represented in the public domain. Ronald Reagan, Jimmy Carter, and Gerald Ford, who had his most extensive participation not as president but as U.S. representative, Michigan in the sixties, made publicized attempts to catalyze the issues more from a motivational impetus. All whom had encounters of their own.

The disclosure and secrecy issues do not own a Genesis date per se. It is recognized by ufology that the modern UFO era had its origin in June 1947 when businessman Kenneth Arnold, also a private pilot, had his Washington State sighting. His description of the flying craft having flight behaviors like "saucers skipping on the water" had a unique resonation with America of that time. The next couple weeks brought more public sighting events, building up to the July 2 and 3 episode that became the Roswell incident.

Up until that July in the Matrix world of disclosure and secrecy, the atmosphere in America was different, indeed far, far different than even the last thirty or more years. World War II had just ended. The entire planet was dressing their wounds. America was extremely fortunate in that not very few bombs had been dropped on our soil. In fact, there was an unprecedented fever of blind faith "in afterburner mode" of the public with our government.

Inherent in the wartime strategy of any active confrontational nation was a behavior that included secrecy. As we are still learning much from even seventy years later, the World War II event theater was no different.

For example, the Manhattan Project, a joint effort by the United States and Canada, ramped up full-scale operations to manufacture inventory of the nuclear elements uranium and plutonium for the upcoming bomb testing. The Tennessee Valley that contains over 4,400 acres of Oak Ridge National Laboratory was over 17 miles long. A single uranium manufacturing plant itself was over a mile long. The workers did not know many of the details beyond what their immediate day-to-day jobs involved except that they were doing it in the service of their country, enough said.

What they did not know was that atomic weapons were being built and tested. Only a select group of nuclear scientists sequestered in the northern New Mexico settlement of Los Alamos and the Roosevelt Administration knew of those plans. So here, many layers of secrecy existed. There were too many examples of such secrecy throughout the war to need to elicit a more favorable conclusion. The risks were too existential, as noted.

Secrecy had to exist or run a risk of compromise in war situational strategy that could cost many lives and economic hardships. In other words, the security of the nation could be at increased risk for the failure of such a program. This notion is so vital to remember as it was and is one of the most significant rationale by similar to businesses, government, and/or military for keeping information, data, and/or knowledge about something to themselves. The term they use for it is *proprietary*.

The U.S. Army enacted the policy of secrecy during a wartime event of alleged encounters with unidentified flying objects over Los Angeles in February 1942. The world was told exactly 1,430 rounds of ammunition were fired that night. Many reports of radar acquisition of object data and mass observations from thousands of civilians, military, police, and professionals were also made public. What, if anything, did the military not tell us about that night? There is no conspiracy rationale argued here. If there were enemy planes in the air, namely Japanese planes, chances are the public would not have been told, at least until well after the war. This is one of many examples of secrecy which was required to lessen any existential risk to the Allied Forces. The stakes were just too high. But why did the Army disclose that exactly 1,430 rounds were fired? Could they have been trusted to identify that precise number?

To be exact, secrecy and disclosure issues have existed throughout humankind's history. At least in times of war, a nation will experience a noted uptick in the graph that involves disclosure and secrecy variables. What about times of peace?

While the mega tonnage of bomb destruction was over in 1945, a *Cold War* had already emerged. This meant more secrecy, more espionage, and more disclosure issues. While civilians of their nations heard posthaste about successes in their government, business, or technology, there existed an incremental nature to the disclosure of those successes. This means that only those advancements that do not leapfrog to a logically small and incremental approach would be made public. In other words, if the advancement was too radical to be fully embraceable, then it may not be disseminated until a more proper time and situation was decided by involved parties, unless a catastrophic event demanded it.

For example, development of the atomic bomb was among the very most secret of projects ever undertaken. Civilians only heard about it after August 6 and 9, 1945. Indisputably, atomic technology was too radical to be introduced to the general public all at once under other circumstances. If those bombs were not dropped, how long would it have been before the public obtained full disclosure?

So there was no single origin of activity that defined an issue of the nature of disclosure in ufology. There were the secrets surrounding most aspects of the Cold War and national security. Many of those secrets involved technology and/or knowledge that, if disclosed, not only may have compromised the public security but also may have upset the paradigm of a "gradual release of incremental technology." The government and military often but not always developed the most advanced technologies of that particular time. After a time, those technologies were gradually integrated into society.

Not to say the government and military always developed the most advanced technologies. For example, in the early 1930s, pioneer Robert Goddard was unrestrained in trying to convince the U.S. Army to adopt his rocket technology. The Army concluded there was no practical use for it.

So with the existing backdrop in the mid-1940s, we know there was motivation and desire to have secrets and sisclosure issues, as history has taught us today about many that existed in World War II and afterward, also that we are learning more and more of them.

Project Sign 12/30/1947–2/11/1949

The modern age of ufology, as noted, was generally agreed upon by most followers of the subject of either side of the argument to have begun in June 1947 with the sightings of pilot and businessman Kenneth Arnold. The lens of the ufology movement became much more focused and expansive after Arnold's pioneering encounter. His observations were very credible, and his descriptive information and data were corroborated so effectively by the public that it is conjectured the term *flying saucer* came directly from Arnold's reporting.

The landscape of ufology study was multiplied immediately after the Arnold sighting. Adding to this multiplicity was that the timing of it coincided with the Air Force's imminent separation from the U.S. Army. For most of the publically disseminated knowledge and information filtered through the Army-Air Force in those times. With all the press and mass media attention devoted to the newest flap or wave of UFO sightings, it was the Army-Air Force that bore the brunt of interaction and explanation with them. This would attach to the upcoming new Air Force when the separation took place.

The then current flap continued four days later when multiple sighting reports originating from Lake Meade, Nevada, to Montgomery, Alabama, and White Sands Proving ground, New Mexico. These sightings were made in the afternoon locally and by Army-Air Force officers and pilots. Witnesses from many other parts of the country reported their own sightings.

On June 21, three days before the Kenneth Arnold sighting, two water patrolmen on duty with other passengers that included a canine in the tributaries of Puget Sound, at Tacoma, Washington, encountered six unidentified objects flying overhead at about two thousand feet. One of the craft was seen to be falling behind and below the other five, losing altitude. The other five ships then circled the apparently damaged one, and one eventually touched up against it and stayed attached for about five minutes.

Without any warning or pre-perceived notion to any of the boat occupants, the damaged ship released a boiling torrent of burning metal from its bottom. The metallic debris, glowing red-hot, rained down all around and into the boat, burning and killing the dog and burning the arms of a couple of the occupants. All the other debris that hit the water caused steam geysers and smoke clouds. Debris that hit land burned briefly on the landing spots before extinguishing.

The patrolmen had tried using their radio to report back to their command base, but it reacted with so much interference no communication could be established. The six ships then headed west out to sea. Next, the radio then returned to normal functionality.[4]

This episode, known as the Maury Island incident, received a lot of publicity much into the future after it occurred for many reasons. A measure of publicity happened because of an alleged encounter between the patrolmen and a first documented report of and with *men in black*. Another reason was there was an abundance of metallic debris that poured down to the earth and waterway. A reported estimate from the patrolmen was around a ton. But there was more intrigue.

Two Army-Air Force officers, a Captain Davidson and Lieutenant Brown, went to Tacoma and met with the patrolmen they had spoken to, so they decided to talk with the Air Force instead of newspapers partly because they had obtained some of the debris that had fallen from the disabled ship overhead. The men in black visit happened to them the next day.

Capt. Davidson and Lt. Brown concluded their investigation and boarded the B-25 Army-Air Force plane back to Hamilton, California, with some of the recovered debris. The plane crashed shortly after takeoff from Tacoma. The pilot and copilot parachuted to safety. Capt. Davidson and Lt. Brown were killed.

*****

The event that is suspected to have changed the existing paradigm of government and military secrecy is the Roswell incident. However, not much of the factual information would become public knowledge until many years after July 1947. There was an initial international flurry of newspaper press and broadcast media in the days at the beginning of the alleged retrieval operation by the U.S. Army-Air Force.

A significant episode that is most relevant to mention here and as it will be also mentioned in a later documentation of the facts and data about the Roswell incident involves the early days of press coverage. The FBI was detected to have been covertly monitoring communications of the press and broadcast media as is proven next.

A teletype operator at Roswell radio station KSWS, Lydia Sleppy, describes the event that took place around 4:00 p.m. on July 7, 1947, with the following statement:

> We were Mutual Broadcasting and ABC. I was the one who did the typing. He (John McBoyle, reporter for KSWS radio station in Roswell) told me he had something hot for the network. I said, "Give me a minute, and I'll get the assistant manager," because if it was anything like that, I wanted one of them there while I was taking it down. John was dictating and [Karl Lambertz, KSWS assistant manager] was standing right at my shoulder when the bell came on [signaling an interruption]. Typing came across: "This is the FBI, you will cease transmitting."
> Sleppy continued by saying:
> "I had my shorthand pad, I told [Karl] that I had been cut off but that I could take it in shorthand and then we could call it in to the network. I took it in shorthand, as John went on to give the story. He had seen them take the thing away. He'd been out there [presumably at the Foster ranch] when they took it away. And at that time,

John said they were gonna load it up and take it
to Texas. But when the planes came in, they were
from Wright Field."[5]

With all due respect to Edward Snowden, his actions in June
2013 of uncovering the National Security Administration's (NSA)
policy of spying on American citizens was not the first time this has
happened. The U.S. government and military agencies have done
this very thing many times throughout history. Here is an example
of such secrecy and intimidation that took place on July 7, 1947.
The teletype dictation memo, the FBI teletype "cease transmission"
message, and the machines they were printed on are on display at
the Roswell UFO Museum in Roswell, New Mexico. Even back in
1947, there were technological capabilities for covert monitoring and
proliferation of maintaining secret activities in place.

The initial days' reports and broadcasts about the Roswell incident were refuted by high-ranking officers such as General Roger
M. Ramey and General Samford. They were the point people tasked
with later explaining to the press that all was mistaken, that a weather
balloon was the object that crashed. Ramey and Samford were to
renew their public popularity during the July 1952 Washington, DC
UFO flap. The public seemed content then to move back into their
daily routines.

To be noted initially about the Roswell incident involves a
library full of data, evidence, testimony, and conjecture. Curiously,
there was nary a mention of the Roswell incident in such publications
as *The Condon Report, Project Blue Book, the UMBRA documents*, and
The Robertson Panel, among others. There is also contained within
that library of information about Roswell more attention than any
other event within the focus of ufology.

So the public got off the government's back quickly in this case.
What was alleged to have followed the July 1947 events concerning
Roswell have been subject of a sometimes acrimonious turn-around
refutation between supporters and non-supporters of disclosure,
secrecy, and other conspiratorial issues of the Feds and the military
complex.

Two events happened in the latter half of 1947 whose existence is supported by a varying level of documentation and testimony. The first was a chronological coincidence. On September 18, 1947, the National Security Act created the U.S. Air Force and began separating its existence away from the U.S. Army. The completion of this process would not occur until the last Transfer Order was signed on June 22, 1949. This can be unanimously proven by the documentation that is in the public record.

The second happened at the same time in 1947 but cannot claim the same achievement of unanimity of proven documentation as the creation of the Air Force. The second event was the alleged creation of what is known in the ufology discipline as the *Majestic 12*, or *MJ-12*, and *Majic 12* as acronyms.

Evidence for both do exist, though much more of the Air Force creation. Not nearly as much exists in the realm of Majestic 12 evidence. The discourse surrounding the debate, though, has lasted until today, so this must be mentioned. As well to mention it for this reason and for the more pragmatic reason that there is some logic with the reliability of these events having happened and were a real fit within the larger landscape of the psychology and sentiment of the government, the military complex, and the structure of society's psyche at that time. While both sets of evidence can be inspected and analyzed by investigators from whatever discipline to draw conclusions from, there is less to study from Majestic 12 and less for both the investigator seeking to confirm its existence and for the refuter or debunker.

In the whirlwind of the Roswell incident, physical evidence was alleged by some to have been transported away from the crash debris field to a variety of locations scattered around the United States. The Air Force had a lot on its administrative plate with simultaneously beginning the separation from the Army and having to answer to the Roswell incident and a growing frequency of UFO sightings and encounters both domestic and abroad, all this within the landscape of growing tensions over the budding Cold War and adaptation activities of a post-war society.

The formation of the Majestic 12 was nowhere near any public domain, so there was no immediacy of information dissemination in existence. What was alleged to have happened was that, in the immediate aftermath of the Roswell infatuation a process was initiated that would bring a group of twelve U.S. government intelligence and military officers and scientists together for an ongoing investigation. In addition to the formal investigation, fact-finding activities regarding phenomena that seem to have no earthly or natural explanation for their existence were established. On September 24, 1947, a letter from President Truman to then Defense Secretary James Forrestal provided the president's approval to proceed with formation of Majestic 12.

That formation activity created the following group of government, military, and scientific minds:

Dr. Vannevar Bush	Rear Admiral Roscoe H. Hillenkoetter
James Forrestal (replaced by	Air Force General Nathan Twining
Gen. Walter Bedell Smith)	
General Hoyt Vandenberg	General Robert M. Montague
Dr. Jerome Hunsaker	Rear Admiral Sidney Souers
Defense official Gordon Gray	Dr. Donald Menzel
Scientist Detlev Bronk	Dr. Lloyd Berkner

It has been verified that Dr. Menzel, Rear Admiral Hillenkoetter, Twining, and Vandenberg maintained lengthy and documented relationships with the ufology subject through the years in their various activities. Vandenberg was known to have been involved in the Roswell incident in that he was the acting Army-Air Force chief of staff that July whose name was in the header of a telegram General Roger Ramey was holding when some significant photos were taken of the alleged debris explained as the real Roswell wreckage on July 8. As noted elsewhere, that telegram was discovered to be a description of the Roswell crash event, debris field, and a proposed method-of-proceeding instruction set.[6]

No documentation regarding direct and detailed Majestic 12 investigations has been uncovered. It is generally believed that

Majestic 12 operated (some say still operates) within the veil of the highest level of government secrecy ever created in the United States. The president, then Truman, was included in this loop and is indirectly referenced by written evidence later in testimony from Air Force Brigadier General Arthur E. Exon, commanding officer of Wright-Patterson Air Force Base in Dayton, Ohio.

The next verified dated documentation came from the discovery of a secret memorandum written by Canadian Engineer Wilbert B. Smith, who worked for the Canadian Department of Transportation. The November 21, 1950, memo recommended Canada form a study project to be called *Operation Magnet* to investigate the UFO phenomenon. Smith's "own discreet inquiries had uncovered that the U.S. government had undertaken classified UFO investigations by a small group headed by Doctor Vannevar Bush."[7]

A set of briefing documents that were discovered in the 1980s, which included the 1947 Truman-Forrestal memo, are more speculative in terms of validity within this debate. A November 18, 1952, briefing document was prepared for the upcoming president, Dwight D. Eisenhower. The briefing detailed the origins of the Roswell crash and subsequent investigations undertaken by the Majestic committee.

Another document discovered a little time after the briefing document and the Truman-Forrestal memo was a letter from Presidential Assistant Robert Cutler to General Nathan Twining, an alleged original member of Majestic 12. The July 14, 1954, letter made continuous references to Majestic 12, also known from then as MJ-12.

There is some potential for valid correlation between MJ-12 and a known existing group in 1954 called the *special group*. President Eisenhower created the NSC 5412/2 Special Group with the same highest secrecy clearance that Majestic 12 was felt to possess. The NSC 5412/2 Special Group included twelve committee members with the president as the thirteenth member.

This reference to thirteen members may have been what Air Force Brigadier General Arthur E. Exon was speaking about when, in 1965, he reported that "a group of high-ranking officers and scientists were somehow involved with UFO studies; nicknamed the Unholy Thirteen."[8]

The term of Project Sign contained one more tragic UFO-related incident and more report bookings. On January 7, 1948, Air Force pilot Captain Thomas Mantell was killed, and his plane was reported to have crashed. The preemptive sighting was made in the afternoon by both civilian and Air Force personnel, including the base commander of Godman Air Force Base. The craft was reported to have been spherical in shape and about 250 feet in diameter.

Three planes sortied after the UFO. Two of them lost contact with the craft. Mantell, however, maintained his tracking and then climbed an altitude too close in pursuit of over twenty thousand feet. This is important because none of the sortied fighters had oxygen aboard. Mantell continued his pursuit while being given radio instructions from the control tower radar that had been tracking the entire episode.

There were numerous other sighting episodes reported throughout the rest of 1948. Many from civilians and some from commercial airline pilots and flight crews in addition to Air Force fighter jet encounters and their base personnel. Many of these reports were classified by Project Sign as "restricted." Use of a "secret" designation was occasionally enlisted. The use of top secret was not used much by Project Sign and would only be picked up as a designation protocol when Project Sign morphed into *Project Grudge*.

This secret designation was used in Project Sign's final report issued as such in February 1949. The groundswell of increased usage and support for the morphing attitude of the government and military complex toward secrecy and disclosure is contained in the recommendations of the last Project Sign report: "Future activity on this project should be carried out at the minimum level necessary on future reports. Future investigations of reports would then be handled on a routine basis like any other intelligence work."[9]

Project Grudge 2/12/1949–12/27/1949

The breathable duration of Project Grudge comprised less than one year's duration. The dearth of the investigation was made clear in that only 244 reports of sightings from dates up to 1949 were studied. The most significant single participant during Project Grudge's duration was J. Allen Hynek. He recorded the findings in statistical format.

Hynek summarized the categories as, "About 32% of the cases, astronomical objects. Another 12%, weather balloons. Some 33% were dismissed as hoaxes or reports that were too vague. A residue of 23% was considered as 'unknown.'"[10]

There was concern even at this time by the Air Force over the variety of problems inherent in pursuing any studies of the UFO phenomenon. In the *Grudge Report* of August 1949 known as, Technical Report No. 102-AC 49/15–100: "Reports of unidentified flying objects are the result of: 1) Misinterpretation of various conventional objects; 2) A mild form of mass-hysteria and war nerves; 3) Individuals who fabricate such reports; 4) Psychopathological persons."

Here were statements that display the same scope of the Air Force's attitude toward the UFO phenomenon and the possible future direction of study potential four years prior to disclosure and secrecy regulations being made permanent within the government and the military industrial complex.

Quote: *The Condon Report* pgs. 852–853: "The investigation of study of reports of unidentified flying objects be reduced in scope, a current collection directives be revised to provide for the submission of only those reports clearly indicating realistic technical applications."[11]

The years 1950 and 1951 were two mostly quiet years in the realm of the disclosure and government secrecy paradigms. There were episodes within the military and national security that became conducive to potential secrecy issues. Those were mostly in relation to testing of thermonuclear weaponry by both the United States and Russia and the emergence of other inductees into the nuclear family, such as France. The landscape of ufology was mainly quiet. As 1951 came into its last quarter, there was a series of encounters that included dogfights with unidentified aerial objects over Fort Monmouth, New Jersey, and were witnessed by the pilots of the T-33 fighters, base and control tower personnel, and an abundance of visiting Air Force officers, as well as the events being captured by multiple radar and other instrumentation.

Correspondence moved between the Fort Monmouth base and the ATIC and other Air Force agencies which led to the eventual appointment of Captain Edward J. Ruppelt to head an effort to remake Project Grudge.[12]

The year 1952 began a yearlong wave of quantitatively historical sightings that only lessened at the end of the year. The U.S. agency responsible for the then current report center was the Air Technical Intelligence Center (ATIC). In total, they received over 1,500 that year. Project Grudge was already in transition when the events of 1952 exploded. A reorganization of Project Grudge which included fiscal infusions (though very short-term and ultimately nonexistent) as well as personnel and a new leader emerged.

Captain Ruppelt was given the authority to originate and amplify a systematic study of the UFO phenomenon. This effort included encouraged input from the public to report their encounters. A dedicated effort to include scientists, engineers, computer programmers, and other professional opinions in the study added a lot of credibility to the ufology concept.

This new motivation attracted an enthusiasm by the press and public in renewed ways. What the Air Force did not anticipate entirely was the influx of so much raw data and study substance. Ruppelt was dedicated and determined to minister to new heightened levels of pragmatism and critical thinking within Project Grudge. He was familiar to past administration paradigms and vowed not to repeat them. Ruppelt made it clear that "open speculation or argument about the origins of unidentified flying objects or the legitimacy of the reports was taboo and ousted several staff members who pushed one theory or another."[13]

It is noted that Ruppelt headed Project Grudge during a particularly exciting and busy time in ufology history in the modern era. So much so that he introduced the community to J. Allen Hynek, then an active Air Force Astronomy consultant, to the position of chief scientific consultant for Grudge. Hynek would then spend fifteen years helping the Air Force perform their functions in reference to the UFO/alien phenomena and ufology.

Project Grudge would lose its official status before 1952 ended. The on-again/off-again Air Force funding turned on again for a brief period. Also the name was changed to *Project Blue Book*. This name would remain until 1969 when the Air Force decided to exit the ufology business, at least for public purposes.

The year 1952 was a landmark year in the annals of ufology not only because of the record wave of encounter reports but also because of this activity the landscape of disclosure and secrecy began to change. The Washington, DC flap of July 1952 caused the Air Force to man up and answer to public anxiety and apprehension by conducting their largest post-event press conference since World War II. Major General John A. Samford, Major General Roger A. Ramey, Colonel Donald L. Bowers, and Ruppelt, among other Air Force brass took turns fielding the long press session.

The direction of the press conference was influenced by the effects the 1951–1952 wave of UFO encounters was having on the Air Force reporting offices and their incapable of obtaining rapid determinations and explanations of why and how these events were happening. The Project Grudge/Bluebook offices were not able to keep up with their hoped for dedicated and thorough study of the casebook that was rapidly multiplying. There were discussions among various government agencies, including the NSA, Air Force, Navy and Army, the Office of Scientific Intelligence, the ATIC, and—more significantly—the Central Intelligence Agency, over growing concerns this wave was negatively affecting their administrations.

On the same day the Air Force Ramey/Samford press conference was held, the CIA Assistant Director of Intelligence Ralph Clark issued a memo announcing formation of a study that would directly influence the formation of the Robertson Panel, a study conducted jointly by Air Force and CIA officials and a panel of well-known and renowned scientists from different disciplines.[14]

The Clark memo referenced a meeting the day before with rocket engineer and scientist Frederick C. Durant that pushed the critical thinking toward the creation of the Robertson Panel. The CIA continued its presence in the pre-panel formation with hints at what future panel recommendations and end points would consist of. An internal CIA paper dated August 19, 1952: "In summarizing this discussion, this whole affair has demonstrated that we arrive at two danger points which, in a situation of international tension, seem to have National Security implications."[15]

A third memorandum to General Walter Bedell Smith, director of Central Intelligence, dated September 11, 1952, was later addressed by the upcoming Robertson Panel: "The study makes no attempt to solve the more fundamental aspect of the problem which is to determine definitely the nature of the various phenomena which are causing these sightings or to discover means by which these causes and their visual and electronic effects may be immediately identified."[16]

The terms were definitely and immediately thematically recurring and crucial to the direction the Robertson Panel would take when they met formally in January 1953. The government and the military were frustrated they could not gain the knowledge to provide conclusive and immediate answers to these phenomena as each episode occurred. The proliferation of the UFO phenomena, which had been coming in massive waves followed by respite periods for the last few years could not be explained or controlled to its satisfaction.

Explanation and control were two of the most central issues facing the government and military for long-term solutions to this increasingly burdensome evolution. Because they could obtain neither, the Robertson Panel was formed and took central stage to provide answers. The four-day conference included Frederick C. Durant, then CIA officer, Edward J. Ruppelt, Samuel Abraham Goudsmit, a nuclear physicist at Brookhaven National Laboratories, associate of Albert Einstein and discoverer of the theory of electron spin in theoretical physics, and J. Allen Hynek, among others.

A point of fact about this panel was that Ruppelt was the first to hint at the existence of such a conference. The Robertson Panel was classified as secret by the government and military. He did make a brief public mention of the Robertson Panel in a 1954 interview for *TRUE Magazine*. Additionally, he mentioned it in his 1956 book, *Report on Unidentified Flying Objects*. Ruppelt did not identify any of the participants or any identification of governmental or institutional involvement due to secrecy and disclosure concerns. Disclosure of the existence of the panel was only made in 1966 thirteen years later.

Upon analysis and discussion of approximately 1% of the over 2,300 cases from 1951–1952 immediately available to them, the panel followed the tone of the concerns expressed earlier by the CIA and Air Force and other agencies in their recommendations. The two central themes focused on the educational aspects of being able to make unidentified flying object sighting and encounter determinations based on known natural phenomena and of allowing for the debunking of episodes when they happen.

The panel's concept of a broad educational program integrating efforts of all concerned agencies was it should have two major aims: training and debunking. The training aim would result in proper recognition of unusually illuminated objects (e.g., balloons, aircraft reflections) as well as natural phenomena (meteors, fireballs, mirages, noctilucent clouds).[17]

The debunking aim would result in reduction in public interest in flying saucers which today evokes a strong psychological reaction. This education could be accomplished by mass media such as television, motion pictures, and popular articles. Basis of such education would be actual case histories which had been puzzling at first but later explained. As in the case of conjuring tricks, there is much less stimulation if the secret is known. Such a program should tend to reduce the current gullibility of the public and consequently their susceptibility to clever hostile propaganda. The panel noted that the general absence of Russian propaganda based on a subject with so many obvious possibilities for exploitation might indicate a possible Russian official policy.[18]

Among the formal recommendations made were:

> The evidence presented on unidentified flying objects shows no indication that these phenomena constitute a direct physical threat to national security, which are attributable to foreign artifacts capable of hostile acts, and that there is no evidence that the phenomena indicates a need for the revision of current scientific concepts.

The continued emphasis on the reporting of these phenomena does, in these parlous times, result in a threat to the orderly functioning of the protective organs of the body politic. The panel recommended that:

> The national security agencies take immediate steps to strip the unidentified flying objects of the special status they have been given and the aura of mystery they have unfortunately acquired. That the national security agencies institute policies on intelligence, training, and public education designed to prepare the material defenses and the morale of the country to recognize most promptly. These aims may be achieved by an integrated program designed to reassure the public of the total lack of evidence of inimical forces behind the phenomenon, to train personnel to recognize and reject false indications quickly and effectively, and to strengthen regular channels for the evaluation of and prompt reaction to true indications of hostile measures.[19]

"True indications of hostile measures" means episodes that affect national security. So from then on, the methods and proliferation of secrecy and disclosure within government and the military complex existed and were acted upon when the situation presented itself. The CIA would issue more memoranda to further develop and solidify the atmosphere of secrecy and disclosure that would prevent public access. They wanted regulation oversight for UFO reporting and a removal of such reporting from the public domain.[20]

It was a predictable if difficult road to traverse for the government and the "military industrial complex," a term President Eisenhower would create and use in his presidential exit address in 1961. Up to this time, before the last couple of years since Project Grudge evolved from Project Sign, there was an atmosphere of frustration that the objective of immediacy in explaining each UFO encounter almost "as it happens" with the same velocity as a voracious news reporter is obsessed with getting his name on the story and getting it published as the "scoop" story was not obtainable. Whatever science was introduced into each situation could not rationally explain the nature of what happened. So explanations of already-known natural phenomena would suffice as the real and rational ones no matter what their level of embodiment within or fit into reality or what they even sounded like when they were broadcast. The explanations had to be immediate and satisfying. Satisfying did not mean "to reality" but alternatively to either "get the public off their backs" and/or to not have to spend a lot of time, money, and resources on them.

Concurrently, with the January 1953 meeting of the Robertson Panel, Project Blue Book scaled back on collecting reports from the public. Eventually, they would stop this procedure altogether. Even though a prevailing thought within the military complex reasoned that even with the annual record of 1,501 reports received from the public, this represented only around ten percent of all true events in the United States. Nevertheless, their motivation to eventually stop receiving reports was based on the redundancy of the new data coming from successive reports.[21]

This was a seminal event that significantly shaped the lens of government and military policy of secrecy and disclosure habits going forward. The backdrop of the Cold War brought the excuse of national security to the forefront. Added to this was the growing public response to more frequent flaps and waves of mass sightings.

Fueling this was the influence the mass media and the entertainment industry had over the public in their own way. Mass media wanted the faster scoop stories to drive sales. The entertainment industry wanted to drive sales also and saw an entry point that took the direction of fiction in their presentations.

Here, technology was exploding with the development of daily live television and radio programming. The public was now able to receive breaking news more immediately than ever before. Did this characteristic of immediacy in daily society of achieving newer, faster, and more sophisticated technologies for news broadcasting harbor the same psychology in the minds of the public as it did the minds of the government and military complex within their own daily situations? Was this a driver of the shift in attitudes toward secrecy and disclosure "down the rabbit hole?"

The new Air Force protocols adopted from the recommendations of the Robertson Panel were implemented by the rest of the government and military agencies starting in 1953. Their implementation included educating the public by giving them scientific explanations for virtually all reported UFO encounters no matter how illogical and by debunking any and all of them by the same rush to judgment fallacy of *post hoc/proper hoc* and "my mind is made up" fallacy of *Aristotle's Principle of Noncontradiction* explanations as logical fallacies. This was done from a new categorized list of probable and possible explanations of events according to an identified category of such an event. This meant, whatever type of sighting or encounter the witness had and identified in his report, it was cross-referenced to the appropriate list of explanations and the pre-printed excuse of a known natural and terrestrial phenomenon would then be the answer.

Another step included gradually eliminating all Project Blue Book funding and withdrew most personnel from their work. Captain Ruppelt eventually left the Blue Book employ in August of 1953. By the end of 1953 Project Blue Book, under new Air Force Regulations, was reduced to a peripheral public relations function with only three employees to conduct it.

By the end of 1953, the Air Force was up to then the popularly-recognized and actively-involved military agency regarding the

UFO/alien phenomenon. Simultaneously, when Captain Ruppelt left Project Blue Book, the Air Force issued *Regulation 200-2*. This regulation dictated new UFO reporting, investigating, and public communicating procedures. The implementation of the reporting function now forced personnel to eventually remove all publicity regarding any UFO encounter from the public domain. Regulation 200-2 would strengthen previous regulations in that personnel were now forbidden to release any data whatsoever.[22] Additionally, the rest of any such anatomy of a UFO sighting report that lay within the investigation and public communication could now be successfully buried within the bureaucracy of the military resting upon the strength of Air Force Regulation 200-2. So much of the regulation was secret information that the regulation itself was secretly kept from the public.

This paved the way for additional pending regulations that were constituted from the U.S. Joint Chiefs of Staff in December 1953. The Joint Army-Navy-Air Force Publication 146 was published. JANAP-146 now made it a Federal crime for any person who had any knowledge of the existence of JANAP-146 to release any information about a UFO report to the public in any way. Only if any military or government authority could make a positive identification and give a satisfactory explanation of a sighting would it then be allowed to provide any information to the public. Secrecy and disclosure protocols were now in full force.

JANAP 146 subtitled *Canadian-United States Communications Instructions for Vital Intelligence Sightings (CIRVIS)* established protocols for communication of sensitive information requiring "defensive and/or further investigation." The criteria for qualification of such information were left to the discretion of the observer. As noted, the existence of JANAP 146 was only acknowledged within the official channels of the services themselves. The release of any information to the public was a crime under the legislation of the Espionage Act and was punishable by a prison term of up to ten years. While (JANAP) 146/CIRVIS was made known to the public via a press conference announcement, its data collection reporting, cataloguing, information analysis, and other protocols were now safe from the public domain under secrecy and disclosure laws and security.

JANAP 146 evolved as three different letters of promulgation (a misuse of the term promulgation that you need to conclude as to degree and scope). The last of these, JANAP 146(C) was published on March 10, 1954.

This letter of promulgation JANAP 146(C) provided "communication instructions for reporting vital intelligence sightings from airborne and waterborne sources, supersedes JANAP 146(B), aircraft (CIRVIS) and all other conflicting instructions."[23]

What (JANAP) 146(C) accomplished was to now integrate all events, whether in the atmosphere or on/under the water's surface where a sighting was or could have occurred, within the reporting process. The regulation mentions that foreign territory was excluded. An abundant use of the phrase "unidentified flying objects and individual surface vessels, submarines, or aircraft of unconventional design or engaged in suspicious activity or observed in an unusual location or following an unusual course" was evident.[24]

While they could never have positively known at the time that any earthly nation was in possession of and using technologies that could display the flight/motion characteristics and behaviors abundantly observed by growing thousands of witnesses, the Joint Chiefs of Staff had to ensure all such sightings and encounters were adequately accounted for in JANAP 146(C). Nevertheless, the final useable JANAP/CIRVIS mechanism was situated and has now existed in a mechanism of secrecy and disclosure for decades.

Other member groups of the public community affected in a very direct and significant fashion were airplane pilots. Yes, they were given a reporting opportunity whenever they observed or encountered anomalous aerial objects. They were not encouraged to do so in any fashion or reward opportunity. They were restricted by law to discuss their encounter at any time afterward with severe penalties as noted. The physicality of the CIRVIS reporting application itself developed a reputation as not being hospitable to the witness participant.

The vast majority of the time when a pilot did report a sighting, it was explained away to them by the litany of excuses the Air

Force used as debunking tactics. This was explicitly demonstrated to discourage pilots from pursuing their reporting. Also the pilot's employer was given full knowledge of their reporting. The term resurrected to describe the sentiment felt by the witness pilot as reasoning not to go forth with a report is professional suicide. This sentiment is felt by the science community to this day.

The net evolution of the implementation of JANAP 146/ CIRVIS was that subsequent reports of UFOs and alien encounters, which was an entirely new phenomenon facing the Air Force and the family of the military industrial complex going forward, dwindled down to a trickle. The mass media press, experiencing a precipitous drop in revenues from stories of this subject, also moved on. Project Blue Book would continue under secret operations until the 1969 wrap-up known as *The Condon Report.*

To succinctly summarize the ramifications of JANAP 146 and CIRVIS,

> The fact is that from 1947 to 1977 and even to the present, the data has flowed directly away from civil aviation and into the restricted arena of the military/intelligence domain via JANAP 146, the CIRVIS reporting system, Project Sign, Grudge and Blue Book and through other data collection programs and systems.[25]

*****

During the 1940s and all the 1950s and 1960s, there were numerous documented episodes. Here are a few of those that have contributed toward a rabbit-hole evolution.

The NICAP agency had been in existence for more than a few years as a library of knowledge about UFO encounters when in 1958 the director, Major Donald Keyhoe, was invited to appear on a CBS Television episode of the *Armstrong Circle Theatre* show to discuss the UFO phenomenon. Keyhoe was a long-time advocate of the existence of the phenomenon. As NICAP director, he was considered an expert on the subject. A handful of other Air Force spokesmen were also scheduled to appear on the show.

The other guests forced Keyhoe to submit his script to them for review. The Air Force instructed CBS to allot the majority of airtime which totaled twenty-five minutes and the rest of on-air time about seven minutes to Keyhoe. They also dictated that the discourse was to be monitored in real time throughout and scripted, not to allow for any unscripted material.

Major Keyhoe's script was heavily edited, and the only excuse that was offered by the Air Force was the script "was too long." The Air Force spokesmen appeared for the first three quarters of the program; Keyhoe in the last segment. Keyhoe insisted on being allowed to reference documents from the Air Technical Intelligence Center (ATIC) from 1948 through 1952 confirmed by Edward Ruppelt during his Project Blue Book term and from the Robertson Panel files in 1953 to communicate conclusions as to the reality of UFOs as intelligently-controlled ships with interplanetary origins. He was denied his request.

In his segment of the show which was broadcast live on January 22, 1958, Keyhoe diverted from his teleprompter and began to make the following statement:

> Now I'm going to reveal something that has never been disclosed before. For the last six months we have been working with a congressional committee investigating official secrecy about UFOs, if all the evidence we have given this committee is made public in open hearings, prove that UFOs are real machines under intelligent control."

CBS Television censored the live broadcast as the receiving television screens all over North America went blank.[26]

The Air Force often added to the landscape of covertness now that they had some criminal authority to utilize with JANAP/CIRVIS and the Espionage Acts in force. On December 24, 1959, the Air Force warned every air base commander in the continental United States about the seriousness of the UFO phenomenon as presented below:

> Unidentified flying objects, sometimes treated lightly by the press and referred to as "flying saucers," must be rapidly and accurately identified as serious USAF business in the ZI (Zone Interior). The phenomena or actual objects comprising UFOs will tend to increase with the public more aware.[27]

This was another is a series of actions the Air Force took to compartmentalize and internalize then future reports they either received or initiated. The criminalization mechanisms offered punctual sanctuary for their operations in this regard. Additionally, all the reports of UFO encounter episodes they received and could not succinctly and swiftly explain were excused away and/or not broadcast as evidenced by the many documents and testimonies that are now in the public domain. A founding member of the alleged original Majestic 12, Vice Admiral Roscoe Hillenkoetter, who was officially visible as former CIA Director from 1947–1950, and a NICAP member in the 1950s, received a copy of the warning and reported the whereabouts of additional copies. He was an advocate of disclosure for quite some time and, after the whirlwind of the Air Force-issued 1954 warnings, made a statement to the *New York Times*:

> It is time for the truth to be brought out in open congressional hearings. Behind the scenes, high-ranking Air Force officers are soberly concerned about the UFOs. But through official secrecy and ridicule, many citizens are led to believe the unknown flying objects are nonsense to hide the facts the Air Force has silenced its personnel.[28]

Throughout this time, the U.S. Congress was not completely unaware of this situation. There was some knowledge of characteristic secrecy perpetrated by the military complex. Senator Barry Goldwater—even with his resume that included terms as air force major general, reserve colonel and chairman of the senate intelligence committee, and an advocate for disclosure—could not crack the code that existed. He was the recipient of a rebuke from Curtis Lemay regarding access to top secret information at Wright-Patterson AFB with severe warning to "never bring this subject up again."

William H. Ayres, Rep., Ohio, was quoted as saying:

> "Congressional investigations have been held and are still being held on the problems of unidentified flying objects. Since most of the material presented is classified, the hearings are never printed." U.S. Senator Richard B. Russell, former chairman of the Senate Armed Services Committee: "I have discussed this with the affected agencies of the government, and they are of the opinion that it is unwise to publicize the matter at this time."[29]

U.S. Representative Walter H. Moeller gave this confident perspective of the American public: "[I have] every confidence that the American public would be able to take such information without hysteria. The fear of the unknown is always greater than the fear of the known."[30]

In 1962, the Air Force issued more orders that oversaw dissemination and/or withholding of information and knowledge from reported UFO episodes. These directive orders applied to the retention of such information pertaining to all printed materials such as books, articles, and scripts, and for either their usage for or additionally applied to separate talks and broadcasts of all kinds. As stated,

> "When the manuscript concerns military subjects, it will be submitted to the office

of information, applies to active duty, retired, civilian employees. By this order, the secretary of Air Force Office of Information must delete all evidence of UFO reality and intelligent control which would contradict the Air Force stand that UFOs do not exist. The same rule applies to A.F. press releases and UFO information given to Congress and the public."[31]

The Project Blue Book Years and The Condon Report

The unofficially official "death knell" of a reporting era, as suspected by some, came with the conclusion of Project Blue Book in 1969. The publication of *Scientific Study of Unidentified Flying Objects*, as directed by Dr. Edward J. Condon, provided an impetus for some groups, most importantly including the U.S. Air Force and all associated involved government and military agencies, to cease most all public dissemination of ufology knowledge. Dr. Condon was poorly regarded by some groups, namely the proponents of the UFO and extraterrestrial hypotheses, and a villainous savior to the skeptics and the data and knowledge accumulators.

There was an abundance of interoffice scandal during the tenure of the Condon study from beginning to the culmination, namely the 1969 publication of a 1485 page hardcover volume that was in effect the final summary of Project Blue Book.

What the *Scientific Study of Unidentified Flying Objects* publication offered was the figurative turnkey to the U.S. government and other affected parties to turn all future reporting and investigation of UFO sightings, encounters, and the newly-emerging subfield of extraterrestrial encounters inward and covert. This was the cause and effect of its importance on disclosure and secrecy.

The summary was unofficially nicknamed *The Condon Report*. A two-sentence synopsis was that the book concluded UFOs and anything construed to have extraterrestrial involvement was not worth investigating in a dedicated scientific manner. Following up with a long-held opinion the Air Force had used for decades that the security and safety of the United States and its citizens were not put in danger by these incidents and reports.

By the 1960s, there were more verbal requests for serious study of UFOs and extraterrestrial encounters from a widespread cross-section of society than ever before. Complaints of lack of attention to, responses of ridicule for, and naiveté from official government channels and Project Blue Book when answering calls for such investigation did not wane as they would have preferred.

Ironically, it was J. Allen Hynek in 1965, still under the Air Force employ and on the side of concerted skeptic, who convinced the Air Force Scientific Advisory Board (AFSAB) to organize a review panel to examine the course of Blue Book. It took over a year, but with help from a Congressional UFO hearing on April 5, 1966 (a rarity for this community), the search was on to complete step 1 of such a project. That would be of finding a group of civilians who were scientists who would enter such a study.

That task was difficult as most members of the scientific community had always been generally unwilling to risk their reputations, careers, and—in extreme situations marriages—for what they felt was no gain, financial, or otherwise. The structure of such scientific communities resided in two arenas. One was private industry and the other was academia.

It took months and a lot of rejections by university communities, but the one that did agree was the University of Colorado. They employed Condon and had an assistant graduate department dean, Robert J. Low, willing to survey research campus opinions for conducting such a study. After months of study, the University of Colorado and Condon agreed to participate because a large amount of sentiment suggested it was for fiscal reasons and that "Who could say 'no' to the Air Force."[32]

Condon became the director of the study, Low became the coordinator and two significant selections, and scientist David Saunders and astronomer Franklin E. Roach became principal investigators, among others. This announcement was about the only that produced mostly positive feedback from both the scientific and public communities. It was the only step in the project that ended with a positive conclusion.

The work began at the end of 1966. There were numerous and often humanly vindictive disagreements over many aspects of the project, including design, human resources, methodology, many types of bias within many of the study elements, and preconceived opinions that translated to an comprehensive incongruity to what the science disciplines call their methods.

Arguments ensued over hiring policies, management, accounting, and uses of monetary funding, use of unqualified staff, numerous committee member disagreements, and a lack of agreement over how to effectively coordinate members' efforts.[33] The early months of the Condon Committee and its actions provided a slippery slope paradigm that was especially revealing in context and punctuated by a public speech Condon gave at the end of January 1967. As such, "The government should not study UFOs because the subject was 'nonsense,' but I'm not supposed to reach that conclusion for another year."[34] Here, Dr. Condon committed a host of logical and communicative fallacies that repeatedly recur in everyday life situations and also in the science communities. These did not attribute well to the context of the study.

Prior to when the speech was given, the committee was receiving a lot of technical and intellectual support from the National Investigations Committee on Aerial Phenomena (NICAP). That support and assistance became significantly more precarious.

Later in July 1967, James E. McDonald, senior physicist at the Institute for Atmospheric Physics and professor in the department of meteorology at the University of Arizona in Tucson, discovered a memorandum written by Robert Low on August 9, 1966, before the University of Colorado accepted the Air Force study proposal. The memo elucidated the response Low gave two university administrator decision-makers in these excerpts:

> One would have to go so far as to consider
> the possibility that saucers behave according to a
> set of physical laws unknown to us. The simple

act of admitting these possibilities puts us beyond the pale, and we would lose more prestige in the scientific community that we could possibly gain by undertaking the investigation.

The tone of Low's memo is clear, in that studying the subject is too dangerous to the career of a scientist because it would uncover the dearth of knowledge we have in all the related subjects. Also the sociopolitical culture of the science communities is such that the amount of ridicule a scientist would encounter would essentially blacklist him or her from a future practice in their profession.

A later excerpt supports this state of affairs and additionally to marginalize the audience they are trying to address and to introduce many levels of fatalistic bias into their research, a methodological taboo:

"Our study would be conducted almost entirely by nonbelievers who, though they couldn't possibly prove a negative result, could and probably would add an impressive body of thick evidence that there is no reality to the observations. The *trick* would be, I think, to describe the project so that, to the public, it would appear a totally objective study but, to the scientific community, would present the image of a group of nonbelievers trying their best to be objective but having an almost zero expectation of finding a saucer."[35,36]

The Low memo was shown to Condon for explanation. Condon rebuked it. It took a few months after the disclosure, but NICAP finally unattached itself entirely from the project in April 1968. The Low memo made the public domain days after the NICAP separation. The U.S. Congress became involved with the Condon Committee by holding a hearing about the matter. Ultimately, Low resigned from the study in May 1968.

The environment of personal politics continued to dominate throughout the rest of the project. Other team investigators such as Roy Craig put their philosophical and personal opinion on the project and its design. Craig subsequently attacked most of the fifty-nine detailed cases describing them as "utter nonsense, highly suspect, and unexplained."[37]

Notwithstanding the actions of the Condon Committee, the Air Force paid the National Academy of Sciences to assess the committee's findings. This was executed before the study was completed. The summary of the NAS concluded, "On the basis of present knowledge, the least likely explanation of UFOs is the hypothesis of extraterrestrial visitations by intelligent beings." And that, "No high priority in UFO investigations is warranted by data of the past two decades."[38]

Reviews of the *Scientific Study of Unidentified Flying Objects* were numerous and came from such sources as the peer-reviewed critiques from the science communities and academia, mass media journals, magazines, and numerous published books by authors who took positions on both sides of the discourse. James E. McDonald, a most highly respected and credible member of the scientific community, complained of the inadequacy and naiveté of the cases selected for study and those not for study. "It represents an examination of only a tiny fraction of the most puzzling UFO reports of the past two decades and that its level of scientific argumentation is wholly unsatisfactory."[39]

Another summary came from Peter A. Sturrock, physicist at Stanford University. He noted the dichotomy of how the parties on the different sides took ironic if not natural conclusions, "Critical reviews came from scientists who had actually carried out research in the UFO area while the laudatory reviews came from scientists who had not carried out such research."[40]

For J. Allen Hynek, *The Condon Report* was the critical mass needed for his polarization from skeptic and nonbeliever to emphatic and unshakeable confirmer of the UFO and extraterrestrial hypotheses. Hynek described the report as "a voluminous, rambling, poorly organized" and wrote that "less than half was addressed to the investigation of UFO reports." And that, "*The Condon Report* settled nothing singularly slanted" and wrote that it "avoided mentioning that there was embedded within the bowels of the report a remaining mystery; that the committee had been unable to furnish adequate explanations for more than a quarter of the cases examined." Hynek contended that "Condon did not understand the nature and scope of the problem." He was studying and objected to the idea that only extraterrestrial life could explain UFO activity. By focusing on this hypothesis, he wrote the report "did not try to establish whether UFOs really constituted a problem for the scientist, whether physical or social."[41]

What the *Scientific Study of Unidentified Flying Objects* by Edward U. Condon accomplished, by all sides of the argument, was to be the enzyme that became the catalyst for the Air Force, government and associated military partners on the subject to successfully bury future ufology data, and studies into the rabbit hole of disclosure and secrecy. This discourse continues.

Top Secret UMBRA Documents

The National Security Agency/Central Security Service (NSA) was introduced as another chapter in the evolution of the modern age of secrecy and disclosure when it was created in November of 1952. (The NSA is the same U.S. government agency that Edward Snowden, then employee of the CIA, became a whistleblower against by disclosing many thousands of classified documents in June 2013.) The NSA was formed and began operating at the highest levels of technology, clandestine communication collection and investigation, and the requisite highest security secrecy and clearance of any U.S. government agency.

This fits into the timeline with which the Robertson Panel was in its final design and implementation stages, the Majestic 12 group was allegedly undertaking an investigation of the landscape of the Roswell Incident, and JANAP 146(A)/CIRVIS was in the design stages of development. Also the United States entered the first thermonuclear bomb design that was successfully tested on November 1,

1952, and Project Blue Book was preparing to turn its open-access policy of UFO reporting and the current level of investigation back inside itself into another clandestine agency. The NSA/CSS is now another U.S. government agency with many of the same missions as the CIA, CIG, FBI, all the agencies of the military complex, and the alleged Majestic 12.

On October 9, 1980, Eugene F. Yeates, NSA Chief, Office of Policy, released a total of 239 classified government documents titled *Top Secret UMBRA* as required to respond to a Freedom of Information Act (FOIA) lawsuit, Civil Action No. 80-1562.[42]

The nature of these documents encompasses many communication-related activities undertaken by the NSA in a long time frame primarily between 1958 and 1979. The emphasis was on UFOs. This was the thrust of the collection of FOIA related requests that went unanswered or denied over the years and culminated in the lawsuit. To obtain a more vital image, some of the details of the Top Secret UMBRA release of 1980 are indicated below.

One of the documents is a critique of an NSA official field investigator's attendance to a UFO symposium conference. About one-third of these released documents involve communications from other government agencies regarding UFO topics and/or events. Some of the documents have titles such as *UFO Hypothesis and Survival Questions* and *UFOs and the Intelligence Community Blind Spot to Surprise or Deceptive Data.*

There are others in the documents that explain an attitude of dismay over the NSA operations and those of other government agency's general lack of ability "to respond correctly to surprising information" and "that release of documents for public scrutiny, for a variety of reasons, would seriously damage the ability of the United States to gather this vital intelligence information."[43]

The most significant difference within its release was that the entire set was heavily redacted for alleged security reasons. Many of the redacted entries described communications by various military and civilian aircraft/marine pilots/captains and crews of unexplainable and unusual sightings and other encounters with unknown craft. Some of these were of single ships; other encounters were with multiple craft. The frustration of investigative teams to accurately identify and conclude the purpose(s) of these observations is another example of the widespread frustration of all government entities to be able to control the UFO issue. Hence, the actions the government took in response to these characteristics at the time and throughout the years were a logical response to internalize more and more of the protocols of secrecy and disclosure and the public domain.

Among the UMBRA documents disclosed in that 1980 decision was this top secret letter written by Albert Einstein and Dr. J. Robert Oppenheimer in June 1947. The "References" section can direct you to the entire original document. Here are excerpts from the original document:

> The possibility of confronting intelligent beings that do not belong to the human race would bring up problems whose solution it is difficult to conceive. International law should make place for a new law on a different basis: "Law among Planetary Peoples." Here, Einstein and Oppenheimer place our civilization's fractured culture (then in 1945, as well as today) into perspective with an overarching need for change if/when this confrontation takes place.

A later excerpt examines the possibility of humankind, whether voluntarily or by a more compelling circumstance, coming to awareness of a durable relationship with ET:

> Now we come to the problem of determining what to do if the inhabitants of celestial bodies or extraterrestrial biological entities (EBE) desire to settle here. It is difficult to predict what the attitude of international law will be with regard to the occupation by celestial peoples of certain locations on our planet, but the only thing that

can be foreseen is that there will be a profound
change in traditional concepts.[44]

Dwight Eisenhower was a most uniquely positioned authority
in the controversy surrounding UFOs and extraterrestrial life in the
early years of the modern era. He was a supreme military commander
and also U.S. president when the office still enjoyed a "need to know"
about things so clandestine. He was a participant in many events of
this nature.

The first one was part of a mass sighting while he was not yet
in political office. It occurred while he was aboard the *USS Franklin
D. Roosevelt* aircraft carrier in the North Sea on September 20, 1952.
This was part of the NATO military maneuvers named *Operation
Mainbrace*. As explained elsewhere here, Operation Mainbrace
turned out to be a composite of mass sightings, data collection from
radar, other electronic equipment and photographic, jet plane chases,
and civilian encounters. The British Defense Ministry attributed the
Operation Mainbrace encounters to be the critical mass needed to
officially recognize the *UFO concept*.

Here were some quotes taken from an article that appeared in
the *New York Post* on June 18, 1997.[45]

> The USS FDR, Sixth fleet flagship, 1952,
> a crew member writes: "We were north and east
> of England with the NATO fleet in the North
> Atlantic. About 1:30 a.m., through the stormy
> rain and lightning, this big blue-white light
> appeared right off starboard bow. It came down
> to one hundred feet of the water and just hung
> there as we cruised by it.
>
> "This UFO was easy to see when the light-
> ning flashed. It then rose straight and left. Four
> of us saw it. Here's the kicker! General Ike, who'd
> flown over by chopper with the Admiral, had just
> come out on the signal bridge wearing pj's and
> robe, looking for coffee.

"We were sitting and making small talk when the bright light came on. We all watched it ten minutes then just stood there staring at each other. After a while, Gen. Eisenhower said, he better go 'check this out' and left. He also told us to 'forget about it for now.'

"Next day and ever after, nothing was ever said about it. I don't know what it was or why it was hushed, but I saw it."

One interesting item about the Operation Mainbrace encounter is that the USS Franklin D. Roosevelt was reported to have had a nuclear weapon on board. While there are many thousands of UFO/UAP encounters that include extraterrestrial interventions, among these are also many thousands of such encounters that happened around military installations all over the world. Often, these are events that occur at sites where nuclear weapons are entrenched. Just a few of the more popular ones include Roswell and Corona, New Mexico, Holliman Air Force Base nearby in the White Sands Missile Range, Malmstrom Air Force Base in Montana, RAF Bentwaters which was part of the Rendlesham-Bentwaters incident in 1980, Nellis AFB in California, and Minot AFB in North Dakota.

The next events allegedly took place during his term as president. Here is a rendition of these events, all which occurred in his first year of office, from the archives.

The First Contact Meeting

Palm Springs, California, February 17–24, at Smoke Tree Ranch where he was staying.

On February 20–21, he was whisked away "to a dental appointment" at 10:00 p.m. Allegedly, Eisenhower "entered a secure hangar at Palmdale field, adjacent to Edwards." Edwards enacted a three-day shutdown off the base. The meeting was with a race of Nordic extraterrestrials. The disc was witnessed to be about one hundred feet in diameter and was housed in Hangar 27. A USAF test pilot witness stated that "five craft, two were cigar-shaped and three were disc-shaped landed." The reason for the meeting with the Nordics was to offer spiritual help only to get us to stop nuclear weapons proliferation. They did not offer us any technology.

Other excuses made as to his absence was that Eisenhower was at church on that Sunday February 21, when he was not a church-goer. And he wasn't a Catholic. There is not much documentation to support this series of events. Eyewitness testimony was more frequent. Among these reports was one from Meade Layne, director of the Borderland Sciences Research Foundation. Layne did corroborate as to the accounts of firsthand witnesses present that day. Other witnesses named various participants present at MUROC/Palmdale as eyewitness testimony. This included military police, Dr. Edwin Nourse (of the Brookings Institute at that time and past chairman of Council of Economic Advisors to the President (1944–1953), and Truman's chief economic advisor, Catholic Cardinal James Francis Macintyre of Los Angeles.

This MUROC meeting with the Nordics did not lead to an agreement.

The Second Contact Meeting

Holloman AFB, February 1955.

President Eisenhower started a quail hunting trip on February 10. He was seen later that afternoon at Secretary of Treasury George Humphrey's estate before disappearing for over thirty-six hours. According to many eyewitnesses, including military and civilians, Ike appeared at Holloman the next morning at 9:00 a.m. to meet the representatives of the grey aliens. This meeting lasted until 5:00 p.m. Meanwhile, back at Thomasville, Georgia, site of the hunting activity, the excuse offered to the press for not seeing the president was "a case of the sniffles."

Here is a more detailed account of the events of February 11th:

The day's events of Friday, February 11, 1955, at Holloman Air Force Base started out as routinely as could be expected due to the excited announcement from a few days earlier of a planned visit from Air Force One and President Eisenhower. Part of the itinerary was included an honor parade which is customary for such dignitaries as generals, heads of state, and the president. It is a little unusual for the short notice of a few days for an honor parade. The base commander, Colonel Sharp, and junior officers, including a Captain Reiner, were polling the base personnel for their possible participation in the parade. If the response was negative, then the respondent would be work scheduled that day. What was highly unusual about this activity was the whole parade was called off only the day before.

At 8:10 a.m., the first words spoken about the day's events came from the Holloman control tower over radio squawk from Air Force One, "Holloman Tower, this is Air Force 7885 ten miles east of Maryhill." Colonel Sharp and many junior officers were in the control tower witnessing the plane's arrival. Tower radar and traffic controllers issued the assigned "runway 13" after giving weather information to pilot Major Bill Draper (Runway 13 means landing on a course of 130 degrees on a compass or from the southeast). Runway 13 was the landing tarmac farthest from the hangars on the base.

At this point, it was estimated that about three-hundred personnel were witnessing the landing, including the clueless tower controllers and radar operations. Air Force One touched down around 8:35 a.m. on Runway 13. Upon landing, new orders were passed around the control tower to "go dark." This meant to turn most of the radar and electronic equipment off. This was on orders from Washington, DC. A few radars from locations up range of Holloman were still on. There were patrols around the area that were not scrambled to go dark.

Control tower personnel were encouraged to conduct their duties like a "business as usual" workday. They were contacted within a couple of minutes after the go dark orders were given from off-base patrols and the other radar sites. Two unidentified objects were detected moving over Range Road 12 with high velocity. Immediately after these reports were broadcast, others came into Holloman stating that the two "blips" had reached Range Road 7. A third "bogie" was spotted behind the first two. All them were approaching Holloman. Control tower occupants now obtained visuals of two round metallic craft followed by a third. The craft had no wings and moved silently toward Holloman and Runway 13.

The first two ships stopped about three hundred feet over the Air Force One craft. They then touched down in front of the plane a few seconds later. The third ship also stopped and hovered for a short time over the tarmac then moved away slightly from the central core of landed craft.

From a separate position, other witnesses observed the same events. A civilian electrician had a more direct view while working on a different part of the base near the electrician's warehouse. This location was behind the cluster of hangar buildings where the control tower would have been to the side of Runway 13 from the worker's vantage point. He spotted Air Craft One taxi from a vector almost in front of him to the northwest. The worker climbed a pole to gain a better vantage point of the president's plane.

The next visual came from behind the witness. He turned around, strapped to the pole, and saw the first round "pie tin like thing" coming right at him a few hundred feet elevated. The flying disc followed the same vector up the landing runway and to the tarmac over Air Force One. It was silent and wobbled slightly in a to-and-fro pendulum motion. This witness reported only the one craft being sighted. He noticed his peers moving away from their positions around the electrician's warehouse. He scurried down the pole on which he was elevated and, therefore, did not witness any other episodes.

At the control tower, reports were made of episodes in the minutes following the congregation. The door to Air Force One opened. Then a door on the round ship opened, and a ramp folded out. The vantage point was such to the witnesses in the tower that only a side of the door and ramp were visible. A person moved out the door from the president's plane. This person walked down the stairs then moved toward the ramp. Nothing else was reported to have been witnessed until about forty-five minutes later when the person reemerged onto the ramp and then moved toward Air Force One.

President Eisenhower was reported to have spent the rest of the day speaking with base personnel in groups of about two hundred and inspecting some of Holloman's active projects as "works in process." In those talks, Eisenhower repeated that, "He was not supposed to be at Holloman that day," as reported by base personnel later. Air Force One departed from Holloman at around 4:45 that afternoon.[46]

President Eisenhower, it was reported, was supposed to have been recovering from a case of the sniffles back at Secretary of The Treasury George H. Humphrey's plantation in Thomasville, Georgia, on February 11. He made an appearance the next day at an autograph session.

Three trusted servicemen from different eras in American history shared the same visualizations and concerns about how human behavior can individually and collectively shape the proliferation of many things. This ranges from the small stage of an individual's knowledge and perceptions to the same knowledge and perceptions of the foreboding immensity of a complex behemoth as the military industrial complex. Inextricably interspersed within and among are the bellows of secrecy and disclosure and the lack of faith and trust inherent when these things become reality and entangled inside.

"You now face a new world—a world of change. The thrust into outer space of the satellite, spheres, and missiles marked the beginning of another epoch in the long story of humankind—the chapter of the space age. We speak in strange terms of harnessing the cosmic energy, of the primary target in war, no longer limited to the armed forces of an enemy, but instead to include his civil populations; of ultimate conflict between a united human race and the sinister forces of some other planetary galaxy" (Address by General Douglas MacArthur to the United States Military Academy at West Point, May 12, 1962) (Eisenhower Ref MacArthur). (May 12, 1962).[47]

A most poignant conjecture in disagreement against a military establishment policy or strategy of withholding disclosure comes again from President Eisenhower in his Farewell speech on January 17, 1961:

> We must guard against the acquisition of
> unwarranted influence by the military industrial
> complex. The potential for the disastrous rise of

misplaced power exists and will persist. Only an alert and knowledgeable citizenry can compel the proper meshing of the huge industrial and military machinery of defense with our peaceful methods and goals so that security and liberty may prosper together.[48]

President Abraham Lincoln had that same idea about one hundred years before Eisenhower and re-quoted here, "I am a firm believer in the people. If given the truth, they can be depended upon to meet any national crisis. The great point is to bring them the real facts."[49]

References

[1] Lincoln, Abraham. Sixteenth U.S. President. Retrieved from: www.lincolnarchives.com/LincolnQuotes.php.

[2] Welles, Orson. 1938. *The War of the Worlds*. Adapted from H.G. Welles's novel.

[3] Jung, C.G. 1978. *Flying Saucers: A Modern Myth of Things Seen in the Skies*. Copyright 1978 by Princeton University Press.

[4] Disclosure Ref Maury: Condon, E.U.1968. The Condon Report: Scientific Study of Unidentified Flying Objects; The Complete Report Commissioned by the U.S. Air Force, chapter 2. Conducted and copyrighted by the Regents of the University of Colorado, 1968. Retrieved June 4, 2016, from: http://files.ncas.org/condon/text/s5chap02.htm.

[5] Disclosure Ref Roswell: 1992 book *Crash at Corona* by Stanton Friedman and Don Berliner, published in the United States by Paragon House.

[6] Disclosure Ref Roswell Ramey: Rudiak, David. 2001. Roswell Proof: What Really Happened. Retrieved June 4, 2016, from: http://www.roswellproof.com/.

[7] Disclosure Ref Majestic Bryan: Bryan, C.D.B. 1995. Close Encounters of the Fourth Kind: Alien Abduction, UFOs and the Conference at M.I.T. Pg. 186. Alfred A. Knopf, ISBN 0-679-42975-1.

[8] Disclosure Majestic Exon: Rudiak, David. 2001. Roswell Proof: What Really Happened. Brigadier General Arthur E. Exon. Retrieved June 4, 2016, from: http://www.roswellproof.com/.

[9] Disclosure Ref Condon: Condon, E.U. 1968. The Condon Report: Scientific Study of Unidentified Flying Objects; The Complete Report Commissioned by the U.S. Air Force, chapter 2. Conducted and copyrighted by the Regents of the

University of Colorado, 1968. Retrieved June 4, 2016, from: http://files.ncas. org/condon/text/s5chap02.htm.

[10] Disclosure Ref Condon: Condon, E.U. 1968. The Condon Report: Scientific Study of Unidentified Flying Objects; The Complete Report Commissioned by the U.S. Air Force, chapter 2. Conducted and copyrighted by the Regents of the University of Colorado, 1968. Retrieved June 4, 2016, from: http://files.ncas. org/condon/text/s5chap02.htm.

[11] Disclosure Ref Condon: Condon, E.U. 1968. The Condon Report: Scientific Study of Unidentified Flying Objects; The Complete Report Commissioned by the U.S. Air Force, chapter 2. Conducted and copyrighted by the Regents of the University of Colorado, 1968. Retrieved June 4, 2016, from: http://files.ncas. org/condon/text/s5chap02.htm.

[12] Disclosure Ref 12: Jacobs, David Michael, 1973. The Controversy over Unidentified Flying Objects. In America: 1896–1973. Copyright David Michael Jacobs, 1973. The University of Wisconsin, 1973. University Microfilms, Inc. Ann Arbor, Michigan.

[13] Disclosure Ref 13: Ruppelt, Edward J. 1956. *Report on Unidentified Flying Objects*, Pg. 114. London: Victor Gollancz, 1956. 2nd, expanded edition New York: Ballantine, 1960.

[14] Disclosure Ref 14: U.S. Central Intelligence Agency. 1952. Memorandum for the Deputy Director/Intelligence from Ralph Clark, "Recent sightings of unexplained objects," July 29, 1952.

[15] Disclosure Ref 15: Unsigned memorandum. 1952. U.S. Central Intelligence Agency. CIA memorandum, unsigned, August 19, 1952, "Flying Saucers." Retrieved June 4, 2016, from: http://www.cufon.org/cufon/cia-52-1.htm.

[16] Disclosure Ref 16: Chadwell, Marshall H. 1952. U.S. Central Intelligence Agency memorandum. CIA memorandum to Director of Central Intelligence from H Marshall Chadwell, September 11, 1952, "Flying Saucers".

[17] Disclosure Ref 17: Report on Scientific Advisory Panel on Unidentified Flying Objects Convened by Office of Scientific Intelligence, C.I.A. January 14–18, 1953. Pgs 19–20. Retrieved June 4, 2016, from: http://www.cufon.org/cufon/ robert.htm.

[18] Disclosure Ref 18: Report on Scientific Advisory Panel on Unidentified Flying Objects Convened by Office of Scientific Intelligence, C.I.A. January 14–18, 1953. Pgs 19–20. Retrieved June 4, 2016, from: http://www.cufon.org/cufon/ robert.htm.

[19] Disclosure Ref 19: Durant, Frederick C. 1953. Memorandum for the Assistant Director for Scientific Intelligence from F.C. Durant, "Report of Meetings of the Office of Scientific Intelligence Scientific Advisory Panel on Unidentified Flying Objects," January 14–18, 1953, February 16, 1953.

[20] Disclosure Ref 20: CIA memorandum to DDI from H Marshall Chadwell, "Unidentified Flying Objects," February 10, 1953; letter from H Marshall

Chadwell to H P Robertson, January 28, 1953; and CIA memorandum for IAC from James Q Reber, "Unidentified Flying Objects," February 18, 1953. CIA memorandum for the record by F.C. Durant, "Briefing of One Board on Unidentified Flying Objects," January 30, 1953, and CIA Summary disseminated to the field, "Unidentified Flying Objects," February 6, 1953.

21 Disclosure Ref 21: Jacobs, David Michael, 1973. The Controversy over Unidentified Flying Objects. In America: 1896–1973, Pg. 112. Copyright David Michael Jacobs, 1973. The University of Wisconsin, 1973. University Microfilms, Inc. Ann Arbor, Michigan.

22 Disclosure Ref 22: Jacobs, David Michael, 1973. The Controversy over Unidentified Flying Objects. In America: 1896–1973, Pg. 112. Copyright David Michael Jacobs, 1973. The University of Wisconsin, 1973. University Microfilms, Inc. Ann Arbor, Michigan.

23 Disclosure Ref 23: U.S. Joint Chiefs of Staff. 1954. JANAP 146(C) Communication Instructions for Reporting Vital Intelligence Sightings from Airborne and Waterborne Sources, March 10, 1954. Retrieved June 4, 2016, from: www.cufon.org/cufon/janp146c.htm.

24 Disclosure Ref 24: U.S. Joint Chiefs of Staff. 1954. JANAP 146(C) Communication Instructions for Reporting Vital Intelligence Sightings from Airborne and Waterborne Sources, March 10, 1954. Retrieved June 4, 2016, from: www.cufon.org/cufon/janp146c.htm.

25 Disclosure Ref 25: Roe, Ted. 2004. Aviation Safety in America: Underreporting Bias of Unidentified Aerial Phenomena and Recommended Solutions. Pg. 7. Revised Edition, July 20, 2004. Retrieved from: http://www.narcap.org/files/narcap_TR-8_2002.pdf.

26 Disclosure Ref Tim Good: Good, Timothy. 1988. Above Top Secret, Pgs. 286–287. Published 1988 by William Morrow.

27 Disclosure Ref Tim Good 2: Good, Timothy. 1988. Above Top Secret, Pgs. 288. Published 1988 by William Morrow.

28 Disclosure Ref Tim Good 3: Unknown author. United Press International. 1960. Air Force Order on "Saucers" Cited. New York Times, February 27, 1960.

29 Keyhoe, Major Donald E. 1973. *Aliens from Space: The Real Story of Unidentified Flying Objects*. Published 1973 by Signet Press, ASIN B000HYOMMG.

30 Disclosure Ref Tim Good 2: Good, Timothy. 1988. Above Top Secret, Pgs. 288. Published 1988 by William Morrow.

31 Disclosure Ref NICAP Bulletin 31: NICAP Bulletin, August 1962. Also Tim Good Ref: Disclosure Ref Tim Good 2: Good, Timothy. 1988. Above Top Secret, Pgs. 289. Published 1988 by William Morrow.

32 Condon Ref 32: Dick, Steven J. 1996. *The Biological Universe: The Twentieth Century Extraterrestrial Life Debate and the Limits of Science*, Pg. 293. Published June 13, 1996, by NY: Cambridge University Press.

[33] Condon Ref 33: Hynek, J. Allen. 1972. *The UFO Experience: A Scientific Inquiry*, Pgs. 192–244. Copyright 1972 by J. Allen Hynek. Published in 1972 by Marlowe & Company.

[34] Condon Ref 34: Clark, Jerome. 1998. *The UFO Book: Encyclopedia of the Extraterrestrial*, Pgs. 593–604. Copyright 1998 by Visible Inks Press.

[35] Condon 4: Craig, Roy. 1995. *UFOs: An Insider's View of the Official Quest for Evidence*, Pgs. 194–195. Published 1995 by University of North Texas Press.

[36] Low memo: Low, Robert. *Robert Low "Trick Would Be" Memo and Transcription.* August 9, 1966. Memo to E. James Archer and Thurston C. Manning. Reproduced by NICAP. Nicap.org/docs/660809lowmemmo.htm

[37] Condon Ref 37: Craig, Roy. 1995. *UFOs: An Insider's View of the Official Quest for Evidence*, Pgs. 228–233. Published 1995 by University of North Texas Press.

[38] Condon Ref 38: Dick, Steven J. 1996. *The Biological Universe: The Twentieth Century Extraterrestrial Life Debate and the Limits of Science*, Pg. 302. Published June 13, 1996 by NY: Cambridge University Press.

[39] Condon Ref 39: McDOnald, James E. 1969. Science in Default: Twenty-Two Years of Inadequate UFO Investigations. December 27, 1969. Retrieved June 4, 2016.

[40] Condon Ref 40: Sturrock, Peter A. 1999. *The UFO Enigma: A New Review of the Physical Evidence*, Pg. 40. Warner Books, 1999.

[41] Condon Ref 41: Hynek, J. Allen. 1972. *The UFO Experience: A Scientific Inquiry*, Pgs. 192–244. Copyright 1972 by J. Allen Hynek. Published in 1972 by Marlowe & Company.

[42] Disclosure Ref 42: Affidavit of Eugene F. Yeates. 1980. Civil Action No. 80-1562 – United States District Court for the District Of Columbia. Dated and created October 9, 1980. Retrieved June 4, 2016, from: www.governmentattic.org/2docs/NSA_YeatesInCameraAffadavit_1980.pdf.

[43] Disclosure Ref 43: David, Leonard. Secret Government UFO File Unveiled: National Security Agency affidavit sheds light on Cold-War efforts. 2005. NBCnews.com. November 16, 2005. Retrieved June 4, 2016, from: Newhttp://www.nbcnews.com/id/10070986/ns/technology_and_science-space/t/secret-government-ufo-file-unveiled/#.V4vDmI-cHtQ.

[44] Oppenheimer, Dr. Robert J. and Einstein, Albert. 1947. Relationships with Inhabitants of Celestrial Bodies. June 1947. Retrieved from: http://www.majesticdocuments.com/pdf/oppenheimer_einstein.pdf.

[45] Eisenhower Ref NY Post: Eisenhower UFO Sighting: An Update. Presidential UFO. Retrieved June 4, 2016, from: http://www.presidentialufo.com/dwight-d-eisenhower/65-eisenhower-ufo-sighting. *New York Post*, June 18, 1997.

[46] Disclosure Ref Eisenhower Holloman AFB: Campbell, Art. 2007. President Eisenhower at Holloman AFB? February 19, 2007. Retrieved 6/4/2016 from: http://www.ufocrashbook.com/eisenhower.html.

[47] Disclosure MacArthur quote 1962: MacArthur, General Douglas. 1962. (Address by General Douglas MacArthur to the United States Military Academy at West Point, May 12, 1962.) (Eisenhower Ref MacArthur) May 12, 1962. (Disclosure MacArthur quote 1962.) www.nationalcenter.org/ MacArthurFarewell.html.

[48] Disclosure Ref Last: Eisenhower, President Dwight D. 1961. Presidential Farewell Speech of Tuesday, January 17, 1961. [DDE's Papers as President, Speech Series, Box 38, Final TV Talk (1); NAID #594599] Retrieved June 4, 2016, from: https://www.eisenhower.archives.gov/research/online_documents/ farewell_address.html.

[49] Lincoln, Abraham. Sixteenth U.S. President. Retrieved from: www.lincolnar-chives.com/LincolnQuotes.php.

CHAPTER 21

Why Would You Become Interested in the UFO/UAP/Alien Hypothesis

If plant life or some subhuman intelligence were found on Mars or Venus, there is no good reason to suppose these discoveries, would result in changes in perspectives or philosophy in large parts of the American public. If super intelligence is discovered, the results become quite unpredictable.

—The Brookings Report[1]

If it is true that a creative society must explore, it is also true that the same society must consider the consequences of its exploration.

—Steven J. Dick, 2007[2]

Key words: ad lapidem fallacy, proof by assertion fallacy, flap, wave, ad populum fallacy, prokaryotic, eukaryotic, exogeology, exobiology

When the United States gained a presence in space in the late 1950s, a significant and growing segment of society shifted some of its attention to the matter of this chapter's subject. The mass media helped instill awareness on the UFO topic by publishing daily news stories. This period in American technology and engineering history was among the most prolific in memory for promotion of the study of the geophysical, astrophysical, and space sciences. There was an untapped encyclopedia of knowledge awaiting discovery, insight, and invention to promote yet further discoveries.

This project helped motivate an increasing number of students to pursue careers in the astrosciences. At least for a generation, from the 1950s to the 1970s, American universities saw increased enrollment in the STEM fields of study. *STEM* is an acronym for *science, technology, engineering,* and *mathematics.*

NASA was the then newly-formed space science section of the U.S. government. It undertook a massive comprehensive project in 1961 to forecast man's potential for future space exploration. Their study, commissioned as the Brookings Institution, itself a social sciences research think tank based in Washington, DC, resulted in a most well-known publication titled, *The Brookings Report.* The Brookings Report was comprised primarily of an anthropocentric dialogue of human's potential and future in space activities. In this text, chapter 20 titled "Why Here...?" discourse on the notion of man's evolutionary propensity for global exploration upon knowledge of the existence of foreigners in those far-off lands is provided. It was one of a large amount of possible motivations for those travelers to risk their lives in the quest for whatever they were seeking.

A couple of sections of the Brookings Report explored the linkup of notions between exploration and the initial contact with other living beings. It touched on speculative ramifications upon such contact based on historical evidential occurrences in man's history: "Anthropological files contain many examples of societies, which have disintegrated when they have had to associate with previously unfamiliar societies espousing different ideas and different life ways."[3]

This statement is an example of potential for the motivation of the governments of the United States and other nations to engage in a disclosure operation, no matter how elaborate, on one end, or improbable it has been known to be dismissed as on the other end. NASA and the Brookings Institution took the fatalistic viewpoint in it. Whether a debate about any societies necessarily facing such dire outcomes upon visitation by another intelligent explorer is not the point at this time.

The report provides two suggestions that factor these two ideas into a research study framework. The first one touches on the disclosure issue: "How might such information, under what circumstances, be presented to or withheld from the public for what ends? What might be the role of the discovering scientists and other decision makers regarding release of the fact of discovery?"[4]

A second recommendation introduces a possibility of such a fatalistic viewpoint the evolution of such events could have on our society. The need for "continuing studies to determine emotional and intellectual understanding and attitudes—and successive alterations of them if any—regarding the possibility and consequences of discovering intelligent extraterrestrial life."[5]

What is significant here is to more deeply engage in how a line of events would be acknowledged and perceived by the recipient society. We are the recipient society is us, the civilization of planet Earth which is so described. The NASA/Brookings study used the United States as a proxy.

The report moves its discussion on what the arrival and discovery of intelligent extraterrestrial beings means to a statement on how leadership could learn from and exploit such knowledge, "The degree of political or social repercussion would probably depend on leadership's interpretation of its own role, threats to that role, and opportunities to take advantage of the disruption or reinforcement of the attitudes and values of others. It would be most advantageous (for leadership) to have more to go on than personal opinions about the opinions of the public."[6]

If governments have been withholding disclosure of information from the public, this statement would fit into an operations protocol quite well. The governments have the evidence of such existence and can then justify their withholding of disclosure to the public on these grounds, among others, while building their case on how and why the public would be incapable of adjusting to such a scenario. With more and more governments currently willing to share what they know, this strategy is embedded within that protocol. Scores of declassified documents from Britain, Brazil, Canada, and even some from the United States, over the years seem to add to that body of evidence.

For people to be able to answer the most general of the associated questions, "Why would you become interested in the UFO/UAP/ extraterrestrial hypotheses?" would probably entail some degree of contextualization or relevant perspective. Human life span on Earth is very long. In our lives, we use a disproportionate amount of time to develop, learn, and mature biologically. We also spend a disproportionate amount of early-life years being educated and trained in what life has to offer. Then we apply that expert knowledge to earn a living. All of us have to invest our time in this way in order to receive the "necessities and trappings" offered by that living. This is another large amount of time devoted to daily activities. Concurrently with this, we use still more time to start and raise a family. Only after the child-raising years do people obtain a more manageable amount of discretionary time for other life pursuits. In summary, a person's life span is devoted to many mandatory pursuits which often take up most of that time.

So if the answer relies on this notion of clock time and how it is used up throughout our lives, then it is concluded that there may not be much time to budget for either a recreational or analytical pursuit of the UFO/extraterrestrial life subject during the years in the middle of our lives.

Of course, interest and motivation are primary factors for anyone who wishes to pursue a subject. Measured interest in a subject lies along a dimension of "not interested," to "recreationally interested," to "profoundly interested." We are often too busy to even make time if we were to be categorized in the "profoundly interested" category.

To answer this "why" question for anyone would entail investigating the other "w's" questions: "what's, when's, where's, and how's." Of those, the "what" questions have the most sweeping and relevant designs. To exemplify, we could ask ourselves, "What would this mean to me?"; "What's in it for me?"; or "Why would I want to?" Would anyone become interested before the ultimate worldwide arrival, introduction, acknowledgment, and acceptance of such existence by being proactive in some way as ufologists the world over have been? Or would it take all those steps up to and including global acknowledgment and acceptance for us to only then become interested? Or would we not even care? Would the subject only be a recreation that proliferates solely within the movie theater? Or is it even too "fictional" to be entertained by? Or do our intuitive precepts compel us to want to look closer at the subject? More on the "what's" momentarily.

We know that from the previous example, the amount of time in our daily lives is of paramount significance for most of us. As in any pursuit only if one is an avid knowledge seeker would the paraphrase, "you can make time for this" be not only appropriate but functional. This means that some of us just do not have that much time for a pursuit unless it is of profound interest.

As it currently stands, there is no global acceptance among all the communities and nations of our civilization which are conclusively maintaining that UFOs/UAPs (unidentified flying objects/ unidentified aerial phenomena) and intelligent extraterrestrial life exist. Those of us, as representatives of the public domain, who have thought about this issue are the most likely to have at least not rejected these hypotheses. According to the findings of abundant current and past statistical surveys, a large majority of us (over 75%) believe intelligent extraterrestrial life exists. Some of those more formally trained in an epistemology of the natural science disciplines, however, may have decided to withhold their conclusions for more evidence of the type which better suits their method of reasoning. Additionally, some of those engaged in the defense and security of their communities' or nations' borders may not be able to offer these conclusions, at least publicly. Many members of those communities have observed and/or experienced a real event in nature that alters their future time commitment to the hypotheses. The information discussed thus far appears quite substantial. We have read about many, many of these situations. We can become knowledgeable about more of them in our own investigations as well as those which will happen in the future. They have been happening for thousands of years through recorded history.

It is of this knowledge which sparks new interest from these individuals. It is also true that the longer one lives, the more chances for a firsthand encounter with such a phenomenon. So in our daily lives, we simply have not been exposed to the subject in any way that captures our attention until that time. A firsthand encounter provides the motivation for many. So older individuals may be collectively more apt to be participants in the subject for this reason.

For younger people who have not had the opportunity of a firsthand experience, interest may not be immediately sparked, or it will lie in a dormant state. Then time commitments to daily life factors take precedence in reasoning for nonparticipation. In this governmental pre-disclosure atmosphere, it appears unless there is some significant directional catalyst toward this end, the dictates of daily life would interfere with developing such an interest.

Global disclosure, to the satisfaction of all communities and nations of our civilization, could permanently and dramatically alter the situation. Participation would change from speculation in a proactive sense to being fully active. Of course, we would have learned many things about the subject along the way so our train of thought may be shaped accordingly. A lot of people would, only after this event (if it is indeed just one special grand disclosure event and not many less profound ones), become more involved. From the noted discussions about every person's individual life experiences which are heavily influenced and shaped by such experiences, as well as cultural perspectives and such, this expectation would be automatic as well as continually dynamic in nature.

The interest level one would acquire is contextualized regardless of being in a pre-disclosure or post-disclosure environment. The state of ufology, at least within the science communities, is still categorized as being pseudoscience. The summation of our life experiences defines their contextualization, as well as ours. We may have gotten caught up in the display of Orson Welles's *War of the Worlds* or any of a plethora of science-fiction movie themes which bias our thinking. These examples shape our contextualization uniquely from one another.

There has been a lot of literature written about possible post-disclosure reaction by the society at large. All this has been written within the umbrella of a pre-disclosure environment. The total amount of time devoted to and the volume of the writing is yet another example of the shift in paradigm from a platform of pseudoscience to one of mainstream science. This has been experienced continually by the scientific community throughout history.

As noted previously, this is how all facets of science have been studied since ancient Greek times 2,500 years ago. The ancient Greek philosophers invented the ontology and metaphysics of philosophy. These descended and branched off into the natural sciences. Since then, the science disciplines developed a method, aka the scientific method, which treats all incrementally small pieces of evidence and research results construct a conclusion over time. All these activities move very, very slowly in a science community paradigm until a revolution takes place in the research community. This revolution redefines and often replaces the old way of doing things. In this example, formal and permanent disclosure of intelligent extraterrestrial life is the revolution up for change. Social constructivist and science philosopher Thomas Kuhn popularized both the term *paradigm* and the study of these science revolutions in *The Structure of Scientific Revolutions* in the 1960s.[7]

Steven J. Dick, former chair of astrobiology at the Baruch S. Blumberg NASA/Library of Congress, further described the scientific discovery paradigm as practiced in science even today as follows: "I've done a book about discovery in astronomy. I made a discovery. It's always an extended process. We have to detect something, you have to interpret it, and it takes a long time to understand it. As for extraterrestrial life, the Mars rock showed it could take an extended period of years to understand it."[8]

The science disciplines have maintained an extensive historical blueprint for this type of idea. If we surveyed the histories of Copernicus, Galileo, Newton, Mendel, Einstein, Tesla, and many other accomplished scientists, the blueprint of science discourse is repeated over and over. There is fluidity within the paradigms of science, but it flows very slowly.

What does this all mean to those of us who have not spent a lot of time being trained in scientific or critical thought? We do practice, at least intuitively, many of the same basic logical and critical thinking processes as any scientist does, much to their disappointment.

Their more formal training, continued practice, and repeated exposure to thinking situations like these help them maintain their acuity in these areas for use in their work and their daily lives. We use the same epistemology in our daily lives, though more informally.

So in answering the question, "What would allow you to become interested in the UFO/UAP/extraterrestrial hypotheses?" we now know of two important impediments. In addition to the amount of time we have to devote elsewhere to a busy life, maybe the confusion we have over the mystique of science, as previously discussed, gets in our way of the understanding and clarity of what is read and seen and heard. Confusion is meant as the process of how science reasons through a particular problem of inquiry and investigation is often rigid, incomprehensible, and ambiguous. Along the way, we since used many fallacious arguments used in articles which we read, information that we obtain and/or article diversions away from a thoughtful, accurate, and truthful investigation of the topic. The resulting writing either deflects our attention or destroys our interest and engagement.

A third influential factor which also diminishes our possible engagement, involves a point of view which is taken, for various reasons, by many scientists today. Many in the science communities maintain that "their mind is made up" on the matter of intelligent extraterrestrial life. They do not wish to be confused by more facts. No amount of future information or evidence will ever change their minds. This error in judgment is known as an *ad lapidem* fallacy. A similarly demonstrative logical fallacy is the *proof by assertion* fallacy. Helping to alleviate this common fallacy is the fact that the number of scientists who maintain this nonbelief about the ET life debate is decreasing over time even as more facts and evidence are collected and made aware to society. This indicates a corresponding shift of the paradigm from pseudoscience closer to achieving acceptance as mainstream science, and *The Humaniverse Guide to Better Reasoning & Decision-Making* supports this track.

A fourth position is the scientist wants more information before he or she draws a conclusion. This is an example of a more responsible and effective use of science protocols. The person is keeping an "open mind" to new sources of facts which are part of reality and nature. It may be a slow process for his decision-making effort, but it does adhere more closely to the practices of scientific thought and discourse.

Maybe we can wade through all this science "muck" and move forward because we have an intrinsic motivation to do so. We might become interested in the UFO/UAP/alien hypotheses via some initial reading exposure to the subject. Or it could be through an actual encounter, introduction from another person, or from our own introduction via entertainment recreation, scholarly introduction, or some other combination.

Often but not always, it is some real encounter we have which sparks this interest, at least within our own minds. All the cases studied in this book involve encounters which are mostly single episodes with a few exceptions. These do not fall within the time frame of a *flap* or *wave*. They are either isolated single events or, in the study of mass sightings, the initial episode in the sequence of such a flap or wave. Flaps and waves, remember, are repeated incidents over a continuing time period of some length. Use of this methodology in *The Humaniverse Guide to Better Reasoning & Decision-Making* is primarily for eliminating potential biases which could contaminate a case study. These are biases that are studied in the psychology and social sciences, in which subsequent episode reports were being unduly influenced by witness knowledge of the initial event and biasing their motivations for reporting their own encounters. This is also to extract hoaxes and the equivalent that could have been conjured up by such witnesses for whatever reason.

Sometimes, the motivational spark is a curiosity which lays dormant in our lives for various reasons. It may be because of our life clock-time factor, or it may be a heightened level of apprehension or fears of personal or professional reprisal we think we would experience because of the subject's controversial nature we hear about in the literature.

Sometimes, our interest is ignited or at least combusted by conversations with other people on the subject. An association with the other person(s) could reduce our level of apprehension and fear of being significantly and discriminately ostracized for our interest in the subject. This is an example of the strength in numbers notion which governs many typical decision-making reasoning in society in more ways than we initially think.

There must also be awareness that there is a type of reasoning associated with the "strength in numbers" which could result in committing a genetic fallacy called an *ad populum fallacy*. This happens often in society, including the science communities. While it is natural and potentially productive to associate with a shared interest, it is the specific decision-making situation in regards to a particular problem which could cause a fallacious decision error of this type. We cannot make a decision about something be reliable only on the premise that "everybody else says so and this is why I agree this is the nature of the ad populum fallacy."

In sum, the potential roadblocks in stimulating individual interest in the intelligent extraterrestrial life debate we first may not have the time to pursue the subject. Second, there may be a level of apprehension present. This could either be social, professional, or personal reservation. Thirdly, the science community mechanism on whom we rely for insight and knowledge to explain how the world and nature exists and works has somehow failed. This could be from opinions heard from science or government that there is not enough evidence to enable or further study the phenomena. Many have been discouraged from hearing dissenting opinions from debunkers who commit numerous logical and argumentative fallacies in their efforts to "explain away" every UFO/ET event without any thoughtful investigation into each case study.

If an interest in the subject has been sparked, could this change any of our perspectives at all? If so, how? More generally and to follow up from a moment ago, *what* would it change about us?

We are investigating a discussion about intelligent life in the cosmos and not microbes or other non-sentient life. There has been a lot of literature written about prospects and pursuit of discovery of non-sentient life-forms such as single-celled bacteria or plants. Most of the popular literature indicates this realization would not have much of an effect on the worldview or the individual. It takes on an importance only slightly greater than the similar discovery of a new species of such life on Earth. There would be no immediate and direct effect on us either individually or collectively as a society. It also would not affect us because its presence, if it indeed somehow traveled to Earth, could not harm us in any way (presumably). We have nothing to *fear* from its existence, either potentially or physically.

In another scenario, presumably, one of man's space projects will show discovery of such non-sentient life-forms on a Mars probe or a lunar probe of some kind. Unless it is decided to try to transport samples back to Earth, it will remain there and out of harm's way. So we would not be changed, at least in the way we live our lives.

When a leap is made up to a level of interest and investigation of intelligent life, the landscape changes, and the individual is now challenged with a more complex state of affairs. In addition to that, the worldview also changes. We could talk about how this could happen in two ways. First, humans could be the ones to discover and make contact with the intelligent life by making the gestures for such as explorers. This would mean for us to either communicate successfully with or make the journey to their home. Metaphorically, we would be the explorers.

Second, the discovery and contact could be made on our home turf as the discovered.

It would be presented with some foundational certainty that, even within these two possible scenarios, our perspectives would be changed in many and different ways. Our brain's limbic system, the neurological center for our emotions, behavior, and sensory input would process this information from the two scenarios. The result would take different outcomes. These outcomes would also depend on our life experiences as inputs to our reasoning.

In the first scenario, humans make the discovery to a physical distance from Earth. If the contact is made via radio message, we may be the transmitters of or the responders to whatever radio communication is involved. Upon contact, we would probably experience excitement, curiosity, relief, but probably not apprehension, anxiety, mistrust, or fear. Here, the physical distance involved that separates both parties of the contact is the determinant.

In the second scenario, we are not the explorers, transmitters, or initiators of the first contact. We are rather the recipients. The most succinct way to describe a potential first reaction is to some other more uncomfortable sociological or physiological responses such as apprehension, anxiety, mistrust, and fear into the situation.

This occurrence has an analog in our earth's history. From recorded documented history, we could explore how the Caribbean natives reacted to first contact with the discovery by the Christopher Columbus's expeditions, or the China treasure explorers of the fifteenth century to the rest of Asia, or the Aztecs and Meso-America to the Herman Cortez explorers, or the Canadian fur traders to Native Americans, or the Jesuits to the Native Americans, among others.

The difference between the explorer and the explored lies in the contextual personal and cultural perspectives of each group. The geography is also a significant primary factor. The dynamics and objectives of the individual's reactions and change to context and perspective will move toward a societal change in perspective, absent any political and/or military intervention.

If we are to be part of an intelligent extraterrestrial life discovery at the present, we may have to be resigned to be the recipient party in this example. Reading through *The Humaniverse Guide to Better Reasoning & Decision-Making* tells us that our society and technology is not capable yet of initiating such contact by means of doing our own exploring to enact that contact.

If our planet would be the recipient of first contact, we know this analog could include looking at it through the lens of the Caribbean natives, Native Americans, or the Canadian fur trappers. Not all of man's past expeditions ended badly for everybody. The Native Americans settled peacefully and profitably with the Jesuits, and initially with the English pilgrims, as did the Canadian fur traders. There were also other situations which did not end very successfully, as we know. This includes many Native American tribes and Meso-American civilizations. We have the advantage of using this body of recorded history and knowledge, and the ability of impressing it onto this scenario for our benefit.

The recipients of the contact, in these examples, were being invaded by explorers whether with prior knowledge of their existence or not. In the human train of thought and experience, the natives' thoughts could have been reflective of one of three things: a) a one-time contact where the explorer later moves on for whatever reason; b) an initiation of further and regular contact with intent of eventual co-habitation; or c) explorer preparation for pre-conceived motivation for a major societal change or an invasion of some kind.

Reviewing chapter 8, "Why Here? Why Not Somewhere Else?" there is a long list of possible motivations for extraterrestrials, if they exist, to wish to explore the Earth. Some, if not most, could accurately fit as explanations into whatever scenario our unique perspectives would allow.

As described in the Brookings Report, "An individual's reactions to such a contact would in part depend on his cultural, religious, and social background, as well as on the actions of those he considered authorities and leaders and their behavior, in turn would in part depend on their cultural, social and religious environment."[9]

A typical individual's reaction would, in turn, be one of a collective reaction that includes our communities. These perspectives would be shaped comprehensively by our life's experiences and is fluid and subject to change. These would also include social contacts, culture, whether or not it has religious foundations, how we perceive our government in general and their reaction to the same and how we perceive everyone else's reaction. Most significantly, we would interact with the reactions of those closest to us.

Our government's reaction would also matter to us as this community is the glue which binds our social structure. Any perspectives might consequently be partly shaped by our government's reactions. The Brookings Report elaborated on such leadership ramifications, "the degree of political or social repercussion would probably depend on leadership's interpretation of (1) its own role, (2) threats to that role, and (3) national and personal opportunities to take advantage of the disruption or reinforcement of the attitudes and values of others."[10]

There is currently a noted chasm between the governments of most of the world's nations and its citizens over the subjects of

UFOs/UAPs/extraterrestrials and disclosure. These examples identify an existential problem when the topic of disclosure and transparency and release of the freedom of information doctrine is raised. In the evolutionary timeline of document disclosure that has been recursive for decades now, here are some observations to analyze:

1. If such documents exist, they are tangible and a representation of some reality. Some of them have already been made available to the public domain but with many heavily redacted modifications. Some of these are noted in the "Disclosure" chapter.
2. As time moves on, more and more documents are being released. England has, within the past ten years, released a second library under its own Freedom of Information Act. France, Brazil, Australia, and Canada have undertaken similar projects recently.
3. How could any government release any documents at all if they did not exist?
4. How, why, when, and where can a person deny their existence when these documents are there in black and white? Many thousands of pages are now in the public domain. Would governmental groups spend that much of taxpayer's finances perpetrating documentary hoaxes? Why would any group of people devote so many resources, including time, to some reality that is so profusely claimed to "not exist?"

All those government leaders and their employees would have been performing more important tasks to occupy their time than fabricating and providing bogus signatures to such documents if the subject of the documents did not exist or was not real. Such make-believe then should have been attacked voraciously by the populace. This surely would have happened in our place of work if such a scenario was perpetrated by our efforts.

Does this notion apply to members of the Advanced Aerospace Threat Identification Program (AATIP) expensed by the United

States Department of Defense in 2007–2012 to the amount of over twenty-two million dollars? The program studied dozens of cases involving unidentifiable aerial craft sightings, dogfights with jet fighters, recordings of these strange events and data collection by mostly military and aerospace witnesses. Also case studies were investigated of incidents at military facilities and nuclear power plants. Some individuals involved in AATIP no doubt either developed a new and highly motivated level of engagement as defined from their job or their timing was right.

Luis Elizondo, military intelligence official, ran the program. His office was in the Pentagon, in the C-Ring, fifth floor. He investigated such incidents as the 2004 sortie of two Navy F/A-18F fighter jets and the carrier USS Nimitz of a "whitish oval object" larger than most commercial passenger jets. It was a bellicose incident, not a hoax or a field practice exercise. Elizondo told the *New York Times* the AATIP program continued to operate past the official 2012 defunding transaction. He worked with the Navy and the CIA for five more years on additional investigations. He resigned his post just recently in October 2017 because of "excessive secrecy and internal opposition."[11]

Three politicians with previous interest in the debate (if these are truly unidentified objects being observed, are they piloted by an intelligence of some kind?) were involved in AATIP. U.S. Senate majority leader Harry Reid heavily influenced formation of the program. His home state is the home of Area 51, an ultra-secret military region which was denied to exist by the government until just a few years ago. Two other U.S. senators, Ted Stevens of Alaska and Daniel K. Inouye of Hawaii, were very vocal in recruiting resources to help start AATIP. All them were influenced by former U.S. astronaut John Glenn, later senator from Ohio, to encourage the Department of Defense to start a study program.

A long-time personal friend of Senator Reid and billionaire entrepreneur Robert Bigelow broke off all relations with membership in the Mutual UFO Network (MUFON) organization to sign on to the study. Bigelow had been a researcher in UAPs for many years prior to this assignment and was a regular contributor to Reid's political campaigns. His company, Bigelow Aerospace, was the front for the research and the funding destination for AATIP. Not only was the program studying UAPs, they also used Bigelow Aerospace facilities as a laboratory/warehouse for testing retrieved and captured metals, alloys, and other materials in the course of events.

Bigelow graduated from AATIP to work with NASA on developing expandable spacecraft for human flight and habitation. He continues to be under contract there today. Bigelow, normally private in these matters, did tell CBS's *60 Minutes* in a May 2017 interview that he "was absolutely convinced that aliens exist and that UFOs have visited Earth." Bigelow told CBS about many case files documenting observations and videos of aircraft which seemed to violate humankind's knowledge about high velocity ship flight and maneuvers which produced no noise or visible characteristics of a propulsion mechanism or exhaust.[12]

One of these investigations Bigelow referenced was initiated from experiences and data collected by the principles in the USS Nimitz sortie in 2004. On November 14 of that year, Commander David Fravor and Lt. Commander Jim Slaight were on a training mission off the coast near San Diego. Their radio operator notified them about the tracking of a "couple dozen of these objects continually for a few weeks and which were now sighted descending from eighty thousand feet altitude."

Fravor and Slaight were ordered to intercept this fleet of unidentified ships, believed to be part of this sighting flap. When they approached for a first visual, Fravor and Slaight described a white ship without wings of any kind. It was about forty feet in diameter and hovering in midair about fifty feet above the ocean surface. The ship was gathering water up from the surface. Videos from both Fravor and Slaight recorded a water swell and a trail rising upward toward and into the craft. After more than a minute, the craft raced toward Fravor's craft and almost collided with it before veering off radically in an opposite direction at acceleration not feasible for any human-built ships. A second sortie was dispatched from the Nimitz and captured more video and radar data before the craft disappeared from view.[13]

Here are some opinions which offer food for thought. Sara Seager, professor and astrophysicist from Massachusetts Institute of Technology, advocates more dedicated study to such manifestations: "When people claim to observe truly unusual phenomena, sometimes, it's worth investigating seriously. What people sometimes don't get about science is that we often have phenomena that remain unexplained."

Luis Alizondo said in his resignation letter in October, "There was an undoubted and crucial need for further detailed study of the many accounts from the Navy and other services of unusual aerial systems interfering with military weapon platforms and displaying beyond-next-generation capabilities."

Harry Reid summed up the project in this way on a *Twitter* feed, "If anyone says they have the answers now (for where the objects came from), they're fooling themselves. We do not know the answers, but we have plenty of evidence to support asking the questions. This is about science and national security. If America doesn't take the lead in answering these questions, others will"[14] (Senator Harry Reid [@SenatorReid] December 16, 2017).

It is said that as a species, we are not yet ready to be the physical explorers in a cosmic scenario, at least in the sense of investigating an astro-geographical domain. All communities in our society are in general agreement with this statement because we do not yet possess the knowledge or technological capacity to be enablers outside our "goldilocks zone." Additionally, these same communities—including the government, military, and the sciences—don't forecast us as being able to achieve this potential anytime soon. No matter if we were able to chart this course and find either intelligent species less than or more advanced than we are in all evolutionary notions, it would take humankind a long time to make that physical journey ourselves.

Therefore, we would have to accept the role of the "recipients" of such first contact. Our perspectives would most probably be shaped through a lens of the *receiver* of the contact like our global ancestors. This notion may have a more profound significance than if we were the explorer or transmitter. First, the surprise element would be unknowingly significant if some other intelligent species were to appear unannounced on our planetary doorstep. We could not anticipate such a first meeting nearly as well as if we were the explorers.

The natives of the Caribbean and Watling Island in the Bahamas were probably more surprised upon first contact than Columbus's expedition were in anticipating the event. Columbus as explorer anticipated such events to transpire. The natives did not, so they were surprised. Next, what each party observed of the other at that time of the first contact helped shape their initial reactions. Did the receivers see the ships the explorers arrived in? Or did they see the explorers coming into shore? How many members of the other party were there? Did either see weapons of the other? How different looking was the other party from themselves? There were language barriers which made the communication more difficult. How much more difficult?

We would reason the exact thought processes upon any such contact. Our impressions would be shaped to a large part by any prior knowledge about all the aspects of what we observed. So our surprise factor may or may not be greater depending on what we can cognitively process about what is happening given what we already knew. Factors here would include "how many of them there were and what kind of vessel they arrived in and what the methodology of their arrival."

Next, we would need a measure of apprehension as is defined differently from that of surprise. This would probably include "how their outer appearances differed from us and by how much, how many of them there were and if they showed any weapons of harm."

Chances are, though, such a first contact would not physically involve anyone individually. Remember that in all the other earthly explorations, it was a group of representatives, whether designated or not, or a lucky select few distinguished members of the native community who were present with the receiving party. Accordingly, the same would probably be true of any such contact encounter in our times. Our interest would only be enhanced via indirect reports we receive from government officials or the equivalent.

Our later impressions and perspectives would be modified accordingly. But it is acknowledged that the possibility our motivation could be initiated by any of these scenarios does exist. Our individual encounter experience, as has been those alleged by the multitude of experiences through history, could consist of only one ship and a couple of occupants. Or there could be a fleet of ships that disembark somewhere here on Earth and establish communications with government or other officials.

These dynamics would be modified again after the initial contact has taken place, and there has been some time to reflect on the facts of the encounter and to analyze them. This initial perspective shift could be investigated in terms of a classic theory of human motivation developed by Dr. Abraham Maslow, a psychologist in the mid-twentieth century.

Dr. Maslow published an article, *A Theory of Human Motivation*, in 1943.[15] In the article, he concluded research on human's basic needs and curiosities as groundwork for daily life. A "needs hierarchy" was created. What is most important to take from this reading is that man's most basic life needs of physiology and safety are a constant activity as positioned at the bottom of his needs hierarchy. Only when these basic needs are satisfied for any individual can he then strive to fulfill those needs higher up on Maslow's pyramid.

Dr. Maslow's theory is applied to our example with consideration to interest level and perspectives regarding the UFO/UAP/ extraterrestrial hypotheses. The more the first contact is threatening to a witness, the more his or her perspective would move toward a heightened emotional level of apprehension and maybe fear. This would play out in an example of whether the explorers somehow threatened our food supply or of a peaceful existence the way we are currently living. So we would be far more interested in the phenomena than if, say the new life discovery was of just microbial or even a plant species.

This is presented to help the reader remember that individual perceptions and interest levels would be modified in slightly different ways by a personal encounter as opposed to learning of such

from others. The impact level of the buildup to an acknowledgment of new life as such is important also. This means a visit from
extraterrestrial explorers, in a dramatic fashion, where they land on
the steps of the U.S. Congress would have a different impact than
if there was a gradual, incremental contact of just one encounter at
a time involving only a few individuals and over a very long period
of time.

Our reactions would be at least a bit different from other people. There would also be a more collective community perspective
that would, in turn, influence our perspectives even more. This may
or may not further influence our individual perspective according
to how we analyze and make conclusions about new facts of the
encounter and its aftermath.

Most of the literature stops at making predictions based on the
perspectives and reactions of the collective society. So what would
change in society under a discovery of such magnitude? It has been
said that if the recognized new life-form(s) were either *prokaryotic* (an
organism whose cells do not have a nucleus) or plantlike *eukaryotes*
(an organism whose cells contain a defined nucleus) the consensus
answer would be "not much." Collectively, we probably would not
change our vacation plans or seek to change our employment situation. Maybe, though, some of us would contemplate a career change;
this would be dramatic. Some of us may have our epistemological
interest in the subject aroused. Some of us who were initially apprehensive about being extroverted with our discussions about alien
life would have a new public emergence. New career paths would
undoubtedly be taken by a lot of people.

Would the discovery provide for a paradigm shift away from
this subject of extraterrestrial life being treated by science as pseudoscience? From historical records, while there are many proponents on
each side of the debate from all walks of life, it appears that it is most
problematic for the scientific professions. Is a reason because there is
too much more to learn about the various scientific principles and
implications of intelligent life-forms, some of which may be more
advanced than our species? Is it more of an inability to own up to
an ignorance of knowledge? Is the reluctance based on anthropocen-

tric or other centric tendencies which have plagued humankind since recorded history? Is it because the current construct of practiced science thought and study need a tune-up?

A discovery of new prokaryotes would sway some younger people into biology and exobiology career and cause a temporary upward spike in revenues of champagne grape growers for those who wish to celebrate such a discovery. Do biologists drink champagne? Maybe not on the job but I am sure here we could provide them a one-time exception.

Would discovery of only new prokaryotic life hinder the evolution of the science train of thought with respect to a shift toward of that of a classification to that of mainstream science? From the standpoint of man being the explorer versus man being recipient of a discovery such as this, it is evident that as explorers, humankind cannot yet do much traveling to other worlds with which to experiment and test these hypotheses. If we cannot travel to these other venues, we can still continue to investigate our own Earth for facts and for learning about new biological life-forms.

If our engagement and perspectives won't change much with an unearthing of new prokaryotic life-forms, as most studies seem to indicate, would a discovery of eukaryotic life be different? The burning question is not of just any eukaryotic life, that of plants and the like, but of the intelligent, sentient species types. How would we react to this discovery? How would science, or government, or society react?

Individually, we would react to both the raw discovery and the context mostly on our own terms. We may also be influenced by society's reaction. Either or both will cause a change in our thinking. The reactions could, of course, be studied according to an application of Maslow's need hierarchy to those components present in the situation. Such components include how benevolent the entire scenario evolved. This could be answered as, "Are my food, security, and belongingness needs to my basic life threatened?" Employment, recreation, and daily life considerations may take on a new dimension as a result.

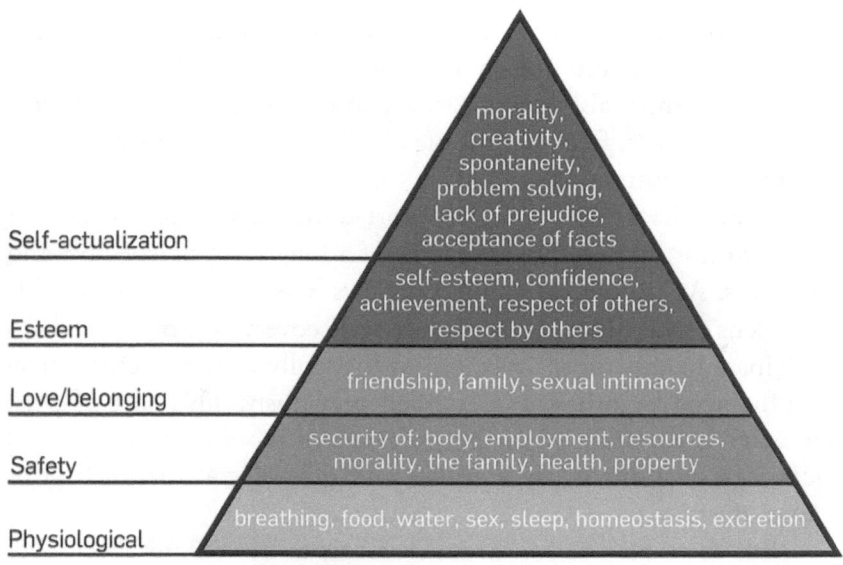

MASLOW NEEDS HIERARCHY

Citation:

Factoryjoe/Maslow's hierarchy of needs. Resized, renamed, and
  copied version of File:Mazlow's Hierarchy of Needs.
  svg. /CC-BY-SA-3.0 (https://creativecommons.
  org/licenses/by-sa/3.0.deed.en)

Our engagement in the subject would probably be affected also. Because extraterrestrial life is just another real part of life and existence in our universe, if we now had knowledge and absolute confirmation of its existence, we may be motivated to learn more about it and how it affects the environment. This is clearly what a scientist seeks to do and to study in his work. So ufology is also a study of the environment around us in the same way radio astronomy and *exogeology* and *exobiology* are. Ufology is a more localized environmental study in geographical scope for now. This is only because of humankind's lack of technological development. We can only study the evidence as it is presented on our own planet and within our physical reach. The other science disciplines extend the definition of environment to include outer space and the cosmos.

The benevolence factor is crucial to consider in both a post-discovery framework and our own reactionary framework. If the intelligent extraterrestrials show no threat to our more basic short-term needs, we are then free to continue our lives and our pursuits with a minimum of restraint.

These hierarchies, being all short-term considerations, do not or should not preclude investigation of a long-term perspective and awareness. An individual's religious views may be moved into a different lens of visualization as a result of discovery. We may decide to participate in this proactively and individually or our decision may be influenced by others as described previously. This sort of pack mentality is certainly what the government would like to see and encourage in terms of their ability and ease of governance by this course of action.

But there may be just as much for us to be anxious about over the government's reactions than what individual level of anxiety we would feel toward the discovery of intelligent extraterrestrials. For example, the government, to whom an analogy of the Maslow's need hierarchy grid could be applied, may become more anxious at even the grid's lower dimension levels of physiological well-being and safety security if these needs are not met. A person's own individual reactions will, thus, be altered accordingly.

This is scary in and of itself because the government has many more weapons at their disposal with which to use to react and maneuver around these possibilities than any of us. The thought of, "If the government has had this knowledge of existence and have not and will not disclose it, are they risking a more tenuous and perilous control of society the further down the road the situation plays out?" Would they be making it only harder on themselves to govern the way they wish, the longer it continues? Do the consequences of nondisclosure become more risky, even riskier than the reasons for continuing to keep the secrets away from society?

The dynamic of these scenarios can take on less benevolent examples which are always possible and cannot be ignored. Who is to say if we have not yet been discovered that an eventual unearthing may not be a harmless one? We only need to investigate man's historical explorations of some of our ancestors' civilizations to conclude that if we are capable of such processes then others can be. We know that this theater can process two such polarized situations. Often, we make predictions, sometimes to a fault, based on historical precedence. If this motivation is egocentric, it can sometimes be wrong to behave in this manner.

If we are to be the receivers and not the explorers in this discussion, we would have no control over whether the explorers are less or more benevolent in nature, at least in the beginning. If the explorers are coming to visit our planet, then we have to resign ourselves to the fact that we could not control their immediate intentions because of the technical advantages they possess. This is in the same fashion the natives of Watling Island, the Bahamas, viewed the Columbus expedition in 1492.

The what-if strategy allows us to acquire knowledge, through simulation, of an epistemology that would manufacture real knowledge to enable us to advance faster than if we did not partake in such exercises. If these exercises can also entertain us, this is additionally beneficial. This is another way interest, and engagement levels in ET life can be ignited or catalyzed.

The takeaway points here are that via this strategy, we could individually and collectively gain more insight into how our perspectives and interest in the UFO/UAP/extraterrestrial hypotheses could be altered. We can gain insight into learning from our own historical perspectives and experiences from being entertained or from information in a more purely scholastic format. As we come to understand more about how nature operates, on philosophical, scientific, cultural, social and spiritual dimensions, positive change in all levels of society at large then becomes more real and attainable. All this means is that, at the end point, our world becomes a better one in which to live.

Here would be a good opportunity to demand more transparency to knowledge disclosure in all the relevant debate areas. *The Humaniverse Guide to Better Reasoning & Decision-Making* has been demonstrating more and more evidence that an abundance of this knowledge is being withheld from society. The Department of

Defense's December 2017 announcement of the Advanced Aerospace Threat Identification Program to study unidentified aerial phenomena is just another case in point and one of many situations which will occur in the future.

Additionally, maybe this notion of a lack of government disclosure and transparency will spark an interest in many of us to wish to further pursue knowledge in this profession. *The Humaniverse Guide to Better Reasoning & Decision-Making* shows that a lot of documentation may be covertly hidden from the public domain. Some written documentation has been released by the world's (not just the United States) governments. The perspective from the entire ufology profession concludes there is much more waiting to be uncovered.

There can be long and drawn-out debates about the notions of the viability and feasibility of any government to successfully cover-up in any harmful way, a conspiracy of this nature and magnitude. One side can argue that, "No government can cover-up a subject forever; therefore, nothing exists."

This is a fatally deceptive argument. If nothing exists, why does the public have access to a sea of documentation from governments all over the world who are disclosing that some things do exist? Why have regular and extensive studies been made part of the public domain and at public expense since at least the 1940s if nothing exists? Why do so many events continue to happen causing evermore studies to be conducted, information withheld, but only fractionally disclosed many years later if nothing exists? Does this all suggest an ocean of information is currently hidden from society?

References

[1] Brookings Institution. 1961. Proposed Studies on the Implications of Peaceful Space Activities for Human Affairs, pg. 103, n. 34. RRO.org (Beau, Jerome). Retrieved 2016-06-02.

[2] Dick, Steven J. 2007. Consequences of Exploration: Learning from History. National Aeronautical and Space Administration. November 22, 2007. www.NASA.gov/.

3   Brookings Institution. 1961. Proposed Studies on the Implications of Peaceful Space Activities for Human Affairs, pg. 183. RRO.org (Beau, Jerome). Retrieved 2016-06-02.

4   Brookings Institution. 1961. Proposed Studies on the Implications of Peaceful Space Activities for Human Affairs, pg. 183–184. RRO.org (Beau, Jerome). Retrieved 2016-06-02.

5   Brookings Institution. 1961. Proposed Studies on the Implications of Peaceful Space Activities for Human Affairs, pg. 183–184. RRO.org (Beau, Jerome). Retrieved 2016-06-02.

6   Brookings Institution. 1961. Proposed Studies on the Implications of Peaceful Space Activities for Human Affairs, pg. 183–184. RRO.org (Beau, Jerome). Retrieved 2016-06-02.

7   Kuhn, Thomas S. 1962. The Structure of Scientific Revolutions. Published 1962 as monograph in Encyclopedia of Unified Science, Mundaneum Institute, the Hague. Published 1962 concurrently by University of Chicago Press.

8   Howell, Elizabeth. 2015. How Would the World Change if We Found Extraterrestrial Life? *Astrobiology Magazine.* January 29, 2015. Retrieved from: http://phys.org/news/2015-01-world-extraterrestrial-life.html.

9   Brookings Institution. 1961. Proposed Studies on the Implications of Peaceful Space Activities for Human Affairs, pg. 184. RRO.org (Beau, Jerome). Retrieved 2016-06-02.

10 Brookings Institution. 1961. Proposed Studies on the Implications of Peaceful Space Activities for Human Affairs, pg. 184. RRO.org (Beau, Jerome). Retrieved 2016-06-02.

11 NY Times Reference 11: *The New York Times*, 2017. "Glowing Auras and Black Money: The Pentagon's Mysterious UFO Program." Published December 16, 2017. Copyright 2017 by The New York Times Publishing Company.

12 NY Times Reference 12: *The New York Times*, 2017. "Glowing Auras and Black Money: The Pentagon's Mysterious UFO Program." Published December 16, 2017. Copyright 2017 by The New York Times Publishing Company.

13 Rosenberg, Eli. 2017. *Washington Post*; Checkpoint. Former Navy pilot describes UFO encounter studied by secret Pentagon program. Published December 18, 2017. Retrieved from: https://www.washingtonpost.com/news/checkpoint/wp/2017/12/18/former-navy-pilot-describes-encounter-with-ufo-studied-by-secret-pentagon-program/?utm_term=.ecbfdb8c3ba3.

14 Doubek, James, 2017. National Public Radio. The Two-way. *Secret Pentagon Program Spent Millions to Research UFOs.* December 17, 2017. Retrieved from: https://www.npr.org/sections/thetwo-way/2017/12/17/571446881/secret-pentagon-program-spent-millions-to-research-ufos.

15 Maslow, Dr. A. 1943. A Theory of Human Motivation. Psychological Review Journal. Volume 50, Number 4. Pgs. 370–396. www.apa.org/pubs/journals/rev.

CHAPTER 22

Prove ET
Ways to Provide Evidence for
Possibility of Existence

The important thing is to not stop questioning. Curiosity has its own reason for existence. One cannot help but be in awe when he contemplates the mysteries of eternity, of life, of the marvelous structure of reality. It is enough if one tries merely to comprehend a little of this mystery each day.

—Albert Einstein, *LIFE Magazine* interview[1,2]

Key words: validity, reliability, credibility, electromagnetic spectrum, frequency, interferometry, bio-natural pacemaker, homopolar machine

A main objective in virtually all learning projects of an investigative nature is to uncover and/or discover facts as evidence. An accumulation of these facts and evidence will be analyzed with the intent of making an inference or deduction (bottom up or top down) about the accuracy of what the investigation is attempts to describe. Other analytical factors include the *validity, reliability*, and *credibility* observed through the lens of the researcher and others. These three factors of validity, reliability, and credibility are discussed elsewhere in this text.

We are also aware that, even though this appears to be a statement only scientists would understand and adhere to, the fact is all of us use this same process in our investigations. At least we should be doing this. No community owns this process.

So far, the discourse presented here has focused on refuting the only two hypotheses presented: those that "UFOs do not exist" and "Extraterrestrial beings do not exist." These discussions are attempts to prove something by trying to disprove it. This is a popular but significantly debated and not a majority-used paradigm in investigative research of the sciences, social sciences, business, political science, education, and so on.

Here are two definitions of the term *proof*, the noun, as it applies to our discussion:

/proōf/: noun – Evidence or argument establishing or helping to establish a fact or the truth of a statement.[3]

[proof]: noun – Evidence sufficient to establish a thing as true, or to produce belief in its truth.[4]

Note both definitions use *evidence*, *establish*, and *truth* as attributes. These attributes should be read and investigated "as is" in the process. Thus, evidence establishes the truth.

The modern era of ufology has entered its eighth decade. This is not to say there were no events involving these phenomena prior to 1947; we know there have been a lot of episodes. Many historically significant ones have been explored. For the most part, it is only within this time frame that humankind has discovered and developed some tools of technology that would permit an investigation of the phenomena and help satisfy some of the requirements for truth; namely the validity, reliability, and credibility that help prove reality. Thus, reality is the equivalent of existence, and truth is in agreement to both.

There has been a great debate involving an information and potential world government cover-up of such that has lasted as long as the modern era of ufology. The actions of individual governments and their military defense complexes, as well as unified efforts of multinational government equivalents, have contributed to the alleged cloaking of knowledge concerning UFOs and celestial beings.

In establishing a relationship between evidence in this realm and technological tools, it is necessary to define the sufficiency of such tools to be useful as well as interest, engagement, and purpose in the UFO and celestial beings realms. We know that, taken as separate communities, a government, which here includes its military forces as defined from direct civilian government operations and the populace are equally interested in the phenomena and what knowledge and discovery that can be gleaned from them. Their objectives for wanting to accumulate and access such knowledge are probably equivalent on some important levels, at least enough to pursue the topic with continued enthusiasm.

When it comes to the sufficiency of being able to accumulate that knowledge, though, for a long time in the last century, the governments have had the upper hand. This was because they had more tools, accoutrements, and authority needed to facilitate a successful program with which to assemble this body of knowledge. They had the fiscal means, the power, and the physical might to be able develop the technological tools needed to make detailed investigations of phenomena which use their own brand of advanced technology. In this way, these tools were and are needed to inspect even more advanced technology. The gap between a "raw" eyesight observance-as-evidence technology versus interpretation and conclusion about the objects processed from utilization of advanced technology need to be narrowed. Access to the tools of the last century gave the governments and military complexes a monopoly over similar efforts that were possible by society in general.

The gap is starting to be narrowed slightly.

As time moves forward and humankind obtains more sophisticated technology tools, that disclosure gap will continue to diminish. The inhibiting factors of the public domain's capacity to acquire these tools will continue to erode. The cost of these tools, as well as future advancements, will accelerate this gap closure. At least in our democratic society, we will now have more and more capability to add to our evidence base and continue to investigate and move toward a conclusion of existence via accessing these tools.

Let's assume the U.S. government has not participated in a secret procedure of the issues to date. The government has repeatedly stated that they have not conducted any type of investigative cover-up concerning any incidents in the modern era of ufology. By now, we have learned about Project Blue Book and the large amount of literature which champion their version of "concerted investigation" regarding the history of cases, as well as others such as the Roswell incident. Basically, let's extend the notion by saying no evidence has been covered up by any community, from government to military to any religious institution, including but not limited to the Vatican, and that no evidence referenced to any of these sources has ever existed.

That would be only be the beginning when it comes to the history and present knowledge of UFO and extraterrestrial cases. There are simply too many past and present events and that continue to happen which warrant a rational and dedicated investigation by whatever community and/or means it takes to do so. We now know about some of the untold number of unusual and documented intrusions into military operations, nuclear weapons installations, both military and civilian airports, government buildings, and civilian site locations as suspected to have been infiltrated by celestial beings.

Here is another recent disclosure which argues against government interest in UFOs and extraterrestrials. In revisiting the December 2017 disclosure by the Department of Defense and the Pentagon that they operated the Advanced Aerospace Threat Identification Program (AATIP) to investigate case studies of unidentified aerial phenomena (UAPs), the author has identified some of the key individuals in the program. The theme was how persons could become motivated and otherwise engaged in the subject. For most of them, interest was more unpredictable and, therefore, not by total choice.

The U.S. military studied the phenomena for over five years from 2007–2012 and spent over twenty-two million dollars in that time alone, according to disclosure documents. The AATIP program's head of operations, Luis Elizondo, maintains that the program, though defunded in 2012, continues to exist to this day. He resigned in October 2017 over protest to lack of disclosure and transparency in the AATIP project. Among the arsenal of high tech tools used in the project were the Navy supercarrier USS Nimitz, aerial high-speed cameras, spectroscopes, F-18 fighter jets, many types of radar technology, and other devices of which we are still unaware. There were dozens of cases investigated by the AATIP, but all were kept secret until now.

Though twenty-first century development of these technologies has introduced more accessibility by public researchers, as we will become aware of next, another crucial argument entirely refutes the government's official proclamations against being interested in the UFO/extraterrestrial hypotheses. Their classic official conclusions have been that "they pose no threat to national security." The AATIP is yet another disclosure admission which defeats this argument, as well as the arguments which state that all sightings and encounters are identifiable and either occur in nature, are man-made, or manifestations of the human imagination psyche in some form or another.

In order to narrow the gap that exists between access by the government and military to technology, arsenal monopoly and societal advances in many technological areas have permitted civilian efforts to step in and start to conduct their own more thorough studies. These new tools come from various areas of natural science. These science subjects include: physics, chemistry, biology, geology, meteorology, and paleoanthropology. Let's get into some of these new techniques by first describing the physical landscape of what these instrumentations would apply to in gathering evidence for proof of their existence.

The anatomy of an extraterrestrial and/or UFO experience would consist of this landscape. The craft used to transport any beings would physically come from and through the outer atmosphere or maybe from below the Earth or sea or other body of water. It may come from a type of extra-dimensional portal. Some natural science knowledge has already theorized existence of more than the three dimensions we live in every day. Albert Einstein and the theoretical physics community have been pursuing incremental additions to this body of knowledge for over one hundred years.

Knowledge of any of the sciences can be used to help better record, provide data for analysis, and draw rational and valid conclusions. Technology that can identify objects that emit and/or reflect light waves, in the visible portion of the spectrum or beyond it, and other forms of radiation waves and/or particles exist now. The Hynek Classification List would record these as CE-1 or maybe CE-2 episodes.

If an object were to land, there is always the possibility it would leave trace chemical and/or geophysical evidence that can be recorded and analyzed with the latest sophisticated instrumentation. Also further physical or other chemical signatures can be detected from a landed object. The Hynek Classification List would record these as CE-2 encounters.

If the object was intelligently piloted, there may be additional situations whereby an experience with the entity could leave more evidence, this time of either a chemical or biological nature that we now have some tools of technology with which to assist in the investigation. The Hynek Classification List would record these as CE-3 situations.

If the experience with the live entity was more involved than just visual, such as some form of communication episode, the Hynek Classification List would record these as CE-4 or higher. All these scenarios can be critically studied with access to some of the more sophisticated tools we now have access to as the average public. Let's now discuss some of the concepts that coordinate with how these tools would operate, what they would measure, and what data they would record for our analysis.

This environmental discussion starts with the atmosphere and bodies of water in relation to how observations can be made and data recorded and measured for later analysis. Humans and instruments can detect and record objects as direct applications of how electrons behave. The *electromagnetic spectrum* is defined as the entire range of wavelengths or frequencies of electromagnetic radiation extending from gamma rays to the longest radio waves and including visible light.[5]

Most types of energy travel in waves. The individual electrons initiate this wave movement by moving up and down within the particular atom. This wave can be envisioned by remembering how a heart monitor portrays the up-and-down wave of a heartbeat. These waves can be longer or shorter depending on what particle is manufacturing the wave. It is the electron's actions within this electromagnetic spectrum that determine the length of a wave cycle.

A portion of this spectrum describes electrons that manufacture waves of a certain cycle that produce light. The length of a particular wave cycle helps to describe the *frequency* that the electrons of that particle move. Frequency in this subject is defined as the entire range of wavelengths or frequencies of electromagnetic radiation extending from gamma rays to the longest radio waves and including visible light.[6] The frequency refers to the number of waves or the number of up-and-down oscillations of a particular wave that occur in a period of time. In most science applications, this unit of time is a second. Also all matter is unique in that every particle that has a particular chemical composition has a different frequency of oscillations. All substances vibrate at different frequencies; the waves travel at a unique number of cycles per second.

Not all light is capable of being seen by humans. Only a small portion of the electromagnetic spectrum is visible to us as light. We have heard of different types of waves, such as gamma rays, x-rays infrared, fluorescent, microwaves, and radio waves, to name a few. These are all examples of waves of electromagnetic radiation. They all travel at different cycles or velocities. It is these unique signatures assigned to all the particles in nature that allow us to see, hear, and help make sense of the environment around us.

This is how we can sight and observe an unidentified object moving through our atmosphere, along the land or through the water. The electrons of the unique chemical substances present are creating these wave signatures that travel through our eyes, ears, and brains and allow us to perceive these manifestations. The instruments of technology can capture these substances and objects and measure them for positive identification for our further analysis.

Some of these instruments have been in existence for almost one hundred years. When radar was invented in 1935, it allowed the operator to transmit radio waves from a ground-based transmitter into the atmosphere. These waves were pointed toward the target it wished to detect. The radio waves would bounce off the target and travel back to a radar receiver located near the transmitter or at least close enough for the detection to appear on the operator's radar screen.

As is most often the case with the evolution of any machine of technology, the improvements that are made are of a very small incremental nature. In the eighty years since, many small improvements were made to radar equipment, so now such instruments can detect the largest variety of objects ever seen objects as far away as the farthest reaches of Earth's orbit and beyond into space.

As a first method of "how to detect ET" is discussed, a sufficient knowledge of the geophysical, chemical, and meteorological attributes of our environment enters into the process of the creation and development of enabling radar technologies. The evolution of the ability to detect and analyze the ground, ocean, and the atmosphere and what is in it at any time progressed in this fashion.

Radar technology was the dominant detection method accessible for a significant part of the twentieth century. Radio Detection and Ranging (RADAR) was a key factor in the outcome of the Battle of Britain between Nazi Germany and England during World War II. The developments in radar technology were incremental and followed a typical path that exemplifies most general processes of how a new technology is evolved. Newer versions were faster, more powerful, and usually smaller than their predecessors. The introduction of unrelated separate technologies would periodically chart a new evolution course.

An example of new and unrelated technology was the introduction of the computer into radar design. Another was the introduction of the concept of *interferometry*. This is a process popularly attributed to the study of astronomy and a good way to help visualize the concept. Multiple telescopes are situated in different locations with a separation of a few yards to thousands of miles away. When all the telescopes are pointed at the same observational target, computers triangulate what each individual telescope views at that time. The result is a far greater resolution and ability to detect more detail than any individual telescope can accomplish.

This method of extracting data applies to the evolution of radar equipment technology. The computer technology allows for existing radar technology to accomplish more at the radar's current sophistication level than could be otherwise obtained.

Generally, what radar is detecting are objects that emit their own unique electromagnetic waves and thus can be identified. As radar equipment evolved along with technology such as computers, inventions like interferometry then allowed for the shrinking in the size of the equipment, the increased power capabilities, and the portability.

By the 1950s, new radars that used all these features were developed. The computer was slowly introduced, and then airborne radar units were invented as the size of workable models decreased. Better microwave transmitters and receivers (microwaves were the preferred form of electromagnetic radiation detection equipment then) paralleled the Cold War Era demand for better equipment. Later another creation, Doppler radar equipment, greatly enhanced detection characteristics. All this was developed by the military and was in widespread use by the 1970s. Doppler radar allowed for detection of a new characteristic of target detection. This is the characteristic of the velocity of the object. Now radar could identify the position, intensity of the object (surface characteristics, etc.), and the velocity as the object moving through the atmosphere, along the ground or in the water.

Computers operated under an analog processing system until the 1980s. This was slowly replaced by digital technology, which greatly increased processing speed and capacity. It was another evolutionary breakthrough that expanded the capabilities of radar installations.

To this point, we have been discussing ways to detect interactions between the atmospheric environment, including the ground and bodies of water, and an object moving through them. The common crossing interaction between radar evolution and the sources of its development involved only the government and military complexes. The overarching deployment of digital computer technology and further miniaturization created the possibility of diffusion of radar technology into the general public. It was now feasible for the public to obtain and use such equipment. Whereas, dedicated and sophisticated analysis of UFOs and associated phenomena were the monopolized milieu of the military now had barriers-to-entry removed for the public to gain access into undertaking their own study analyses.

The notion of technology diffusion is a very important process that occurs in societies all over the world. As noted, computers allowed for the eventual diffusion of radar detection technology into societies' reach. The science inherent in the understanding of how nature behaves and the physics, chemistry, and geophysics behind the concepts of electromagnetic radiation have also led to creation of such other technologies as GPS, many applications in astronomy, as noted, and the current structure of weather detection and forecasting. Knowing how light the various types of wave transmission and such types of radiation react and travel in our universe is central to understanding some of the ways UFOs and extraterrestrials can be detected.

We can now start to imagine how a type of radar equipment could be used to detect objects in the earth's environment. Beams of wave energy are transmitted into the atmosphere, along the ground, or into the water to gather information for analysis. The physical process of aiming this radar equipment at a target has a parallel with other detection equipment. Such detection by the use of photography was the grandfather technology used in this type of investigation. A camera could be pointed into the atmosphere and record images of UFOs and other objects. This technology also was the original equipment available to mainstream society and earlier ufology research investigations.

Needless to say, the literature was full of controversy when it came to photographic technology. Here, the individuals in the debunking community did not need to establish their own credibility in their critiques despite their lack of detailed investigation and lack of utilizing many facts that led to their use of fallacious "hasty generalizations." Photographic evidence was inherently flawed, at least with each UFO incident taken in isolation and without argument of an accumulation of many separate UFO sightings as a larger body of evidence. An individual photograph of one UFO sighting did not have a compelling enough evidence to convince enough people that existence was reality.

In continuing to use the epistemology of the electromagnetic spectrum, let us now focus on a small portion of it. Traditional radar technologies, especially the earlier ones, detected and inspected readings predominately from radio and microwave sources. Earlier evolutions of photography could only depict and detect waves in the visible light spectrum portion of the entire electromagnetic spectrum. Until the last few decades, only visible-light photography was within practical reach of the general public, including most ufologists.

When we observe a printout of "sine waves" or look at them on an oscilloscope, we see many repetitions of the wave as it travels across the paper or on the screen. The closer the waves appear together, the higher the frequency of the waves. This means that more frequent cycles of the wave occur in one second of time than waves that are farther apart.

Waves of the highest frequency occur at the shortest end of the electromagnetic spectrum. These are gamma rays. Waves at the lowest frequency are at the other end. These are radio waves. The visible spectrum is a little closer to the highest frequency end where the gamma rays reside. Just to the shorter side of the visible portion are the ultraviolet waves, followed by x-rays, then gamma rays. Going in the other direction from the visible portion are infrared, then microwaves, then the various forms of radio waves. These microwaves and radio waves are those that carry the various audio and video communications, namely radio and television, through our atmosphere. It is the existence of this portion of the entire spectrum that allows us to listen to the radio, watch television, and use radar equipment.

Electromagnetic waves are all forms of energy. So radio waves, television, microwaves, visible light, x-rays, and gamma rays are all forms of energy. In science, this is referred to as electromagnetic *radiation*. When any matter in nature gives off light waves it is giving off a form of energy that radiates from the surface of the matter. The radiation, as we see, does not have to be that of visible light. It is the task of various instruments to be able to detect the radiation specific to the measuring capabilities of that instrument. Thus, cameras detect radiation in the visible light portion of the electromagnetic spectrum, radar in the various radio "bands" of radiation, and so on.

Moving toward a discussion on what detection technologies are currently accessible to ufologists, we can describe some of the types of radar equipment systems that have been developed recently. First, *active radar* refers to a system that includes a radio wave transmitter and a separate receiver. The transmitter shoots out the radio waves toward the object in the sky or water and then waits for the waves to bounce off the object and return to the receiver that is positioned some distance away from the transmitter. Triangulation mathematics is used to obtain distance and size measurements of the object. One other system, *passive radar*, operates like active radar but has no active transmitter. A passive system only detects ambient radio signals that radiate from nearby transmitters. Examples of passive radar include radio and television signals and cell phone tower emissions.

Most people are familiar with the weather forecasting system called *Doppler radar*. Named after Austrian physicist Christian Doppler in 1842, it was discovered that light waves shift their colors according to whether the underlying matter that are emitting the waves are traveling toward or away from the observer. This provided opportunity to invent many technologies that have helped many fields from astronomy to medicine and meteorology.

Doppler analysis breaks down light waves according to whether the underlying matter or object is moving toward or away from an observer. A redshift means the waves of light have decreased in frequency and increased in wavelength. The number of waves per second has decreased which automatically means the length of each wave has increased. Another way to perceive this is to note that the wave line is being stretched out. So the redshift means the matter or object that is emitting the waves is moving *away* from the observer. Consequently, a blueshift occurs when the frequency of wave per second has increased and the length of each wave has decreased. The wave line here is being crunched together. Blue is on the shorter end of the visible light portion of the spectrum; red is on the longer end of the electromagnetic spectrum.

From this analysis, we can see why the Doppler effect was a revolutionary discovery for many fields including astronomy, medicine, and meteorology. None of these fields would have had the ability to provide nearly as much knowledge without it.

Another technology that has been influenced by knowledge of the evolution of electromagnetic spectrum is the Global Positioning System (GPS). This is simply a sophisticated radar system in which many earth-orbit satellite transmitters and receivers interact with these waves of energy continuously to provide accurate location information anywhere on Earth or beyond. Like Doppler radar, Global Positioning System technology was only developed by the government and the military complex a few decades ago and available to society since around the year 2000. They are both now readily available to any ufologist who wishes to use these as detection equipment.

So on many basic levels it is the natural behavior characteristics of electromagnetic radiation that govern how we can observe things around us. As most matter emits a variety of electromagnetic radiation and any particular substance has unique signature electromagnetic radiation characteristics, it is possible to detect, analyze, and identify many substances. We are most interested in identifying unidentified aerial or submerged objects so the expected sources of detectable radiation come from metallic ship surfaces.

We can now introduce some of the technological advances in instrumentation that ufologists currently use or are planning to introduce in the near future. The first of these merges the technology as a tool with "a way to use it" method. Here is an example of a system that has been used for a time by private individuals.

In the 1990s, during the Gulf War in Iraq, cable television news networks started using a dedicated camera station whereby a twenty-four-hour/seven-day-per-week live camera was in operation. Autonomous from any human operator, the cameras could allow news shows to broadcast either real-time or recorded videos of missiles flying toward Baghdad. This was a safe technique to use and captured anything within the camera's field of view. The method merged tool with strategy to provide a working design. The camera platform became a real tool for capturing potential unidentified phenomena.

Unfortunately, the only waves these cameras could capture were the visible-light waves. Eventually, as other instrument technologies diffused into mainstream society, more access to other types of electromagnetic wave detection became possible. Instruments such as infrared detectors, microwave receivers, and equipment that could capture virtually any wavelength of radiation eventually became available to anyone at an increasingly-decreasing cost.

Any dedicated instrumentation platform would have to consider, in addition to the electromagnetic spectrum, environmental characteristics as a way of acknowledging more evidence with which to investigate unidentified aerial phenomena (UAPs) or UFOs. Some environmental characteristics that are measurable in this respect include ambient temperature, air pressure, magnetic properties (not electromagnetic properties per se), and gravitational characteristics, to name a few. The instrumentation with which to capture this data is now accessible to the general public at a cost that makes this feasible only within the last generation. These tools are also the most sophisticated in humankind's history and can detect the smallest differences in ambient measurement. These capabilities allow for such investigation as gravitational anomalies that could not be detected before.

Any object with mass can be detected moving through the airspace (or water) that is captured by readings from this instrumentation. Optical cameras, fitted with lenses that can capture the entire sky on a wide-angle basis or zoom into telephoto settings, can record either still or video in the visible light spectrum. With the right type of sensor, these cameras can now record in multiple bands from fluorescent to microwave. Microwave spectrometers are dedicated to capturing movement and structure within the microwave band. Geiger counters can be synthesized into recording equipment to capture waves in the x-ray, alpha, beta, and gamma ray radiation bands. Various types of radar and radio wave detection technology are already known.

These instruments can also detect ambient temperature, wind and air pressure changes, and some changes in gravity due to the chemical reactions taking place. In addition to this equipment allowing for capture of these anomalies, more popular weather station instruments such as thermometers, anemometers, and barometers respectively are also part of some of the UFO detection platforms. Magnetometers can detect subtle changes in magnetic fields that can also be caused by objects who exhibit powered-flight characteristics. Gravimeters can inspect changes in local gravity. Sound and ultra-sound equipment can investigate audio wave differences within the event theater.

To summarize, dedicated detection programs can measure visual, audio, electromagnetic, temperature, air (or water) pressure, wind speeds, magnetic, and gravitational aberrations for great distances. Here are a few of the contemporary projects focused on these designs.

UFODATA

UFODATA, short for "UFO Detection and Tracking"[7] is a not-for-profit research organization that at the time of this writing is in the construction phase of the first platform that will use the technology just described. Their prototypical station will maintain an electromagnetic unit and an environmental unit. A computer station will direct the functionality and record the data or at least make it transformative from the source measuring instrumentation. The UFODATA station will be unmanned but able to communicate all investigative readings off-site for analysis.

The electromagnetic unit will consist of various imaging and recording equipment to acquire data of objects moving through its ambient airspace. Each piece of hardware will obtain data only within specified electromagnetic wave bands. A microwave spectrometer and Geiger counter will also be onboard the platform to perform their specific functions of investigating microwaves and x-ray, alpha, beta, and gamma ray radiation respectively. This dedication allows for more sensitive and precise measurements by all equipment. The environmental unit will consist of typical weather station devices in addition to a magnetometer that will record subtle differences in the ambient magnetic field as the object moves through its space.

Peripherally, around this suite will be high speed video optical cameras to perform its visual functionality.

The UFODATA platform is planned to be a twenty-four-hour/seven-day active station, derived from the concept of the cable television Gulf War real-time video platform of the 1990s. A team of volunteer scientists and science education professionals from different disciplines constitute the research portion of the project. Fundraising is being done from a crowd-funding platform.[8,9]

UFOTOG:

A second type of unidentified aerial phenomena surveillance platform will be known under the series title, UFOTOG, an acronym for "UFO Photography." The current iteration in the series is UFOTOG II. UFOTOG will be a comprehensive electromagnetic spectrum-capturing platform similar in design to the UFODATA system.

This technology will concentrate, at least initially, on various instruments each of which is most sensitive in measuring a unique narrow and precise portion of the spectrum. There will be the requisite high-speed visual imagers, then a separate ultraviolet camera, infrared imager, gamma ray detector, x-ray detector, spectrometer, magnetic anomaly detector, and a separate electromagnetic field detector. Any object moving through its electromagnetic spectrum-capturing range will be detected and data recorded measuring a large variety of both electromagnetic and environmental characteristics. UFOTOG II will also supply a direct link to communication and radio transmitters and receivers for data transfer such as cellular networks, GPS satellites, the SPOT technology utilized in backcountry regions that lack continuous connection with cellular or GPS reception.

The UFOTOG II will be solar-powered and require no other power source for its operation. Battery backup will also be on board. This system will also be a dedicated twenty-four-hour/seven-day operating platform with full remote control capabilities. A further design feature will be a dome exterior housing unit that will be remotely operated and used when inclement weather conditions prevail. The station site will be automated and entirely controlled remotely. Computers that will be transformative in linking to the source instrumentation will collect all data for further analysis.[10]

Breakthrough Listen

This next effort is not driven by any dedicated detection platform technology per se but rather a funding effort for others to tap into for their own projects. Project aims are for more classic pursuits similar to what Search for Extraterrestrial Intelligence (SETI) has been pursuing.

Yuri Milner, a billionaire investor from Russia, announced in July 2015 creation of an endowment fund with one hundred million dollars in capital for this purpose. This funding, known as Breakthrough Listen awards, will be made to research teams in order for them to obtain and fund telescope time at many observatories in the United States and Australia. Additionally, data processing and analysis will be incorporated into the scientists' own searches for anomalous radio signals captured in the cosmos that could infer intelligent sources for those transmissions.

This endowment, scheduled to last for ten years but with assurances from Milner to be renewed, will assist researchers who are conducting experiments similar to the projects undertaken by the SETI Institute. Milner is seeking to help astronomers and scientists detect "an artificial signal not explainable by science."

Milner, a trained physicist, became financially successful by investing in many Silicon Valley start-up companies. He has had Stephen Hawking assist him in the ramp-up of Project Breakthrough Listen. Funding will be available to anyone with a viable plan design in place. Milner has a tempered perspective about the state of life in the cosmos: "The universe is not teeming with life, but we're probably not alone."[11]

Project Hessdalen

Erling Strand is professor of computer science at Ostfold University College in Kråkerøy, Norway. He is a founder of Project

Hessdalen, a continuing effort to learn the answers to long-time sighting waves of anomalous lights and patterns that have occurred for many decades over the region in south and south-central Norway. Strand is also on the board of the UFODATA project.

Project Hessdalen was founded in 1983. Strand has been leading numerous research teams in field experiments and also for UFO-Norge and UFO-Sweden projects. The technology design features used in Project Hessdalen are especially prototypical of the noted UFODATA and UFOTOG systems. These dedicated studies began in 1984 and have employed the investigations of hundreds of scientists through the two decades and more since its inception.

This part of Norway is sparsely populated. Starting in 1981, continual observations by citizens as well as the science, government, and academic communities have documented lights "larger than automobiles" moving through the skies at many different altitudes and in many different forms and velocities—some fast, some slower, and some stationary. Some moved directly over and touched houses. The lights have been shown in many different colors and usually of a solid one or two color arrangement.

What is of importance here is the experimental design and instrumentation used since Project Hessdalen's inception. This is the "grandfather" of dedicated scientific research of unidentified aerial phenomena. A large variety of instrumentation has been employed in its history. First, cameras with gratings that allow for the capture of waves in multiple portions of the electromagnetic spectrum. Second, seismographs to detect atypical sound waves as evidential for the phenomena have been used. Additionally, magnetometers, Geiger counters, infrared viewers, lasers, and spectrum analyzers have all been utilized to record data with both electromagnetic and environmental characteristic.

This application of what a scientist may describe as a pure form of investigation has not been able to confirm the true reality of the nature of the Hessdalen lights, though many hypotheses have been formulated and tested. All have been recycled after new knowledge was analyzed.[12,13]

Materials and Chemical Physical Evidence Analysis

Another substantial type of examination offering meaningful reflection in the ET hypothesis comes from the study of actual artifacts and materials left by unidentified craft and/or its occupants. Among the variety of physical artifacts or manifestations that have not been discussed but that are capable of being inspected include: ground artifacts or debris, chemical traces, consequential effects on vegetation and wildlife, physiological outcomes to witnesses, effects of gravity on ambient objects or their environment, and other tangible substances deposited in either the airspace, water surface, or landscape.

There is some documented history of this type of investigation in the ufology literature. A workshop series was organized in Tarrytown, New York, on September 30–October 3, 1997, which brought teams of scientists and UFO researchers together for the first of two rounds of workshops. These activities introduced the subject of material evidence. Next, a substantive conference to discuss specific case studies, identify problematic areas within the evidence domain, and offer determination for future investigations was held November 28–30, 1997, in San Francisco. The LSR Fund—originated by the Laurance S. Rockefeller Trust and encouraged by Laurance Rockefeller, of the Rockefeller Empire, and who had an ongoing interest in the subject until his death in 2004—funded the conference.

Peter A. Sturrock, PhD—scientist, astrophysicist, and professor emeritus at Stanford University—led a panel of nine scientists, including Charles Tolbert, professor of astronomy at the University of Virginia, Dr. Yervant Terzain of Cornell University, Dr. Harold E. Puthoff, director of the Institute for Advanced Studies in Austin, Texas, and David E. Pritchard, professor of physics at the Massachusetts Institute of Technology, among others. The workshop included eight ufologists, among them Jacques Vallee, Mark Rodeghier, Richard F. Haines, and J-J Velasco.

The workshop was not conducted to determine theoretical conclusions as to the overarching objective of determining the existence of UFOs or extraterrestrials. It was arranged to provide recognition that this was a subject that warranted further and much more con-certed scientific investigation than it was being given was the general consensus of the conference.

A few cases that were discussed in those workshops were exemplary for detailed discussion. Three cases will be highlighted here to show some of the techniques that were and can be used in future study. The first two were discussed at the workshop. This is especially significant because there have been fifty years of techno-logical developments since some of these incidents that allow for the most rational, detailed, and informative study of the phenom-ena to date.

On January 8, 1981, an unidentified craft landed on a farm close to a village in France called Trans-en-Provence. Among the types of artifacts left at the scene included physical trace substances, landing marks, and ground indentations and damaged vegetation. Eyewitnesses to the incident also reported visual and audio charac-teristics throughout the encounter.

French gendarmes, some of whom were also witnesses, arrived on the site almost immediately and began to collect samples from all around the event theater. Investigators from a division of the French National Center for Space Studies (CNES), the equivalent of NASA, called GEPAN arrived that same day after the gendarmes began gath-ering evidence. The GEPAN investigators joined in collecting more samples.

GEPAN released a detailed case report in 1983 called "Technical Note No. 16, Inquiry 81/01." Some of the report indicated ground disturbances did not disappear or erode for months afterward. The ground around the landing site was heated to over six hundred degrees Celsius causing much nuclear irradiation and disturbances to the vegetation. The disruption affected plants that maintained their biochemical phenotypes (outward appearances) but whose biochem-ical constitution were altered to produce a signature characteristic of very old specimens. Further ground depressions from the landing gear of the craft did not fill in. The ground atop the landing site became devoid of vegetation and became hydrophobic, unable to absorb water.[14,15]

In his science research paper, *Physical Analyses in Ten Cases of Unexplained Aerial Objects with Material Samples*, author Jacques Vallee attests to a problematic issue in UFO trace evidence investigations: "The primary concerns have to do with inaccuracies in data gathering, lack of information about exact dates and times, lack of detailed, critical field investigation, and failure to provide an irrefutable chain of evidence in the collection, transportation, and examination of the samples."[16]

Regarding the Trans-en-Provence case, none of Vallee's concerns were ever problematic. This was a textbook process of how scientific and forensic investigations should be conducted. There are many more of these examples in the ufology literature. The takeaway points here are that chemical laboratory analysis of physical evidence has developed for many decades, does work and continues to evolve to provide ever more sophisticated techniques, and is a key contributor in building a case book of proof going forward.

On December 17, 1977, two eyewitnesses observed a flying craft crash into the countryside near a dike at Council Bluffs, Iowa. They saw "a bright flash" then flames which rose ten feet high from the ship. When they arrived at the crash site minutes later, they reported a big portion of the dike grid was engulfed by molten metal which emanated red-orange waves and ignited the vegetation around it.

The police and fire departments and many other new witnesses arrived at the scene moments after the first two witnesses. Among these new observers were Kenny and Carol Drake and their nephew Randy James, all natives to Council Bluffs. What they encountered was a huge blob of this molten orange-glowing metal running down the sides of the dike parallel to a road and railroad track. As the various police and fire department officials collected samples by using glass bottles that could withstand the searing temperatures of the metal, other civilians also collected samples.

The Drakes waited until some of the pieces cooled enough before acquiring some pieces of "blue crystals" about three inches square. Another witness who did not reconnoiter until early the next morning was astronomer Bob Allen. After inspecting the area Allen collected many additional pieces of debris. Formulating a plan of scientific investigation, Allen took it upon himself to take some of the samples to a local metal manufacturer, Griffin Pipe Co., Iowa State University and the Foreign Technologies division of Wright-Patterson Air Force Base for analysis.

Both Griffin and Iowa State U. concluded some the metal could have been a very high grade of iron or nickel and chromium. Other samples additionally identified substances made of highly-refined carbon steel.[17] Colonel Charles H. Senn replied that the debris could not have originated from space because there was no evidence of any impact crater from a meteor. This would seem to be corroborated by all the witnesses who stated in their reports that a physically-defined oval-shaped object dispensed the refuse and was not traveling at a high speed; in fact, it was hovering more or less in place at the time of the episode.[18]

This sequence of events was what piqued the interest of Jacques Vallee. He studied the incident and became intrigued enough to include the Council Bluffs event in the Tarrytown and San Francisco workshops. After a detailed study of all the unique elements was collected, Vallee conjectured that the debris may have originated from a liquid metal electrical propulsion system or a similar nuclear-powered engine apparatus. The identification of the variety of metallic substances present in the samples also lead one to determine a possibility of the "slag" being generated from the power plant of a *homopolar machine*. A homopolar motor is a booty current electric motor with two magnetic poles, the conductors of which always cut unidirectional lines of magnetic flux by rotating a conductor around a fixed axis so that the conductor is at right angles to a static magnetic field. This could explain why some of the chemical analyses indicated presence of some substances readily found on Earth.[19,20]

The depth of the investigation included testing four different hypotheses adhering to the scientific method criteria. What was studied but concluded out of possibility were an impact from a meteor from outer space, debris that broke off and fell from a man-made aircraft, and various hoax scenarios. Both the space meteors and the aircraft equipment hypotheses were eliminated as all parties contacted for analysis of various parameters of the situation dismissed them, including the Air Force from Wright-Patterson and Eppley Air Field. Various meteorological and geophysical examinations were included in the study. To summarize, testing from many different science disciplines was conducted and analyzed.[21]

The Council Bluffs case is another example of a proper, comprehensive, and dedicated research study that passes all of Jacques Vallee's criteria for a correct investigation as noted. The central theme of this chapter applies here in that, even back in the 1970s, there was at least limited access within the public domain for materials and chemical examinations of trace evidence from UFOs. There was a detailed, specific, and thoughtful research study design in place that was followed by multiple tests conducted by various independent teams from different scientific communities. Now that access just grows stronger as knowledge is accumulated and new and cheaper technologies and techniques are developed and implemented; as time goes forward, new cases will be better scrutinized by this new knowledge and technology.

On May 19, 1967, amateur prospector Stephen Michalak was searching quartz mines near Falcon Lake, Manitoba, Canada. He observed two oval-shaped craft descend to near his vantage point and hover for a few minutes, glowing a bright red. One of them landed near him. The landed craft began emitting a toxic ulphur odor and made loud whirring noises while changing colors almost continuously.

Michalak moved to within a few feet of the ship. He noticed the exterior hull was shiny and of a polished glass texture that is ultrasmooth. After observing the craft and sketching it for a few minutes, he touched the side of the ship with his hand enclosed inside a heavy miner's glove. The hand part of the glove disintegrated immediately. At that instant, the ship rose off the ground and prepared to fly off. The exhaust apparatus was now pointed at Michalak. An emission of hot gas and air scalded him and burned off his clothes. The ulphur smell came back at that time, along with an odor of burning electrical insulation and burning vegetation on the ground around him.[22]

Michalak was treated for grid-like third-degree burns on his chest and torso. He had recurring episodes of nausea and headaches for weeks afterward. This type of physiological evidence is evidential of real injuries to the bodies of their victims. The types of burns to Michalak's torso were determined by medical physicians to be caused by applications of heat from some source(s) that were problematic in discussing any known manmade and available instigators.

There are many other case studies where physical evidence included physiological episodes like severe burns, ulcers, and head and bodily trauma throughout the ufology literature. As the medical sciences develop new technologies that lead to new discoveries and additional knowledge, more of these analyses will provide more accurate and determinate conclusions.

Hill Star Map

The Barney and Betty Hill 1961 experience created a new level of awareness in the landscape of extrabiological entity experiences. Multiple factors were present that added to this platform. Some of them were noted earlier, and one was mentioned. When Betty Hill reported the existence of a "star map" that was shown to her while on board the Zetas's ship, little could be learned from the depiction. The map was not labeled, and all of Betty's requests for details were rebuffed by the captain. So what she and the world were left with was a basic spatial map of a section of the local star group.

Betty was told our solar system was depicted on the map. She was reproducing a depiction of a three-dimensional wall chart. The details showed relative size of the stars within that local region. Distances were also spatially accurate, as would later be proved. The Zetas's relationship to each of the stars on the map was also spatially and categorically detailed. For that time, Betty Hill was left with the map and not much else.[23]

Her conception could be thought of in this way. If we were handed a three-dimensional map of planet Earth, minus the labels of topographical and contrived identifications and delineations of the various countries and nations, we probably could not say much about the inhabitants of this planet, but we would know the person showing us the map was a native inhabitant (as Betty Hill was told by the captain). We would know planet Earth was inhabited and that intelligent life existed there.

Enter astronomer Margaret Fish. She became intrigued upon first hearing of the Hill incident and after reading the Jacques Vallee book *Anatomy of a Phenomenon*. Previously, in 1966, the inventory of types, quantity, and accessibility of evidence were far less than nowadays.

Being a teacher and amateur scientist, Fish knew when evidence could present itself and be presented as advantageous for pursuing more knowledge. She made a connection with Betty Hill and obtained the star map. The next two years were spent collecting sufficient data with which to begin a detailed investigation.

Marjorie Fish determined early on that there were no star catalogues which accurately depicted the region of local space the Zetas claimed as their habitat. There were some undiscovered candidates on the map as of 1966. The astronomical field maintains a centralized and globally-recognized archive of star catalogues. Named the Gliese Catalog, annual and interannual editions are published and disseminated for the astronomical communities.

It took until the end of 1969 for Fish to run a simulation model that accurately included most of the star candidates. She constructed dozens of simulation maps in eight years hence. There were other considerations besides just matching stars on the map with a picture or visual investigation of the night sky. Because the driving force behind this was an alleged discovery of life, there had to be some accounting for existence and identification of the right type of stars that could harbor life-forms. This took a highly significant amount of Fish's investigation but was vital to accurately determine if the candidate stars were viable matches for such harbor for life, especially intelligent life.

Finally, in 1972 and after the latest Gliese Catalog was published, Fish successfully located the last three stars. She simulated the depictions a few more times to confirm her investigation and evidence. In February 1973, she was confident she had identified the location of what would be called Zeta Reticuli 1 & 2. The Zeta Reticuli system is a binary G type star pair situated about thirty-nine light-years from Earth. They are visible to the naked eye only in very dark skies and in the southern hemisphere within the Reticulum constellation.

Marjorie Fish made her announcement in 1974, and the astronomy community peer-reviewed her work. Almost all of it was laudatory and confirmatory. What Betty Hill observed was an accurate depiction of the local Star Group in which our solar system was part.

The striking features of the Fish investigation were that Betty Hill could not have conjured any of this up as has been conjectured by skeptics and debunkers. No one on Earth knew of the Zeta Reticuli binary system and the accurate placement of the local region because no comprehensive astronomical maps existed showing all these systems until the Gliese Catalog of December 1969. Stars are discovered then vetted by the astronomical community and verified before publication. This is an example of good science at work.

Another striking feature was that Betty Hill was told by the captain that the three-dimensional map he showed to her contained the trade routes the Zetas used. If all the candidate stars were trade-route destinations, that would mean some form of life would exist in those places.[24,25]

Forty-three years of astronomical knowledge, technology innovation, and development have accrued since Marjorie Fish's Zeta Reticuli discovery confirmation in 1974. Over 3,500 exoplanets have been confirmed to exist, and many thousands more are in the "pending confirmation" category to date.

It should be not only feasible but almost empirical and rudimentary now to be able to conduct an earnest investigation to discover the existence and some characteristics of exoplanets within the Zeta Reticuli binary system. Telltale geophysical and chemical signatures are now accessible by humankind technology that can provide many answers and additional confirmation, if they do in fact exist, about life capabilities and possibilities on any such discovered planets.

Additionally, this methodology of design can be extended into the area of many other investigations. This particular paradigm contains requirements for equipment whose access may be either restricted for use or problematic for acquisition. The spectroscopic analysis required to conduct such investigations requires telescopic and imagery hardware and software that are beyond the fiscal means of many in the general public. But it is also to say that we have the technology now with which to investigate these inquiries in a sensible and rational way. The key is to get interested parties together and on the same page. In this paradigm, we are "going to them" instead of "waiting for them to come to us." Humankind is the explorer, not the explored.

Thus far, we have searched for ways to enable the sciences to study questions about the existence of other intelligent life in the universe and close to Earth. The survey of the natural sciences, with distinct and exciting capabilities to provide more and more compelling evidence, continues with contributions from the biological and exobiological disciplines.

The exciting and exponential increases in biomedical technology has allowed for an entry into investigations of providing evidence for the existence of ET. This technological revolution has become more accessible within the public domain than at any time in humankind's history.

The feasibility of obtaining such biomedical evidence and recognition seems more and more likely through the lens of DNA inspection. As noted, Deoxyribonucleic acid (and RNA – Ribonucleic acid) is nature's most intricate and definitive fingerprint for biological life in the universe. Since its proven discovery in 1953, the development for measurement and study of the double helix structure that pictorially defines DNA has increased exponentially. As the body of knowledge increases, the capability and achievement of obtaining new knowledge increases in a higher-than-linear function. This is because there are more facts with which to base new discoveries. These new discoveries will arrive at a faster and faster pace than before.

For DNA analysis, this notion of exponential change can be exemplified this way. When the human genome was first mapped in 2003, the cost of the project was millions of dollars. Today, anyone can get his or her individual genome mapped and analyzed for less than five hundred dollars. Global databases have catalogued the genomes of many thousands of life-forms. As the cost has decreased exponentially, the knowledge about this natural biology has likewise increased.

Recall the Peter Khoury incident of 1992. The alleged encounter with a Nordic-type and an Asian-type EBE's left DNA evidence that he saved with the potential for later analysis. It took a large handful of years in order for some rational analysis to become possible for at the time the human genome project had not yet even started, though it was planned.

Scientist and researcher Bill Chalker accepted the case and formed a team of genetic researchers to study and test the saved samples. DNA has been able to survive for millennia which allows for later replication, mapping, and almost unlimited dedicated testing and analysis from many different techniques. Recent successful replications and mapping include ancient Egyptian mummies and frozen corpses of ancient European bodies and the manufacturing of beer taken from the restored and replicated DNA from ancient yeast samples from over 2000 BCE.

The scientists at the Anomaly Evidence Group in Australia tested mitochondrial DNA from the Khoury samples. Mitochondrial DNA, as opposed to nuclear DNA, allows for investigation of parental lineage, specifically on the maternal side. These samples, as having come from a female, could provide answers to the lineage of that female's partner.

What was found at the time were five distinct anomalies in gene segments. The data from these segments statistically calculated to a probability of that DNA originating from an earthly donor as being virtually zero. When more layers, such as the results of an investigation of the four recorded humans with all five anomalous markers were uncovered, the probabilities of a definitive human origin were reduced even more dramatically. Additionally, only a few DNA segments of the female partner sample had been examined to that time in 1999.[26] The Human Genome Project had not even been close to being completed by then.

Eighteen years hence and thanks to that notion of the exponential expansion of technology development and knowledge, we now have access to the complete genome sequencing catalogues, techniques to provide gene therapy treatment for many diseases, and potential for the genetic cure of diseases. An example of this cure potential is concerned with the implantable pacemaker. A research team at the Cedars-Sinai Heart Institute in Los Angeles, California, is—at the time of this writing—testing a prototype of a *bio-natural pacemaker*. This technique proceeds by genetically-modifying normal pig heart natural pacemaker cells to create the natural pacemaker cells all human hearts possess and need for the heart organ to produce its beating normal sinus rhythm. This would represent a cure for abnormal heart function, whereas the pacemaker is only a treatment.[27]

It is now possible for us to launch the most vigorous studies of DNA samples that will offer more knowledge and answers than ever before. DNA and forensic analysis have been part of the lexicon of the law for decades now and is the most sophisticated contemporary technique used to determine guilt or innocence of an alleged defendant. There have been no questions of unreliable and invalid evidence in any judgment handed down in any such case involving DNA investigations. Viable and conclusive sampling has been obtained from even the smallest and most nuanced organic substances deposited by an involved party.

When the hypothesis of proving or disproving the existence of extraterrestrial biological entities is entertained, this body of knowledge and technique is among the most powerful of forensic strategies that is now accessible within the public domain. The knowledge that DNA can survive for millennia introduces an exceedingly enormous inventory of archaeological artifacts into the analysis landscape. Additionally, studies of new current and future encounters will possess these new powerful tools with which to deploy. Many new investigations can be postulated and attempted that offer potential for providing a correspondingly larger body of evidence that could lead to further disapproval of the two hypotheses at hand.

References

[1] Ref 1: Einstein, Dr. Albert. 1955. "Old Man's Advice to Youth: 'Never Lose a Holy Curiosity.'" *LIFE Magazine* (May 2, 1955). Volume 38, number 18, p. 64.

[2] Ref 2: NOVA Television network. 2005. "Einstein's Big Idea," June 2005. From the memoirs of William Miller, an editor, quoted in *LIFE Magazine*, May 2, 1955; *Expanded*, p. 281. Copyright 2005 by NOVA Television. Website copyright 1996-2016 by WGBH Educational Foundation. Retrieved August 28, 2016, from: http://www.pbs.org/wgbh/nova/einstein/wisd-nf.html.

[3] Ref 3: Definition of Proof. Google. Retrieved August 28, 2016, from: https://www.google.com/#q=proof+definition

[4] Ref 4: Definition of Proof. Dictionary.com. Retrieved August 28, 2016, from: http://www.dictionary.com/browse/proof.

[5] Ref 5: Definition of Electromagnetic Spectrum. *Merriam-Webster*. Copyright 2016 by Merriam-Webster Incorporated.

[6] Ref 6: Definition of Frequency. *Merriam-Webster*. Copyright 2016 by Merriam-Webster Incorporated.

[7] Ref 7: Byrd, Deborah. 2015. Scientists to monitor skies for UFOs. Human World I Science Wire I Space. November 2, 2015. Copyright 2016 by Earth Sky Communications Inc. Retrieved August 28, 2016, from: http://earthsky.org/space/scientists-ufodata-monitor-skies-ufos.

[8] Ref 8: UFODATA. 2015. Technology. Copyright 2015 by UFODATA Project. Retrieved August 28, 2016, from: http://www.ufodata.net/technology.html.

[9] Ref 9: UFODATA. 2015. Board Members. Copyright 2015 by UFODATA Project. Retrieved August 28, 2016, from: http://www.ufodata.net/team.html.

[10] Ref 10: Rik Johnson. 2015. What Is UFOTOG II? Mutual UFO Network, Colorado MUFON Chapter, Issue 132, February–March 2015, p. 1–5. Copyright 2015 by MUFON.

[11] Ref 11: Vella, Matt. 2015. Yuri Milner: Why I Funded the Largest Search for Alien Intelligence Ever. *Time Magazine*, July 20, 2015. Copyright 2015 by Time, Inc.

[12] Ref 12: Project Hessdalen. 2016. Copyright 2016 by Project Hessdalen.org and Ostfold University College. Retrieved August 28, 2016 from: http://www.hessdalen.org/index_e.shtml.

[13] Ref 13: Strand, Erling, MSc.EE. 2016. Project Hessdalen 1984. Final Technical Report. Copyright 2016 by Project Hessdalen.org.

[14] Trans-en-Provence Ref 14: Berliner, Don. 1995. Unidentified Flying Objects Briefing Document (The Best Available Evidence). Retrieved August 28 2016 from: https://archive.org/stream/pdfy-uHw9zGct4dsyqXcU/Unidentified%20Flying%20Objects%20Briefing%20Document%20%5BThe%20Best%20Available%20Evidence%5D_djvu.txt.

[15] Trans-en-Provence Ref 15: Trans-en-Provence Physical Trace Case. 2011. Copyright 2011 by ufoevidence.org. Retrieved August 28, 2016, from: http://ufoevidence.org/cases/case110.htm.

[16] Trans-en-Provence Ref 16: Vallee, Jacques F. 1998. Physical Analyses in Ten Cases of Unexplained Aerial Objects with Material Samples. Published in the Journal of Scientific exploration, March 1998, Volume 12, Number 3, p. 374. Copyright 1998 by Society for Scientific Exploration.

[17] Trans-en-Provence Ref 17: Vallee, Jacques F. 1998. Physical Analyses in Ten Cases of Unexplained Aerial Objects with Material Samples. Published in the Journal of Scientific exploration, March 1998, Volume 12, Number 3, p. 374. Copyright 1998 by Society for Scientific Exploration.

[18] Council Bluffs Ref 18: Jerrett, Greg. 2004. Mystery at Big Lake Fire in the Sky. Council Bluffs Daily Nonpareil, May 16, 2004. Copyright 2004 by The Daily Nonpareil LLC.

[19] Trans-en-Provence Ref 19: Vallee, Jacques F. 1998. Physical Analyses in Ten Cases of Unexplained Aerial Objects with Material Samples. Published in the Journal of Scientific exploration, March 1998, Volume 12, Number 3, p. 374. Copyright 1998 by Society for Scientific Exploration.

[20] Council Bluffs Ref 20: Definition of Homopolar Motor. Retrieved August 28, 2016, from: https://en.wikipedia.org/wiki/Homopolar_motor.

[21] Trans-en-Provence Ref 21: Vallee, Jacques F. 1998. Physical Analyses in Ten Cases of Unexplained Aerial Objects with Material Samples. Published in the Journal of Scientific exploration, March 1998, Volume 12, Number 3, p. 374. Copyright 1998 by Society for Scientific Exploration.

[22] Michalak Ref 22: The Stephen Michalak Encounter at Falcon Lake. Copyright 2011 by ufoevidence.org. Retrieved August 28, 2016, from: http://www.ufoevidence.org/cases/case376.htm.

[23] Ref 23: Fish, Marjorie E. 1974. Journey into the Hill Star Map. Presented at the 1974 MUFON Symposium. Reproduced by NICAP. Retrieved August 28, 2016, from: http://www.nicap.org/reports/hillmap.htm.

[24] Ref 24: Fish, Marjorie E. 1974. Journey into the Hill Star Map. Presented at the 1974 MUFON Symposium. Reproduced by NICAP. Retrieved August 28, 2016, from: http://www.nicap.org/reports/hillmap.htm.

[25] Ref 25: Dickinson, Terrance. 1974. The Zeta Reticuli Incident. *Astronomy Magazine*, December 1974. Volume 2, Number 12. Copyright 1974 by *Astronomy Magazine*.

[26] Ref 26: Chalker, Bill. 1999. 1988, DNA Sample from Khoury Abduction Raises Big Questions. Published in International UFO Reporter, Spring 1999. Copyright 1999 by Center For UFO Studies.

[27] Ref 27: Pacemaker Ref – Ng, Yi-Di. 2016. Can a genetic fix replace the pacemaker? Published 2016 by *Cosmos Magazine*. Retrieved November 10, 2016, from: http://cosmosmagazine.com/technology/can-genetic-fix-replace-pacemaker.

CHAPTER 23

The Final Chapter

*Let the future tell the truth and evaluate each one according
to his work and accomplishments. The present is theirs;
the future, for which I have really worked, is mine.*

—Nikola Tesla[1]

Key words: placatory, mystique of science, textual criticism

Events of a technologically inexplicable nature are happening,
lots of them. A large number of people have borne witness to sightings and experiences of high strangeness. Perspective and interpretation do not matter. The reality and truth is through history millions
of people have seen, heard, or otherwise experienced in profound
ways manifestations of things that have proved difficult to explain.

Twenty-three chapters of material have been presented here.
Through this text, we have had the opportunity to wear a variety of
hats, the hats of a forensic researcher, a scientist, and a juror.

As noted in the "Introduction," there is a lot of similarity in
the way people from all walks of life conduct their critical thinking,
methodology, reasoning, and decisions. Consider now how it is for
the forensic researcher, the juror, and the scientist. They all use the
same methodology to answer the questions and hypotheses presented
to them.

The only difference is that the forensic researcher and the juror
consider evidence to decide answers based on a situation involving
input from human beings; the scientist uses a similar methodology,

but that format involves only studying to explain how something in nature and the environment works. On many levels, this is like the methodology of an engineer when the project at hand is design and construction of a new tool of technology. The engineer puts a lot of thought and study into how his apparatus will work.

When the question is proving the existence of UFOs and ETs, the ways of science and engineering can offer invaluable input into some of the proof required. But the usefulness and vigor these professionals carry into the investigation often prove inadequate when questions of existence and reality and not just how some event in nature could occur and operate. This is especially true and becomes more magnified when the most sophisticated knowledge of science and engineering cannot explain how some of these events may occur or how some of these objects and beings can exist. The technologies are beyond what they can explain. Are we expecting too much from scientists and engineers here?

In his book, *Society and Technological Change*, sociology professor emeritus and author Rudi Volti explains a transparent characteristic of science, technology, and their communities that points to this incapability: "Whereas science is directed at the discovery of knowledge for its own sake, technology develops and employs knowledge in order to get something done. The content of the knowledge may be rather similar, but different motivations underlie its pursuit and application."[2]

This suggests a scientist or engineer may not be interested in truth or reality but just in discovering knowledge for its own sake. Anthropologist and science sociologist Bruno Latour wrote in his book, *Science in Action: How to Follow Scientists and Engineers through Society*, the following:

> "Much of the prestige accorded to science is the result of its supposed purity; science is thought to be an intellectual venture free from political, organizational, and economic constraints." But as a lot of recent literature has analyzed and concluded contrary to what this *mystique of science*

> was self-constructed and self-proclaimed to be, Latour continues, "Science has been described and analyzed as a social construction. Scientific inquiry is not a disinterested, fact-driven search for truth but a human creation shaped by cultural patterns, economic and political interests, and gender-based ways of seeing the world."

A significant and comprehensively conclusive example of this discovery can be found in another of Latour's books *Laboratory Life: The Construction of Scientific Facts*, as noted earlier in *The Humaniverse Guide to Better Reasoning & Decision-Making*. Latour offers this summary: "Successful scientific outcomes may have more to do with negotiation, the support of designated authorities, and resonance with prevailing attitudes than theoretical elegance or experimental evidence."[3]

Volti makes an additional *placatory* (conciliatory) reference to the science and technology disciplines by offering this insight: "Science and technology are dynamic enterprises that build on their past successes, but they also make profitable use of their failures. An inadequate scientific theory may lead to the formulation of a better one, and a collapsed bridge is likely to provide valuable lessons that help to prevent future failures."[4]

When all these thematics become part of the script for a lot of determinations, in addition to the UFOH and ETH hypotheses, it often translates down to the level of insufficient explanation given by whatever party is providing the commentary. These examples of determination could come from the government or military representative, the scientist, the engineer, or the debunker.

So where does all this lead in our and the general public's relationships with and credibility toward scientists and technologists? Maybe this statement provides a good starting point: "We should hug our nearest scientist for the work that they do but be a bit skeptical and equally open-minded as to what, how, and what amount and type of proof and commentary he is offering in his explanations."

Because we know that some of the mystique of science that defines and is defined by its quintessential quest of knowledge for its own sake that discovery and explanation of things beyond the "nature of something" limits some of the usefulness of their conclusions. This moves into the realm of reality and existence. Reasons for this include the methodology and the motivation of whoever is providing the commentary.

For the juror, because he or she investigates situations that intrinsically involve human beings as intelligent beings and not just things in nature, he or she is better equipped to add an extra dimension to the methodology and analysis. This added dimension takes the form of other types of reasoning, such as the abductive reasoning.

So humans or other intelligent life beings like earthly animals or perhaps others from outside of our planet, if they exist, have input and influence on the local environment when they interact with and within it. Not every phenomenon in nature is caused only by an exhibit of natural forces. Picture a scene like this: a flying craft moves overhead. It is the pilot's maneuvers that introduced the craft into the scene in the first place. The input from the life-form is causing the craft to fly overhead. Another example of the intrusion of a living being into a scene that influences the environment would be this: if all the animals from a local zoo were to escape into one of our neighborhoods, it would not take an extensive and full-blown scientific study or methodology to tell you they have done so. Observations of this series of events and perhaps capture of one or more of the animals by one, a few or many people would allow the investigation to conclude, rather obviously, that these animals exist and are roaming around freely. The witness(es) can see, hear, and smell them as they move through your front yard. The thought processes that the social sciences, judges, jurors, and forensic scientists use in their professional careers would provide the correct functionality of a preponderance of the evidence in conjunction with the natural scientific methodology on all basic levels.

If UFOs/extraterrestrials are real craft/live beings and not fallaciously excused away just for the sake of it, in a knee-jerk reaction without any thought or study, then shouldn't the scientific methodology now used to debunk the events also fully utilize these other methodologies used in the law and social sciences? A blatant example of this point is the "lenticular cloud" excuse used to explain away a UFO. No matter that the cloud is, most of the time, far larger than what the witness is trying to report. These are assaults on people's intelligence derived from such debunking exercises. No matter that uncalculated thousands of such witnesses were, are, and will be police officers, military officers, scientists, and many other reputable societal authorities shouldering a burden of a lot of at-risk personal reprisal.

The methodology of the social sciences takes on an added burden of having to account for the consequences of intelligent life-forms that are part of many extraterrestrial encounters. The field of psychology uses such concepts as mass hysteria and pareidolia, from your earlier reading, as explanations that parallel the process of insulting the witness's intelligence. These are more reasons why a more diverse form of methodology should be collectively used in UFO and extraterrestrial investigations. Included should be the other forms of reasoning and analysis not yet used in ufology investigations.

There is, as recognized by many famous thinkers in history, including Sir William of Ockham (creator of Occam's Razor), more than one method to answering a question or solving a problem. Some argue that nothing short of dissecting an extraterrestrial will provide sufficient proof of their existence. Sir William would conclude that this is overly burdened with complexity in its procedural footprint. What is being asked is, "Do they exist?" The controversial ramifications of the dissection advocacy are correctly placed in its logical sequence after answering the first question.

Chapter 1: The Hypothesis

From reading all the quotes made by famous and accomplished scientists from different time periods spanning over a century, we can see that many polarized perspectives reside within the scientific communities. This is correct. It is part of what the peer review is to those communities. The result if the discovery of new knowledge.

Other scientists critique the study and try to replicate the results. This becomes the basis for most of those perspective issues and is how science is practiced.

There are a lot of examples of this paradigm. These polarized perspectives often explode into full-scale social war among those groups. Hal Hellman's book, *Great Feuds in Science: Ten of the Liveliest Disputes Ever*, and the book coauthored by Stanton Friedman and Kathleen Marden, *Science Was Wrong: Startling Truth about Cures, Theories, and Inventions "They" Declared Impossible*, explain how these wars have been occurring since ancient times. They are about proclamations, personality clashes, egos, politics, and money. What is most vital is if combatants can ever reconcile at least most of the differences for the science communities?

Chapter 2: Critical Thinking and the Scientific Method

We are asked to have confidence in the claims scientists provide. We are forced to have faith in what they say because we tend not to have the time to sufficiently research and verify the facts ourselves. See the "Why Would You Become Interested in the Hypotheses" chapter for a fuller appreciation of this notion. We are also being asked to accept their credibility prima facie without question. Another factor is the underlying knowledge needed to fully understand the facts behind their claims may be beyond what we understand. So we are often asked to trust their answers.

This does become problematic for many reasons and is proof that we should respect them for their work but continue to be skeptical until we have been able to study a sufficient number of research results. In her *Scientific American* article, "Doing Good Science: Evaluating scientific claims (or do we have to take the scientist's word for it?)," author Janet D. Sternwedel lists five ways to evaluate the credibility of a scientist or any other refuter, debunker, or other commentator.

The first strategy for effective evaluation of scientific credibility is to reflect on the logical structure of the answer or argument they offer. A second evaluation point, if time permits, is to look at the references they cite in their answers. The researcher's credibility rises if the references have been cited correctly. Actually, this is also what a good, credible journalist does in his or her work. Verifying references is a significant part of a journalist's job.

Third, there should be some study to determine if the scientist is biased in his or her response or argument. Bias takes many forms, and this may be harder to discern, according to Sternwedel. Intuition is one of them and can be a very useful skill.

The fourth strategy level of a credible investigation is to analyze any areas the argument may have been fabricated or falsified because our intuition and skepticism peaks in situations where the claims are "too good to be true" in a critical sense.

Often, examples of falsification of data are broadcast after research journalists have exposed them. We would have no significant access to any of the research because it is done covertly inside laboratories. In this way, we have to trust the journalist's credibility for accuracy and reality. The last credibility check is actually an extension on the fourth. When facts are actually kept secret and away from our inspection, it makes our extension of credibility more tenuous. Eventually, if that knowledge is kept from us for so long, we will come to distrust the claimant, and all levels of credibility are lost.[5]

This appears to be occurring with the fluidity present in our level of trust in the government over the UFOH and ETH and evidence claims.

Chapter 3: What Is Science?

It is possible to succeed within the practice and paradigm and also be a bit skeptical and simultaneously be open-minded to whatever facts and evidence are uncovered while upholding the disciplines principles. Dr. John E. Mack was one of the scientists who practiced his profession according to rules within the methodology:

> "When I first encountered this phenomenon, I had very little place in my mind to take this seriously. The idea that we could be reached

by some other kind of being, creature, intelligence that could actually enter our world and have physical effects as well as emotional effects was simply not part of the world view that I had been raised in. I came very reluctantly to the conclusion that this was a true mystery. In other words, that I—I did everything I could to rule out other sources."[6]

As we know, the scientific method is a process whereby a practitioner enters an investigation with observation, evidence, and hypothesis. Most often, the researcher tries to find enough additional evidence with which to disprove it though there is no universal commandment to structure the study as a falsification effort only a generally agreed upon ideology. There are many examples of the alternate methodology used by many other researchers who try to prove the main or null hypothesis.

This point is not important here. What is important is that Dr. Mack, as any scientist should strive to be, was conducting his investigations skeptically but with an open mind to accept any evidence without bias. He was also very thorough in his investigations that lasted for over forty years. Dr. Mack was originally a skeptic in the realm of alien abductions. His decades of work proved him to adopt an alternative advocacy.

Other successful researchers who made this adaptation include Dr. Stephen Hawking and J. Allen Hynek. For a long time, Hawking was skeptical of humankind's prospects for interstellar travel. In an interview with television host Larry King in 2010, he said, "Time travel used to be thought of as science fiction, but Einstein's theory of general relativity allows the possibility that we could warp space-time so much that you could fly off in a rocket and return before you set out. Unfortunately, it is likely that the warping would destroy the spaceship and maybe the space-time itself."

His evolution of thought has taken another mainstream course since then as evidenced by this quote:

Of course, it is possible that UFO's really do contain aliens as many people believe, and the government is hushing it up. If the government is covering up knowledge of aliens, they are doing a better job of it than they do at anything else. "To my mathematical brain, the numbers alone make thinking about aliens perfectly rational."[7,8]

Hynek went into his decades-long affiliation with the U.S. government and the Air Force as a confirmed skeptic of the UFOH and the ETH. Many quotes exist from his early years refuting and discrediting away UFO and extraterrestrial encounters. Early evidence of the shift in Hynek's views appeared to the general public in 1953 when Hynek wrote an article for the April 1953 issue of the *Journal of the Optical Society of America* titled "Unusual Aerial Phenomena," which contained what would become perhaps Hynek's best-known quoted statement:

"Ridicule is not part of the scientific method, and people should not be taught that it is. The steady flow of reports (UFO), often made in concert by *reliable* observers, raises questions of scientific obligation and responsibility. Is there *any* residue that is worthy of scientific attention? Or if there isn't, does not an obligation exist to say so to the public—not in words of open ridicule but seriously, to keep faith with the trust the public places in science and scientists?"[9]

The case that cemented Hynek's change of thought came in the March 1966 Michigan UFO flap. Hynek traveled there to perform the whirlwind investigation appropriate to the protocol of Air Force regulations. His press conference enraged the public and all the eyewitnesses when he dismissed the entire affair as that of the natural phenomena known as "swamp gas." A better example of the use of disconnect and ridicule between science and the public in the realm of the UFOH and ETH is hard to find.

There are many examples of situations where, without performing any substantive study or analysis, famous scientists make proclamations which turn out to be entirely wrong. Refer to the Friedman-Marden book, *Science Was Wrong: Startling Truths about Cures, Theories, and Inventions "They Declared Impossible"*, for more examples. It is full of such proclamations across history that exemplify many scientists' "rush to judgment" without any investigation. An example is of Lord Kelvin's 1903 proclamation that, "Heavier than air flying machines are impossible," Lord Kelvin's critical thinking clearly did not follow the protocols of science methodology.

Chapter 4: Inductive and Deductive Reasoning

Consider this quote from J. Allen Hynek: "It reminds me of the days of Galileo when he was trying to get people to look at the sunspots. They would say that the sun is a symbol of God; God is perfect; therefore, the sun is perfect; therefore, spots cannot exist; therefore, there is no point in looking."[10]

Reasoning requires a constant readjustment of any individual's analysis, reflection, and often conclusion because most everything in nature is constantly evolving. Proof of this notion was noted in the discussion of entropy earlier. Entropy's propensity for the behavior of matter of an atomic nature (all matter in the universe) toward disorder and chaos is one of the basic thermodynamic properties of the natural sciences, chemistry in particular. This leads to the requirement of the researcher to maintain a skeptical but open-minded to changes or entropic circumstances in the evidence obtained in any investigation. This includes case studies of the UFOH and ETH.

Another famous quote from the late Stanford astrophysicist and professor emeritus Peter A. Sturrock reinforces this practice:

> On the other hand, patterns that appear consistently in data derived from several independent sources are far more significant than a pat-

tern that shows up in only one source. "Strong" facts of this type can be obtained only be careful cataloging of data from as many reliable sources as one can find. After a catalog has been compiled and patterns supported by the weight of evidence in the catalog have been established, one can then begin the comparison of evidence and hypothesis. (An outstanding example of this process is the construction of the Hertzsprung-Russell diagram in astrophysics, which provides the crucial test for any theory of stellar evolution.) This procedure is complex, calling for a careful organization of theoretical work and data reduction.[11]

In all situations where hypotheses of inquiry are studied, the use of critical thinking, rationality, reasoning, and an investigation of the validity, reliability, and credibility of the subject matter are required while using an open mind. This is often difficult to achieve. We are human. This flexibility often helps in correct decision-making. The natural sciences have used this ideology successfully many times. So have the legal sciences, social sciences, and forensic sciences. Some of these use other forms of reasoning such as abductive and so on, as noted earlier. The study of ufology could adopt these new strategies as well as the others to form a more relevant protocol that performs better with contemporary society and knowledge.

An example for the deployment of additional reasoning strategies could assist in the mystery surrounding the ancient Egyptian practice of mummification. We have the knowledge of "how" and "why" procedures. As all ancient societies were more holistic than we are today, the Egyptians had a well-versed understanding and belief of the afterlife and reincarnation. The process was the apparatus and the mummified bodies were portals and conduits toward those ends. Incidentally, all of the ancient cultures appear to have possessed this technology as evidenced from a study of their historical records.

The main questions here are, "Where did they obtain this knowledge and when?" The earliest known examples of the mummification process came from the Neolithic Chinchorro peoples of Peru and Chile in South America around 5000 BCE. Remember that the entire epistemology and schematic procedure of mummification was at a level of technological sophistication that was exceedingly out of place with any parallel of such sophistication of the technological advancement of life in that era. The practice of successful mummification is yet another example of oopart of ancient times that even today we find exceedingly difficult to accomplish even with our advanced technology and knowledge of the subject.

Instead of relying solely on the "experimentation and replication" of the laboratory testing intervention, let's introduce some other ways of thinking out of the box, such as abductive reasoning or causal reasoning. They are being used in the social and legal sciences already. Even classical literature captured the essence of some of this strategy when the trials of Sherlock Holmes were penned by Sir Arthur Conan Doyle. The case study historical record is replete with cases of failed reasoning which resulted in incorrect conclusions for research inquiries. The most effective and desirable, at least for most people, outcome is to use what works and uncovers the correct answers. It is the only path to truth and knowledge.

Chapter 5: The Yin and Yang of the Science and Ufology Argument

The disciplines of the natural sciences are not currently able to address issues of nature or inquiry beyond those that occur in three dimensions. As we have seen elsewhere, answers to questions concerning matters involving the fourth and additional dimensions, the ones that Einstein and other theoretical physicists and cosmologists have been hard at work researching for centuries now, are very elusive and create even more questions for each one that is answered.

A large part of this example and more generally any such matter of science inquiry demands the invention of new measuring instruments and other technologies capable of addressing the questions being asked and before they can be studied. In the theoretical physics field, the CERN Hadron Collider in Switzerland was needed before any research that could result in the discovery of the Higgs-Boson particle in theoretical physics could be undertaken. Johannes Kepler and Galileo were not able to make any of their planetary discoveries until Tycho Brahe developed the instrumentation and techniques for night sky observation.

If an unidentified aerial object suddenly appears and/or suddenly disappears under either high acceleration or simply "vanishes into thin air," the episode instantly becomes one that is beyond humankind's technology with which to already have manufactured and duplicated. It is also beyond science's capability to study properly according to the methodology paradigm. This is because the phenomenon cannot be duplicated in an experimental laboratory and the instrumentation needed to measure data does not exist. Therefore, the event also cannot be replicated by their peers.

I have discussed many examples where a daring scientist steps out of the paradigm box and reasons out a new discovery. It may take a very long time for the discovery to be accepted in the mainstream due to these very realities and the current methodologies. This slows down Ray Kurzweil's exponential technology development principle into a representation and imagery more like that of a linear graph.

So the clash of ideology between mainstream science inquiry methods and ufology continues under this scenario. Science refutation has either dismissed the UFOH and the ETH as pseudoscience or non-testable under their design. Ufology has tried and called for repeatedly to study UFO and extraterrestrial experiences under these guidelines as an appeal for their acceptance, but yet to no avail. If there would be an allowance for adoption of new more social science-structured study methodology into an entire new form of inquiry investigation that uses both, maybe the two disciplines can mend some of their differences.

It appears that, with so, so many experiences on record documented through history that, like I stated at the beginning of this chapter, something is going on. The linearity of science protocol and practice tends often to enact painfully slow discovery investigations. Ufology has moved closer to a mainstream use of testing instrumentation, as evidential from the "Ways to Prove ET" chapter. Maybe a synergy of the protocols of all the philosophical ways of thought, including new ones, the natural sciences and the social sciences exemplified by the science inquiry of the law and the juror concept could be utilized to make for a more effective and fluid future protocol.

Instead of calling for just a tune-up or a scientific method that has been practiced using protocols designed 2,500 years ago, maybe adoption of a separate newer format reasoned from additional utilization of critical thinking and out-of-the-box thinking is needed. This would then allow for more discoveries of real things that happen in nature and the environment. Additionally, it could make for new and lasting productive epistemological partnerships between science and ufology and between learning new knowledge and how we get there. As Carl Sagan paraphrased from the original quote by mathematician and science scholar Pierre-Simon LaPlace, if, "extraordinary claims require extraordinary evidence," then an extraordinary method with which to study this evidence is needed. Collection of facts and data in this construct would make for a more correct and faster discovery and affirmation of new knowledge and answers to those claims.

Chapter 6: Who Has Seen Them?

If UFO experiences are a psychosocial syndrome, science has no substantial explanation for most. There have been, through history, more or less arbitrary and ill-placed assignments of such explanations as hypnagogia, mass hallucinations, mass hysteria, character pathology, mind tricks, contagious sightings, the Tetris Effect, and the crowd effect to the case at hand in an illogical and unstudied fashion. This course of action reinforces the proof that most UFO and extraterrestrial experience studies are not given any kind of worthy or valid critique.

We have read about some of these situations. The Roswell incident, the Phoenix Lights, the Washington, DC 1952 flap, and the March 1966 Michigan swamp gas flap are all examples of where such psychosocial excuses were added to the official answers to causes of the phenomena people experienced during those encounters.

Many rebuttals to those who have had an encounter of any kind go no deeper than use of a phrase that seems to sound like it would fit into their explanations. A few others try to explain by citing a factor or concept from the subjects of cognitive science and cognitive psychology.

Both of these subjects study the human brain. The main difference is that cognitive scientists study the clinical aspects of brain function; whereas, cognitive psychology heavily utilizes the humanistic factors that participants bring to the behaviors being studied.

The Stanford University *Encyclopedia of Psychology* readily admits that cognitive science and cognitive psychology to a somewhat lesser degree is replete with a lack of knowing that a lot of the underlying principal concepts are definitively correct and representative of nature; that most of it is either not correct and/or is therefore conjecture and subject to future revisions as more certainty is obtained.

> Cognitive science has unifying theoretical ideas, but we have to appreciate the diversity of outlooks and methods that researchers in different fields bring to the study of mind and intelligence. Although cognitive psychologists (on the other hand) today engage in theorizing and computational modeling, their primary method is experimentation with human participants.

The many critics of cognitive science and cognitive psychology cite the inability to predict with consistency what particular human behavior will occur given a particular situation. In other words, "The claim that human minds work by representation and computation is an empirical conjecture and might be wrong. Philosophical critics have claimed that this approach is fundamentally mistaken." The refutation factors include: physical environments, consciousness, emotional, social, mathematics, and the dynamic systems of the mind unique to the individual. Also the unique life experience set that imprints our brain's neurons that are different from everyone else's.[12]

What this means is these fields are in their infancy in attempting to acquire a sophisticated and advanced understanding of the human mind. It may be a mistake to place such final rebuttals on an inquiry about any of the people who have had such encounters as UFO sightings or experiences. A premature explanation using this analysis without an examination of any other facts is probably not correct when seeking truth and reality for the UFOH and the ETH.

Chapter 7: Man's Technology Explosion

Consider this quote from Carl Sagan: "Who are we? We find that we live on an insignificant planet of a humdrum star lost in a galaxy tucked away in some forgotten corner of a universe in which there are far more galaxies than people."

An example of this was the continuing proliferation of the idea that the earth was the center of the universe. In all the ancient astronomy and astrology maps of the solar system and indeed the universe, Earth was depicted as the "center of it all." When Copernicus finally convinced the world the earth was not the center of the solar system, it almost cost him his life. Even though astronomer and mathematician Claudias Ptolemy made numerous attempts to update his solar system models with Earth at the center, humankind's anthropocentrism was evident 1,400 years earlier.

Copernicus's "paradigm revolution" led to more revolutions soon after. The advancements of this era were an example of science and technology evolving at a faster rate than a linear fashion. As noted, new knowledge led to new inventions in instrumentation which led to new science. This process repeated itself. Johannes Kepler followed Copernicus with new discoveries about the movement of planetary bodies in space. Tycho Brahe invented a lot of instrumentation which helped astronomers discover even more at a faster pace. Giordano Bruno also followed Copernicus's cosmological discoveries. He was ultimately martyred because he would not recant his views for society, which were actually true. Finally, Galileo Galilei was also almost martyred for his own radical "pseudoscientific" discoveries which also lent support for the heliocentric universe.

So humankind has maintained a long history of practicing anthropocentrism. All these scientists and the more contemporary scientists like Einstein, Tesla, Sagan, Hawking, and such dared to challenge mainstream science with their knowledge, data, and manner of thinking. Because of their "outside the box" thinking, scientific revolutions in their subjects happened.

This is also true when the subject turns to the UFO and extraterrestrial hypotheses. The manner with which the many millions of case history reports by people from all walks of life at different times in history have been not studied properly necessitate a call for application of the same dedicated scientific methodology preached and practiced by the science world for over 2,500 years.

Many people cannot reason that extraterrestrial life can mean more than an invisible microbe that we may yet find on Mars or Europa or Pluto. Maybe it is because individual microbes do not have spaceships with which to travel to Earth. Because we have not discovered anything microbial outside of Earth's atmosphere is evidence enough of humankind's lack of technological sophistication and the need to retreat from an all-encompassing doctrine of anthropocentrism when discussions about extrabiological life takes place.

What these people need to do is practice what they preach. It is a call for commitment to detailed and thorough inspections of the myriad of archaeological evidence accumulated as artifacts from unidentified objects and possible evidence left by forms of intelligent life through the millennia. Ufology researchers should not continue to be deterred by the ad-hoc and offhand rush-to-judgment answers offered by those who fail to practice what they preach.

The alleged EBEs who have visited Earth must have arrived in more advanced spaceships than we are capable of constructing at this time. There may be more than eighty-five different alien visitors which all have different forms and levels of technology. An accomplishment such as this would make them more advanced than humankind.

It is also a direct attack on the notion of anthropocentrism. This argument is not some form of New Age conspiracy theory. We just have to get down and practice what is preached about how to and how deeply to investigate hypotheses and not to dismiss them as pseudoscience without the study that they are due.

Chapter 8: Why Here?

John Salter, Jr. was the alleged experiencer of the Wisconsin 1988 incident with his son, John III, as noted earlier. The following is an excerpt from his report:

> Consider two families on the same block: the nice, orderly family on one end and the trashy family on the other. The children of the latter move to the tricycle stage and slowly start moving down the sidewalk toward you. Alarmed, you have three choices: blast them into oblivion, but you can't because of theology and ethics; you can wait until they come to your yard and try to deal with things then; or you can take a deep breath, go out and down to their end of the block, and very carefully play the role of social worker; help 'em grow up.[13]

The experience of the Salters was explained to be not only a benevolent but productive one. John Salter, Jr. documented the occurrence of his many physiological changes such as the immediate healing of both new and old bodily injuries and medical conditions. We explored an expanding series of scenarios where exploration and visitation could be both eventual and part of humankind's propensity to partake in these treks. It seems obvious who the humans would be characterized as in Salter's preceding quote.

In his article, *Where Are They? Why I Hope the Search for Extraterrestrial Life Finds Nothing*, futurist Nick Bostrum tries to uphold and answer to the reasoning for theoretical physicist Enrico Fermi's proclamation of the same title name. The issue is significantly problematic as he responds to the point of "why Earth has not been visited" or, in his terminology, use of the paraphrase, the Great

Silence. While Bostrum maintains his belief that the Great Silence is still happening, he finds a synergy in the reasons other intelligent life beings would explore the galaxy with those of humankind as experienced all over Earth. These reasons are the crux of the rebuffs one could offer in this area.

Bostrum claims, "There are a number of considerations that make this (exploring the galaxy) less plausible explanation of the great silence." Bostrum explains that Terran life "manifests a very strong tendency to spread"; that humankind has performed this exact activity in both form and scope; it is the resources that other civilizations could covet; those civilizations could change their minds shortly after their initial decision not to visit or explore Earth; and/or it could be us eventually that does the exploring.[14]

This is precisely one of the matters we studied earlier. One further aspect that does not generate as much discourse is the prospect and reality of Earth being a strategic planet within our intragalactic neighborhood. One of the takeaway pieces of evidence from the alleged Betty-Barney Hill experience was the conversation Betty had with the leader while discussing the star map. The leader mentioned the trade routes depicted on the map. This is logical with any intelligent life-form that has to live in the three dimensions as we do. Space-faring civilizations living in our universe must use tangible materials to maintain the structures they live in or the ships they navigate to make contact and actively pursue their trade activities. From our myopic and rudimentary perspective, humankind can say that at least our solar system is perceived to be a harsh haven for life support. Earth, therefore, seems unique and atypical compared to our other planets.

Not only would the trade of tangible materials be an end point for them wanting to visit us, but also some of these civilizations may, even with their advanced knowledge and technology, still maintain a perspective of bellicose behavior. Because we cannot yet detect with overwhelming probability planets that definitely support and harbor life, we do not know whether some advanced civilizations have their own federation like those that have been depicted to exist in Hollywood. Maybe there are defense, and maybe trade and sociopolitical alliances among some of these alleged nations that we are unaware of yet. Earth may be strategically, spatially, and temporally situated in the middle of one of these theaters of operation in which war, trade, or other friendly activities occur with regularity. Once we are better able to travel beyond our moon's lunar landscape, we may quickly gain knowledge of some of these activities.

Chapter 9: How to Move My UFO

Humankind's status of trailing behind the technological development of extraterrestrial civilizations is consistently referred to in various forms of literature. These assertions frequently indicate it would take millions or even billions of years to catch-up and attain the sophistication level of our alleged visitors.

Looking back to his TED Talk, "The Accelerating Power of Technology," Ray Kurzweil attributes one community involved in commentary about ufology, namely the scientific community, with thinking and constructing ideas in a linear fashion. What this means is the accused science constituency does not account for the new layers of knowledge and discovery that are laid sequentially atop the existing base and become useful for making even newer insights and inventions.

This can be visualized like an inverted pyramid of knowledge. The time it takes for new inventions and technologies, in this digital civilization, to create new knowledge becomes less. This means that if you visualize this upside-down pyramid shape you will see new layers, containing more and more knowledge than the ones below it. This new knowledge is obtained in less time due to the speed with which digital technologies can process and advance them. Examples of this way of thinking and evolution of both the technology itself and the exponentially shorter changes in their time frame include the development of computer memory, DNA genetic technology, and the increasing velocity, in one specific length of time, with which a celestial body moves toward a black hole, to name a few.

There are many examples of this notion both in nature and in humankind's development of technology throughout history. The notion of exponential growth does not suggest there is only one path the graph takes. Any dimension that is being measured on this scale is primarily making projections into the future. Consequently, there can be a turn, or a few, in the road that this projection could take. This could eventuality be influenced by any factors, not the least of which are societal factors such as sociopolitical, fiscal, etc., which are unique to that civilization. There are many alternatives that humankind will uncover that will answer the main question asking how to "move his UFO."

In this way, it is giving us some credit for having the ability to evolve our civilization in an exponential or at least something more than a nonlinear process. We can decide for ourselves whether humankind is billions, millions, thousands, or hundreds of years behind the technology of extraterrestrial explorers. Maybe Gene Roddenberry got it right on the timeline of *Star Trek* that twenty-third century Earth will look and be like the television show depicts and that those technological complexities and capabilities will, in fact, exist.

Chapter 10: Ancient Texts of High Strangeness

We have studied how ancient texts were subject to a common characteristic of misinterpretation and altered interpretation from their original meaning when translated over a long period of time and by many different translators and mindsets. Consider this classic classroom thought and memory experiment. All the students in the class are sent out into the hall except for one. The teacher tells the one remaining student a factual story. A second student is called back into the classroom where the first student relays the story as accurately as possible. Then a third student is called back into the classroom and the second student then repeats the procedure and so on until the last student is told the story. The last student then retells the story as best he or she can for accuracy. The end result of the experiment is an entirely recreated version of the original story with many facts distorted or entirely omitted.

The results invariably appear to implicate the human thought process of cognitive dissonance. This is because humans are gener-

ally incapable of maintaining one hundred percent accuracy or anything close to it, unless one possesses an eidetic memory. This occurs
when firsthand knowledge is passed to a second recipient, then to
a third recipient, and so on. This study on accuracy is conducted
by a branch of textual scholarship called *textual criticism*. Textual
criticism is a branch of literary scholarship that studies changing
interpretations of older passages, scripts, and other linguistic forms
of communication.

This phenomenon is not really observed when inspecting
ancient glyphs and other archaeological artifacts and paraphernalia.
A reason for this is because these are the original objects, akin to the
original information the first student in the experiment possesses.
The archaeologist who discovers and analyzes these originals is making a firsthand translation. A firsthand translation is unencumbered
by many layers of temporal translations, like written religious script
has been manifested throughout our history. They are the original sources of semiotics and can therefore be literally interpreted
far more accurately as firsthand communication than ancient texts.
Ancient texts have been reinterpreted by many scholars during their
own eras. The new versions include a characteristic imprint of the
current culture, lifestyle, political and societal paradigms, views and
life experiences the scholar lived through as he began interpreting
the texts. As noted, this translation exercise includes changing words,
phrases, and entire passages. The overall result is the permanent alteration of the original. Unless the original scripts themselves are still
available, the original meaning of these as intended by the original
authors can be entirely lost.

Therefore, it can be a challenge to trust the accuracy of more
modern textual scripted translations of the ancients. Another problematic area is with our own modern interpretation into those original versions of text, glyph, symbolic, and other original linguistic
sources of passage. It is wrong to use our own modern perspectives,
cultural influences, and life experiences to analyze ancient writings. The life experiences of people living today are different from
those who wrote the original source. People had a holistic worldview at that time; whereas today, we are far more microscopic in our

vision—a form of myopia. Our microcosmic interpretation is often not an accurate proxy for those of the ancients for whatever motive the investigator acts upon. It would be a fatally biased and flawed activity. Proper application of the special training archaeologists and anthropologists receive will effectively eliminate these biases.

Holism can be defined as the theory that parts of a whole are in intimate interconnection, such that they cannot exist independently of the whole, or cannot be understood without reference to the whole. This is, thus, regarded as being greater than the sum of its parts. Holism is often applied to mental states, language, and ecology.

A significant phrase within this definition is the "intimate connection" and "cannot exist independently of the whole." The act of being holistic means to keep the "whole" in mind when discussing a topic. Two examples of this would be when talking about global warming or religion. How often do you listen to or are involved with a conversation where only a couple of countries are mentioned (in the global warming topic) or a couple of religions when all them should be included within the conversation.

When applied to the UFOH and ETH of ancient times, it is especially critical, for accuracy sake, to get the interpretation right. This notion can also be applied to and taken from the classroom memory exercise. If the investigator has access to the firsthand communication—whether it is script, glyphs, artifacts, structures, or other paraphernalia—he is in a better position to discover and analyze facts accurately. Additionally, he can be far more confident that the reliability of this original body of evidence is at its highest level.

Chapters 11 and 12: Ancient Structures of High Strangeness

The ancients sure knew how to build gigantic structures that were built to last—far better than we appear to be able to do today. Have a conversation about the Roman Colosseum with any of the U.S. professional sports teams and stadium owners and ask them why they feel the need to have a new stadium every twenty years. Or ask any of the skyscraper owners in Dubai how long they think their buildings will last.

Now consider the tools that are used to build any of our modern stadiums or skyscrapers. Compare them with the tools that either were or should have been used to build the ancient megaliths and monoliths. What would make this discussion even only a fraction more cogent is the discovery of more of the tools with which our ancestor societies built their durable structures.

Another final thought of very many to consider stems from the Nazca phenomenon. We have discussed and know that sophisticated tool sets may not have been needed to construct the Nazca Fields of structural glyphs. This notion does fly somewhat in the face of the preceding analysis. But the Nazca Lines, as they are commonly called, have other uninvestigated attributes that could constitute forensic evidence for the future.

Remember that the Nazca Fields encompass a land area of hundreds of square miles. Most of the glyphs, among the world's very largest, are many hundreds of meters long and wide. Taking all these characteristics into consideration and studying what the motivations of the Nazca natives were for creating such astonishing ancient architecture would support the assertion that there are parallels between the cognitive thinking of the Nazca and the native peoples of Indonesia in the twentieth century whose communities were known as the Cargo Cults.

Contrast and compare these motivations of an ancient people from thousands of years ago with those from recent history and even today, where some of these Cargo Cult communities still exist! We know the reasons that the Cargo Cults were trying to reestablish contact with the soldiers from the American Army Air Force, the U.S. Navy, and British and Allied military forces after World War II ended. For them, the reasons were for food, supplies, and shelter. What or more precisely who were the Nazca Fields made to attract and for what purpose?

Chapter 13: Ancient Paraphernalia of High Strangeness

If archaeologists have been able to uncover many venerable texts, artifacts, hunting clovis-tipped spears, and tool sets from the

Acheulian Period (seven hundred thousand to four hundred thousand years ago) and the even earlier Oldowan Period (over eight hundred thousand years ago), then why isn't there a much larger museum tool inventory from much more recent periods of only up to about twelve thousand years ago?

We know that a lot of ancient civilization superstructure complexes are still buried underground. Up to ninety percent of the Mayan city-states of Palenque, Chichen Itza, Tikal, as well as most of the Meso-American, Peruvian, and Bolivian architecture are still unseen. Also a treasure of suspected artifacts around the Great Sphinx, Giza, Gobeckli Tepe, Baalbek, and other Chinese and Indian cities are being continuously unearthed or recovered from underwater. Even Stonehenge and Avebury are conjectured to have more material paraphernalia buried underneath their super structures. For such large dimensional objects to have been built, any engineer today would admit that the tools needed to build them would have also been larger and more durable.

Chapter 14: Physical Evidence of UFOs

Discussions about trace evidence and the UFO and extraterrestrial hypotheses generally do not include any forensic inspection regarding other witnesses who were present on site at the time of the encounter. Animals are living organisms that interact with humans. From our pets to farm animals and even those in the wild are capable of noticing and reacting to unusual manifestations from UFOs and EBEs.

Refer back to the Delphos, Kansas, incident of November 2, 1971. The case file reports explain significant aberrant behaviors exhibited by all the animal life around the scene of the incident. Ronald Johnson, the human observer, was accompanied by his dog and sheep within the barnyard at the time of the encounter. For two weeks after the UFO encounter, the dog would anxiously try to get back inside the Johnson house around nightfall when it never before exhibited this characteristic. It ripped two doors down in the process. At about the same time each day, the sheep would leap over the fences and run wildly around the compound. All animals would continuously stay far away from the area of ground where the "ring" of white powder trace evidence was left by the alleged landed ship. Even wildlife was noted to have avoided the area.[15]

There are sensory receptors in the different animal species whose range are unique to many live organisms. We know dogs have a greater sense of hearing and smell than humans. This is because of the physiological attributes and phenotypes they have that we do not. The animal kingdom is rife with these characteristics. It is possible that these animals can sense things about the Delphos compound grounds that humans cannot.

For example, when humans cannot perceive a feature in nature—such as being able to see infrared light, magnetism, DNA "in-situ" or in real time, or hear above 20000 kHz—we build instruments with which to detect and capture a perception of these phenomena. An invention process where new equipment is developed and implemented with the ability to detect and record these unperceivable manifestations could be undertaken.

Remember that in the Delphos case and many thousands of other documented cases animals were interacting directly in the environment around the event theater before, during, and after. Some unusual things were and are occurring within those natural settings. The animals are interacting and reacting to these episodes for a long time afterward. Maybe this suggests a new initiative into the design, implementation, and committed study of these formerly untouchable and discarded phenomena be established.

Chapter 15: What Scientists Are Saying

As the evidence has been accumulating to offer refutation against *The Humaniverse Guide to Better Reasoning & Decision-Making* hypotheses, let's maintain the perspective that through our particular lens of science, jury member, judge, and forensic investigator, we are asked to study and analyze only what is being hypothesized.

What are being hypothesized are two statements. The first asks about the nonexistence of airborne and submerged objects that are alleged to occur and navigate in ways only possible by intelligent control. The second is the nonexistence of intelligent beings that may be the impetus for the existence of the objects stated in the first hypothesis. What the science and philosophical principles that originated from over 2,500 years ago are asking modern science to do is to prove or disprove them by testing and replication protocols in a laboratory setting by any of a number of peers who conduct similar experiments.

How can they replicate and test the existence of the objects stated in the first hypothesis? The only way would be to build proxies and experiment with them. So let's say they are somehow able to build and test-fly some of them. All this says is that, "It is possible for humans to build ships that display some of the characteristics that have been observed." Also the current state of affairs has dictated that the ships that possess this sophisticated level of operational design and flight characteristics involve knowledge the sciences do not seem to currently possess. They are incapable of proceeding in this fashion except in a painfully slow paradigm, almost something less than linearly, to a level of some form of logarithmic scale. So they are not testable. This is the excuse given that explains that either they definitively do not exist or untestable.

How can they test and replicate the substance of the inquiry stated in the second hypothesis in a laboratory? It may be necessary for many to dissect if they wish to maintain the laboratory setting and replication design. Researchers are directed by the methodology to seek as many data as they can. Maybe they could test the "aliveness" of them if they in fact introduced themselves to us and allowed for us to experiment on them. That would be a turn of alleged events in the extraterrestrial abduction/experiencer scenario!

What is missing is the acceptance and inclusion of a social science fashioned forensic body of investigation that also discovers facts and data that are used by juries to decide on a claim of hypothesis of the acts of a human being. The science of the social sciences or the law does not seek to insist on the use of laboratories to replicate experiments on a human being to determine the person's guilt or innocence. But it does seek to find answers to the existence and reality of situations in which humans, as living beings, are accused to have had a significantly influential activity in. The discovery of new particles in theoretical physics or a new drug can be aided by laboratory experimentation. It is much harder to uphold that experimentation protocol on the cause of determining the existence of life-forms or the ships they use.

If some animals escape from the zoo one day and hundreds or thousands of people saw them move through their front yards, it would be regarded as obvious that they exist, and everyone is seeing the same phenomenon.

Chapter 16: What People (a Lot of Them) Are Observing: The Public

A topic that is perhaps a bit more esoteric and less pragmatic than the mainstream study of ufology is related to crop circles. It is surely popular enough for mention in the high strangeness of unexplained phenomena and also here.

Documented sightings of such manifestations go back at least to the 1500s. The discovery incidence of such formations rose exponentially only in the latter half of the twentieth century. The complexity and sophistication of the manufactured imagery also has risen in a similar exponential fashion. There is no generally agreed upon explanation for why crop circles continue to appear. Targeted sites have included every continent, but a majority of them have been laid down in northern Europe. Who is making these formations is a mystery, at least in part.

Some of the crop circles that appeared in England in the 1980s were discovered to have been perpetrated hoaxes. A few engineers came forward demonstrated how they were able to manufacture a basic formation locally in just one night. Other study simulations confirmed that basic formations are within the capacity of humans and basic wooden tools to manufacture. But this is only a small part of the story.

As mentioned previously, the sophistication and wide global distribution of formations are two explanations that the majority of crop circles are beyond human capability. Such designs have included fractal mathematical imagery that are over an acre in area and the precision of the designs are such that no tools available to those "weekend warriors" can accomplish this at all, much less in one night or without being detected.

There have been more recent investigations that have utilized science and technology instrumentation and testing of crop circles. Leaders in this research include biophysicist William C. Levengood, Nancy Talbot of BLT Research Team Inc., and physicist Eltjo Haselhof.

Dr. Levengood has published many accomplished scientific research papers testing the crop circle phenomenon.[16,17,18,19] His conclusions about various biophysical, geophysical, and chemical characteristics testing of worldwide formations indicate patterns which include the application of intense heat and temperature instantaneously to such crops in which parts of the crop structure, the nodes, literally explode. Photographic evidence of some crop formation episodes show spheres of light maneuvering over the field at the precise time the imagery shows decimation and bending of the crop victims to produce the design known as the crop circle. This imagery is shown to manifest almost instantaneously or within a few seconds of the beginning of the episode. This video-taped evidence is evidential to support proof of the findings of Dr. Levengood.

Eltjo Haselhof, of the Dutch Centre for Crop Circle Studies, replicated Levengood's test experiments with his own research in 2014. His findings included node length expansion of crop victims formed in a concentric fashion. This means that the node lengths, which help cause the requisite bending of the crops but not their destruction, are most pronounced within the inner area of the formation. As the observation is made going outward from the center core, the node lengths decrease. Haselhof's findings corroborate with other research in that the manifestation of the studied crop circles, and crop circles in general, is caused by a short intense application of high temperature and/or of extreme electromagnetic radiation administered to the crops in a highly sophisticated algorithmic design. The advanced technology needed to and displayed in these incidents is not within humankind's possession or achievement.[20]

While the ultimate truth and reality of who or what has created all the crop circles still remains unanswered, save for the identified hoaxes, it is evident that the perpetrators use a highly-advanced technology instrumentation to produce them. Some research has moved in a different direction to explore the question, "What do the circles mean?" Are they messages to be decoded? Are they to be analyzed as a group to detect similar messages? Is humankind the intended audience for these?

Chapter 17: What People (a Lot of Them) Are Observing: The Especially-Trained Public

In his speech and paper at the 2001 MUFON Symposium Proceedings, Ryan S. Wood lectured on the Cape Girardeau 1941 UFO crash and retrieval of the remains of an alleged extraterrestrial. Local fire and police departments have archived logbooks documenting the activities of that time in April 1941 that invoked and involved the services of Baptist minister Reverend William Huffman. He was asked by officials on the scene, which included the U.S. Army and the FBI, to administer a last rights to the victim of the ship that crashed. There were many representatives of the "especially trained public" on site that night.

Wood summarized his study of the case in this way: "The conclusions and implications are more predictable and widely supported outside this case. Namely, ETs are real; certain factions of the government have been hiding this fact since at least 1941; and the involvement has been a matter of grave concern to several presidents."[21]

Chapter 18: What People (a Lot of Them) Are Observing: The Military:

There has been a strategy that the military agencies of the U.S. government have used to summarize its position in a lot of press releases regarding UFOs and ETs. It is that they have not kept secret files on UFOs and the intelligent life beings that navigate them through the skies and under its oceans. It would be evidently asserted that this strategy is not truthful because they have surely paid a lot of attention to them in history. From the "What People Are Observing...The Military" chapter, it seems conclusive that UFOs (and later to be known as UAPs) have repeatedly punctured the fabric of reality and of Earth's atmosphere and hydrosphere so often as to lose count. They successfully knocked out the U.S. Air Force's nuclear warheads and their delivery systems at Malstrom Air Force Base and many other hundreds of similar armament at various times and all over the globe since the 1940s.

While the U.S. Navy, if they had ever been involved in the subject of UFOs/UAPs and ETs in an official capacity, maneuvered this strategy "under the radar," the Air Force has been the pivotal point for the lion's share of publicity in these areas of inquiry. This strategy degradation even achieved new higher levels of protocol in the late 1960s when the Air Force Academy published and assigned the textbook, *Introductory Space Science*, for its course titled "Physics 370."

According to then Public Information deputy Director Major Stewart Kilpatrick, about twenty-five students regularly took the physics course. The Volume II version of the book contained a chapter on unidentified flying objects. Chapter XXXIII was fourteen pages long and was in use in its particular script from 1968 to the fall semester 1970. This script included intensive use of the UFO moniker, and all subject matter was enriched by abundant attributions to the general ufology subject. Quotes from that edition such as: "Fifty thousand virtually reliable people have reported sighting unidentified flying objects. This leads us with the unpleasant possibility of alien visitors to our planet or at least alien controlled UFOs."

The 1968 version of chapter XXXIII closes with an affirmation of the urgent need for renewed and extensive investigation into the subject of UFO and alien study. There were as many epistemological and philosophical takeaway points as technological ones. The academic perspective included a desire to teach students that when in an investigation, when there are streams of facts and data that seem either refutable or contradictory that the correct course was to maintain an open mind and open reasoning and seek more facts and data. This is, of course, the same protocol that is used in the appropriate application of research study in any endeavor.

When this edition was published and first used in 1968, the famous *The Condon Report* study was just getting started. After *The Condon Report* was published later in 1969, the Air Force Academy took the existence of that publication and some of its recommendations to task. Because some of the summary concluded that UFOs, "posed no threat to national security and, therefore, is not deemed to warrant further study," the academy was compelled, officially or not, to revise chapter XXXIII of the textbook. For the fall 1970 semester, the newest edition of *Introductory Space Science* toned down the intensive phrasing of UFOs and actually started using the unidentified aerial phenomena acronym UAP as a descriptor instead of the traditional UFO. Parts of the conclusions set forth in *The Condon Report* became text for this new chapter.[22]

The mainstream inclusion of an entire chapter in an upper-level collegiate physics course to ufology was a curious undertaking by the Air Force. There must have been more to the reality and substance of the matter than what was portrayed by the litany of excuses offered by all the refutation communities at that time and even through to today. The chapter XXXIII script was also official acknowledgement from another reliable and credible source that more dedicated study must be undertaken of the UFOH and the ETH. Who is more credible here, the U.S. military or the debunking community?

Chapter 19: What People (a Lot of Them) Are Observing: Plus a Lot More

The alleged extraterrestrial experiencer unequivocally has a most impenetrable case study in all the accounts of case studies in ufology. Harvard University psychiatrist and professor John E. Mack saw his career track move through the methodology from skeptic to advocate. He studied hundreds of cases of what were then called alien abductions and alien encounters. Among his patients were the sixty-two grammar schoolchildren at the Ariel School in Ruwa, Africa, in 1994 who had their own encounter, as noted earlier. His revelations concerning this incident were vital to the understanding of his conversion from doubter and skeptic to believer. Statements made by him capture the essence of some of this dimension: "Just how literally to take this is one of the most interesting and complex aspects of this. There are aspects of this which I believe we are justified in taking quite literally. UFOs are in fact observed, filmed on camera at the same time that people are having their abduction experiences."

He also feels that:

> People, in fact, have been observed to be missing at the time that they are reporting their abduction experiences. They return from their experiences with cuts, ulcers on their bodies, triangular lesions, which follow the distribution of the experiences that they recover, of what was done to them in the craft by the surgical-like activity of these beings. All that has a literal physical aspect and is experienced and reported with appropriate feeling, by the abductees, with and without hypnosis or a relaxation exercise.[23]

Mack laments the shortcomings of mainstream science to have full ability to effectively study episodes within this phenomenon using their methodology. According to Mack,

> Insofar as the abduction phenomenon does enter the physical world, it can be studied by the traditional methods of science. But even here, it has been difficult to obtain data that would satisfy the scientific community that something extraterrestrial or otherwise strange has taken place. What are we to make of this consistent elusiveness of the sort of physical evidence that would satisfy scientists? Is it that the phenomenon itself is redolent of a kind of tricksterism

that mocks our technology and the literalness of
minds, which require material proof before they
believe anything really exists?[24,25,26,27,28]

Time and time again, scientists speak out about the need for
reform in the mystique of science that is shackling their pursuit of
knowledge, reality, and truth. Mack is another of the enlightened scientists who recognized the need for more dedicated and multimodal
ufology research, as evidenced in this quote from a letter from col-
league Paul Bernstein to John Mack: "Perhaps, the greater weakness
in the field of abduction research is not inadequacies in methodology
so much as the paucity of courage motivating a sufficient number of
people to enter and explore its risky territory."[29]

Chapter 20: Disclosure: An Overview of What We Don't Know

Reflecting upon the "Disclosure" chapter, there exists much
knowledge which supports the notion that a lot of things both UFO
and ET are being kept secret. Disclosure implies either a form of
transparency or one of secrecy. We all know that many of the agencies
that have governed the thousands of societies in humankind's history
have had a propensity to hide facts and information for whatever
reason. Among the rebuttal attempts offered throughout the ufology
literature and elsewhere are those that proclaim the government, the
U.S. government or any other, is not capable of keeping things secret.
The excuse of "there is nothing to disclose" has also been recycled
both overtly and covertly to the point of dysfunction with reality.

Consider some examples of the fallaciousness that these excuses
portray in their dissemination. *The Washington Post* published a
December 2016 article titled "Scientists Are Frantically Copying
U.S. Climate Data, Fearing It Might Vanish under Trump" (presi-
dent-elect at the time). The article explains the mounting of a "gue-
rilla archiving event" by the frantic science communities to copy and
store as much geophysical global climate data and facts as possible
before Trump took office. This was all over the fear that the information would disappear from any domain, public or otherwise. The

data is damning evidence for the reality that Earth's biosphere is experiencing severe global warming. The science communities have concluded that humankind is a leading cause of this phenomenon.[30]

Here is a recent example of scientists who are self-proclaimed to be rational, logical, reasoned, and analytic professionals performing a task to alleviate the fears, real or otherwise, of a pending disclosure and secrecy catastrophe. Some members of these communities are the same people who have proclaimed elsewhere and in other forums that the U.S. government does not hide information or hold secrets when answering to an analysis of the UFOH or ETH hypotheses.

A second example of the epidemic of flawed critiquing happened on July 8, 1947, at the height of the Roswell incident. As noted earlier, the transmission specialist responsible for broadcasting the reports of the UFO crash and possible bodies recovered from Roswell worked at radio station KOAT in Albuquerque, New Mexico. Her name was named Lydia Sleppy. She was teletyping the story transcript received from Colonel William Blanchard's office at RAAF 509th Operations Office at Roswell Army Air Field via local Roswell radio station KSWS to the Associated Press for syndication.

In the middle of her transmission, Lydia Sleppy received an urgent transmission from the Dallas, Texas, office of the Federal Bureau of Investigation. The communicate ordered her to "cease transmitting this story immediately." She complied with the FBI order, and the station later found other means to disseminate the information relayed from the Roswell Air Force base.

There are additional points inherent in this event besides being ordered by a government agency to "stop talking." It is another case of blatant disclosure and secrecy censorship. Why and how was the FBI monitoring this in the first place? The FBI has sworn for decades that they do not get involved with the UFOH or ETH hypotheses. The FBI was monitoring all U.S. news services at that time and before that time during World War II and even before that. After that, when the FBI was not as active, it was the National Security Agency (NSA) or any of a host of other government organizations doing it. But the explanations from debunkers of the UFOH and ETH continue to proclaim that the government does not do this and never has. They are offered as reasons both of these hypotheses are false.

A third example of the incorrect rebuttals offers and supports further proof of the second one. In 2013, whistleblower Edward Snowden disclosed the existence and proliferation of exactly these types of covert activities when he made the NSA protocols public that effectively intercepted the electronic communication and financial transactions of every American citizen or others present and communicating such on U.S. territory and overseas.

How can a reasoned, rational person offer these as refutations and who still wishes to be regarded as or are held up to be credible?

Consider this quote from Carl Sagan:

> "We have designed our civilization based on science and technology and have also arranged things so that almost no one understands anything at all about science and technology. This is a prescription for disaster. We might get away with it for a while, but sooner or later, this combustible mixture of ignorance and power is going to blow up in our faces."

Turning away from the immediate concerns about science and technology in their generic forms and toward its applications to and involvement in government and its secretive agencies that develop and use them, the disclosure discourse among and between the American public and its government has not been perfect for some time. It appears many of America's Founding Fathers and early presidents had the vision to forecast such eventualities.

Consider this quote from U.S. President James Madison:

> "A popular government, without popular information, or the means of acquiring it, is but a prologue to a farce or a tragedy, or perhaps both. Knowledge will forever govern ignorance; and a people who mean to be their own governors must

arm themselves with the power which knowledge gives."[31]

Scientists often practice their duties and obligations toward the public, whether implied or stated, for the good of the public. To punctuate the James Madison quote, Albert Einstein said this, "The right to search for truth implies also a duty; one must not conceal any part of what one has recognized to be true."[32]

Chapter 21: Why You Would Become Interested in the UFO and ET Hypotheses?

If the discipline of ufology wants to dedicate itself to an objective of recognition of a larger research community, it could take some pedagogical steps to provide entry points for citizens of all ages. The process of igniting people's attention and interest may involve a multistep approach. First, prospective investigators would need to be given access to more information and knowledge about what technology developments and forensics are represented by the study of ufology. Second, use these as exploration and motivational strategies to expand and use this knowledge in other areas of science, engineering, and the social sciences.

Both of these steps will embody a layer of coursework in the numerous diverse fields attributable to both the UFOH and the ETH. Again, there are many epistemological elements that require a basic working knowledge of some currently-known principles of chemistry, physics, engineering science, geophysics and meteorology, forensic science, and the social science disciplines which include anthropology, archaeology, and psychology. This landscape is painted on a background which is made of some knowledge of critical thinking, reasoning, and the hypothesis-theory development process.

A more profound study of the social and forensic sciences should be a requirement as many evidence artifacts are documented from observations made by other people. As mentioned previously, frequently, a historical perspective must be learned and sustained often as different cultures through history saw and operated differently within their own world. This readily translates down to the texts, artifacts, structures, and paraphernalia they left behind. Knowing about the principles of holism when studying the ancients would spark discovery of many new answers which add to the knowledge-base and move down the correct road toward confirmation or durable refutation of the UFOH and the ETH and especially its aftermath.

It is a menu of these entrees that should be introduced to all people, students, and adults alike. The nature of a meaningful and durable first-contact disclosure would automatically ignite curios-ity, motivation, and engagement for all our civilization no matter their age, culture, life experience, or history. Such contact would also probably reintroduce people to the notions of open-mindedness and holistic worldviews.

Chapter 22: Prove ET: Ways to Provide Evidence for Possibility of Existence

Another example of how the debunking discourse leads to prob-lematic issues of reliability and credibility can be found in the dis-cussion involving crop circles. For decades in the twentieth century, thousands of reports came in globally introducing new crop circle episodes. The pattern of quantity accelerated after the 1960s. There was a wide range of sophistication in the final products; the circle designs took on a life of extreme sophistication. All initial explana-tion from the responder community dismissed all them as some type of natural phenomena of atmospheric or other cause.

In 1991, two British citizens came forward with a public announcement that they were responsible for all the crop circles. They demonstrated their technique for making an exemplary crop circle to authorities. They did show an ability to duplicate basic replicas of the rough design of at least some of them. Many government and science professionals fallaciously inferred that these pranksters must have been the creators of all the circles. The refutation community committed many basic logical fallacies in this case, which included the most popular *hasty generalization*, also known as jumping to con-clusions. An additional *invincible ignorance* fallacy, where "my mind is made up; nothing will change that," is present and argued in many of these cases.

The fatal fallacies can be attributed to all these "refutations" for many reasons. First for the conclusion of one party committing to all the crop circle creations since the 1970s is fatally flawed because there were documented reports from over twenty-eight countries compiled through the early 2000s. Second, none of their replicas contained samples that contained any characteristics of the types of crop samples discovered and studied by Dr. William C. Levengood and Eltjo Haselhof, discussed earlier (chapter 16). This was yet another case where scientific evidence was not considered.

Third, none of the explanations are provided with any details about any type of investigation of factors regarding any crop circle case. For instance, these debunkers gave no detailed description of what happened within any of the circle fields from a micro-perspective. These explanations did not study any of the individual stalks of vegetation and other close-up characteristics of any of the environment within the theater of operations. The commentary was only from a macro-perspective from a vantage point high above the field in question and offered no investigative evidence whatsoever.

There is a lot of evidence in these crop circle cases that is simply being ignored. Rather, many fallacious arguments are substituted for any consideration of these facts. It has historically been the overarching operational characteristic of the critic's paradigm of irrationally and incorrectly dismissing these events based on a total lack of investigative study. It is situations like this that have caused many dedicated ufology researchers to repeatedly declare a call for renewed and truthful undertaking of a new paradigm for research studies.

There is also continuing technological advancement in the astronomy field that opens up new and exciting opportunities for ways to provide evidence for extraterrestrial life. Discovery of new planets and detection of telltale signatures of atmospheric chemical attributes to a possible existence of life-forms are two methods in use now for over a decade. Additionally, a particular study could be undertaken to test the hypothesis of the Zeta Reticuli system harboring such an intelligent civilization that Betty Hill met in her alleged encounter.

Emerging as perhaps a superior measure of technological advancement and ways to provide evidence is an accounting of all the divisions of the super discipline of genetic biosciences. Some of these have been discussed as central themes earlier in this text. Another path of possibility occurs when reports are disseminated that some hybrids are walking among us on Earth already. Some of them have been introduced to us in public forums. Now that we have achieved the technological status of meaningful investigators, would it not best serve the UFOH and ETH to offer to obtain DNA samples from some of them for analysis?

Additionally, an exceedingly overflowing inventory of biological artifacts exists in all the world's museums which beg for an opportunity for exploratory DNA testing. Why not let the forensic sciences conduct investigations to take samples of skull and body fragments and from other archaeological biological artifacts and try to recover some DNA identification patterns? We have already taken DNA from some yeast samples from ancient Egyptian remains and have successfully manufactured the type of beer they drank over 2,500 years ago!

Additionally, UFO and ETH investigations should be undertaken to uncover facts about the evolution of government and military disclosure. It appears the stream of knowledge and facts are being uncovered and disseminated with a significant degree of regularity and this phenomenon has been occurring for some time. Current efforts are being made by appropriate officials in the United Kingdom, with their National Archives UFO/UAP website and Ministry of Defense site, France and their Geipan and their UAP file disclosures, the Library and Archives-Canada, Denmark's Air Force, New Zealand's Defense Force, Russia's Navy UFO library files, and Chile's government release of their UFO studies on a scale equivalent to the U.S. Air Force's Project Blue Book from the 1950–1960s.

The United States even has a new entry into the area of information dissemination. We have seen some details from the AATIP's December 2017 proclamations about the Department of Defense and the Pentagon's on-going ten-year study of UAP experiences and encounters by more military officials and trained personnel. The eight dozen case studies have recorded data, radar, video, audio, and other spectrographic and electromagnetic phenomena. The AATIP program is collecting and scientifically analyzing the data. Their reflections so far equate to the call for further detailed study to continue. A prime motivation for the entry into AATIP is recognition to the possibility for threats to national security. We have seen that official government position on this dimension has long been the reverse; they have published many reports indicating there had been no threat to national security in all the studies since the 1940s. Thus, there are many dynamics in the notion of public newly formed access to additional data and knowledge collected by governments worldwide. The archive continues to grow.

Final Thoughts

It is hoped that you, the reader, have taken away some thoughts and insight to reflect upon as you near the end of *The Humaniverse Guide to Better Reasoning & Decision-Making*. This is not meant to sound fatalistic! Here, the end of the book is just the beginning of what is hoped to be a future "humaniverse" that embraces a rational, meaningful, and determined study of the UFO hypothesis and the extraterrestrial hypothesis. The future nomenclature would probably initially redefine the UFO hypothesis as the unidentified aerial phenomenon hypothesis or UAPH.

As the fields of science and ufology continue their debate over the definitions, existence, usefulness, and admissibility of existing UFO and ET evidence into the court of rational and reasoned analysis and consideration, it is hoped that you can see and appreciate both perspectives. It is also hoped that your lens of inspection, consideration, and reflection has been focused more clearly by reading *The Humaniverse Guide to Better Reasoning & Decision-Making*.

It is through my lens that the field of ufology is still in its formative years. It seeks to find its way through the landscape of the want for acknowledgment, elimination of the fear of ridicule, and lack of respect. Ufology is also imploring and entreating others for the long-awaited commencement of good investigation and meaningful answers; most importantly, a respect for reality and truth free from any biased contrivances of the mind. The many different societal communities have had their influence on these factors over the years. Science does not acknowledge ufology, the government ridicules it, and the media do not respect it. The field of ufology, for these reasons, may be experiencing an identity crisis of sorts. Its many contributors are pursuing the search for knowledge in a loosely-coordinated and a similar problematic fashion which besets the mainstream sciences.

Is everything in nature capable of being investigated via the current scientific practices? Does every specimen have a requirement to be physically experimented upon before one can conclude a simple acknowledgment of its existence? The experiments can be used to gain more knowledge beyond the simple fact of its existence, like how the phenomenon works, breathes, and otherwise operates within its environment. Can the methodology investigate and conduct experiments with things that are influenced by a fourth dimension? It seems theoretical physics is having problems in this area by its inability to prove a lot of Einstein's theories, over one hundred years later, even with the existence of the Large Hadron Collider and many other laboratories from around the world solely dedicated to explore these questions.

Can the same conclusions be made about the works and life of Nikola Tesla? He obtained over seven hundred patents for inventions during his lifetime. He won the "electricity war" with Thomas Edison. He sparred often and then collaborated just as often with Einstein on many projects undertaken in the societal environment of world war. To the shock of his family, upon his death in 1943, it took only hours until the FBI had entered his apartment and confiscated all the files and project details he had worked on for decades never to surface again. These mysteries are still being investigated.

If, in the 1900s and 1920s, we were not ready to accept Tesla's advancement of wireless electrical power transmission, are we—as the meta-community of public citizens—ready yet one hundred years later? Back then those technologies seemed futuristic and too sophisticated for society to adopt successfully. Are they now?

This way of thinking parallels that of reflecting on the UFOH and ETH. The modern era of ufology began in 1947, as informally but generally agreed upon. Has the measurement of our level of awe over the UFOH and ETH lessened, remained the same as or more than the measurement of the awe over Einstein's and Tesla's futuristic inventions? Are we, seventy years later, more capable and willing to adopt a more serious study and examination of the UFOH and ETH? Or are these scenarios manifestations of a government wanting to make our decisions for us for whatever reasons?

Carl Sagan reworked his famous phrase, "extraordinary claims require extraordinary evidence," from his reading of Simon-Pierre LaPlace's 1812 treatise *Théorie analytique des probabilités*. In it, LaPlace wrote, "The weight of evidence for an extraordinary claim must be proportioned to its strangeness." He may have fashioned his quote from philosopher David Hume, who wrote in 1748, "A wise man proportions his belief to the evidence," and "No testimony is sufficient to establish a miracle, unless the testimony be of such a kind, that its falsehood would be more miraculous than the fact which it endeavors to establish." A more contemporary interpretation in 1978 by sociologist Marcello Truzzi states, "An extraordinary claim requires extraordinary proof."[33,34,35]

The same urgency of requirement exists for an entirely bias-free investigation and access to the extraordinary tools and all existing evidence that exist in the UFOH and ETH. Practice of the same level of voracity all other scientific methodologies presently employ, and advocate is both required and demanded if its lofty accreditation by society is any consideration. Use of the other investigative methodologies that you have learned about, if they are useful to the effort, must be employable without controversy. Discovering the existence of and learning about entities of extraterrestrial origin would be the most enormously fantastic learning experience in humankind's history. If we could only be sure of having all access to all that knowledge and the unbiased thinking that seems is damning humankind's progress toward meaningful science, technology, and society.

LaPlace and Sagan forgot one thing: these claims also demand extraordinary study in order to properly address them and prove their existence. They are notably worthy of it. Would you demand the same of the many questions and problems in your life?

References

[1]   Quote Ref 1: On patent controversies regarding the invention of radio and other things, as quoted in "A Visit to Nikola Tesla" by Dragislav L. Petković in *Politika* (April 1927); as quoted in *Tesla, Master of Lightning* (1999) by Margaret Cheney, Robert Uth, and Jim Glenn, p. 73, ISBN 0760710058 ; also in *Tesla: Man Out of Time* (2001) by Margaret Cheney, p. 230.

[2]   Volti Ref 2: Volti, Rudi. 2014. Society and Technological Change, Seventh Edition, p. 64. Copyright and published 2014 by Worth Publishers.

[3]   Latour Ref 3: Latour, Bruno. 1987. *Science in Action: How to Follow Scientists and Engineers through Society*. Copyright 1987 by Bruno Latour. Published 1987 by Harvard University Press.

[4]   Volti Ref 4: Volti, Rudi. 2014. Society and Technological Change, Seventh Edition, p. 64. Copyright and published 2014 by Worth Publishers.

[5]   Critical Thinking Ref 5: Sternwedel, Janet D. 2011. Doing Good Science: Evaluating Scientific Claims (or do we have to take the scientist's word for it?) Scneitific American. September 30, 2011. Retrieved December 17, 2016, from: https://blogs.scientificamerican.com/doing-good-science/evaluating-scientific-claims-or-do-we-have-to-take-the-scientists-word-for-it/.

[6]   What Is Science Ref 6: NOVA Online. 1996. Kidnapped by UFOs? Interview with John Mack, Psychiatrist, Harvard University. Copyright 1996 WGBH, Boston, Massachusetts.

[7]   What Is Science Ref 7: Cofield, Calla. 2015. Stephen Hawking: Intelligent Aliens Could Destroy Humanity but Let's Search Anyway. Space.com, July 21, 2015. Retrieved December 17, 2016, from: http://www.space.com/29999-stephen-hawking-intelligent-alien-life-danger.html

[8]   What Is Science Ref 8: The Telegraph. 2010. Stephen Hawking: Alien Life Is out There, Scientist Warns. April 25, 2010. Retrieved December 17, 2016, from: http://www.telegraph.co.uk/news/science/space/7631252/Stephen-Hawking-alien-life-is-out-there-scientist-warns.html.

[9]   What Is Science? Ref 9: Clark, Jerome (1998). The UFO Book: Encyclopedia of the Extraterrestrial. Visible Ink. p. 305. ISBN 1-57859-029-9. Published by: Omnigraphics, Inc.; 2 Sub edition (June 1, 1998).

[10]   Inductive & Deductive Reasoning Ref 10: Hynek, J. Allen. 1977. *Newsweek Magazine*, November 21, 1977. Volume 128, number 19, p. 97.

[11] Inductive & Deductive Reasoning Ref 11: Sturrock, Peter A. PhD. 1999. The UFO Enigma, Pg. 39. Copyright 1999 by Peter A. Sturrock. Published 1999 by Warner Books Inc.

[12] Who Has Seen Them Ref 12: Stanford University. 2014. Cognitive Science. Stanford Encyclopedia of Philosophy, p. 2–10. Originally published September 23, 1996, and revised July 11, 2014. Retrieved December 17, 2016, from: https://plato.stanford.edu/entries/cognitive-science/.

[13] Why Here? Ref 13: Salter, John R. Jr. 1992. Extraterrestrial Contact and Human Social Justice Sensitivity and Action, p. 12–13. Copyright 1992 by John R. Salter Jr.

[14] Why Here? Ref 14: Bostrum, Nick. 2008. Where Are They? Why I Hope the Search for Extraterrestrial Life Finds Nothing. Published May/June 2008 by MIT Technology Review, p. 72–77.

[15] Chapter XXI Physical Evidence Ref 15: Case Report. 2011. Delphos, Kansas Landing Ring. Copyright 2011 by ufoevidence.org. Retrieved December 1, 2016, from: www.ufoevidence.org/cases/case192.htm

[16] What People...Public Ref 16: Levengood, W.C. 1994. Anatomical Anomalies in Crop Formation Plants. Physiologia Plantarum, March 24, 1994, Volume 92, Issue 2, p. 356–363.

[17] What People...Pubilc Ref 17: Levengood, W.C.; Burke, John A. 1995. Semi-Molten Meteoric Iron Associated with a Crop Formation. Journal of Scientific Exploration. 1995, Volume 9, Number 2, p. 191–199.

[18] What People...Public Ref 18: Levengood, W.C.; Talbot, N.P. 1999. Dispersion of Energies in Worldwide Crop Formations. Physiologia Plantarum, April 1999, Volume 105, Number 4, p. 615–624.

[19] What People...Public Ref 19: Levengood, W.C. 1958. Instability Effects in Vortex Rings produced with Liquids. Nature, June 14, 1958; Volume 181, Number 6, p. 1680–1681. Retrieved December 17, 2016, from: http://www.iccra.org/levengood/CircleScans/1958-Nature-instability%20Effects%20in%20Vortex%20Rings.pdf.

[20] What People...Public Ref 20: Haselhoff, Eltjo H.; Boerman, Robert J.; and Bobbink, Jan-Willem. 2014. An Experimental Study for Reproduction of Biological Anomalies Reported in the Hoeven 1999 Crop Circle. Journal of Scientific Exploration, January 2014.; Volume 28, Number 1, p. 17–33, 2014 0892-3310/14.

[21] What People...Especially Trained Public Ref 21: Wood, Ryan S. 2001. The First Roswell: Evidence for a Crash Retrieval in Cape Girardeau Missouri in 1941. MUFON Symposium Proceedings of November 11, 2001. Copyright 2001 by Ryan S. Wood.

[22] What People...Military Ref 22: Carpenter, Major Donald G.; Therkelson, Lt. Colonel Edward R. 1968. Introductory Space Science, Volume II, p. 455–468. Published 1968 by United States Air Force Academy.

23 What People...Plus a Lot More Ref 23: NOVA Online. 1996. Kidnapped by UFOs? Interview with John Mack, Psychiatrist, Harvard University. Copyright 1996 WGBH, Boston, Massachusetts.

24 What People...Plus a Lot More Ref 24: Mack, John E. M.D. 1999. *Passport to the Cosmos*, pp. 26–27. Copyright 1999 by John Mack. Published by Three Rivers Press.

25 What People...Plus a Lot More Ref 25: Sturrock, Peter, et al. 1998. "Physical Evidence Related to UFO Reports," proceedings of workshop held at the Pocantico Conference Center, Tarrytown, New York. September 29–October 4, 1997. Published in *Journal of Scientific Exploration*, Vol. 12, no. 2, pp. 179–229.

26 What People...Plus a Lot More Ref 26: Jung, C.G. 1959. "On the Psychology of the Triskster-Figure." *Collated Works of C.G. Jung: Archetypes and Collective Unconscious*. Vol. 9, part I. Princeton, New Jersey: Princeton University Press.

27 What People...Plus a Lot More Ref 27: Nisker, Wes "Scoop." 1990. *Crazy Wisdom*. Berkeley, California: Ten Speed Press.

28 What People...Plus a Lot More Ref 28: Radin, Paul. 1956. *The Trickster: A Study in American Indian Mythology*. New York: Philosophical Library.

29 What People...Plus a Lot More Ref 29: Mack, John E. M.D. 1999. *Passport to the Cosmos*, pp. 40. Copyright 1999 by John Mack. Published by Three Rivers Press.

30 Disclosure Ref 30: Dennis, Brady. 2016. Scientists are frantically copying U.S. climate data, fearing it might vanish under Trump. *Washington Post*, December 13, 2016. Retrieved December 17, 2016, from: https://www.washingtonpost.com/news/energy-environment/wp/2016/12/13/scientists-are-frantically-copying-u-s-climate-data-fearing-it-might-vanish-under-trump/?utm_term=.7affb62ec432.

31 Disclosure Ref 31: Madison, James; in a Letter to W.T. Barry. 1822. The Founders Constitution, Chapter 18, Document 35. August 4, 1822. Retrieved December 17, 2016, from: http://press-pubs.uchicago.edu/founders/documents/v1ch18s35.html.

32 Disclosure Ref 32: Einstein, Dr. Albert. 1992. National Academy of Sciences, National Academy of Engineering, Institute of Medicine, and National Research Council, 1992 Report to Congress (Washington, DC: National Academy of Sciences, 1992), p. 1.

33 LaPlace Ref 33: Théorie analytique des probabilités, 1812, «... Plus un fait est extraordinaire, plus il a besoin d'être appuyé de fortes preuves; car, ceux qui l'attestent pouvant ou tromper ou avoir été trompés, ces deux causes sont d'autant plus probables que la réalité du fait l'est moins en elle-même.

34 LaPlace Ref 34: Hume, David (1748). An Enquiry Concerning Human Understanding, chap. 10.4. http://www.davidhume.org/texts/?text=ehu#10.

35 LaPlace Ref 35: Marcello Truzzi, On the Extraordinary: An Attempt at Clarification, Zetetic Scholar, Vol. 1, No. 1, p. 11, 1978.

APPENDIX A

Shapes of UFOs

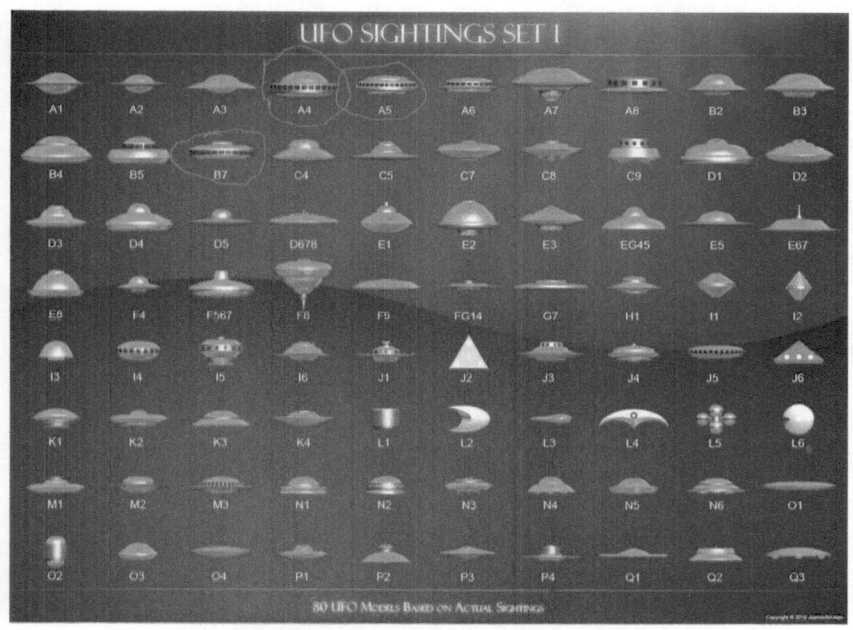

APPENDIX A: SHAPES OF UFOs

Citation:
Artwork courtesy of Jason Christiansen/JasonsArt.com.
    Published June 25, 2014.

Types of Reasoning and Fallacies Dictionary

The following is a compilation of *reasoning* frameworks. The original source creation of the general ideology for reasoning and fallacies is attributed to the Greek philosophers, including Thales, Aristotle, Socrates, Plato, etc.

Reasoning Frameworks

Abductive: The process starting with observations and developing an explanatory hypothesis which is the simplest and most probable. See the "Inductive and Deductive" chapter.

Analogy: Use of similar attributes to other designs as basis for argument. This is one of the oldest and most popular forms of reasoning in existence.

Causation: Using causes (independent variables) to explain effects (results)

Comparative: Differentiating two or more explanations, attributes, or other things against each other.

Conditional Reasoning: Explicit use of "if...then" in the explanation process.

Counterfactual: Analyzing alternative explanations that ignore factual representations of a previous occurrence in real life.

Deductive and Inductive: See the "Inductive and Deductive" chapter.

Intuition: Subjective theory that ignores logic and factual explanation in favor of a "gut feeling." A controversial form of reasoning in that formal critical thinking methodologies cannot verify testability according to its parameters of study.

Syllogistic: Explicit statement of conclusion from premises. Created by Aristotle. Also known as *traditional logic*.

Fallacy Frameworks

In our culture, everyone has utilized elements of at least some of these reasoning types, as well as having been either victim or initiator of some of the following list of *fallacies*. This list is not all exhaustive. The list includes examples of logical, common, and syllogistic fallacies found in everyday life. If the wording structure proves problematic, replace the operative premises/conclusions with letters. For example: All x are y. All z are x. Therefore, some z must be y.

The list is sprinkled with further examples. Common fallacies represented in the book are denoted by "*"

Ableism: Exploiting a weaker opponent's argument by victimizing the opponent in some way instead of the argument. Also known as *the con man's fallacy*.

Accent Fallacy: When the argument's meaning is misinterpreted by nuances on particular words or passages.

Accident Fallacy: A universal application to a general rule or argument when there exists rational exceptions to the rule.

Ad Fidentia: A fallacious personal attack on the person's confidence. This is very common in historical debate across all learning disciplines. See the *ad hominem* fallacies.

Ad Hoc Rescue: Fabricating less-than-valid excuses to maintain confidence in our belief.

Ad Hominen (Abusive): Making an extraneous personal attack to deflect the argument away from the facts.

Ad Hominem (Circumstantial): Inferring that the person's argument is prejudiced due to bias or purposed to conjecture a particular viewpoint.

Ad Hominem (Guilt by Association): When the argument is discredited due to it being attributed or associated with another person of similar persuasion.

Ad Hominem (Tuquoque): When the person making the argument is accused of supporting the argument's premises or claims.

Ad Numeram: Argument that asserts because others participate or align with that view that it justifies your use of it.

Affective: Argument uses justification of "feelings" as legitimate proof of conclusion.

Affirmative Conclusion from a Negative Premise: Use of a premise with a negative connotation to prove a positive conclusion.

Affirming a Disjunct: When an argument with two or more premises that is connected with the word "or" is used, the false claim that, if one is true the other must be false.

Alleged Certainty: Wording a conclusion as absolute without the factual premises to prove it as certain.

Affirming the Consequent: If the consequent is spoken to be true, the antecedent must be true. If A then B. B is true. Therefore, A is true.

Alphabet Soup: A confusion technique that masks a fallacious argument.

Alternative Advance: When a point is argued using two different reasons which mean the same, just worded differently.

Amazing Familiarity: A fallacy used often in fiction stories where premises are so fantastic so as to cause doubt in their validity.

Ambiguity Fallacy: When a premise is used in the argument that contains many definitions and ultimately does not support the conclusion.

Amphiboly: Use of a statement with multiple meanings without specifically defining it.

Anecdotal: Arguing from personal experiences instead of formal reasoning.

Anonymous Authority: Often used in journalism, an information source that is not identified.

Anthropomorphism: The assignment of human attributes to other non-human objects or events.

The Appeals Fallacies: Commonly titled as "Appeal to xxxxxxx," where the "xxxxxxx" is the attribute under fallacious attack. These are mostly plays on an emotional appeal. Use the attribute's definition and synergy with the argument to complete the explanation. This group includes appeal to:

Accomplishment, Anger, *Authority, Celebrity, Closure, Coincidence, *Common Belief, Common Folk, Common Sense, Complexity, Consequence, Definition, Desperation, Equality, Emotion, Extremes, Faith, Fear, Flattery, Force, Heaven, Ignorance, Improper Authority, Intuition, Motive, Nature, Normality, Novelty, Pity, Popularity, Possibility, Probability, Ridicule, Self-evident, Spite, Stupidity, the Law, the Moon, Tradition, Trust.

The Arguments Fallacies: Similar in construct to the Appeals Fallacies, the "Argument from yyyyyyy," where the "yyyyyyy" is the attribute under attack. Also similar to the Appeals Fallacies this group plays on emotional factors. This group includes Arguments from:

Consequences, Emotive Language, Fast Talking, Gibberish, Inertia (stay the course), Motives, Personal Charm, Pigheadedness, Repetition, Selective Reading, Age, Fallacy, False Authority, Hearsay, Ignorance, Incredulity, Silence, of the Beard, to Moderation, to the Purse, Avoiding the Issue.

Base Rate Fallacy: Using information deemed irrelevant in the argument.

*Begging the Question: One of the premises is used as an assumption of the conclusion.

Biased Sample Fallacy: Using information from a suspected biased sample pool to prove a conclusion.

Blind Authority Fallacy: Asserting an argument is true based entirely on the authority making the argument.

Bribery: Argument persuasion by some form of payoff.

Broken Window Fallacy: An argument that ignores opportunity alternatives or unseen alternatives.

Bulverism: Combines circular reasoning and genetic fallacy. Attributions of the arguer's personal nature to the alleged falseness of the argument.

But: Use of the term but to negate a commonly-accepted argument.

Causal Reductionism: Falsely narrowing the situation to a single reason when many valid reasons exist.

Cherry Picking: Favorable selection of certain evidence to support the argument while ignoring other evidence.

Circular Definition: Defining an argument by stating that argument in the definition.

Circular Reasoning: See *Circular* definition above. A similar train of thought where the conclusion is used as evidence in the premises, which continues around to again support the conclusion, thus completing the fallacious circle.

Commutation of Conditionals: Reversing the antecedent and consequent.

Complex Question Fallacy: A statement that implies an action and similar to *Begging the Question* is coercing the audience to accept the statement as fact.

Confirmation Bias: Handpicking only proof points to support your argument though knowing (or not) that other points exist.

Conflicting Conditions: A self-contradicting argument stream.

Confusing Explanation with Excuse: Using a fact to justify validity of the same fact.

Cum Hoc Ergo Propter Hoc: One premise causes the other but without any demonstrable proof.

Currently Unexplainable Fallacy: Arguing that a statement is unexplainable because it currently cannot be sufficiently explained to the satisfaction of all parties to the argument.

Conjunction Fallacy: Choices that are argued to be distinctive alternatives when in reality they are members of the same set of choices.

Conspiracy Theory Fallacy: When an argument is asserted as false because of the claim that evidence support is hidden by efforts of two or more people.

Contextomy: Altering or removing information to distort the true meaning.

Definist Fallacy: Defining a concept in a unique way to make the argument more defensible.

Denying a Conjunct: Stating that if the antecedent or consequent are true then the other conjunct must be false and vice versa.

Denying the Antecedent: If the antecedent is made to be not true, then it is concluded that the consequent is also not true.

Denying the Correlative: Stating alternatives when none exist.

Disjunction Fallacy: X is more likely than (x *or* y).

Distinction Without a Difference: Claiming that an argument is different from another when in reality they are the same.

Dogmatism: False declaration of arguments as unassailable or uncontestable.

Double Standard: The bias of using different standards of measure to support or refute two arguments when the same standard should be used.

Ecological Fallacy: Falsely using statistical data to infer characteristics of an individual from group data. Person A is in group 1. Group 1 has characteristic Z. Person A has characteristic Z.

Etymological Fallacy: Assuming a current meaning has the same as its historical one. A is defined as 2. A used to be defined as 1. Therefore, 2 means 1.

Equivocation: Using the same term in more than one context misleading the audience.

Exclusive Premises: Use of two negative premises.

Existential Fallacy: Use of two or more absolute premises (all or none) but which forces the conclusion as a non-absolute (some). All x are y. All z are x. Therefore, some z must be y.

Extended Analogy: If one is like a second and one of those is like a third, then the other is also like the third.

Failure to Elucidate: When an explanation of some premise or conclusion is made deliberately confusing.

Fake Precision: Misusing mathematical data to make a statement appear allegedly true (or false).

Fallacy of Undistributed Middle: Similar to the *Existential Fallacy*, all A's are C's. All B's are C's. Then all A's are B's.

Fallacy of Composition: Arguing that a whole is true because some part of the whole is true.

Fallacy of Division: See the *Fallacy of Composition*. The Fallacy of Division is the exact opposite.

Fallacy of Every and All: A form of the *Equivocation Fallacy* where specific use of the quantifiers "all, some, every, and/or none" with different meanings that confuse both the argument and the arguer.

Fallacy of Four Terms: Its use in three-term syllogisms which disqualifies the argument on this basis.

Fallacy of Opposition: An assault on the arguer only because they disagree with your argument.

False Attribution: Using a fabricated not relevant or otherwise biased statement for argument support.

False Compromise: Arguing that only a middle course is correct and refuting any extreme argument.

False Conversion: A formal fallacy where the terms of either the premises or conclusion are reversed and used as proofs. All B are C. Therefore, all C are B.

False Dilemma: When it is argued that only a certain amount of options exist but, in reality, more than that exist. Synonymous with *Bifurcation*.

False Effect: An argument asserting an inaccurate effect from a cause.

Fantasy Projection: Arguing that facts that are in reality subjective or fantasized incidents must be accepted as truth.

Far-Fetched Hypothesis: Arguing to a conjectured conclusion while ignoring other more pragmatic explanations.

Faulty Comparison: Arguing for desirability of something by relating two premises when, in reality, they are not related. Most often, arguments comparing two or more premises or objects are fallacious in their nature.

Free-Speech: Resort to proclamation of the First Amendment to support your argument.

Gadarene Swine Fallacy: Arguing that because one is not aligned with all the others in a group that it must be that one "outlier" that is in the wrong.

Galileo Fallacy: The argument that a premise or conclusion must be valid if for no other reason than the premise or conclusion is ridiculed, forbidden, or derided.

Gambler's Fallacy: In an argument of chance, the mistaken belief that the conclusion is dependent on previous outcomes.

Gaslighting: Deliberately distorting facts or argument points.

Genetic Fallacy: Arguing a truth that is based on the source of its premises.

Hasty Generalization: Arguing for a conclusion based on a small or no inspection of the sample or evidence when more evidence exists.

Having Your Cake: Arguing in a new direction when your first argument is not clear.

Hedging: Amending an argument to avoid other evidence while simultaneously assuming the amended argument is identical to the original. Argument M is made. This is disproven. Argument N is then made and argued to be identical to Argument M.

Historian's Fallacy: Arguing a position in the present that incorrectly assumes that the same information was available those in the past that taking the same position as the one you are arguing for.

Homunculus Fallacy: Argument X is explained by reason Y, but in reality, Y depends on X.

Hot Hand Fallacy: In games of chance, the belief that, just because you win a few times in a row, that it will continue.

Hypnotic Bait and Switch: Substituting a false argument or condition after stating one or more true statements.

Hypothesis Contrary to Fact: Arguing from a weak position about either what happened in the past or what is predicted to happen in the future.

Identity Fallacy: An argument evaluated only on the arguer's physical or societal identity.

If by Whiskey: Often seen in political discussions, an argument based on the arguer's opinions and stated to appear that it supports both sides of an issue.

Illicit Contraposition: Switching values of two premises of a categorical proposition, which results in an invalid argument. All X are Y. Therefore, no non-Y are non-X.

Illicit Major: An argument where the major condition is distributed in the conclusion but not in all the premises. All X are Y. No Z are X. Therefore, no Z are Y.

Illicit Minor: Similar to the *Illicit Major* but involves minor premises. All X are Y. All Y are Z. Therefore, all Z are X.

Illicit Substitution of Identicals: Arguing to confuse the knowledge of something with knowledge of it under many different names or definitions.

Incomplete Comparison: An assertion of proof that is impossible to be challenged. Used often in business and advertising. Product P is "the best" but no argument is offered as to "what."

Inconsistency: Two or more statements that cannot be absolutely true or false and that do not align with the argument.

Inflation of Conflict: A form of black-and-white reasoning. Either the argument is absolutely true (or false) or nothing at all can be concluded.

Insignificant Cause: Asserting that a single statement (factor) is the sole proof (cause) of the argument.

*Jumping to Conclusions: Arguing a conclusion without learning about the argument itself. Similar to *Knee-Jerk Reaction*.

Just because Fallacy: Arguing that there need not be any reason for a conclusion.

Just in case Fallacy: Ignoring more probable support of an argument in favor of a "glass is half empty" argument.

Kettle Logic: Arguing from many positions in hopes of accurately supporting one position. Analogous to taking many shots at a target at once in the hope of hitting the target once.

Least Plausible Hypothesis: Arguing from less supportable positions than from one with more evidence.

Limited Depth: Strongly arguing your position without offering any explanation.

Limited Scope: Arguing nothing about a conclusion except the premises that are offered to explain the conclusion.

Logic Chopping: Ignoring the main substantive arguments in favor of insignificant details.

Ludic Fallacy: Incorrectly arguing that models of proof can be taken from one conclusion and aligned as proof of other conclusions.

Lying with Statistics: Making arguments while using inaccurate data.

Magical Thinking: Arguments based on superstition.

McNamara Fallacy: When both quantitative and qualitative evidence exists but only quantitative arguments are made.

Meaningless Question: An argument that cannot be rationally responded to.

Misleading Vividness: Arguments from ignoring reliable evidence in favor of sensational events and reasoning.

Missing the Point: Making wrong conclusion about a set of premises.

Moralistic Fallacy: When a conclusion is based only on "what ought to be" instead of "what is."

Moving the Goalposts: Arguing that evermore points must be responded to even after the main arguments have been successfully refuted.

Multiple Comparisons Fallacy: When an argument of a false comparison becomes a significant one.

Naturalistic Fallacy: The opposite of the *Moralistic Fallacy.* A conclusion of "what ought to be" based only on "what is."

Negating Antecedent and Consequent: After switching the antecedent and consequent in the premises, they are not transposed in the conclusion. If B, then C. Therefore, if not-C then not-B.

Negative Conclusion from Affirmative Premises: Arguing from both premises being positive but the conclusion is negative.

Nirvana Fallacy: Arguing an unrealistic or nonsensical conclusion over a realistic one.

No True Scotsman: When a universal conclusion (an all or nothing) is refuted, the argument turns to specific conditions (not all or none but only one or a few).

*Non Sequitur: When the argument for the conclusion does not logically flow from the premises.

Notable Effort: Arguing that the effort is enough reason to accept the conclusion.

Overextended Outrage: A form of stereotyping where a rare instance of evidence is argued to be the normal to incite emotional disagreement.

Oversimplified Cause Fallacy: The argument of reducing many causal factors to a single one. A form of "either this one is, or none are."

Overwhelming Exception: When an argument has so many exceptions so as to destroy its efficacy.

Package Deal Fallacy: Arguing that premises that are grouped together must always be grouped. If not, then fatal negative effects will occur.

Poisoning the Well: An ad hominem attack on the opponent's personal or factual attributes in advance of the discourse.

Political Correctness Fallacy: Arguing an overly corrective position that attributes two or more groups or ideas being equal. A polar position of the *Overextended Outrage Fallacy*.

Post Designation: Like the *Multiple Comparisons Fallacy*, drawing conclusions without stating what the expectations were before the conclusion was made.

Post Hoc: Part of a group of common and logical fallacies. T follows S. Therefore, T is caused by S.

Prejudicial Language: Connecting appeals to emotional terminology or dialogue to the acceptance of the argument.

Profanity: Use of unrelated strong unacceptable language to support an argument; often in conjunction with other fallacies.

Proof by Intimidation: Intentionally making an argument confusing to understand so as to gain acceptance by intimidation. Related to *Proof by Assertion*.

Proof Surrogate: An argument masked as proof of a conclusion when no proof is being offered.

Prosecutor's Fallacy: Arguing that a low probability of a false occurrence from a population falsely assumes a similar low probability of a particular example of a false occurrence.

Proving Nonexistence: Arguing for proof of nonexistence rather than asking to prove the existence. If x cannot be proved, then prove that it doesn't. Absence of evidence is not evidence of absence.

Psychogenetic Fallacy: Stating that an argument is invalid due to inferring that a psychological reason is given for the argument. This is different from:

Psychologist's Fallacy: The argument presupposes that his observations are unbiased.

Quantifier-Shift Fallacy: Every P has a related Q. Therefore, there is some Q related to every P.

Questionable Cause: Incorrectly arguing that one event caused another just because they are often associated.

Rationalization: Arguing from false premises because the real premises may be more problematic but still true.

Red Herring: A deliberate argument that seeks to divert attention from the original argument in the hopes of that argument being abandoned.

Reductio ad Absurdum: An argument that takes a premise and develops invalid absurd conclusions by asserting contradictory proofs of the same premises.

Reductio ad Hitlerum: Arguments that share any analogy to Adolf Hitler or the Nazi Party. The *Appeasement Fallacy* is an example.

Regression Fallacy: Arguing for a nonnatural causation of a phenomenon when none exists except for the reasons due to natural causes.

Reification: When a non-tangible idea or belief is argued to have physical or tangible reality or basis.

Relative Privation: Arguing for a better appearance of a conclusion by using a "best case" or "worst case" comparison.

Repetition: Deliberately repeating a statement or conclusion automatically makes it true.

Retrogressive Causation: Arguing for the source of an effect to alleviate the effect of that source. A causes or is the source of B. To alleviate B, do more of A.

Retrospective Determination: Arguing that the inevitability of an action occurring is just because that action has been undertaken.

Righteousness Fallacy: Arguing that because of good intentions only that their facts are truthful and ignores basis of reality.

Rights to Ought Fallacy: Arguing for one's actions based on their rights to do that action.

Scapegoating: Unfairly targeting an unpopular conclusion because the conclusion is easy to argue against.

Self-Righteousness Fallacy: See *Righteousness Fallacy*.

Selective Attention: Attending to only designated specific parts of an argument while ignoring other relevant argument parts.

Self-Justification: Falsely approving your behavior to support an argument by comparing it to others who demonstrate bad behavior.

Self-Sealing Argument: A conclusion that is possible only if no argument can be successfully brought against the conclusion.

Shifting of the Burden of Proof: Making a conclusion then arguing that the opponent proves the opposite of that conclusion.

Shoehorning: Forcing evidence into an argument.

*Slippery Slope: When an insignificant argument or premise can be entered into a more significant argument or premise which could affect the conclusion.

Social Conformance: Threatening to agree with the argument or be socially ignored or uninvited.

Special Pleading: Similar to the *Double Standard Fallacy*.

Spiritual Fallacy: Arguing that a conclusion is caused by spiritual explanation so as to make it impossible to logically prove.

Spin Doctoring: Presenting a deceptive argument so as to convince the opponent to accept your argument.

Statement of Conversion: Accepting a conclusion of conversion without knowing any reasons for the conversion.

*Stereotyping: How we generalize an argument on the basis of inferring a generality from one specific example.

Stolen Concept Fallacy: Arguing for truth of something you are trying to disprove.

*Strawman Fallacy: Arguing a conclusion by misrepresenting or falsifying the person's actual argument.

*Style Over Substance: Use of embellishing language or emotional appeal as major argumentation strategy.

Subjectivist Fallacy: Falsely claiming something is true for one person and not for another when it is actually true of all.

Subverted Support: Arguing for a conclusion in which there is no evidence that the argued premises have actually occurred. Similar to *Begging the Question*.

Sunk-Cost Fallacy: Arguing that additional investment is necessary to offset what was already spent while ignoring the overall costs.

Suppressed Correlative: Fallaciously redefining one or more mutually exclusive premises to limit the possible conclusions.

Survivorship Fallacy: "Dead men do not tell tales."

Texas Sharpshooter Fallacy: Falsely concluding a situation by acknowledging only the similarities of the premises while ignoring differences.

Tokenism: Incorrectly accepting an insignificant courtesy as sufficient substitute for the complete or real item.

Traitorous Critic Fallacy: Arguing by attacking the opponent's favoritism toward the opposition view or group rather than the true nature (reasons) for the disagreement.

Two Wrongs Make a Right: Mistakenly justifying a wrongful action against an opponent only because the opponent would do the same.

Type-Token Fallacy: Use of a word or phrase that is unclear in an argument whether it applies to an abstract idea or a tangible "token" object.

Unfalsifiability: Arguing that a conclusion is true or false when the conclusion cannot be tested. A major precept in scientific thought.

Use-Mention Error: Confusing use of a word in an argument when the word is used to describe a thing and the thing itself.

Weak Analogy: Use of an analogy that detracts, rather than enhances, support of a conclusion.

Willed Ignorance: Rejecting all arguments because they do not want to change their belief. "My mind is made up, do not bother me with the facts."

Wishful Thinking: When the unfounded wishes or hopes of someone dominate factual evidence or arguments.

Wrong Direction: The cause and effect parameters are reversed in an argument.

References

Changing Minds. 2016. Types of Reasoning. Copyright 2002–2016 by Changing Works. Retrieved July 29, 2016, from: http://changingminds.org/disciplines/argument/types_reasoning/types_reasoning.htm.

Damer, T. 2009. *Attacking Faulty Reasoning: A Practical Guide to Fallacy-Free Arguments*. (sixth ed.) Wadsworth. ISBN 978-0-495-09506-4.

Dowden, Bradley 2010. "Fallacy." *The Internet Encyclopedia of Philosophy*. ISSN 2161-000. https://www.iep.utm.edu/fallacy/

Hansen, Hans, "Fallacies," *The Stanford Encyclopedia of Philosophy* (Summer 2018 Edition), Edward N. Zalta (ed.), URL = <https://plato.stanford.edu/archives/sum2018/entries/fallacies/>.

Khan, Salman. 2018. *Fallacies: Formal and Informal Fallacies*. Khan Academy. Retrieved August 28, 2018, from: https://www.khanacademy.org/partner-content/wi-phi/wiphi-critical-thinking/wiphi-fallacies/v/formal-informal-fallacy

Palomar College. 2018. *Table of Fallacy Categories*. Bruce Thompson; Philosophy Department-Palomar College. Accessed at: https://www2.palomar.edu/users/bthompson/Table%20of%20Fallacies.html

Rational Wiki. 2018. *Logical Fallacy*. Retrieved August 28, 2018, from: https://rationalwiki.org/wiki/Logical_fallacy

University of Arizona. 2018. *Logical Fallacies*. Retrieved August 28, 2018, from: http://www.u.arizona.edu/~shunter/logic.txt

University of Texas-El Paso. 2018. *Master List of Logical Fallacies*. UTEP English Department.

Wilson, W. Kent. 1999. *Formal Fallacy*. In Audi, Robert. *The Cambridge Dictionary of Philosophy*. (second ed.) Cambridge University Press. ISBN 978-0-511-07417-2.

About the Author

Keith A. Seland, EdM, MBA, has been a science writer and researcher in the disciplines of science education and ufology for eleven years. *The Humaniverse Guide to Better Reasoning & Decision-Making* was inspired by the lack of meaningful connection in society today between the wonderful applications of reasoning, science thought, its practitioners and an estranged public, and to help fill this void.

www.ingramcontent.com/pod-product-compliance
Lightning Source LLC
Chambersburg PA
CBHW020717180526
45163CB00001B/9